Methods of Correlation and Regression Analysis

Linear and Curvilinear

Methods of Correlation

Mordecai Ezekiel

> *Head, Economics Department*
> *Food and Agriculture Organization*
> *of the United Nations*

Karl A. Fox

> *Head, Department of Economics*
> *and Sociology,*
> *Iowa State University*

THIRD EDITION

and Regression Analysis

Linear and Curvilinear

New York · John Wiley & Sons, Inc.

London · Sydney

FIFTH PRINTING, JUNE, 1966

Library of Congress Catalog Card Number: 59-11793

Printed in the United States of America

Preface to Third Edition

Thirty years have elapsed since the original edition of this book was written—years of political tensions and upheaval and of enormous progress in technical development. This last has been reflected in changes in some of the examples cited—from an automobile with two-wheel brakes in the 1920's to the orbit of an earth satellite in the late 1950's, and from methods for using hand calculators and card tabulators to those for electronic computers. Despite this technical progress, the basic elements of correlation analysis continue unchanged. The major emphasis, however, has shifted from correlation to regression, and the wide range of uses of the method in varied fields has led to many specialized applications or modifications. This is especially true in econometrics. Here the long controversy over mutually inter-correlated variables has finally produced an effective simultaneous-equation method for dealing with situations where single-equation solutions are inadequate; but apparently such situations are relatively infrequent.

In this third edition the senior author has been fortunate in securing the collaboration of an associate who has made distinguished contributions in these newer aspects of the field, particularly in their application to problems of actual research. The new chapter on simultaneous-equation solutions, Chapter 24, is one of his contributions. He is also responsible for the extended treatment of the analysis of variance in relation to regression problems (Chapter 23), the modernization of the chapter on standard errors in multiple regression (Chapter 17), and a complete revision of the treatment of error formulas for time series (Chapter 20), as well as for many other contributions throughout the text.

In the revision, methods of determining regressions by algebraic equations have been given first consideration, and graphic approximation methods have been treated second, with due consideration of the limitations of each. Modern terminology, especially in the econometric field, has been recognized and used where appropriate, and the presentation of sampling and of confidence intervals has been modernized.

The order of presentation has been rearranged and grouped under seven major sections. We hope this will make the general development clearer both to students and to teachers. Modern methods of calculating correlation and regression constants are outlined in a new chapter, and methods of using electronic calculators for this purpose are briefly treated. The chapter on uses of correlation analysis has been recast and extended to cover newer fields in which the method is now widely used, as well as recent work and developments in fields long familiar. Examples from other countries, and also from the international field in which the senior author has been working this past decade, have been introduced here and elsewhere in the book. The treatment of standard errors and their meaning for statistics derived from small samples has been materially revised. Introduction of adjustments for correlation coefficients and indexes to remove bias due to limited degrees of freedom has been deferred until a late chapter in the book (Chapter 17).

Despite these innovations, the general simplicity of expression and explanation has been retained as far as possible. Mathematical derivations have been relegated to the technical appendix and many of the more obvious ones have been eliminated; the notation has been kept as simple as possible, and only a modest level of mathematical training is assumed for the reader.

As in previous editions, full attention is given to multiple *curvilinear* and *joint functional* regressions, essential for adequate treatment of many problems in both natural and social sciences. Many standard statistical texts still largely ignore the use of non-linear regressions as practical working tools. In recognition of the increased emphasis given to regression analysis in general, and the unusually full treatment of non-linear regression, the title of the book has been changed to *Methods of Correlation and Regression Analysis* with the subtitle *Linear and Curvilinear.*

This book is in part an exposition of standard statistical methods with no attempt to give proofs or to show their mathematical derivation. In substantial part, however, it is based upon procedures first developed by the senior author or by colleagues associated with him. In such cases chapter references indicate the professional papers in which the methods were first presented and proved and other important papers relating to these methods.

The authors would like to express their deep appreciation to many fellow workers in many lands who have contributed to this revision by supplying suggestions, criticisms, materials, or illustrations, and to the many students who through the years have called attention to errors in the previous printings or editions. Our special thanks are due to John H. Smith for many helpful suggestions on the entire manuscript, to Martha N. Condee for calculating the examples used in Chapters 13 and 24, and to J. P. Cavin and R. J. Foote for making their computing facilities available for this purpose. We hope readers will again call our attention to any new errors of computation or of type setting that may have slipped in in this revision and in the many new examples introduced.

In the first edition the senior author acknowledged his debt "to the spirit of research with which the Bureau of Agricultural Economics was imbued by the broad vision of Henry C. Taylor." The junior author owes a similar debt to the research environment that was maintained in the Division of Statistical and Historical Research of that Bureau under the leadership of O. C. Stine and J. P. Cavin, and to such former colleagues there as R. J. Foote, Harold F. Breimyer, and C. Kyle Randall, who contributed to its high standards in applied research during the current decade. His new colleagues at Iowa State University, including T. A. Bancroft, George Snedecor, and Emil Jebe, have helped him to appreciate the shift in emphasis from correlation to regression (and to the analysis of variance) that has taken place in a number of other sciences in addition to economics. The two authors are, of course, jointly responsible for the particular emphasis adopted in this edition and for such errors and imperfections as may exist in it.

We wish to thank all our helpers in Ames, Rome, and at Wiley & Sons in New York, for their part in making this new edition possible.

MORDECAI EZEKIEL
KARL A. FOX

Rome, Italy
Ames, Iowa
August, 1959

Preface to First Edition

This book is not intended to cover the entire field of statistics, but rather, as its name indicates, that part of the field which is concerned with studying the relations between variables. The first two chapters are devoted to a brief review of the central elements in the measurement of variability in a statistical series, and to the essential concepts in judging the reliability of conclusions. These chapters are not to be regarded as a full statement, but instead as brief summaries to clarify the basic ideas which are involved in the subsequent development.

No attempt is made in the body of the text to present the mathematical theory on which the art of statistical analysis is based. Instead, the aim throughout has been to show how the various methods may be employed in practical research work, what their limitations are, and what the results really mean. Only the simplest of algebraic statements have been employed, and the practical procedure for each operation has been worked out step by step. It is believed that the material will be readily comprehensible to anyone who has had courses in elementary algebra.

Although the examples which are used in presenting the several methods are drawn very largely from the author's own field of agricultural economics, the methods themselves are explained in sufficiently general terms so that they can be applied in any field. In addition, two chapters are devoted to a discussion of the types of problems in a great many different fields of work to which correlation analysis has been successfully applied, and to research methods and the place of correlation analysis in research. It is hoped that this presentation will assist research workers in many fields to appreciate both the possibilities and the limitations of correlation analysis, and so gain from their data knowledge of all the relations which so frequently lie hidden beneath the surface.

Where the methods presented are the well-established ones developed by the fathers of the modern science, mainly the English statisticians, no attempt is made to prove or derive the various formulas. On a few crucial points, however, or where derivations not generally accessible are involved, the derivations of the formulas are shown in notes in the technical appendix, in the simplest manner possible.

The methods presented in this book, insofar as they constitute an advance over those previously available, represent largely the joint product of a group of young researchers in the Bureau of Agricultural Economics of the United States Department of Agriculture during the past decade. The new methods include (a) the application of the Doolittle method to the solution of multiple correlation problems, greatly reducing the labor of obtaining multiple correlation results, and making feasible the use of multiple correlation in actual research work; (b) the development of approximate methods for determining curvilinear multiple correlations, and, more recently, very rapid graphic methods for their determination; (c) the recognition of "joint" correlation, and the gradual development of methods of treating it; and (d) by extensive use in actual investigations, concrete demonstration of the possibilities of these methods in research work. These recent developments in correlation analysis are as yet largely unavailable except in the original articles in technical journals. One object of this book is to present them in organized form, and with such interpretation that their significance and application may be fully understood.

During the last two decades, the English statisticians "Student" and R. A. Fisher have been developing more exact methods of judging the reliability of conclusions, particularly where those conclusions involve correlation or are based on small samples. These new methods have as yet received but little recognition from American statisticians. They are presented here as simply as possible, and the discussion of the reliability of conclusions gives them full consideration.

So many persons have helped in the years during which this book has been growing that it is difficult for me to enumerate them all. First of all I should like to mention Howard R. Tolley, from whom I received my introduction to statistics, and with whom it has been a constant joy to work. I give him credit for much that is included here. The very order of presentation reflects that which he worked out for his classes. In a very real sense this book is a product of the spirit of research with which the Bureau of Agricultural Economics was imbued by the broad vision of Henry C. Taylor. John D. Black was the first to point out some of the undeveloped phases of statistical analysis, and then aided with encouragement and counsel in their solution. Bradford B. Smith aided in the beginning of the new developments, and his vivid imagination and logical mind have been a constant help. Among others who have collaborated in various stages, or who have independently worked out various phases of the problem, may be mentioned Sewall Wright, Donald Bruce, Fred Waugh, Louis Bean, and Andrew Court. Susie White, Helen L. Lee, and Della E. Merrick have given intelligent, conscientious, and loyal assistance in the clerical work in the development and testing of each new step.

In the preparation of the book itself I have had generous and willing help. Dorothea Kittredge and Bruce Mudgett have given the very substantial assistance of a detailed reading of the entire text, and many improvements in presentation and in material are due to their suggestions. For two terms the mimeographed manuscript has been used as a text in the United States Department of Agriculture Graduate School, and the members of the class have helped me in working out the illustrations, in clarifying the text, and in eliminating errors. R. G. Hainsworth, who prepared the figures, deserves credit for the excellence of the graphic illustrations. O. V. Wells helped in computing many of the illustrative problems, and Corrine F. Kyle in verifying the arithmetic. For the laborious and exacting work of typing the preliminary stencils, the many revisions, and the final manuscript, and for her care, patience, and suggestions, I am indebted to my mother, Rachel Brill Ezekiel; and for editing the manuscript and helping in the lengthy task of proofreading, to my wife, Lucille Finsterwald Ezekiel.

To all these, and to the many others who have helped me in the development of this work, I take this opportunity of expressing my obligation and my gratitude.

For any errors in the statements made and in the theories advanced, I alone am of course responsible. Although the text has been checked painstakingly, it is hardly to be hoped that a publication of this character will appear without some errors creeping in, in mathematics, in arithmetic, or in spelling. When such errors, or any ambiguities of statement, are noted by any reader, I would be very grateful if he would inform me of them.

MORDECAI EZEKIEL

Washington, D. C.
April 20, 1930

Contents

SECTION VI

Miscellaneous Special
Regression Methods

SECTION VII

Uses and Philosophy of Correlation
and Regression Analysis

SECTION I

Introductory Concepts

CHAPTER I

Measuring the variability of a statistical series

Statistical analysis is used where the thing to be studied can be reduced to or stated in terms of numbers. Not all the undertakings that rely on measurements ordinarily employ statistical analyses. In surveying, physics, and chemistry, for example, the particular thing being studied can usually be measured so closely, and varies over such a small range, that the true value can be established within narrow limits. But even in these fields, the modern work on atomic subparticles has involved the use of statistical concepts. In fact, the statistical concept of true value owes its existence to the reproducibility of measurements in fields like these.

In many natural sciences, the problem to be studied can be simplified by the use of controlled experimental conditions, which permit the influence of various factors to be studied one at a time. In such sciences, statistical methods can be used to plan experiments in such a way as to make the conclusions most reliable with a minimum of effort, and they can be used to measure the interrelations in sciences like astronomy, where the phenomena can be observed but not controlled.[1]

In the social sciences, there are fewer opportunities for the use of controlled experiments. Such sciences have to rely on statistical analysis, both to judge the importance of observed differences and to untangle the separate effects of multiple factors. Statistical analysis is used in the study of occurrences where the true value or relation cannot be measured

[1] W. G. Cochran and Gertrude M. Cox, *Experimental Designs*, 2nd ed., John Wiley and Sons, New York, 1957.
R. A. Fisher, *The Design of Experiments*, 5th ed., Oliver and Boyd, Edinburgh and London, 1949.

directly or is hidden by other things. The numerical statement of the occurrence or of the relationship cannot be obtained directly from the original or "raw" figures. Instead, the data must be analyzed to determine the values desired.

The special need for analytical methods in the social sciences has been clearly stated by an eminent Englishman, as follows:[2]

Causation in social science is never simple and single as in physics or biology, but always multiple and complex. It is of course true that one-to-one causation is an artificial affair, only to be unearthed by isolating phenomena from their total background. Nonetheless, this method is the most powerful weapon in the armory of natural science: it disentangles the chaotic field of influence and reduces it to a series of single causes, each of which can then be given due weight when the isolates are put back into their natural interrelatedness, or when they are deliberately combined (as in modern electrical science and its applications) into new complexes unknown in nature. This method of analysis is impossible in social science. Multiple causation here is irreducible.

The problem is a two-fold one. In the first place, the human mind is always looking for single causes for phenomena. The very idea of multiple causation is not only difficult, but definitely antipathetic. And secondly, even when the social scientist has overcome this resistance, extreme practical difficulties remain. Somehow he must disentangle the single causes from the multiple field of which they form an inseparable part. And for this a new technique is necessary.

The Arithmetic Average. The basic forms of statistical analysis concern the organization of quantitative information as a basis for drawing inferences. Some of the basic work involves averaging and classifying data. Thus, if a person were studying the yield of corn in one year in some area, say a county, he might talk with 20 farmers picked at random and obtain figures, such as those in Table 1.1, showing the yield of corn each farmer had obtained.[3]

---

[2] Julian Huxley, The science of society, *Virginia Quarterly Review*, Vol. 16, No. 3, pp. 348–65, summer, 1940.

[3] "Picked at random" means so selected that, for each observation, there is just as great a chance of any one farm in the universe (as here the county) being selected as of any other farm. One way of making a random selection would be to put slips with the names of all the farmers in the county into a bowl, mix the slips thoroughly, and then have a blindfolded person draw out slips one at a time, repeating the mixing before each new drawing. Data would then be obtained from the n farmers represented by the slips so drawn. A sample so selected is known as a "random sample" or an "equal probability sample." (See V. G. Panse and P. V. Sukhatme, *Statistical Methods for Agricultural Workers*, pp. 36–40, Indian Council of Agricultural Research, New Delhi, 1954.) These two terms, "universe," meaning the whole group of cases about which one is interested in finding out certain facts, and "sample," meaning a certain number of those cases, picked at random or otherwise from all those in the particular universe, are both used frequently in statistical work, and should be clearly understood.

Table 1.1

YIELDS OF CORN OBTAINED BY 20 FARMERS*

Farmer	Yield	Farmer	Yield	Farmer	Yield	Farmer	Yield
	Bushels per acre		*Bushels per acre*		*Bushels per acre*		*Bushels per acre*
1	39	6	43	11	39	16	43
2	35	7	30	12	45	17	41
3	48	8	38	13	36	18	47
4	40	9	40	14	33	19	38
5	37	10	39	15	41	20	42

* In making a table such as this, the actual values may be "rounded off" to any desired extent. In this case they are rounded to the nearest whole bushel. For example, 43 bushels represents any report of 42.5 bushels or more, and up to but not including 43.5 bushels. If the original reports were secured to the nearest tenth bushel, this might be indicated by writing 42.5–43.4 instead of 43; or if secured to the nearest hundredth bushel, by writing 42.50–43.49.

In performing arithmetic calculations on rounded off data, the results may always have a certain range of inaccuracy due to the effects of rounding. See Appendix 3, Note 7.

The most natural first step in reducing such a series of observations to more usable shape is to find the arithmetic average—to add all the yields reported and divide by the number of items. The 20 reports total 800 bushels, or an average of 40 bushels.[4] This provides a single figure into which one characteristic of the whole group is condensed.

[4] Bushels are used here to represent any other quantity in which one might be interested in a particular case. If we let X' represent the number of bushels reported by farmer 1, X'' the bushels reported by farmer 2, X''' the bushels by farmer 3, and so on, we can then represent the sum of all the reports by the expression ΣX (read "summation of the X's"). Similarly, if we use n to represent the number of observations we have obtained and use M_x to represent the *average* (or *mean*) number of bushels for all reports we can define the *arithmetic mean* by the formula:

$$M_x = \frac{\Sigma X}{n} \tag{1.1}$$

This formula can be applied to anything we are studying, no matter whether X means bushels of corn, inches in height, degrees of temperature, grade in a school examination, distance of a star, height of a flood, or any other measurable quantity; or whether there are 2 cases or 2 million. This is thus a perfectly general formula which can be applied to any given problem. As statistics is a study of general methods, so stated that they can be applied to particular problems as desired, it will be necessary to use many general formulas of this sort. The student should therefore familiarize himself with the definitions given above and with the way they are used in formula (1.1), so that he will be able to understand and use each formula as it occurs.

But the average is not the only characteristic of the group that might be of interest. The average would still be 40 if every one of the 20 farmers had had instead a yield of 40 bushels per acre; yet the mean of 20 reports each of 40 bushels would certainly be more reliable than the mean of 20 reports ranging from 33 to 48 bushels, even though both did have the same average.

Classifying the Data. One way of showing the differences in the individual reports is to arrange them in some regular order. If the farmers interviewed have simply been visited at random, and not selected so that those visited first represent one portion of the county and those visited later another portion, the order in which the records stand has nothing to do with their meaning. As a first step to seeing just what the data do show, they can be rearranged in order from smallest to largest, as shown in Table 1.2.

Table 1.2

YIELDS OF CORN ON 20 FARMS, ARRANGED IN ORDER
OF INCREASING YIELDS

Bushels per acre

33	38	40	43
35	38	40	43
36	39	41	45
36	39	41	47
37	39	42	48

It is now easier to tell from the series something about the group of reports. One can now see that only 1 farmer had a yield of less than 35 bushels per acre, and only 2 had more than 45, so that 17 out of the 20 had 35 to 45, inclusive. The series shows, too, that 10 of the farmers had less than 40 bushels of corn per acre and 10 had 40 or more, so that the figures 39 and 40 mark the middle of the number of yields reported. If we divide each half into halves again, we see that 5 men had yields of 37 bushels or less, 5 had yields of 43 bushels or more, whereas 10 men— half of those reporting—had yields of 38 to 42 bushels, inclusive. This tells something about how variable yields were from farm to farm in the area from which the reports were secured—half the reports fell within this 5-bushel range.[5]

[5] In statistical terminology, the figure that divides the number of reports into halves—39.5 in this case—is termed the *median*; and the figures that divide the numbers into quarters—37.5 and 42.5—are termed the *lower* and *upper quartiles*. The difference between the two quartiles, within which the central half of the reports fall, is termed the *interquartile range*.

Even as rearranged in Table 1.2, the 20 reports still constitute a large tabulation. If there were several hundred, such a listing would be so unwieldy that it would be difficult to use.

Frequency Tables. The records can be studied more easily if, instead of writing "39" three times when there are 3 farmers with 39 bushels each, we simply show that each of 3 men reported 39 bushels. Similarly, instead of putting "40" down twice, we can show that 40 bushels were reported by 2 men. If this operation is performed for all the reports, the data can then be assembled into what is known as a "frequency table." The result is shown in Table 1.3. It gives the frequency, that is, the number of times that each yield of corn was reported.

Table I.3

FREQUENCY TABLE, SHOWING NUMBER OF TIMES EACH YIELD
WAS REPORTED, BY INDIVIDUAL BUSHELS

Yield of Corn	Number of Times Reported	Yield of Corn	Number of Times Reported
Bushels		*Bushels*	
33	1	41	2
34	0	42	1
35	1	43	2
36	2	44	0
37	1	45	1
38	2	46	0
39	3	47	1
40	2	48	1

In preparing such a frequency table, spaces are put in for all yields (such as 34 bushels) for which no reports were received, but which lie between the largest and the smallest report, to show clearly that no such yields were reported.

Table 1.3 is an improvement on Table 1.2, but it is still pretty long—and if the lowest yield had happened to be 25, say, and the highest 70, it would have been longer still. For that reason it is frequently desirable to group the reports, not only for a yield of a specified number of bushels, but also for yields within a certain range of bushels. This is illustrated in Table 1.4, which gives the number of reports for groups covering 3 bushels.

The presentation is now condensed enough so that it can be readily understood. It is easy to see that most of the reports fell around 35.5

Table 1.4

FREQUENCY TABLE, SHOWING
NUMBER OF TIMES EACH YIELD WAS
REPORTED, BY 3-BUSHEL GROUPS

Yield of Corn	Number of Times Reported
Bushels	
32.5–35.4	2
35.5–38.4	5
38.5–41.4	7
41.5–44.4	3
44.5–47.4	2
47.5–50.4	1

to 44.4 bushels and that more fell near 40 bushels than anywhere else. Of course, the 3-bushel group is purely arbitrary, and any other convenient "class interval," as it is called in statistical terminology, could have been used. Thus, if a 5-bushel class interval had been selected, the convenient groups 29.5–34.4, 34.5–39.4, 39.5–44.4, and 44.5–49.4 bushels could have been established, giving frequencies of 1, 9, 7, and 3 for the four groups. Just what class interval makes the most satisfactory table for any given set of data depends upon how the data run and how much detail it is desired to show.[6]

Measures of Deviation

The Mean Deviation. Table 1.4 shows, in fairly compact form, the way that the several individual reports fall on each side of the average value. For some uses, however, it is desirable to have a single figure which expresses the variation or "scatteration" of the whole group of reports, in just the same way that the arithmetic mean expresses the average yield of the whole group.

One way in which the tendency of the group to scatter either far from or close to the mean can be measured, is by finding out how far, on the

[6] Where there is a tendency for the reports to be grouped around certain values, such as 5, 10, it is desirable to take the class intervals so as to make these values fall in the middle of the groups. Thus, with a concentration on whole numbers ending in 5 or 0, the groups 2.5–7.4, 7.5–12.4, 12.5–17.4, etc., may be used.

average, each report lies from the mean. Table 1.5 illustrates the way in which this can be done.

Table 1.5

COMPUTATION OF MEAN DEVIATION

Original Report	Mean	Report Minus the Mean
Bushels	*Bushels*	*Bushels*
39	40	−1
35	40	−5
48	40	8
40	40	0
37	40	−3
.. *		
Total		60†

* The remaining 15 reports are not shown in this table, though included in the total.
† The plus and minus signs are disregarded in making this total.

$$\text{Mean deviation} = \frac{60 \text{ bushels}}{20} = 3 \text{ bushels}$$

In computing the mean deviation, the plus and minus signs are disregarded in adding up the individual differences from the mean.[7]

The new figure, 3 bushels, is the *mean deviation* of all the reports. It

[7] Before writing the general formula for the mean deviation it is first necessary to have some way of writing *any* deviation. Using X to indicate any given report, as before, and M_x to indicate the arithmetic average of all such reports, the small x will be used to indicate the deviation of each report from the mean of all, thus:

$$X - M_x = x \tag{1.2}$$
$$X' - M_x = x'$$
$$X'' - M_x = x''$$

and so on.

Parallelling the previous notation, $\Sigma|x|$ (read "summation of all the small x's") is used to indicate the sum of the values such as x, x', x'', etc., taken without regard to sign.

The mean deviation is then defined:

$$\text{Mean deviation} = \frac{\Sigma|x|}{n} \tag{1.3}$$

It is necessary to disregard the signs in taking this sum, as otherwise the sum would be zero.

shows that the 20 individual reports differed from the mean yield of 40 bushels by an average of 3 bushels each. This furnishes a single figure which expresses how much or how little the individual yields differed from the average yield. If the group of 20 reports were being compared with another group of 20, all of 40 bushels each, the *mean deviations* of the two sets would indicate at once a striking difference in their make-up. The second set would have a mean deviation of 0, as compared to the 3-bushel mean deviation for the first set.

Whereas the arithmetic average is a measure of the central tendency of a group of reports, the mean deviation is instead a measure of the "scatteration" of the individual reports—of their tendency to lie near to or far from the central value.

The Standard Deviation. The variation of a group of reports around the mean of the group can also be measured by another statistic which has certain advantages from a mathematical point of view. (A statistic is the value of a given coefficient computed from a sample.) This measure is based on the deviation of each report from the mean, just as is the mean deviation. After the individual deviations are computed, each one is squared, and then added. This process is shown in Table 1.6.

The sum of the squared deviations is then divided by the number of items included in the group, and the square root of the result is computed. The computation is as follows:[8]

$$\frac{288}{20} = 14.4$$

$$\text{Standard deviation} = \sqrt{14.4} = 3.79 \text{ bushels}$$

[8] The letter s is used as the sign for the standard deviation computed from a sample. Using x to represent individual differences from the mean, as before, x^2 for the square of each of such deviations, and Σx^2 for the sum of all such values, the standard deviation is defined mathematically by the formula

$$s_x = \sqrt{\frac{\Sigma x^2}{n}} \tag{1.4}$$

Where the arithmetic average is a fraction, so that computing each individual deviation and squaring it would take much arithmetic for accurate work, the standard deviation may be computed more easily by the following formula:

$$s_x = \sqrt{\frac{\Sigma X^2}{n} - M_x^2} \tag{1.5}$$

Here the original X values are squared instead of the deviations from mean, or x, values. It can be readily demonstrated algebraically that the two formulas give identical values for s_x.

along the bottom of the chart and the number of reports falling in each group along the sides.[11]

Besides showing the number of reports included in each 3-bushel group by the height of the continuous line, the position of the mean in about the center of the group of reports is indicated, and likewise the number of reports included within a range of both one average deviation and of one standard deviation on each side of the mean.

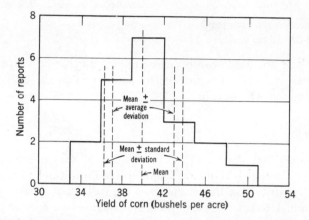

Fig. 1.2. Frequency distribution of corn yields, and range above and below the mean included by average and standard deviations.

If a similar chart were made for a very large number of observations of a variable normally distributed, it would have the shape shown in Figure 1.1.

Statistics and Parameters. This chapter has defined three measures of a series of data, the arithmetic average or mean, the average deviation, and the standard deviation. These were obtained from a sample of cases (yields on individual farms) selected at random from all the cases in the given universe (the county). If records were collected for the entire population of the universe (all the farms in the county) as in a census, only one fixed value could be obtained for each of these measures. These

[11] Mathematically, the quantities which are measured from left to right and shown along the bottom of the chart, as the bushels of corn are here, are called the *abscissas*, whereas the quantities which are measured from bottom to top and shown along the sides, as the number of reports are here, are called the *ordinates*. Since any point in the chart can be located by telling how far it is from the left side and how high it is from the bottom, these two items tell exactly where any particular point in the figure should fall. Thus, the line for the group from 38.5 to 41.5 bushels has for ordinate the height 7 farms, and the abscissas of the ends of the line are 38.5 and 41.5 bushels. The ordinate and abscissa, taken together, are called the *coordinates* of a point.

true fixed values for the entire universe are designated *parameters*. The values calculated from a sample drawn from the universe can only be estimates or approximations of the true parameters. Such estimates are called *statistics*. To make this difference clear, we will use a different notation for the values calculated from a sample and for the fixed (and generally unknown) true values for the universe, as follows:

Measure	Statistic from the Sample	Parameter for the Universe
Arithmetic average (or mean)	M_x	μ_x
Standard deviation	s_x	σ_x

Latin letters are used to represent sample values, and corresponding Greek letters are used to represent the parameters in the universe from which the sample was drawn.

Summary

This chapter has shown (1) how a series of measurements of any variable, such as the yield of corn from farm to farm, can be classified into a frequency distribution that shows how the individual reports are distributed from high to low; (2) how an arithmetic average may be computed that shows the value around which all the reports center; and (3) how the variation of the individual reports from the average may be summarized by computing the average deviation or the standard deviation, which serve as indicators of the variability of the items included in the particular series. Although these measures, especially the arithmetic average, are frequently of value for themselves alone, they are discussed here because it is necessary to know how they are computed and what they mean before the next proposition to be discussed can be fully understood.

Note 1.1. Where the number of observations is large, the standard deviations can be computed more readily from a grouped frequency table than from the individual items. This process is illustrated in the tabulation shown on p. 13.
The standard deviation is then calculated from the grouped data by the formula

$$s_u = \sqrt{\frac{\Sigma(d^2F)}{n} - \left[\frac{\Sigma(dF)}{n}\right]^2 - \frac{c^2}{12}} \qquad (1.6)$$

Substituting the values shown in the tabulation

$$s_u = \sqrt{\frac{33}{20} - \left(\frac{1}{20}\right)^2 - \frac{1}{12}} = \sqrt{1.65 - (0.05)^2 - 0.0833} = 1.25$$

In making this computation, any convenient group may be selected as the assumed

Yield	Number of Reports, F	Deviation from Assumed Mean, d	Extensions	
			dF	d^2F
32.5–35.4	2	−2	−4	8
35.5–38.4	5	−1	−5	5
38.5–41.4	7	0	0	0
41.5–44.4	3	+1	3	3
44.5–47.4	2	+2	4	8
47.5–50.4	1	+3	3	9
Sums	20	...	+1	33

mean, and the deviations of the other groups (d) in class-interval units calculated as departures from it. This method assumes that all the cases in each group fall at the center of the group. With most variables, with a tendency toward a normal distribution, the average of the items in each group will fall somewhat nearer the center of the distribution than the midpoint of the group, so the use of this method tends to give too large a value for the standard deviation. The correction $-c^2/12$ (called "Sheppard's correction" after its originator) makes an approximate allowance for this tendency. The c of the formula stands for the number of units of d in each class interval. Where a unit of 1 is used for each class interval, as in this problem, the correction becomes simply $-\frac{1}{12}$, to be applied to s_u^2. In this case, failure to use Sheppard's correction would increase s^2 by 5%. If a larger number of groups is used, the effect would be less and therefore this correction is usually ignored in practice.

In computing the standard deviation from a grouped frequency table, the s calculated will be in terms of the units in which d is expressed. In the illustration, each unit in d —one class interval—represents 3 units in X, since the yields were grouped in 3-bushel classes. The standard deviation computed in terms of class intervals, s_u, is therefore only one-third as large as is the standard deviation in terms of X. The latter may be calculated from the former by multiplying s_u by the number of units in each group. That is, $s_x = $ (units of X per class interval) s_u. In this problem $s_x = 3(1.25) = 3.75$.

The resulting value, 3.75, found by the short-cut method, is seen to be nearly the same as the exact value of 3.79 bushels, previously found by the longer method. The greater the number of cases in the sample, and the more nearly normal the distribution, the more time will the short-cut method save, and the more nearly will its approximate result agree with the exact value found by the longer method.

CHAPTER 2

Judging the reliability of
statistical results

Almost without exception, the object of a statistical study is to furnish a basis for generalization. In a case like that discussed in the preceding chapter, for example, no one would be likely to visit 20 farms scattered all over a county simply for the purpose of finding out what the yield of corn was on *those particular farms*. Instead, he might be studying the yield on those farms as a basis for determining what the average yield of corn was for all the farms in the county. Stated in statistical terms, he would be finding out what was the average yield in a *sample* of farms, picked at random, with a view to judging what was about the average yield in the *universe* in which he was interested, that is, all the farms in the county.

Of course it would be possible to visit all the farmers in the county, find out exactly what yield each one obtained, and so get an average of all the yields in the whole county. But this process would not only be expensive but also in most cases would be a pure waste of time and energy. We need only take a large enough sample by a well-designed sampling method to satisfy ourselves to any desired degree of confidence concerning the actual average for all the farms of the county. In this case, 100 records may enable one to determine the average yield quite as accurately as is necessary. Obtaining records from all the several thousand farmers in the county might add nothing to the usefulness of the results.

Before considering ways of finding out how many records would be needed in any given case, we might well discuss a little more fully what the process of statistical inference involves. Really, all that we do is to examine or measure a certain group of objects, and *infer* from the size or measurement of those objects, or from the way those objects behave, what will be the size of other objects of the same sort, or how other objects of the same kind will behave. Thus, statistical inference is a special case of the logical process of *induction*, whereby we reason from particular facts about particular objects to *general* conclusions as to what will be the facts for all objects of a given class. Now of course we do not really know what the particular facts are for any particular object without actually examining that individual object. All that we can do is to separate certain groups of objects we know to be alike in one or more particulars, and then assume that they will be alike in other particulars too, even though we do not examine every one to prove it.

Assumptions in Sampling. The basic assumptions upon which the theory of sampling rests apply both to the way the sample is obtained and to the material being sampled. With respect to the material sampled, the assumption is that there is a large universe of items subject to more or less uniform conditions, in that throughout the universe the individual items vary among themselves in response to the same causes and with about the same variability. With respect to the selection of sample, the values must be so selected (1) that there will not be any relation between the size of successive observations, that is, that the chances of a high observation being followed by another high observation will be just the same as of a low or a medium observation being followed by a high observation; (2) that the successive items in the sample are not definitely selected from different portions of the universe in regular order, but are simply picked at random so that the chance of the occurrence of any particular value is the same with each successive observation in the sample; and (3) that the sample is not picked all from one portion of the universe, but that the observations are scattered through the universe by purely chance selection.[1] Where these assumptions are fulfilled, the sample is designated a "random sample," and its reliability can be estimated by the methods now to be described.

Taking up the question of how reliable a statistical average really is, we must first consider, "What is the meaning of *reliable?*" If we are interested in corn yield, for example, it is obvious that a perfectly reliable sample would be one whose average agreed exactly with the average yield in the county. But if we are interested in knowing the average yield to

---

[1] Where the items are so selected as to represent different portions of the universe, it may be called a "stratified sample"; where they are all selected from one portion of the universe, it may be called a "spot" sample.

Where the universe is not completely uniform, a stratified sample tends to be more reliable than a random sample, and a spot sample tends to be less reliable than a random sample. See G. U. Yule and M. G. Kendall, *Introduction to the Theory of Statistics,* 14th ed., pp. 533–539, and P. V. Sukhatme, *Sampling Theory of Surveys with Applications,* pp. 83–137, Iowa State College Press, Ames, and Indian Society of Agricultural Statistics, New Delhi, 1953, for formulas as to the reliability of stratified samples.

within 1 bushel, then for that purpose the sample would be sufficiently reliable if its average was almost certain to come within 1 bushel of the average for the whole county.

Variations in Successive Samples. Suppose that 20 farms had been visited at random, with the results presented in Chapter 1. If we wanted to find out how near we could expect the average from that sample to come to the average for the county as a whole, we might try taking another sample—visiting 20 other farms at random, and getting the average yield for those 20. If the average yield of the second sample differed from the average of the first sample by, say, 3 bushels, we should know that both could not come within 1 bushel of the true average; if, however, the average of the second sample came within a half bushel of the first average, we should be inclined to place more confidence in both of them. If we repeated the process several times over, and all the different samples had averages falling within 1 bushel of each other—say between 39.0 and 40.0 bushels—then we should feel pretty certain that the average yield for the county as a whole was about 39.5 bushels.

Let us suppose that 15 more samples had been taken, each of 20 farms selected at random, and that when we tabulate the 16 averages from the 16 different samples, we have the 16 values shown in Table 2.1.

Table 2.1

AVERAGE YIELD OF CORN IN ONE COUNTY, AS DETERMINED
BY 16 DIFFERENT SAMPLES OF 20 FARMS EACH

Sample	Yield	Sample	Yield
	Bushels per acre		*Bushels per acre*
1	40.0	9	40.3
2	37.5	10	38.9
3	39.3	11	39.3
4	40.6	12	38.0
5	39.8	13	39.2
6	41.1	14	40.9
7	38.3	15	39.1
8	39.6	16	40.4

Although the 16 averages range all the way from 37.5 bushels for the smallest to 41.1 bushels for the largest, we can see that most of them fall around 39 or 40 bushels. This is even more evident when we arrange the 16 reports in a frequency table, as shown in Table 2.2.

Although there is some tendency for the averages to cluster around 39 and 40 bushels, still there are several below 38.5 and several above 40.5. The average for the whole group is 39.5 bushels, and the standard deviation is 1.00 bushel.

The fact that the standard deviation of the group of averages is 1 bushel tells us one thing about the way they scatter, from what we already know about the meaning of *standard deviation*. It tells us that about 68 per cent

Table 2.2

FREQUENCY TABLE SHOWING THE NUMBER OF TIMES VARIOUS AVERAGE YIELDS WERE OBTAINED OUT OF 16 SAMPLES, BY ONE-HALF-BUSHEL GROUPS

Yield of Corn	Number of Averages in Group	Yield of Corn	Number of Averages in Group
Bushels		*Bushels*	
37.5–37.9	1	39.5–39.9	2
38.0–38.4	2	40.0–40.4	3
38.5–38.9	1	40.5–40.9	2
39.0–39.4	4	41.0–41.4	1

of them will fall in the range between one standard deviation below the mean of all the averages and one standard deviation above the mean. In this particular case, the mean is 39.5 bushels, and the standard deviation is 1 bushel, so the interval of one standard deviation above and below the mean covers the range from approximately 38.5 bushels to 40.5 bushels. Checking this against the array of averages shown in Table 2.2, we find that this interval does include 10 out of the 16 cases, or fairly close to the proportion expected.

Now let us go back to our single original average of 40 bushels, based on visits to the original 20 farms. What we want to know is how reliable that one average is. Stated another way, how much is that average likely to be changed if the study were made over again—if another sample of the same size were taken?

In Tables 2.1 and 2.2 we have seen how it might actually work out if we *did* do the study over several times. We have seen that, if the new averages did fall as shown in those tables, two-thirds of the new averages would fall within an interval of 2 bushels. Furthermore, those figures showed that *all* the different averages fell within a range of 4 bushels (37.5–41.5). But those conclusions were obtained only *after* getting 15 more samples of 20 cases each, and making 15 new averages, one for each sample. Is there any way to find out how much the original average is likely to vary from the true average without going to all the work of taking a number of new samples?

Estimating the Reliability of a Sample

If we could estimate the extent to which the averages from new samples would be likely to vary, *without ever getting the new samples*, then we should know something more about how much faith we could put in the particular average that we had already. For example, if in the present case we knew that, if we did go out and get a large number of new averages (such as those shown in Tables 2.1 and 2.2), those new averages would have a standard deviation of 1 bushel, this fact would tell us at once *something* about how much our one average was likely to be different from the real average on all the farms. For example, we should know that about 68 per cent of the sample averages would lie in an interval of 2 bushels (one standard deviation on each side of the mean of the samples). The one particular average that we had obtained might be any one of all those in a distribution like that shown in Table 2.2. If we assume that the mean of a large number of samples would coincide with the true average, then the chances would be about 68 out of 100 that our average was one of the averages falling within *1 bushel* of the true mean. If, on the other hand, we knew that the standard deviation of a group of new averages would probably be, say, 5 bushels, then we should know that we had only about 2 chances out of 3 of the mean of any one sample of 20 cases coming within *5 bushels* of the true average. Obviously, when an average has 2 chances out of 3 of coming within 1 bushel of the true average, it is much more reliable than if it had 2 chances out of 3 of coming within 5 bushels of the true average.

Whether we can judge how much confidence we can place in a given average depends, therefore, on whether we can tell what would be the standard deviation of a number of similar averages, computed from random samples of the same number of items drawn from the same universe. If we could tell exactly what that standard deviation would be, we should know how much faith we could put in the average we had —we should know what the chances were of its being changed by more than a given amount if the study were made over, or if a very large sample were taken. Even if we did not know *exactly* what the standard deviation of the whole group of similar averages would be, it would be some help if we knew approximately what it would be, or if we had a minimum or maximum value for its size, so that there would be some measure of how much trust to place in the particular average.

Computing the Standard Error. Fortunately, it is possible to estimate with some degree of accuracy what the standard deviation of a whole series of averages is likely to be, if each average is computed from a sample of the same size and drawn at random from the same universe. Even so, the ability to make such an estimate is a tremendous aid to statistical investigators, for it affords some check on the dependability of results without going to the expense that would be involved in repeating every sample 15, 20, or more times, to make sure that a reliable result had been obtained.

The method for computing the estimated standard deviation of the average involves just two values. These are (1) the standard deviation of the items in the universe from which the sample was drawn; and (2) the number of items in the sample. We do not know the standard deviation of the items in the universe, however, and can only estimate it from the standard deviation of the items in the sample. It has been determined that an unbiased estimate of the standard deviation in the universe can be made by adjusting the standard deviation observed in the sample as follows:[2]

$$\text{Estimated standard deviation of the universe} = s_x\left(\sqrt{\frac{n}{n-1}}\right)$$

In the sample considered in Chapter 1,

$$= 3.79\sqrt{\frac{20}{19}} = (3.79)(1.026)$$

$$= 3.89$$

[2] Using the symbol s_x as before to mean the standard deviation observed in the sample and $\bar{s}$ to represent the estimated standard deviation in the universe from which the sample was drawn, we can define the sample value adjusted for bias as

$$\bar{s}_x = s\sqrt{\frac{n}{n-1}} \tag{2.1}$$

It can more readily be computed by the equation

$$\bar{s}_x = \sqrt{\frac{\Sigma x^2}{n-1}} \tag{2.2}$$

Actually, it is the value of the estimate of σ^2 which is unbiased; and $\bar{s}_x$ is the square root of that unbiased estimate.

The two equations are identical, as may readily be proved by combining equations (2.1) and (1.4).

When equation (1.5) is used, $\bar{s}$ may be computed

$$\bar{s}_x = \sqrt{\frac{\Sigma X^2 - nM_x^2}{n-1}} \tag{2.3}$$

The mean from a sample has no bias, so the value adjusted for bias is identical with the sample value, M.

The standard deviation of the group of averages may next be estimated by dividing the estimated standard deviation in the universe by the square root of the number of cases in the sample. Thus, for our original sample of 20 farms,[3]

Standard error of the average

$$= \frac{\text{estimated standard deviation of items in the universe}}{\text{square root of the number of cases in the sample}}$$

$$= \frac{3.89 \text{ bushels}}{\sqrt{20}}$$

$$= \frac{3.89 \text{ bushels}}{4.47}$$

$$= 0.87 \text{ bushel}$$

In comparison with the 15 other averages, all shown in Table 2.1, we see that the standard deviation of all 16 averages was a trifle larger in this case than we estimated it was likely to be—1.00 bushel, as compared to 0.87 bushel expected. It has already been noted that where a number of repeated samples are actually taken, this may easily occur. Since the observed standard deviation will vary from sample to sample, the estimate of the standard error of the average will also vary. Even so, this estimated "standard deviation of similar averages" is an exceedingly useful figure. Such an estimated standard deviation for an average (or any other statistic) is called the *standard error* of that average (or other statistic). It serves as a standard measure to give some indication of how much such a sample may give results that vary from the true facts of the universe, solely as the result of chance fluctuations in sampling.

[3] Here the symbol s denotes the standard deviation as before, the subscript x indicates that it is the standard deviation of the individual items that go to make up our sample, and the subscript M indicates that it is the standard deviation of the means which is to be computed, thus:

$\bar{s}_x$ = standard deviation of the items in the universe, estimated by equation (2.1), (2.2), or (2.3)

s_{M_x} = estimated standard deviation of the group of averages if similar samples were repeated = *standard error* of the mean of X

The standard error of the mean is then given by the formula

$$s_{M_x} = \frac{\bar{s}_x}{\sqrt{n}} \tag{2.4}$$

Here, just as in the previous formulas, n stands for the number of items in the original sample—the same items as those from which s_x was computed.

Confidence Intervals for Large Samples. It has been shown that the *means* of successive random samples are distributed in a manner very close to that of a normal curve, even if the distribution of the original observations is far from normal.[4] Where the sample size is large (30 observations or more) and is drawn under conditions of simple sampling, the interval $M \pm s_M$ will include the true mean in 68 per cent of all such samples. The interval $M \pm 2s_M$ will include the true mean in 95 per cent of such samples, and the interval $M \pm 3s_M$ in 99.7 per cent. These ranges are called *confidence intervals*, and indicate for any given sample about how much confidence we can place in the average (or other statistic) derived from that sample as an approximation to the universe average (or other parameter).

Confidence Intervals for Small Samples. When the sample size is smaller than 30 observations, however, the distribution of averages from successive samples is somewhat different from the normal curve. The proportion of samples in which a stated confidence interval, such as $M \pm 2s_M$, will include the true mean is *smaller* and the proportion of sample means departing from the true mean by more than $\pm 2s_M$ is larger. The smaller the number of observations, the more serious the difference. The exact distributions of averages (and other statistics) from small samples have been worked out and tabulated for different sample sizes.

This correction shows for specified confidence limits around the statistic ($\frac{1}{2}$, 1, $1\frac{1}{2}$, 2, etc. times the standard error computed from the sample), what the probabilities are that those limits will include the true value for the universe. These proportions are given in Table 2.3 and in Figure 2.1.[5]

Table 2.3 shows the proportion of samples of each given size which yield confidence intervals of each type that actually include the parameter for which they were constructed. Thus, if the sample is large, and we state that the true average lies within one standard error of the computed average, we should probably be right for 7 out of 10 of such statements.

[4] M. G. Kendall, *The Advanced Theory of Statistics*, Vol. I, p. 180, 1949.

[5] Table 2.3 applies as stated only in the case of statistics such as the arithmetic average, which are computed from the original data by the determination of a single constant. Where the computation of the statistic involves simultaneously determining two constants from the original data, $n - 1$ should be used for the number of observations in the sample. This applies to the coefficient of regression. Where the computation of the statistic involves simultaneously determining a large number of constants, say j in number, from the original data, then $(n - j + 1)$ should be used for the "number of observations" in Table 2.3 or Figure 2.1. Thus, for a coefficient of partial regression, $b_{12.345}$ obtained from a sample of 20 observations, 5 constants are involved, so 16 would be used as the number of observations in using Table 2.3 to judge the reliability of the computed value. (Subsequent chapters will explain the meaning of the new coefficients mentioned here.)

(The exact proportion expected is 683 out of 1,000.) If there were 20 observations in the sample, and we made the same statement, we should be right 67 times out of 100. But for samples with only 2 observations, such a statement would be right only 50 times out of 100, on the average.

Table 2.3

PROPORTION OF SAMPLES OF EACH GIVEN SIZE THAT YIELD
SPECIFIED CONFIDENCE INTERVALS THAT ACTUALLY INCLUDE
THE PARAMETER*

Confidence Intervals	Probability (P) for Samples with n Equal to						
	2	4	6	10	16	20	30 or more
$M \pm 0$	0	0	0	0	0	0	0
$M \pm 0.50 \, s_M$	.295	.349	.362	.371	.376	.377	.383
$M \pm s_M$	.500	.609	.637	.657	.667	.670	.683
$M \pm 1.5 \, s_M$	.626	.769	.806	.832	.846	.850	.866
$M \pm 2.0 \, s_M$	.705	.861	.898	.923	.936	.940	.954
$M \pm 2.5 \, s_M$	.758	.912	.946	.966	.975	.978	.988
$M \pm 3.0 \, s_M$	.795	.942	.970	.985	.991	.993	.997
$M \pm 3.5 \, s_M$	.823	.961	.983	.993	.997	.998	.9995
$M \pm 4.0 \, s_M$	.844	.972	.990	.997	.999	.999	$\cdots$

* Based on article by "Student," New tables for testing the significance of observations, *Metron*, Vol. V, No. 3, pp. 105–120, 1925.

The estimated standard error of 0.87 bushel from our single sample of 20 cases with an average of 40.0 bushels, would therefore tell us that 67 per cent of such samples would have averages which fell within an interval of ±0.87 bushel of the true mean. If our sample was a true random sample, we should then have 2 chances out of 3 of being right if we assumed that the confidence interval of 39.13 to 40.87 bushels included the true average yield for all the farms in the county.

It is evident from Table 2.3 that very small samples have much less reliability than samples of even moderate size. Thus, a sample of 30 cases has a confidence interval of $\pm 2s_M$ for $P = 0.95$; but for a sample of 10 cases, we see from Table 2.3 (or Figure 2.1) that we have to take a confidence interval of $\pm 2.25s_M$ to have the same probability; for a sample of 6 cases, $\pm 2.6s_M$; and for a sample of 4 cases, $\pm 3.2s_M$. (The selection of 30 cases as the dividing line between "small" and "large" samples is somewhat arbitrary; but the distribution of averages from samples of 30

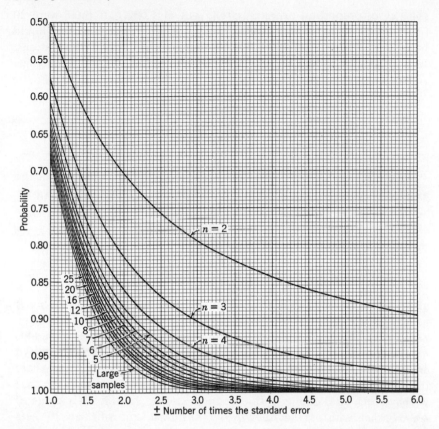

Fig. 2.1. The proportion of random samples in which the interval of the observed mean plus and minus the stated multiple of the standard error computed from the sample, will include the true mean, for samples of stated numbers of observations.
(To apply to coefficients of regression, see footnote 6.)

cases is so close to the normal distribution which underlies the last column of Table 2.3 that the differences are generally disregarded.)

From equation (2.4) it is clear that our estimates of s_M would vary from sample to sample, as each sample would give us a different value for $\bar{s}_x$. The *standard error of the standard error*, stated in relative terms, depends solely upon the number of cases in the sample. It is computed as follows:[6]

[6] Using σ_s/σ_x to represent the standard error of the estimated standard error, we can define it by the equation

$$\frac{\sigma_s}{\sigma_x} = \frac{1}{\sqrt{2(n-1)}} \tag{2.5}$$

Relative standard error of the standard error

$$= \frac{1}{\text{square root of two times (number of cases in sample} - 1)}$$

For our sample of 20 cases

$$\frac{\sigma_s}{\sigma_x} = \frac{1}{\sqrt{2(20-1)}} = \frac{1}{\sqrt{38}}$$

$$= 0.162$$

With very small samples, even the estimate of the standard error of the average is subject to a wide zone of uncertainty. With 4 cases, its own standard error is 41 per cent of the value computed.

Meaning and Use of the Standard Error

It is good statistical practice, whenever an average is cited, to give with that average its estimated standard error, so that the reader will know about how significant that average is, and will not be led into using it to make comparisons or to draw conclusions that are not justified by the number of observations which are summed up in that average. This may be done either by showing the average followed by ± its own standard error, or with its standard error shown below it in parentheses; and where the sample is small, by stating the number of reports. Thus, in the case we have been considering, with the single sample showing an average of 40.00 bushels with a standard error of 0.87 bushel, and with only 20 cases in the sample, the correct statement is to say "the average yield has been shown by the sample to be 40.0 ± 0.87 bushels (20 cases)." If a similar sample from a different area has shown the average yield to be 38 ± 2.0 bushels (20 cases), the reader would know that there was a fair chance that the true average yield in the second area was really as large as in the first area in spite of the difference in the two sample averages.

The greatest value of the standard error does not lie in merely indicating how near the sample value may come to the true value for two samples out of three. Mathematicians have determined for large samples that 19 out of 20 (95.45 per cent) of the samples will give averages which fall within *two* standard deviations of the mean, 369 out of 370 (99.73 per cent) will usually fall within *three* standard deviations of the mean, and all but one case out of 16,667 samples (99.994 per cent) will usually fall within *four* standard deviations of the mean.

As shown in Table 2.3, when there are less than 30 observations in the sample, the tendency of the computed standard error to be misleading is greater for high odds than it is for lower odds. Thus, with samples of 20 cases, 6 samples out of 100 will give averages differing from the true average by more than twice the computed standard error, and 7 samples out of 1,000 will miss the true average by more than three standard errors. This last is three times the proportion of such failures which would occur in the long run with samples of over 30 observations. With very small samples, the failures for high odds occur even more frequently. Thus, for samples with only 4 observations, 14 samples out of 100 will differ from the true mean by twice the computed standard error, and about 6 out of 100 will differ by three times the standard error, on the average. As already explained, it is therefore necessary to take wider confidence intervals, in multiples of the standard error, to attain even approximately the same probabilities with smaller samples.

Interpreting the Standard Error in the Illustrative Problem. We can interpret the statement that the average yield in the area studied was 40 ± 0.87 bushels in any of the following ways:

1. If we state that the true mean lies within one standard error of the observed mean (between 39.13 and 40.87 bushels, in this case) each time we use a sample of this size, we shall be wrong in our statement 1 time out of 3, on the average.

2. If we state that the true mean lies within two standard errors of the observed mean (between 38.26 and 41.74 bushels) each time we use a sample of this size, we shall be wrong in our statement 1 time out of 17, on the average.

3. If we state that the true mean lies within three standard errors of the observed mean (between 37.39 and 42.61 bushels) each time we use a sample of this size, we shall be wrong in our statement 1 time out of 135, on the average.

4. If we state that the true mean lies within four standard errors of the observed mean (between 36.52 and 43.48 bushels) each time we use a sample of this size, we shall be wrong in our statement only 1 time out of 1,250, on the average.[7]

Comparing these conclusions with the 16 samples shown in Tables 2.1 and 2.2, we see that 2 of those samples did fall outside the limits given by twice the estimated standard error. If we had been so unlucky as to have got the worst one of these as our single sample, instead of the one we actually did get, then we should not have hit the average even if we had used a range of twice the computed standard error as that within which we expected the true average to lie. On the other hand, every one of

[7] These odds are calculated from tables carried to more decimal places than those shown in Table 2.3.

the averages fell within the range covered by three times the standard error. Even if, in picking our single sample, we had been unfortunate enough to draw the poorest one of the lot—the one which gave an average yield of 37.5 bushels—and had used a range of three times the standard error, we should have been correct in our statement as to the range within which we expected the true average to lie. Then we should have concluded that the true mean lay somewhere between 34.3 and 40.7 bushels, which would have been wide enough to include the real mean. Of course, if we had taken four times the standard error, we should have been almost absolutely certain of including the true mean in the stated range, with only one chance in over 1,000 of being wrong.

For any given size of sample, each of the confidence intervals cited corresponds to a particular probability, P. However, as P varies with sample size, many statisticians first choose the value of P that they are willing to "act on" in a given investigation and then select the corresponding confidence interval, which, in general, will not be an exact multiple of the standard error. Probabilities of 0.95 and 0.99 are most commonly used—their use implying that the investigator is willing to accept the one chance in 20, or the one chance in 100, that the universe value lies outside the specified range. Logically, the "probability of error" $(1 - P)$ one is willing to accept should vary with the importance of the actions that might be based on the results of the investigation. For the general run of statistical problems, and with fair-sized samples, it would seem safe to regard three times the standard error as about the largest extent to which the conclusions might be out *solely because of the chances of getting* an *unusual sample* in random sampling—the probability of error being held between 0.01 and 0.001.

In view of the possibility of the standard error itself being in error, however, the number of observations should always be stated, as well as the standard error of the statistic, particularly where the sample is small.

Bias in Sampling.   The figure as to standard error tells nothing at all of how much error there may be because of *bias* in sampling. Thus, if in taking our sample of 20 farms, we had visited only the largest farms with the most prosperous-looking buildings, we should be very likely to get a sample which was not representative of *all* the farmers in the county, but simply of the better ones, and so might get an average yield, considerably above the true average for the county. Even if we selected our farmers only to the extent of including those who were most willing to give us the figures we wanted, we might have a badly biased sample, as usually the best farmers and the most intelligent ones are most willing to answer such questions. The only theoretically sound way to avoid bias is to define the universe of interest very carefully in advance of the field investigation and then draw our sample in such a way that every item in the universe actually has an equal chance of being chosen in the sample. (In *stratified* random sampling the universe may be divided into two or more parts, but within each part each item should still have an equal probability of selection). In the corn yield example, this might require that we have a list of all farms in the county on which corn was grown, or that we number each square mile of land in the county, draw a random sample of these numbers, and measure the corn yield on every farm lying wholly or partly within each of the square miles appearing in the sample. If the expense of obtaining a complete list of farms or drawing an area sample seemed too great for the purpose at hand, we might select our 20 farms in a less formal way, still trying consciously to have them represent the different yield levels around the county in about the true proportions. However, in this case, we must depend largely on common sense and on other knowledge of the situation we are studying, and not on statistical computations, to tell us whether or not our sample is really representative of the universe we want to study. Thus, we might compare the average size or value of the farms in our sample with the averages for all the farms in the county, as shown by the census reports, to see whether they were representative or not in these respects. All that the computed standard error can tell us is about how closely the sample statistic is likely to approach the average (or other characteristic) *of the group it does actually represent*—whether that group is the one we meant it to represent or only a part of that group. This caution must always be kept in mind in using samples: Computed standard errors tell us how far our results may be off solely because of the chance of getting a poor sample with a limited number of cases; but they do not tell us how far we may be off because of a *biased* sample, which is not a fair selection from the universe we wish to study.

Deciding on the Size of Sample Necessary to Obtain a Stated Reliability. One other application of the standard-error formula remains to be mentioned. The way in which this formula can be used to estimate the reliability of the average from a given sample, when the number of cases is known, has already been explained. The same formula can be used to determine how large a sample would have to be taken in order to secure results within any previously assigned limits of accuracy.

Thus it has already been shown that the records from 20 farms could be used to say that the true average yield lay somewhere between 37.39 and 42.61 bushels, with about one chance in 135 of that statement's being wrong. How many farms would one have to visit to state the same average yield to within one bushel, with the same chance of the statement's being wrong? The same formula which was used to determine the standard error of the average can be turned around to answer this question also.

If we know that we want to get an average reliable to within one bushel, for a range of three times its standard error, then we know that the standard error of that average would have to be only one-third of a bushel. We may also assume that when we take our larger sample, the standard deviation of the yields on the individual farms will be found to be not very different from what it was in our sample of 20 cases, and so use the same standard deviation as we did before.

Taking the relation which was used in computing the standard error before, we have:

$$s_M = \frac{\bar{s}_x}{\sqrt{n}}$$

In the new case we have the required standard error given, $\frac{1}{3}$ bushel; we are assuming that the estimated standard deviation for the universe from our larger sample will be 3.89 bushels, just as it was from our sample of 20 cases. Substituting these values in our equation, and using n'' to represent the number of cases required in the new sample, we then have

$$\frac{1}{3} \text{ bushel} = \frac{3.89 \text{ bushels}}{\sqrt{n''}}$$

When the terms are shifted around, this becomes

$$\sqrt{n''} = \frac{3.89 \text{ bushels}}{\frac{1}{3} \text{ bushel}} = 11.67$$

Hence

$$n'' = 136.2$$

We therefore conclude that if a sample of 136 reports were obtained, we should probably get an average yield which would not differ from the true average yield for all the farms by more than one bushel in more than one sample out of several hundreds of such samples. If any other limit of error was set, we could similarly determine how many reports would probably be necessary to satisfy that limit.[8]

[8] Because the probability (P) associated with a confidence interval of three standard errors is larger for a sample of 136 cases than for a sample of 20 cases, we could hold the probability of being wrong by more than one bushel at one chance in 135 ($P = 0.993$) by choosing somewhat fewer than 136 cases. Thus, from Table 2.3 it appears that in large samples $P = 0.993$ for a confidence interval of about $M \pm 2.7 s_M$. This would lead to an acceptable s_M of about 0.37 bushels and a necessary sample size of about 110 cases. This refinement is superfluous in most practical situations; if the smaller sample size is 30 or more and if the acceptable probability of error is at least 0.01, the method of calculation shown in the text is sufficiently exact.

Standard Errors for Other Measures. This whole discussion has been in terms of determining how closely it was possible to approximate the *true average* from the *average shown by a sample*. In exactly the same way standard-error formulas have been worked out indicating how closely it is possible to approximate the true values of other parameters (such as standard deviations, for example) from the values for those measures determined from a sample. These are shown and interpreted in much the same way as are the standard errors of averages; they will be referred to in subsequent chapters.

Universes, Past and Present

Any statistical measurement relates to something that is already past by the time the measurement can be analyzed. Thus our records of the yield of corn obtained must relate to some crop that has already been harvested. Yields for a crop still growing could only be forecasts, and could never be precisely accurate until the crop was harvested and was weighed or measured. Yet human beings cannot live in the past. If we are planning an agricultural control program, for example, and wish to estimate how many acres will produce a given total bushelage, we shall be dealing with future years. We can do nothing to change the past. Only the future can be affected by our actions. When we take the average yield for a past year as our "universe" to be studied, what we are really interested in knowing is usually something about the yield most likely to be secured in one or in a series of years in the future.

Analysis of what has happened in a succession of years in the past may help us to make a better estimate of the future. Such analysis may show a steady upward trend, or a variation from year to year with rainfall, or other variations whose cause we do not know. But before we can project the past trends into the future, we must try to understand what caused them, and judge whether those causes will continue to operate. These judgments are not a matter of statistical analysis as such but must be based upon scientific and technological study of all the forces at work. Thus a steady upward trend in cotton yields might reflect a rising price of cotton in the period studied, and a resulting increase in the quantities of fertilizer applied per acre. But equally well it might reflect a steady decrease in the total acreage (due to crop control or other causes) and a concentration of the remaining acreage on the better lands. Or it might reflect the gradual adoption of improved strains. A forecast of whether the upward trend would continue into the future would be materially different in the three cases. Besides the statistical facts, it would involve study of what had been happening in cotton production in the area, and non-statistical judgments as to whether the increase in price or the limitation of acreage or the improvement in seed was likely to continue.

Whether we are dealing with the statistical characteristics of people or of crops or of prices or of atoms, the real universe for which we wish to estimate is the universe of future events. Our ability to forecast those events will differ widely from field to field. Presumably the characteristics of atoms or of chemical compounds will be less subject to change than will those of crops or prices. In each case, however, the statistical information gained from the study of past samples must be tempered by other knowledge of the situation, based on study and analysis which may be quite non-statistical in nature. When we move from the facts of the past to forecast the unknown universe of the future it is not the statistics but the statistician who is on trial. Unless he mixes an ample measure of anthropology or agronomy or economics or other appropriate scientific information with his statistics—plus a liberal dash of common sense— he may find his analysis of past events a detriment, rather than an aid, in judging the future. Some of the issues introduced here are discussed further in Chapters 17, 20, and 26.

Summary

This chapter considers the question of how far statistical results derived from a selected "sample" drawn from a universe can be used to reach general conclusions as to the facts of the entire universe.

The confidence which can be placed in any statistic computed from a sample, say an average, depends upon how closely that average is likely to come to the true average of the whole universe. One way of determining that would be to collect additional samples, each of the same size. From the way the averages from each of these different samples varied one could judge how near the average from any one sample was likely to come to the true average. For samples which meet the conditions of simple sampling, another much more rapid way is to estimate the *standard error* of the average, which gives a basis for estimating how much confidence can be placed in the observed average. With large samples, the true average will probably be within twice the standard error from the observed average for 21 samples out of 22, and within three times the standard error 369 times out of 370. Where the number of observations in the sample is less than 30, the possibility of error is larger, as is indicated by Table 2.3.

The same formula can be used to estimate how large a sample must be taken to attain any desired degree of confidence in the final average.

The estimated standard error does not take into account bias in selecting the sample, but only indicates the limits of confidence that can be placed in the result even when an "honest" random sample is obtained.

After the values in the universe have been estimated from the facts shown by the sample, the statistician must still remember that that universe is a past universe. In applying that knowledge to problems of future action, he must give due allowance to the fact that the as yet unborn universe of the future may never be identical with the past and dead universe from which his sample was obtained.

CHAPTER 3

The relation between two variables, and the idea of function

Relations are the fundamental stuff out of which all science is built. To say that a given piece of metal weighs so many pounds is to state a *relationship*. The weight simply means that there is a certain relationship between the pull of gravity on that piece of metal and the pull on another piece which has been named the "pound." We can tell what our "pound" is only by defining it in terms of still other units, or by comparing it to a master lump of metal carefully sheltered in the Bureau of Standards. If the pull is twice as great on the given piece of metal as it is on the standard pound, then we say that the lump weighs 2 pounds. If, further, we say it weighs 2 pounds per cubic inch, that is stating a composite relationship, involving at the same time the arbitrary units which we use to measure extent or distance in space and the units for measuring the gravitational force or attracting power of the earth.

Relations between Variables.   Besides these very simple relationships which are implicit in all our statements of numerical description—weight, length, temperature, size, age, and so on—there are more complicated relationships where two or more variables are concerned. A variable is any measurable characteristic which can assume varying or different values in successive individual cases. The yield of corn on different farms is a variable, since it may differ widely from farm to farm. So is the length of time which a falling body takes to reach the earth, or the quantity of sugar that can be dissolved in a glass of water, or the distance it takes for an automobile to stop after the brakes are applied, or the quantity of milk that one cow will produce in a year, or the tensile strength of a piece of metal, or the length of time it takes a person to memorize a quotation. In contrast to these *variables* there are other numerical values called *constants*, because they never change. Thus one foot *always* contains 12 inches; one dollar *always* is equal to 100 cents; and a stone *always* falls 16 feet in the first second (under certain specified conditions). Science, of any sort, ultimately deals with the relation between variable factors and with the determination, where possible, of the constants which describe exactly what those relationships are.

The variables which have been mentioned may be used to illustrate the way in which changes in one variable can be related to changes in another. Thus the length of time which a falling body takes to reach the earth varies with—that is, is related to—the distance through which the body has to fall. The quantity of sugar which can be dissolved in a glass of water varies both with the size of the glass and the temperature of the water. The distance it takes for an automobile to stop after the brakes are applied varies with the speed at which the car is traveling when the brakes are applied, the area of braking surface on the drums, the area of tire surface on the road, how tightly the brakes are applied, how much the car weighs, the kind of road, and so on.

Then when we come to variables like the production of milk, or the time required to memorize a quotation, we find the situation still more complicated. How much milk a cow will produce varies with her age, breed, inherent ability, and the richness of the milk, and with the kind, quality, amount, and composition of the feed she receives, the way she is stabled and cared for, and many other similar factors. The time it takes to memorize a quotation may be affected by its length, the subject's age, sex, training, fatigue or freshness, his familiarity with material discussed, and his interest in the topic. The strength of a piece of metal will be affected by its size, shape, composition, heat treatment, temperature, and so on.

Yet it is precisely with relations between complex variables that many statistical studies must deal. Science deals in large part with the relations between variables, and with the parameters that express these relations.

The statistical methods which may be used to handle such problems can best be understood if presented first for the simplest cases, and then expanded to cover the more complicated ones. Suppose a physicist made some experiments to determine the relation between the distance a body has to fall and the length of time it takes, by dropping a marble different distances and measuring the time it takes to reach the ground, and obtained the results shown in Table 3.1.

Looking over these figures we see that there is some sort of general relation between the two columns. As the distance increases, the time increases also. But that is not uniformly true. In one case the distance increased without there being any increase in the recorded time; in some other cases the recorded time was not the same even though the distance was unchanged.

Table 3.1

RELATION BETWEEN DISTANCE A MARBLE DROPS AND TIME
IT TAKES TO FALL

Distance Traveled	Time Elapsed	Distance Traveled	Time Elapsed
Feet	*Seconds*	*Feet*	*Seconds*
5	0.6	20	1.1
·5	0.5	20	1.1
5	0.6	20	1.2
10	0.9	20	1.1
10	0.8	25	1.2
10	0.7	25	1.3
15	1.0	25	1.2
15	0.9	25	1.3
15	1.0		

Graphic Representation of Relation between Two Variables. We can get a better idea of what the relation is if we "plot" it on cross-section paper, so that we can see graphically just how the time does vary with

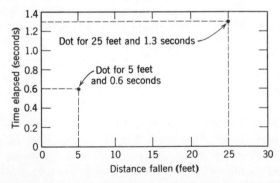

Fig. 3.1. Method of constructing a dot chart. Time elapsed is the dependent variable, and distance is the independent variable.

the distance. Figure 3.1 illustrates the way this is usually done. The units of one variable, in this case the distance to be traversed, are measured off from the left, starting with zero in the lower left-hand corner and counting over toward the right. The units of the other variable, in this case the time elapsed, are measured off from the bottom, starting with zero and counting up toward the top. If negative values are present, then the counting is started with the *largest negative* value, decreasing from left to right or from bottom to top, until zero is reached and the positive values begin to appear.

Where one variable may be regarded as the cause and the other variable as the result, it is customary to put the causal variable along the bottom. In this case with the measurements made by dropping the marble known distances and measuring the time elapsed, it may be said that the differences in distance traversed cause the differences in time elapsed. Distance, therefore, is measured in the horizontal direction, and time in the vertical. There is no particular reason for plotting data just this way except that this is the customary way of doing so. Some relations of this sort can be reversed, so that either may be regarded as cause and either as effect.[1]

Having laid off the chart in the way indicated, we next "plot" the individual observations. The way this is done is illustrated in Figure 3.1. The first observation was that it took 0.6 second for the marble to fall 5 feet. This is indicated on the chart by counting over to the 5-foot line from the left of the chart, and then counting up along that line until 0.6 second is reached. A dot is placed on the chart at that point. As indicated, this dot is at the *intersection* of the line starting from the "0.6 second" at the left of the chart and extending parallel to the "0-second" line, with the other line starting from "5 feet" at the bottom of the chart and extending parallel to the "0-foot" line. Similarly, the last observation, 25 feet in 1.3 seconds, is indicated by a dot where the horizontal line representing 1.3 seconds crosses the vertical line representing 25 feet.

Entering a dot for each individual observation in the same way, we get the chart shown in Figure 3.2. This figure now gives a visual representation of the way in which the length of time changes as the distance traversed changes. Such a chart is known as a "dot chart" or a "scatter diagram."

But even this figure does not show the *exact* relation between the distance and the time. Both the first and the second trials were for exactly the same distance, yet the time was slightly different. Obviously that difference in time could not have been due to the difference in distance between the two, because there was no difference. The investigator must therefore assume that some outside cause, perhaps the accuracy with which the time was measured, may have been responsible for these slight differences. It will be noted, too, that when the different observations are plotted as in Figure 3.2, they come close to all lying along a continuous curve, but do not fall exactly on it. If we are willing to assume that all the

[1] For a more extended discussion of this point, see pp. 47 and 48.

differences between the different observations at the same point along the curve are due solely to extraneous factors, we can estimate the true effect of the distance, by itself, by averaging together the several observations as to time taken for each of the several tests for the same length of fall. A continuous curve drawn through these averages would then indicate the way in which the duration of fall varied with the distance, *on the average* of the cases studied. Although it might not hold true for

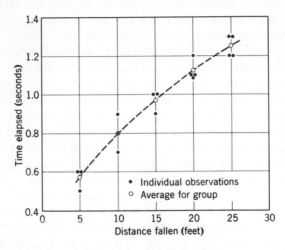

Fig. 3.2. Relation of distance a marble falls to time elapsed in falling, as shown by individual observations and curve of average time.

any one individual case, as we have just seen, still it does indicate *about* what the time will be for any given distance. For practical purposes we may say that under given conditions the time a body takes to fall *is determined* by the distance which it has to fall, as shown by the curve.

Expressing a Functional Relation Mathematically. The relation shown by the curve in Figure 3.2 is what mathematicians call a *functional* relationship; the time it takes a body to fall *is a function* of the distance which it has to traverse.[2] This means that for any particular distance-fallen, there is some corresponding time-required. The term "function" means that there is *some* definite relation between the two variables, number of feet and number of seconds, but it does not tell just *what* that relationship is. When, however, it is said that time is a function of distance according *to the curve shown in the figure*, then the statement has been made specific. The curve shows, for the distances shown on the

[2] Using Y for time and X for distance, we state this mathematically

$$Y = f(X)$$

chart, how long it will take a body to fall, on the average of a series of trials.

In this particular case the function is defined only by the graphic curve. It may also be stated as a mathematical expression

$$Y = \tfrac{1}{4}\sqrt{X}$$

using X for distance in feet and Y for time in seconds. This equation corresponds to the curve. If any value of X is substituted in it, and then the value of Y determined, that will be the value of Y—time in seconds—corresponding to that particular value of X—distance in feet—as shown by the curve in Figure 3.2. This equation is therefore *the equation of the function*, since this simple mathematical expression tells just as much about the relation between the two varying quantities—time and distance —as does the entire curve in the figure.

The way this equation can be used may be illustrated by two examples. Suppose a marble falls 16 feet; how long should it take to fall? The value of X would then be 16; substituting this value in the equation, we have

$$Y = \tfrac{1}{4}\sqrt{16} = \tfrac{1}{4}(4) = 1$$

This gives a value of 1 for Y, which means that it would take about 1 second to fall. Suppose again a bomb were dropped from an airplane 10,000 feet above the ground. How long would it take to reach earth? The value of X is then 10,000; substituting this value in the equation, we have

$$Y = \tfrac{1}{4}\sqrt{10,000} = \tfrac{1}{4}(100) = 25$$

The result means that it would take about 25 seconds for the bomb to fall.[3]

It is evident that the equation goes much further than does the graph of the curve. The latter gives the relation between distance and time only for the distances which are shown on the chart. The equation, on the other hand, gives the relation for any distance whatever, no matter what it may be. It is possible to state this aspect of the *law of gravity* in an equation only because physicists have studied this relation in the past

---

[3] Outside causes, such as friction with the air, may make the time of fall slightly different from the calculated time; therefore with so long a fall as this the time might differ perceptibly from the theoretical time given by the equation. This equation gives the time required *when no influence other than gravity* is taken into account. Obviously a marble would fall in air much faster than a feather—the resistance of the air has very little influence on the speed of the marble and a great deal of influence on the speed of the feather. In a vacuum they would fall at the same rate.

and determined exactly how the one quantity varies with the other. Having found that the same relation between the two variables held through their entire range of observation and having worked out on logical grounds a good reason why that relation should hold, they have felt safe in coming to the conclusion that it will continue to hold even beyond the range of the experimental verification.[4] Where only a graph of the function is available, on the contrary, only the relation within the stated range is known. The graph does not tell, of and by itself, the direction or shape the curve would take if extended beyond the limits determined by the experiments.

Now if instead of the relation we have just been discussing we consider the relation between the quantity of sugar which can be dissolved in a glassful of water and the temperature of the water, we have quite a different problem, and yet one that is similar in many aspects. If we start to determine it experimentally, we must first make sure that the quantity of water with which we are working is the same in every trial; then we must measure accurately both the temperature of the water and the amount of sugar which could be dissolved in it. Many other similar factors which might possibly influence the result would have to be considered before even the exact plan of the experiment could be drawn up.

Once the experiment had been run the numerical results would probably be somewhat similar in character to those in the gravity test. If the data were plotted on a scatter diagram like Figure 3.2, it would be found that the data fell in the general shape of a curve, but that very few of the dots fell exactly on the curve, some lying above and some below the continuous line which could be drawn about through the center of them. Again we might conclude that these slight differences from exact agreement were due to factors other than the temperature of the water—to slight experimental errors in the quantity or temperature of the water, or to slight errors of measurement in determining the quantity of sugar—and be willing to conclude that the line drawn through the center of the series of observations showed the *real* effect of differences in temperature on the quantity of sugar dissolved, when extraneous influences were removed. This again would be a *functional* relation. The curve would express the relation between changes in temperature and changes in quantity of sugar, showing for any given temperature exactly how much sugar could be dissolved. It might then be possible to determine a type of equation which

---

[4] It should be noted that for very great distances, say 10,000 miles, the formula might need to be modified, since then the pull of the earth would be less than it is at the surface. The equation holds true only for those distances from the earth within which its pull is practically a constant.

would accurately specify the function by a mathematical formula, similar to that discussed for the gravity example, if the logical type of relation between the two variables could be worked out.[5]

Determining a Functional Relation Statistically. In the two cases which have been discussed the relation between the two variables was sufficiently close so that by taking proper experimental precautions other influences which might affect the result could be largely removed and a series of observations obtained sufficiently consistent with each other so that the exact nature of the relation could be readily determined. In many other types of relations this cannot be done so easily. It is with this type of relation that statistical methods really become important.

If we were making a traffic study in a given city, for example, we might wish to know what would be the safe speed limits to permit on different streets. In that connection we might need to know in what distance an automobile could be stopped when traveling at different speeds, so that by comparing this distance with the width of the different streets and the length of view at intersections we could judge how fast machines might be able to travel without risk of collisions at street intersections. One way to determine what is the relation between speed and stopping distance would be to make a number of tests in different portions of the city,

---

[5] Some logical foundation *is* needed before a mathematical equation to a curve can be of any more value than merely the chart which graphs the curve. Thus in the gravity example it is evident that the farther a body falls, the faster it falls; in every successive instant the speed it has already attained is increased by the effect of the continued pull which is added to it. Purely mathematical investigations of the relation between such constantly growing magnitudes and the variable with which they grow have enabled physicists to determine the general mathematical *type* to which the relation must conform. Then, knowing the type of the curve, it is relatively easy to determine the value of the constants (such as the "$\frac{1}{4}$" of the equation $Y = \frac{1}{4}\sqrt{X}$) which makes the general equation applicable to a given specific case. This is done by using experimental results, such as those given in Table 3.1, to calculate the constants for the specific type of curve which has been determined upon.

An algebraic equation which expresses the relation logically expected between or among two or more variables is sometimes called a "model" of the relationships. Such a model is a mathematical expression of the hypothesis according to which the observed data will be examined to see whether or not the facts support the hypothesis, and to determine the value of the statistics. Sometimes a model will consist of two or more equations. (Note further discussions in Chapters 24 and 26.)

Not all functional relations can be subjected to this type of logical analysis, however, and it is sometimes impossible to tell what sort of equation the results should really follow. In that case any mathematical curve "fitted" to the data has no more special meaning than the graphic curve drawn through the center of the observations; both are merely empirical descriptions of the relations, and both are limited in their interpretation to the range of the particular data upon which they are based. This fact will be discussed more fully later on.

taking different types of machines and different drivers chosen at random so as to get a representative sample.[6] Let us suppose that as the result of such a series of tests we obtained the series of observations shown in Table 3.2.

Table 3.2

RELATION BETWEEN SPEED OF AUTOMOBILE AND DISTANCE REQUIRED FOR STOPPING AFTER SIGNAL, AS SHOWN BY 63 OBSERVATIONS

Speed When Signal is Given	Distance Traveled After Signal Before Stopping	Speed	Distance	Speed	Distance
Miles per hour	*Feet*	*Miles per hour*	*Feet*	*Miles per hour*	*Feet*
5	2	21	39	28	84
10	8	26	39	27	57
10	17	25	33	30	67
10	14	24	56	16	34
8	9	18	29	18	34
16	19	25	59	8	8
17	29	27	78	5	8
12	11	25	48	5	4
9	5	21	42	13	15
7	6	25	56	14	14
7	7	30	60	8	13
9	13	29	68	9	5
4	4	17	22	14	16
5	8	16	14	8	11
13	18	13	27	35	85
15	16	12	21	40	110
18	47	12	19	39	138
19	30	26	41	31	77
20	48	28	64	35	107
21	55	29	54	22	35
36	79	30	101	40	134

[6] The problems involved in choosing a sample that will both properly represent the universe, and measure the relations in the sample so as to best judge what are the true relations in the universe, are discussed at greater length in Chapters 17, 18, and 20.

It is apparent from the table that there are great variations in the distances which different cars or different drivers required to stop, even when traveling at the same speed. This is shown even more clearly when we make a dot chart of the data in just the same way that was illustrated in Figure 3.2. The graphic comparison between speed and distance-to-stop, shown in Figure 3.3, reveals that there is only a general agreement between the different tests. There is certainly some relation between the two

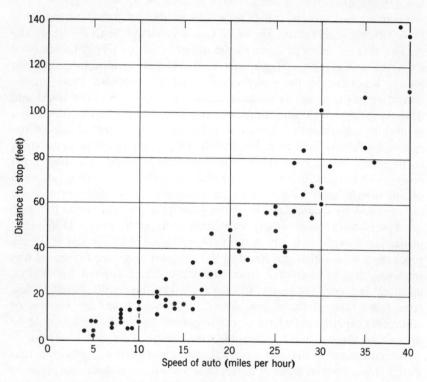

Fig. 3.3. Relation of speed of automobile to distance it takes to stop, as shown by individual observations.

variables, but it is vague and uncertain in comparison with the relatively sharp and clear-cut relation shown in Figure 3.2.

There is no difficulty in understanding why the relation is not more definite. The stopping distances may be influenced by many differences from car to car. Some cars will have brakes in adjustment and others brakes well worn; some cars will be nearly empty and others heavily loaded; some will have tires worn smooth and others new tires; some will have average brakes and others power brakes or exceptionally large ones, as on sports cars. In addition, the drivers differ. Some are experienced drivers, some inexperienced; some strong and some unable to press the brakes fully down; some with almost instantaneous reaction to our signal to stop, some with faltering or lagging response; some bright and wide awake, others tired and unobservant; some calm and steady, others nervous and erratic. Finally the conditions of the tests might be different —some on concrete pavement, others on asphalt; some on up-grades, some downhill.

There are two ways by which we might go about deciding exactly what these varying observations showed. One way would be to divide up the data so that the effect of some of the different factors would be removed from the results. Thus if we separated the observations into different groups according to the make of car, and then each of these groups according to the model or the year made, the relation between speed and distance for any single subgroup would no longer be affected by differences in braking equipment so far as engineering design went. Most of the remaining factors, however, would still be present to affect the results, so that even within each subdivision the records would still show great diversity in the relation. Only if we continued the process of subdivision of our sample until we got down to successive observations of a single car operated by a single driver at the same place, would we be likely to get observations more nearly consistent with each other. Differences in the promptness with which the driver responded to the signal, in the preciseness with which the speed at the moment of giving the signal was observed, and in the force with which the driver applied his brakes, all might influence the result, so that even then the results would be less consistent—"the curve be less definitely defined"—than in a series of laboratory experiments where *all* the important outside variables could be definitely controlled and so prevented from affecting the results obtained.

Should the entire mass of observations be analyzed as suggested, that would give a great number of different sets of relations, each one showing how long it took a given car to stop when driven by a given driver, when traveling at different speeds. But this great number of different curves might not be suitable to answer our question. They might be so different from curve to curve that it would seem that there was no real general relation between speed and distance. A new car with power brakes, driven by an experienced driver, might stop in its own length at the same speed at which an old car with brakes badly worn, driven by an inexpert driver, might require a far greater distance. Obviously neither one of these extremes would be typical of the general relation; but what would be typical? Even the less extreme cases might show great variations among themselves, so that it would be almost impossible to pick from the great diversity of curves one or a few that would serve as a basis of judgment for our problem.

A second way of going about it would be to try to determine some sort of *average* relation between speed and distance. In that case we should admit that there were great differences from the average in individual cases, yet should feel that the average would serve as a general indication of the relation, even though we were aware it would not be true in every, or perhaps even in any, individual case. If we knew *nothing* about a car except the speed at which it was moving, that average relation, however, would serve to give us the best guess we could make as to how far it would go before stopping.

Where the relation between two variables is clear and reasonably sharply defined, as in the experimental cases discussed, it is not difficult to determine the average relationship, since the relation for individual cases and the average relation for all cases are nearly identical. Where the relation is not so well defined, however, and especially where many other relations are involved in addition to the particular one which is being studied, it is by no means easy to determine exactly what the true relationship is. A considerable body of statistical methods has been developed to treat this problem, and is presented in the balance of this book.

Summary

A statement of the change in one variable which accompanies specified changes in another is known as a statement of a *functional relation*. A functional relation may be stated either graphically by a curve, or algebraically by a definite equation. Although functional relations may be readily determined from experimental conclusions where all influences except the one being studied are held constant, many problems cannot be studied by such methods. The statistical methods of *regression analysis* may be used to study functional relations where experimental methods are not satisfactory.

CHAPTER 4

Determining the way one variable changes when another changes: (1) by the use of averages

The problem stated in Chapter 3 was to determine how many feet automobiles traveling at a given speed require for stopping. It involves determining the *average* extent to which one variable changes when another variable changes. Stated mathematically, the problem is to find the functional relation between speed and distance—the probable distance required to stop with any given initial speed. Of the many different ways of doing this, the simplest, and the one which would suggest itself most naturally, would be to classify the record into groups, placing all of one speed in one group, all of another speed in another group, making as many groups as there are different rates of speed recorded, and then *averaging* the different distances for all the cases in each group. This would then give an average distance for stopping for each given rate of speed in the series of records. Table 4.1 shows this operation carried out.

Where there were only single observations, this fact has been indicated by placing the average—the single report—in parentheses.

The averages in the last column of Table 4.1 show quite specifically how the distance required for stopping tends to increase with speed. But the increase is not uniform. The cars at 13 miles per hour averaged a greater distance than those at either 12 or 14, and the cars at 26 a shorter distance than those at 25.

If the successive averages from Table 4.1 are plotted and connected by lines, both the general increasing tendency and the irregular change from group to group are easily seen. Figure 4.1 shows this comparison.

Do these differences between the different group averages have any real significance? Is there any reason to think that this very jagged line is the *true* average relation between speed and distance? We can consider

Table 4.1

COMPUTATION OF AVERAGE DISTANCE REQUIRED FOR STOPPING AFTER
SIGNAL, FOR DIFFERENT INITIAL SPEEDS

Speed when Signal is Given	Different Distances Noted at That Speed*	Average Distance for That Speed
Miles per hour	*Feet*	*Feet*
4	4	(4)
5	2, 8, 8, 4	5.5
7	6, 7	6.5
8	9, 8, 13, 11	10.3
9	5, 13, 5	7.7
10	8, 17, 14	13.0
12	11, 21, 19	17.0
13	18, 27, 15	20.0
14	14, 16	15.0
15	16	(16)
16	19, 14, 34	22.3
17	29, 22	25.5
18	47, 29, 34	36.7
19	30	(30)
20	48	(48)
21	55, 39, 42	45.3
22	35	(35)
24	56	(56)
25	33, 59, 48, 56	49.0
26	39, 41	40.0
27	78, 57	67.5
28	64, 84	74.0
29	68, 54	61.0
30	60, 101, 67	76.0
31	77	(77)
35	85, 107	96.0
36	79	(79)
39	138	(138)
40	110, 134	122.0

* Data taken from Table 3.1.

that from two points of view; the logic of the relation and the statistical basis of the differences. Logically the differences are quite nonsensical. If a given machine can stop in 40 feet when it is going 26 miles an hour, of course it can stop in at least the same distance when going 25 miles per hour, and probably something less. It certainly would not take 49 feet, as the table shows. Then from the statistical point of view the groups are entirely too small to show definitely how many feet on the average

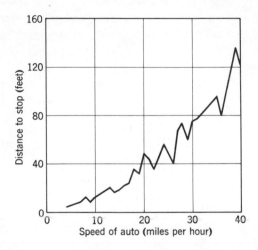

Fig. 4.1. Relation of speed of automobile to distance it takes to stop, as shown by averages of small groups.

it takes to stop at any one selected speed. Even the largest groups have only 4 cases, whereas we have seen in Chapter 2 that 10 to 25 cases may be required as a minimum to give an average of much reliability. Computing the standard error for the average from the 25-mile group, it comes out 5.8 feet. With only 4 cases, however, Figure 2.1 shows that we have to take a range of 1.15 times the standard error to make the observed value come within that range of the true value in 2 samples out of 3.[1] There is thus 1 chance out of 3 that the mean shown by our sample lies more than 6.7 feet above or below the true mean. The average for this group of records might therefore be written 49.0 ± 6.7 feet. When we say that the average distance required for stopping when traveling 20 miles per hour (for all automobiles in town, say) is between 42.3 feet and 55.7 feet, we are likely to be wrong in 1 out of 3 such statements,

[1] The standard error is computed from the standard deviation of the 4 reports at 25 miles, using equation (2.4). This gives a value of 5.8. Figure 2.1 shows that for 4 reports a range of 1.15 times the computed standard error must be taken to secure a reliability of 0.67, so the confidence limit, with $P = 0.67$, is ±(5.8)(1.15), or ±6.7.

on the average. With the average from one of the *largest* groups showing as little reliability as this, it is quite clear that the zigzag variation from average to average has no real meaning.

Does that mean that in spite of the relationship we can *see* in Figure 4.1 that we can get no reliable statistical measurement of the relation? That is overstating the case a little; all that we have determined so far is that the line of averages, the irregular function shown in Figure 4.1, has but little statistical meaning *just as it stands now.*

We might be able to make the results more reliable by basing our averages on a larger number of reports. But that would be a long and expensive process. Isn't there some way we can find out something more just from the records we have?

Another way of making the conclusions more stable would be by combining the records so as to give fewer groups, but with more cases in each group. So far we have been working with 29 different groups, one for each of the 29 different speeds measured. If instead we group them into a few groups—say 6 or 8—we shall have considerably larger groups to work with.

Independent and Dependent Variables. The question might be asked whether the groups should be made on the basis of the rate of speed or of the distance required for stopping. (In preparing Table 4.1 we used the rate of speed without discussing the matter.) That comes back to the question of what we really want to find out. Do we want to know the *average speed at which machines were traveling* when it took them, say, 20 feet to stop; or do we want to know the *average distance* machines took to stop when they were traveling at a given speed? Obviously, the thing we are going to set is the speed limit, and we are merely interested in the distances for stopping as one factor to guide us in deciding what the speed limit should be. We therefore want to know the effect of *speed* upon *average distance*, and not the reverse. For that reason we shall classify our records on the basis of speed, and then average all the different distances for the cars traveling at that speed.

The same question is met with in nearly all problems where the relation between two variables is to be dealt with. It is always necessary to think over the problem carefully, and decide which variable we are going to regard as the independent or *causal* variable, and which one as the dependent, or *resultant.* Thus if we were relating variations in tobacco yields to applications of fertilizer, obviously the differences in fertilizer would be the cause and the differences in yield the result, so we would sort our records according to the differences in fertilizer. Other relations may not be so clear cut. If the size of stores were being related to profits, it might be as logical in some situations to consider that the more successful men were able to afford the largest stores as to consider that the larger stores returned the greater profits. Careful consideration of the facts in each given case is necessary to clarify exactly what is the particular relation involved.

As shown later (pages 74–80 and 475–476), it is frequently impossible to say which variable is the cause and which is the effect. Yet one may wish to regard one variable as the one whose values are given or known. This is then called the *independent variable* and plotted as the abscissa. The second variable will then be regarded as the one whose values are to be related to, or estimated from, the values of the known variable. This is then called the *dependent variable*, since it is treated as *depending upon* the given values of the independent variable. It is sometimes desirable in particular problems to consider first one variable and then the other as independent.

Groups of Larger Size. To return to our automobile problem. Since the speeds varied up to 40 miles per hour, and we have 63 reports to deal with, we might try breaking them up into 8 groups and see what kind of averages that will give us. Using groups covering a range of 5 miles per hour each, we can group the records and determine the averages for the 8 groups thus formed, getting the results shown in Table 4.2.

Table 4.2

SPEED OF AUTOMOBILE AND DISTANCE REQUIRED FOR STOPPING AS SHOWN BY GROUP AVERAGES

Speed when Signal Is Given	Number of Cases	Average Speed	Average Distance
Miles per hour		*Miles per hour*	*Feet*
Under 7.5	7	5.4	5.6
7.5–12.4	13	9.6	11.8
12.5–17.4	11	14.9	20.4
17.5–22.4	9	19.8	39.9
22.5–27.4	9	25.6	51.9
27.5–32.4	8	29.4	71.9
32.5–37.4	3	35.3	90.3
37.5 and over	3	39.7	127.3

These averages can then be plotted and connected by straight lines, just as were the averages in Figure 4.1. In constructing Figure 4.2, which shows the result, it is necessary to use the average speed as well as the average distance-for-stopping in locating each point. This is because each of the average distances, as shown in Table 4.2, represents not one speed, but several different speeds thrown together. The circles in Figure 4.2 represent the several group averages plotted this way. The first one is located at the intersection of the lines for 5.4 miles per hour and 5.6 feet, the second at 9.6 miles per hour and 11.8 feet, and so on for the remainder.

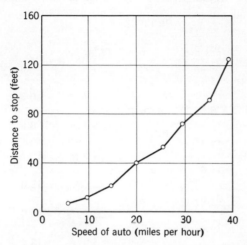

Fig. 4.2. Relation of speed of automobile to distance it takes to stop, as shown by averages of large groups.

When the group averages of Figure 4.2 are connected by straight lines the relation between speed and distance is shown much more satisfactorily than it was in Figure 4.1. The line in the new figure shows a fairly continuous relation between speed and distance. But on close examination even the relation shown in this last figure is not found fully satisfactory. If we compute the change in distance-for-stopping for each change of 1 mile in speed, we find that the conclusions are somewhat erratic. Between the first two averages, the change in speed from 5.4 to 9.6 miles per hour, an increase of 4.2 miles per hour, is accompanied by a change in distance from 5.6 to 11.8 feet, or an increase of 6.2 feet. Between 5.4 and 9.6 miles per hour, therefore, the distance-for-stopping apparently increases 1.5 feet for each increase of 1 mile per hour in the speed of the machine. Similar computations for all the other groups are shown in Table 4.3, carrying out just the same process.

The increase in speed from group to group varies between 4 and 6 miles per hour, and the increase in distance tends to increase, but irregularly. This irregularity is more clearly shown in the last column of the

Table 4.3

COMPUTATION OF CHANGE IN DISTANCE FOR EACH CHANGE OF 1 MILE
IN SPEED, FOR GROUP AVERAGES

Speed when Signal Is Given	Average Speed	Average Distance Required for Stopping	Increase in Speed	Increase in Distance	Increase in Distance per 1-Mile Increase in Speed
Miles per hour	*Miles per hour*	*Feet*	*Miles per hour*	*Feet*	*Feet*
Under 7:5	5.4	5.6			
			4.2	6.2	1.5
7.5–12.4	9.6	11.8			
			5.3	8.6	1.6
12.5–17.4	14.9	20.4			
			4.9	19.5	4.0
17.5–22.4	19.8	39.9			
			5.8	12.0	2.1
22.5–27.4	25.6	51.9			
			3.8	20.0	5.3
27.5–32.4	29.4	71.9			
			5.9	18.4	3.1
32.5–37.4	35.3	90.3			
			4.4	37.0	8.4
37.5 and over	39.7	127.3			

table, in terms of increase in distance for each 1-mile-per-hour increase in speed. This varies erratically, jumping from 1.6 to 4.0 between the second and third pairs, dropping back to 2.1, rising again to 5.3, and then dropping sharply, and rising in the last pair to 8.4.

This same variability in the rate of change can be seen directly from Figure 4.2 by noting the steepness of the several portions of the line. The irregular and zigzag character of the line shows the same fluctuations that the computations in Table 4.3 show. Simply by examining this chart closely it would have been possible to tell about the unsatisfactory character of the conclusions without taking the time to calculate the exact rates.

Are the irregularities shown in Table 4.3 and Figure 4.2 of any importance, or are they due simply to the possibilities of variation in using so small a sample, just as were the differences in Figure 4.1 and Table 4.1? Is it really true that an increase in speed has a larger effect upon the distance required for stopping between 15 and 20 miles per hour than between 10 and 15?

Reliability of Group Averages. The answer involves a consideration of the statistical basis upon which our conclusions are based. These last results were calculated from the average speed and average distance for the several groups of records; obviously they can be no more reliable than those averages are themselves. In measuring the reliability of those averages by the methods already discussed, we must estimate the standard errors which will tell us about how much confidence we can have in each figure.

The next step, therefore, is to calculate the standard error for each of the 8 averages of distance. The computation, which is exactly the same as that used before, based on equation (2.4), is shown in Table 4.4.

Table 4.4

COMPUTATION OF CONFIDENCE INTERVALS FOR AVERAGE DISTANCES
REQUIRED FOR STOPPING

Grouped by Speed	Number of Cases, n	Average Distance Required for Stopping M	Standard Deviation, $\bar{s}$	Standard Error of Mean, s_M	Confidence Interval for* $P = 0.67$	Confidence Intervals for $P = 0.67$
	(1)	(2)	(3)	(4)	(5)	(2) + (5)
Miles per hour		*Feet*	*Feet*	*Feet*	*Feet*	*Feet*
Under 7.5	7	5.6	1.56	0.59	±0.6	5.0– 6.2
7.5–12.4	13	11.8	5.08	1.41	±1.5	10.3– 13.3
12.5–17.4	11	20.4	6.82	2.55	±2.7	17.7– 23.1
17.5–22.4	9	39.9	8.81	2.94	±3.1	36.8– 43.0
22.5–27.4	9	51.9	13.43	4.48	±4.8	47.4– 56.4
27.5–32.4	8	71.9	14.96	5.29	±5.7	66.2– 77.6
32.5–37.4	3	90.3	14.89	8.68	±11.4	78.9–101.7
37.5 and over	3	127.3	15.33	8.95	±11.8	115.5–139.1

* These values are obtained by multiplying the standard error by the values given in Table 2.3 and Figure 2.1. Interpolating from Figure 2.1, these are found to be; for $n = 3$, 1.31; for $n = 7$, 1.08; for $n = 8$, 1.08; for $n = 9$, 1.07; for $n = 11$, 1.05; for $n = 13$, 1.04.

Comparing the several averages with their respective confidence intervals, as shown in the last column of Table 4.4, we find that if we made the same number of observations over again a number of times and used the same grouping, there is 1 chance out of 3 that we might find that the true distance for the second group was less than 10.3 feet or more than 13.3

feet; and so on until for the last group it might be under 115.5 feet or over 139.3 feet. With this wide possible variation in the sample average from the true values, it is quite evident that the real facts have not yet been measured accurately enough to justify detailed computations of the differences in the slope of different portions of the line. By changing any one of the averages as much as has been indicated, the slope of the line would be very materially changed.

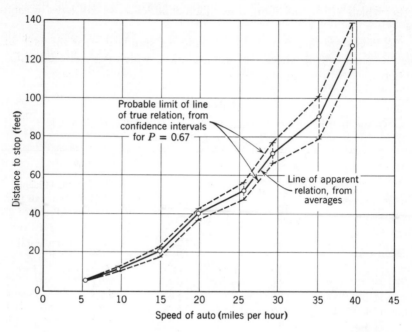

Fig. 4.3. Relation of speed of automobile to distance it takes to stop, as indicated by the confidence intervals.

Range Within Which True Relation May Fall. The extent to which reliance may be placed in the relationship between the two variables as shown by the 63 observations which we have to deal with may be judged from Figure 4.3. Here the actual averages have been plotted, and lines drawn connecting them, just as before. The confidence intervals (for probability of 0.67) are then indicated by placing a short bar above and below each average. Connecting these by dotted lines then indicates the zone of confidence intervals around the line of averages. Figure 4.3 indicates that there is a rather wide zone within which the true relation may fall, especially for the higher speeds, even when we make that zone as narrow as we have in calculating for the moderate probability of being right

2 times out of 3. But we still do not have any definite measure of the general relation between speed 'and distance.

If a smooth continuous line or curve were determined, either by appropriate mathematical processes or by freehand smoothing, and were drawn through the successive group averages, that would give a far better measure of the relationship. (It is left for the student to try drawing a smooth continuous freehand curve through the line of group averages, and to see how many times it falls outside the confidence zone. For this exercise, Figure 4.3 should be replotted on a larger scale, using the data of Table 4.2, and the last column of Table 4.4.)

We have seen how increasing the number of cases included in a single group increased the dependence which would be placed in that group. However, even by reducing our 63 cases to 8 groups we have not been able to get a consistent and satisfactory statement of the relation. Is it possible that by handling all the data as a single group we could get a better result? One way of doing this would be to average all the speeds and all the distances together. But that would only tell us what was the average distance required for stopping and the average speed. What we want to know is what distance is most likely to be required *at any given speed*, and the treatment just suggested would not give us that.

There is another way, though, of determining the relation while considering all the records together. If we are willing to assume that an increase of one mile per hour in the rate of speed will increase the distance required for stopping by exactly the same number of feet, no matter how rapidly or how slowly the machine is already moving, then we can determine this relation for all the data as a whole. On this basis a straight line can be used to represent the relation. All that we have to do to determine a straight line which will come as near as possible to representing the relation as shown by all 63 individual observations. (A straight line might also be drawn by eye, as an exercise, and the results compared with those of the preceding exercise.)

Summary

The change in one variable with changes in another may be approximately determined by grouping the records according to the independent variable and determining the corresponding averages for the dependent variable. Unless a very large number of observations is available, however, the functional relation shown by the successive averages will be irregular and inconsistent, owing solely to sampling variability. For that reason some method is needed for measuring the functional relation for the group of records as a whole. The simplest way in which this can be done is by assuming that the relation can be represented by a continuous straight line. Methods of determining such a line will be considered in the next chapter.

Note 4.1. It is always possible to reverse the dependent and the independent variables. Thus the data presented in Figure 3.2 might have been plotted with time as the independent variable and with distance fallen as the dependent. A curve might then have been drawn in to show the average distance which a body can traverse for a given time of fall. Similarly, the data charted in Figure 3.3 might have been charted with distance as the abscissa and speed as the ordinate. These two alternate ways of stating the relation generally yield somewhat different functions if errors of measurement or "outside" sources of variation are at all large.

Simple Regression, Linear and Curvilinear

CHAPTER 5

Determining the way one variable changes when another changes: (2) according to the straight-line function

There are several ways by which a straight line can be determined to show the functional relation between two variables. One way would be simply to place a ruler over the chart along the several group averages, or to stretch a black thread over them, and draw the line in by eye so as to fall as nearly as possible along them. Although no two persons would draw their lines exactly the same, still this method might give fairly satisfactory results where only a rough measure was wanted. However, it would often be advantageous to determine a particular straight line that could be exactly duplicated by other persons and that would qualify in some sense as the best possible straight line for expressing the relationship in this particular set of data. We shall therefore use the exact regression method of determining the straight line. But first we must consider the meaning of a straight line.

The Equation of a Straight Line

The determination of what this line will be consists in finding the *constants* for the *equation* of the line. Just as we have already seen (Chapter 3) that the curve showing the relation between the distance a body has to fall and the time it takes can be expressed by the relation,

$$Y = \tfrac{1}{4}\sqrt{X}$$

so any straight line can be expressed by the relation[1]

$$Y = a + bX \qquad\qquad (5.1)$$

Figure 5.1 illustrates the meaning of a and b in this formula. When the value of X is zero, b times X is zero and Y is equal to a. This constant, a, therefore, gives the height of the line (in terms of Y or vertical units) at the point where X is zero. This is indicated at the left edge of the chart.

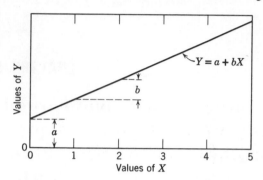

Fig. 5.1. Graph of the function $Y = a + bX$.

From the same equation, every time X increases one unit, Y increases b times one unit, since Y is computed as a plus b times X. The difference of the height of the line (measured in Y units), between the point where X is 1 and where X is 2, is therefore b units of Y, just as indicated on the chart. And this continues to hold true for every unit change in X, whether from 1 to 2, or from 0 to 1, or from 99 to 100.

The meaning of these constants in the *equation of the straight line*, as equation (5.1) is known, may be illustrated more concretely by taking some actual values for the constants a and b, and seeing how the line would look then. If we take 3 for a, and 2 for b, the equation would then read:

$$Y = 3 + 2X$$

Figure 5.2 shows the line for which this is the equation. Thus if X is taken as zero, the value of Y is found to be

$$Y = 3 + (2 \times 0) = 3 + 0 = 3$$

And 3 is therefore the Y value corresponding to the X value zero.

---

[1] Written this way, the equation is a perfectly general one which can be applied to the relation between *any* two variables, by calling one of them Y and the other one X. The symbol Y in the equation simply represents *the number of units* of the variable we designate as Y, whatever that may be, acres, dollars, pounds; and the symbol X likewise represents the number of units of the variable we designate as X.

Similarly if X is taken as 10,

$$Y = 3 + (2 \times 10) = 3 + 20 = 23$$

And the Y value corresponding to the X value of 10 is therefore 23. All other values of Y which may be computed for values of X within the range shown in Figure 5.2 will similarly be found to lie exactly on the same line.

Figure 5.2 illustrates again the meaning of the constants a and b. When X is zero, the value of Y is three units above zero, as indicated,

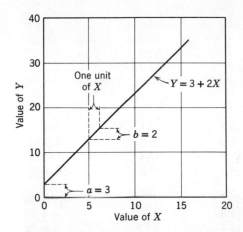

Fig. 5.2. Graph of the function $Y = 3 + 2X$.

and for every unit increase in X (say from 5 to 6) the value of Y goes up 2 units. This is exactly the same thing as shown in Figure 5.1, except that there no definite values were assigned to a and b, whereas here they have been given exact numerical values.

It is quite apparent from Figures 4.2 and 4.3 that the relation between auto speeds and distance required for stopping could not be well represented by a straight line. We will therefore use another example to illustrate this technique. This case is taken from hydrology, where it is important in planning hydroelectric, flood-control, or irrigation projects, to know the past history of stream flows. In one case a dam was being planned on the Kootenai River at Newgate, B.C., near the point where it crossed the Canadian border, but stream records there began only in 1931. Records were available for a longer period at Libby, Montana, further down the stream. How could the flow at Newgate be estimated from that at Libby? Since we wish to estimate the flow at Newgate from that at Libby, we must regard the volume at Newgate as the *dependent variable*, and that at Libby as the *independent variable*.

Table 5.1

WATER FLOW AT TWO POINTS ON KOOTENAI RIVER, IN JANUARY,*
AND CALCULATIONS FOR FITTING A STRAIGHT LINE

Year	Newgate, B.C., Y	Libby, Mont., X	X^2	XY
	Units of 100 cfs			
1925		42.0		
26		24.0		
27		38.0		
28		49.4		
29		24.6		
1930		24.2		
31	19.7	27.1	734.41	533.41
32	18.0	20.9	436.81	376.20
33	26.1	33.4	1,115.56	871.74
34	44.9	77.6	6,021.76	3,484.24
1935	26.1	37.0	1,369.00	965.70
36	19.9	21.6	466.56	429.84
37	15.7	17.6	309.76	276.32
38	27.6	35.1	1,232.01	968.76
39	24.9	32.6	1,062.76	811.74
1940	23.4	26.0	676.00	608.40
41	23.1	27.6	761.76	637.56
42	31.3	38.7	1,497.69	1,211.31
43	23.8	27.8	772.84	661.64
Totals	$\Sigma Y = 324.5$	$\Sigma X = 423.0$	$\Sigma X^2 = 16,456.92$	$\Sigma XY = 11,837.32$
Means	$M_y = 24.96$	$M_x = 32.54$		

* Source: *Extending Stream-Flow Records*, U.S. Department of the Interior, Geological Survey, Water Resources Branch, pp. 7, 8, September, 1947.

The relevant data are shown in the first two columns of Table 5.1, together with data for six earlier years at Libby. (These Libby observations are actually available still further back.) When data for 1931 to 1943 are plotted on a dot chart, as shown in Figure 5.3, a marked relation between the two variables is evident, one that can apparently be represented by a straight line. The next step is to determine the straight line that best describes the relation. We will therefore use X to stand for flow at Libby, and Y for flow at Newgate, each in hundreds of cubic feet per second (cfs).

Thus when we write the equation $Y = a + bX$, we shall be using that as shorthand for

Flow at Newgate (in 100 cfs) $= a + b$ (flow at Libby, in 100 cfs)

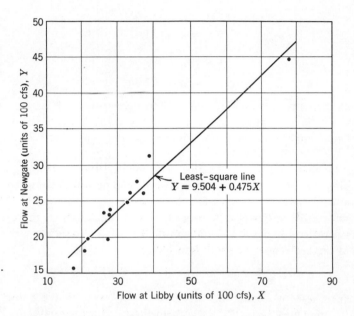

Fig. 5.3. Dot chart of observations of water flow at Libby and Newgate, and straight line fitted to them.

To give this equation definite meaning we must determine the numerical values for a and b, just as in our preceding illustration we had to assume numerical values for these constants before the graph had any definite meaning for us.

The "Observation Equations." One way of finding the values is by regarding each one of our original observations (Table 5.1) as an algebraic equation itself. Thus the 1931 observations, 27.1 at Libby and 19.7 at Newgate, would be written

$$19.7 = a + b(27.1)$$

putting the 19.7 in the place of Y in the equation, and the 27.1 in place of X.

Similarly the next observation, 18.0 at Newgate and 20.9 at Libby would be

$$18.0 = a + b(20.9)$$

and so on right through the last observation.

Bringing all these different equations together would give a series looking like this:

$$19.7 = a + 27.1b$$
$$18.0 = a + 20.9b$$
$$26.1 = a + 33.4b$$
$$\cdots$$
$$31.3 = a + 38.7b$$
$$23.8 = a + 27.8b$$

(The middle equations are omitted here to save space.)

Since we have 13 observations of both variables, we have in all 13 observation equations, each containing the two unknown constants a and b.

Now by the rules of simple algebra, any two independent equations containing the same two unknown constants can be solved simultaneously to obtain the numerical values for those constants. One way to find the values of our a and b would be to pick two of the equations representing our observations and solve them simultaneously. Suppose we take the first and the last ones; we shall then have:

$$a + 27.1b = 19.7 \quad \text{and} \quad a + 27.8b = 23.8$$

Solving these two equations simultaneously, we find the values $a = -139.01$, and $b = 5.86$. But in getting these values we have used only 2 out of the 13 observations, and also 2 whose values of X were very close together. Would we get the same result if we used another pair? Let us try the second observation and the next to the last. Then we will have

$$a + 20.9b = 18.0 \quad \text{and} \quad a + 38.7b = 31.3$$

These two equations, solved simultaneously, give the values $a = 2.39$, and $b = 0.747$, which are certainly far different from the first set. Apparently the values obtained by this method would depend upon the particular pair of observations selected, perhaps varying with each pair.

If we work out estimated values of Y for selected values of X according to these two solutions, we get results as follows. According to the first set,

$$Y = -139.01 + 5.86X \quad \text{so when } X = 20, \ Y = -21.8;$$
$$\text{and when } X = 30, \ Y = 36.8$$

But according to the second set,

$$Y = 2.39 + 0.75X \qquad \text{so when } X = 20, \ Y = 17.4;$$

$$\text{and when } X = 30, \ Y = 24.9$$

If we should then plot the two calculated points for the first of these equations, and connect them by a straight line, we should find that that line also passes through the two dots which represent the two observations from which the values were calculated. The same would hold true for the second line and the second pair of observations. We could compute as many different lines as there are different possible pairs of observations *not* lying on the same line. Fitting a straight line to two points, as we have done here, is simply equivalent to drawing a line to pass through those two points. This can be seen by plotting the two lines just calculated on Figure 5.3. (This is left as an exercise for the student.) Quite clearly no single line could pass through all the different points. If we computed more lines by this process of using selected pairs of points, we should just get a larger variety of different lines. And the closer together the points were, with respect to the values of X, the greater the differences in the slopes of the lines would tend to be.

Fitting the Line by "Least Squares." If we are going to use a mathematically determined straight line at all, what we need is one which represents all 13 observations, instead of any particular pair of them. No one line can *exactly* fit all 13 observations, for, as we have just seen, the line which would exactly agree with the first and last would not agree at all with the second and next to the last. What we shall have to find is some compromise line which will come as near as possible to agreeing with all the 13 observation equations, even though it does not exactly agree with any one. Mathematicians have derived a method of obtaining such a line by the use of what is known as the "method of least squares." This method takes all the observations into account, giving each of them an equal weight in determining the result—a line such that the sum of the squares of the departures of Y from the line will be as small as possible. It also has certain other mathematical properties which make it of great value in handling problems of this sort.

The equations upon which the process is based are derived by the use of calculus. The method itself, however, is simple and can be used by anyone with a knowledge of simple algebra.

Computing the Extensions. The individual observations are listed as already shown in Table 5.1. Each X item is squared and entered in the column headed X^2; and each X item is multiplied by the accompanying Y item, and the product is entered in the column headed XY. All the items in each column are summed (excluding those for the years before 1931), giving the totals at the foot of each. The symbol ΣX represents the sum of all the X items; ΣY represents the sum of all the Y items; $\Sigma(X^2)$ represents the sum of all the X^2 items, and $\Sigma(XY)$ represents the sum of all the products of XY. The means of X and Y are also calculated, and entered in the final line.

SOLVING THE EQUATIONS. We next proceed to find the values of a and b by using the following formulas:

$$b = \frac{\Sigma(XY) - nM_xM_y}{\Sigma(X^2) - n(M_x)^2} \tag{5.2}$$

$$a = M_y - bM_x \tag{5.3}$$

In using these formulas the value of b is determined first, then it is used in the next formula to determine the value of a.[2]

Using the values for ΣX, ΣY, $\Sigma(X^2)$, and $\Sigma(XY)$ given in Table 5.1, in equations (5.2) and (5.3), we find the values of b and a to be:

$$b = \frac{\Sigma(XY) - nM_xM_y}{\Sigma(X^2) - n(M_x)^2} = \frac{11{,}837.32 - 13(32.54)(24.96)}{16{,}456.92 - 13(32.54)(32.54)} = \frac{1{,}278.74}{2{,}691.85} = 0.475$$

$$a = M_y - bM_x = 24.96 - (0.475)(32.54) = 9.504$$

[2] If both X and Y had been stated in terms of deviation from their mean values (just as was done when the standard deviation s was computed in Table 1.6) they would have been denoted by the symbols lower-case x and lower-case y. The product in the fourth column of Table 5.1 would then have been designated xy, and its sum, $\Sigma(xy)$. The correction factors used in equation (5.2) are simply to change the sums of the original extensions, $\Sigma(X^2)$ and $\Sigma(XY)$, to what they would have been if computed instead from the deviations from the mean. That is to say,

$$\Sigma(XY) - nM_xM_y = \Sigma(xy) \qquad \text{and} \qquad \Sigma(X^2) - n(M_x)^2 = \Sigma(x^2) \tag{5.4}$$

Equations (5.2) and (5.3) are only another way of stating the "normal equations," which can be solved simultaneously to give the values for a and b. These equations are

$$na + (\Sigma X)b = \Sigma Y$$

$$(\Sigma X)a + (\Sigma X^2)b = \Sigma XY$$

When these two equations are solved simultaneously, the results are

$$b = \frac{\Sigma(xy)}{\Sigma(x^2)} \qquad \text{and} \qquad a = M_y - bM_x \tag{5.5}$$

The method by which this line is fitted rests upon the assumption that the scatter of the individual observations around the fitted line will approximate a normal distribution. If one or two observations are exceedingly erratic as compared to the others, so that the scatter of the observations around the line will be very skew, this method of fitting may be unsatisfactory.

The equation for the straight line, as thus determined by all the observations, is therefore

$$Y = 9.504 + 0.475X$$

This line is called the *line of best fit*, since it is the line which gives, for all the observed values of X, values of Y which come as near as possible to agreeing with *all* the different Y values observed.[3]

ESTIMATING Y FROM X. We can now take any given flow at Libby we wish and estimate from the equation what would be the most probable flow at Newgate on the basis of the straight-line relationship.

If 2,090 cfs (the flow in 1932), is taken, X will be 20.9. Substituting this value in the equation gives an estimated value of Y, designated Y'.

$$Y' = 9.504 + 0.475(20.9) = 19.43$$

So the expected flow at Newgate, for 20.9 at Libby, would be 19.4. In 1932, the actual flow was 18.0—slightly lower than estimated. Similarly, if we take the unusually heavy flow in 1934, of 77.6 at Libby, and substitute in the equation, we get the result

$$Y = 9.504 + 0.475(77.6) = 46.36$$

This estimate compares with an actual flow of 44.9 that year—again somewhat lower than estimated. If we similarly estimate Y for each year from the equation, using the corresponding values of X, we get results shown in Table 5.2.

Subtracting each Y' from the actual Y gives the differences which are designated z. The z's have a mean of practically zero (the slight difference from zero is due to rounding off in the calculations). We can now plot the line of relationship on Figure 5.3, by using any two Y' with the corresponding values of X. This line, shown as a solid line, indicates the estimated value of Y, Y', for any value of X within the range shown.

Interpreting the Linear Equation. Just what does the line of least squares tell us? What is the meaning of the fitted line $Y = 9.504 + 0.475X$? What do the constants of the equation—the statistics 9.504 and 0.475, which we have calculated from the values given in our sample of 13 cases—really tell?

The first of these statistics, the value for a, is the height of the line when $X = 0$. It indicates that a flow of 9.5 units might be expected at Newgate

---

[3] If the differences between each of the actual observations and the estimated values given by this equation are computed, squared, and summed, that sum will be smaller than it would be if any other straight line were used. Since this method determines the line with the smallest possible squared deviations, the line is known as the *least-squares* line, and the method of computing it is known as the *method of least squares*.

Table 5.2

WATER FLOW AT TWO POINTS, AND FLOW ESTIMATED BY LINEAR
EQUATION

Year	Libby, Mont., X	Newgate, B.C., Y	Estimated by Equation, Y'	Residual, $Y - Y' = z$
		Units of 100 cfs		
1925	42.0		29.5	
26	24.0		20.9	
27	38.0		27.6	
28	49.4		33.0	
29	24.6		21.2	
1930	24.2		21.0	
31	27.1	19.7	22.4	−2.7
32	20.9	18.0	19.4	−1.4
33	33.4	26.1	25.4	0.7
34	77.6	44.9	46.4	−1.5
1935	37.0	26.1	27.1	−1.0
36	21.6	19.9	19.8	0.1
37	17.6	15.7	17.9	−2.2
38	35.1	27.6	26.2	1.4
39	32.6	24.9	25.0	−0.1
1940	26.0	23.4	21.9	1.5
41	27.6	23.1	22.6	0.5
42	38.7	31.3	27.9	3.4
43	27.8	23.8	22.7	1.1

even when there is no flow at all at Libby. Since Libby is downstream
from Newgate, this seems to be an absurd result. The statistic *a* therefore
has no meaning of and by itself in this particular example, beyond placing
the height of the line as a whole for the range within which it does have
meaning.

The statistic for *b*, on the other hand, is always meaningful. It indicates
the difference in *Y* for every difference of 1 unit in *X*, on the average of all
the observations, and within the range covered. In this example the value
of 0.475 indicates that between 17.6 and 77.6, each increase of 1 unit in *X*,
that is, each increase of 100 cfs, at Libby, is associated with an increase of
0.475 units, or 47.5 cfs, in the flow at Newgate. This kind of interpretation

of the value of b can always be made, and is one of the most important results obtained by determining the statistics for the straight line.

Range Within Which the Estimates Are Meaningful. If we estimate the flow Y when $X = 200$, it comes out 107. But 200 for X is almost three times as large as the highest value reported during the entire period from 1931 to 1943, on which we based our statistics. Just as we have seen that the estimate of Y' for $X = 0$ at far below the smallest of the actual observations gives an irrational result, we must use caution in estimating what the relation would be for a value of X far beyond the range of actual observations. We know that within the observed range of X, from roughly 18 to 78, the straight line represents the relation fairly well. Beyond that range, we cannot be sure it will still apply. Chapters 17 and 18 give more exact bases for estimating the confidence with which we can use particular estimates.

Only within the range covered by the original observations of X can an estimating equation of this type be used with confidence. The only exception would be in a case where there is some logical reason, based on other knowledge, to believe that the linear relation would hold true beyond the observed range.

CONFIDENCE INTERVALS FOR THE STATISTICS. The statistics for a and b calculated from a sample of observations are not necessarily the true parameters for the entire universe. Instead, the statistics will vary from one sample to another. If, for example, it had been possible to use records of water flow for one hundred years at the two points (and if the conditions in the universe had shown no change during that period) it would have been possible to compute statistics for a and b which would have been far more reliable. Just as standard errors and confidence intervals can be estimated for means, so they can be estimated for a and b, by methods given in Chapter 17.

CONFIDENCE INTERVALS FOR THE ESTIMATED VALUES. Similarly, the estimates of $Y(Y')$, made from the equation, are subject to possible error. The standard error of estimate based on the standard deviation of the residuals, $\bar{s}_z$, serves as one indication. The standard error of estimate for the straight line is given by the equation

$$\bar{S}_{y \cdot x}^2 = \frac{\Sigma z^2}{n - 2} \tag{5.6}$$

In this case, $\bar{S}_{y \cdot x} = 1.79$. From Figure 2.1 we see that with 13[4] cases

---

[4] Since there are two constants involved in the estimate in this case, a and b, we must deduct one observation, i.e., use 12 instead of 13, in reading the values from Table 2.3 or Figure 2.1.

we must multiply the calculated standard error by about 1.05 for $P = 0.67$, and by 2.2 for $P = 0.95$. Accordingly, in this case we would expect two-thirds of the new estimates made by the calculated equation to fall within a confidence interval of $Y' \pm 1.05S_{y \cdot x}$, or ± 1.88 of the true values, and nineteen out of twenty within an interval of $\pm 2.2S_{y \cdot x}$, or ± 3.93.

In using these confidence intervals, we are assuming that a given value of X might have been associated with any of a number of values of Y and that these would be distributed about an average value of Y which, in the universe of all possible Y's, would be associated with the given value of X. Thus, we might write the 95 per cent confidence interval for Y *given* the 1928 value of X as 33.0 ± 3.93, or from 29.1 to 36.9.[5] If we could select a large number of years in each of which X stood at the 1928 level, we should expect to be right 95 times out of 100 if we assumed that the range $Y \pm 3.93$ included the average universe value of Y which was associated with an X value of 49.4. This assumes that there was no change in the universe during the period. The residuals in Table 5.2 suggest a slight upward trend with the passage of time, so this assumption may not be completely accurate in this case.

FITTING A LINE BY SEMI-AVERAGES. There is a simpler method of fitting a straight line, which is useful for preliminary studies, and where there are no extreme cases or single wide departures from the apparent line, especially at the extremes (as with the observation for 1934 in the present example).

In using this method, we divide all the observations into two approximately equal halves, according to the values of the independent variable X, and take the averages of X and Y for each group. From Table 5.1 we see that the first group should be from 17 to 28 (7 cases), and the second 32 and above (6 cases). Calculating the averages for each group, we obtain values as follows:

$$\text{Lower group, } M_x = 24.09 \qquad M_y = 20.51$$
$$\text{Upper group, } M_x = 42.4 \qquad M_y = 30.15$$

Setting up these values in equation form as before,

$$20.51 = a + 24.09b$$
$$30.15 = a + 42.4b$$

and solving the equations simultaneously, we get the values $a = 9.14$, and $b = 0.47$. The corresponding line, $Y = 9.14 + 0.47X$, can also be plotted in Figure 5.3. Since the values of a and b here are very close to

---

[5] More exact procedures for estimating the reliability of a single estimate are given on pp. 319 and 320 of Chapter 19.

those determined for the line by least squares, the two lines will lie very close to one another.

The line of semi-averages has no exact basis for its fit, and will vary somewhat with the exact composition of the groups selected, and does not necessarily make the average of Y' coincide with the average of Y. We can see the effect of this lower accuracy when we compute the estimated Y' for 1942 by the semi-average line, with $X = 38.7$, $Y' = 27.2$, and $z = 4.0$. This year showed the largest error by the least-squares line, but that was only 3.3, which is substantially smaller.

Methods of estimating standard errors or confidence intervals for the line fitted by semi-averages have not been developed, and it is therefore not used where reliable results are desired.

Usefulness of the Straight Line.   The straight line is a type of relation of very great importance and usefulness. It is one of the simplest functions to fit and to explain, and for that reason it is very widely used. Equations (5.4) and (5.5), which are used in determining the constants of the equation, are therefore of great importance. The student of analytical statistics should become thoroughly familiar with the methods of determining the constants of the equation and should understand thoroughly both the meaning and the limitations of this type of analysis.

Determining the constants for an equation for a given set of observations is called *"fitting" the equation to the data*. Because the linear equation is one of the simplest of all equations to "fit," it is widely and frequently used. In many cases no other possible relation is even considered. Actually, however, the linear equation is very limited in its logical meaning. By its very nature, it can represent only a situation where the change in the dependent variable for a unit change in the independent variable would be expected to be just the same regardless of how large or how small the independent variable was; i.e., where the regression line has the same slope throughout. This is a very precise and narrow relation. In many cases, the line which theoretically would be expected would have a changing slope as the value of the independent variable changed. Unless there is a good logical reason to expect the linear equation to represent truly the situation present, fitting a straight line can be regarded only as an empirical exercise, with no meaning to the constants obtained beyond the purely formal one of specifying the straight line that most nearly represents the observed data.

Summary

To express a functional relationship by a straight line, the constants may be determined arithmetically by the methods of *semi-averages* or of *least squares*. The least-squares line gives the *line of best fit* under the assumptions of that method: a normal distribution of the observations around the line and the reduction of the squared residuals to a minimum. Estimates of the dependent variable may be made according to the linear function for any value of the independent variable. Only within the range which includes the bulk of the independent values does this estimate have meaning, however; and only then if the straight line gives a satisfactory expression of the observed relation, either empirically or logically.

Note 5.1. Just as a straight line can be fitted to show the average flow at Newgate for each given flow at Libby, so another straight line can be fitted if the variables are reversed. In that case the flow at Libby would be regarded as the dependent or *Y* variable, and that at Newgate would be regarded as the independent or *X* variable.

CHAPTER 6

Determining the way one variable changes when another changes: (3) for curvilinear functions

A straight-line equation is frequently a fairly good empirical statement of the relation between two variables even when the true relation is more complex than the straight line can portray. Yet it may be just as important to know the exact or approximate "form" of the relationship as it is to have an empirical statement of it. For that reason it is necessary to consider other ways of expressing a relationship than the straight line.

The automobile-stopping case (Figures 4.1 and 4.2) showed that the relations could not be well expressed by a straight line, especially below 15 miles per hour, and above 35. The latter might be very important for the purpose of the whole study.

The real difficulty involved would have been the assumption that the straight-line function applied. That would have assumed that an increase of one mile in the speed of the car increased the distance required for stopping by the same number of feet, no matter how fast the car was already traveling. When we examine Figures 4.1 and 4.2 closely, we see that this is not correct; the line of averages slants up slowly at first, then tends to rise more steeply as the speed is increased, until it has the steepest slope at the highest speed. It is therefore incorrect to assume that we can express the slope of the line by determining the average increase in stopping distance for an increase of one mile in the rate of speed; for *the increase in stopping distance is not the same regardless of the rate of speed, but tends to become greater as the rate of speed increases.* Only if our way of stating the line can express that fact too will it sum up all our observations with sufficient accuracy.

What is needed is some general way of stating the relation between speed and distance, similar to the general relation expressed in the straight-line formula, yet expressing *a changing slope* instead of the uniform slope shown by the straight line.

Different Types of Equations

In the same way that it is possible to represent relations mathematically by a straight line, it is possible to represent them by curves of various types. We have seen how the equation $Y = a + bX$ can be used to represent any straight line by determining the proper values to be assigned to the constants a and b. There is practically no limit to the different kinds of curves which can be similarly described by mathematical equations. The equations of a number of curves which are useful in statistical analysis of the relations between variables are:

$$Y = a + bX + cX^2 \tag{a}$$

$$\log Y = a + bX \tag{b}$$

$$\log Y = a + b \log X \tag{c}$$

$$Y = a + b \log X \tag{d}$$

$$Y = \frac{1}{a + bX} \tag{e}$$

$$Y = a + bX + cX^2 + dX^3 \tag{f}$$

$$Y = a + bX + c\left(\frac{1}{X}\right) \tag{g}$$

Each of these equations can be used to represent a certain type of curve. Thus type (a) is the equation of a parabola. If we take certain values for the unknown constants a, b, and c, substitute them in the formula, work out the values of Y for various values of X, and plot them just as we did before, we will see the sort of curve this equation can be used to express. Thus if we take 1 for a, 0.5 for b, and -0.1 for c, the equation will read:

$$Y = 1 + 0.5X - 0.1X^2$$

When the value of X is 0, Y will be 1, obviously. When X is 1, Y will be

$$Y = 1 + 0.5(1) - 0.1(1^2) = 1.4$$

Performing this operation for other values, we obtain

X value	Y value
0	1.0
1	1.4
2	1.6
3	1.6
4	1.4
5	1.0
6	0.4

Plotting each of these values on cross-section paper and drawing a smooth curve through the several points, we have the curve shown in Figure 6.1, center top section. This discloses one characteristic of this type of curve—the curve is always symmetrical on both sides of the highest point—the point where it stops going up and starts to turn down (as half way between $X = 2$ and $X = 3$ in this case). The value of Y when $X = 2$ is the same as when $X = 3$. When $X = 1$ it is the same as when $X = 4$ and, for $X = 5$, Y is the same as when $X = 0$. The curve could be cut into halves at the point of turning downward, one of which would be the reverse of the other. Besides this characteristic symmetry, this curve has another peculiarity—it has one, and only one, change from moving upward to moving downward, no matter what values are assigned to a, b, and c, or how far it is carried out. For the equation shown, the curve reaches its highest point when $X = 2.5$. As shown in Figure 6.1, the curve continues downward on both sides of this point, no matter how large the positive or negative values of X become. Thus if $X = 100$, $Y = -949$, or if $X = -100$, $Y = -1,049$.

If the value of b were negative and c were positive, the curve would then be concave from above instead of convex and would be symmetrical with respect to its lowest point.

Because of the characteristics mentioned, this type of curve is not very satisfactory to represent many types of relations. It does have great flexibility, in that many differently shaped curves can be represented by some particular segment of the parabola; but on the other hand the parabolic shape itself is so simple that many times the real relation between the variables cannot be represented by it.

The characteristics of a number of other types of simple curves are also illustrated in Figure 6.1. In each case an equation of the type indicated has been assumed, and the values of Y corresponding to values of X

have been computed as has just been done for the simple parabola. Then plotting these computed values gives the curves shown. Thus type (*f*), the cubic parabola, is seen to have one maximum point and one minimum point and one point of inflection (the point where the curve changes from

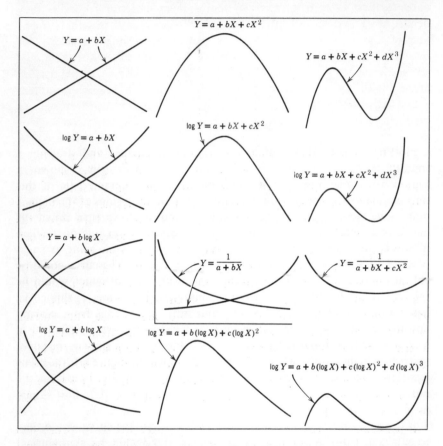

Fig. 6.1. Curves illustrating a number of different types of mathematical functions.

concave from above to convex, or *vice versa*). No matter what values are assigned the constants in this equation, it can have only the single inflection and the two points of maxima and minima. Of course the particular data to be represented might fall anywhere along the entire course of the curve —if only a single change from positive to negative slope were required, the point of inflection in the cubic parabola might lie beyond the extremes of the data, and so not show at all when the fitted curve was plotted for the range covered by the data.

Figure 6.1 also illustrates curves of types (b) to (e), as well as some others not given special designations here. In each case where the log of Y is used in place of Y, it is evident that the previous curve has been modified as if by compressing the ordinates nearest zero and stretching out the ordinates farthest away from zero, stretching them more and more as they depart more and more from zero. This process transforms the straight lines of $Y = a + bX$ to a curve concave from above when $\log Y = a + bX$ is used; or, when $\log Y = a + bX + cX^2$ is substituted for $Y = a + bX + cX^2$, it can sharpen the top of the bend if b is positive, or the bottom of the dip if b is negative. Similar results are found with the cubic parabola.

Similarly, when $\log X$ is used in place of X, the previous curves are modified as if the abscissas were compressed near zero, and stretched out in the higher values. This changes the straight line of $Y = a + bX$ to a curve for $Y = a + b \log X$, convex from above when b is positive and concave from above when b is negative. The parabolas are similarly transformed, making the slopes different on each side of the bend in the simple parabola or on each side of the inflection in the cubic. The effect is to move the first "hump" or "dip" in nearer to the zero abscissa and to stretch out the remainder of the curve (including the second bend, in the case of the cubic parabola).

When logarithms are used for both X and Y, the effect is to modify both sets of coordinates in the manner previously described. The curve $\log Y = a + b \log X$ may have either a concave or convex bend if b is positive, but is always concave from above if b is negative. Similar modifications are noted in the case of the simple parabola.

In any event it should be noted that the curves whose equations contain logarithms retain some of the same characteristics as those with similar equations without logarithms. Thus the linear equations (with only a and b) *never* change from a positive to a negative slope; the simple parabola *always* has one such change, if carried out far enough; and the cubic parabola always has two such changes. In addition, it should be noted that a variable can be stated in terms of logarithms only if it has no negative values. Whereas the other functions can express negative values as readily as positive ones, the logarithmic curves always become asymptotic as they approach zero—that is, they tend to flatten out and to run almost parallel with the axis. This is because a logarithm cannot be obtained for a negative number. No matter how small a logarithm becomes, the corresponding antilogarithm is still positive, even if only a very small decimal fraction.

The hyperbola [type (e)] shown just below the center of Figure 6.1 also is peculiar in that it can become asymptotic as it approaches both the X axis and the Y axis, even if one or both of the variables are in negative values.[1] However, the values of X and Y which it approaches are not the zero values, as with the logarithmic curves, but special values which vary in each particular case and depend upon the value of the constants a and b in the equation. Still more complex curves of the same hyperbolic type may be obtained by including higher powers of X, such as

$$Y = \frac{1}{a + bX + cX^2}$$

Still other curves may be represented by hybrid equations, which combine two or more of the simple types described thus far. Thus type (g) is a compound of a simple linear equation and a simple hyperbola. This is sometimes useful to represent curves which cannot be represented by the simpler types. The choice of an equation to represent a particular set of data, however, depends upon logical analysis as well as upon the empirical ability of a given equation to represent the relation found.

The Logical Significance of Mathematical Functions. There has been frequent reference previously to the question of whether an equation did or did not express "the real nature" of a relationship. To know when we would be justified in using a simple freehand curve, and when we should go to the additional work of determining an equation for the curve, we must understand the logical bases for different types of equations, so that we can judge whether or not any particular type of curve can logically be expected to express the relation in any given set of observations.

The Linear Equation. Many relations are so simple that ordinarily we would not think of expressing them mathematically. Thus, if a train is traveling 45 miles an hour, the distance traveled is equal to the time multiplied by the speed. Using t for the time in hours, d for distance, and s for speed, the relation is obviously $d = st$.

This is a simple straight-line relation. Now, if, in addition, the train were a miles away from a given station at the beginning, after t hours of additional travel away from the station it would be D miles away, where $D = a + d = a + st$.

This is now expressed in the usual form for the straight-line equation,

---

[1] There are three types of simple hyperbolas which are frequently useful in curve fitting:

$Y = \dfrac{1}{a + bX}$ is an equilateral hyperbola, asymptotic to a line parallel to the X axis;

$Y = a + b\left(\dfrac{1}{X}\right)$ is an equilateral hyperbola asymptotic to a line parallel to the Y axis;

$\dfrac{1}{Y} = a + b\left(\dfrac{1}{X}\right)$ is an equilateral hyperbola asymptotic to lines parallel to both axes.

$Y = a + bX$. This equation is therefore the one to be used when it can logically be expected that each unit change in X is accompanied by a corresponding change in Y, regardless of the size of X. Such a mathematical expression of the logical relationship, which is expected to exist in a problem under study, is called a "model" of the relationship.[2] Thus in computing the distance the train has traveled we are assuming that it continued to travel at a definite rate, say 45 miles an hour, the whole way, and traveled the two-hundredth mile just as fast as the first mile. But if we were dealing with something where the change in Y was not the same for different values of X, the linear model would no longer be satisfactory. For example, an airplane on a long-distance flight has to carry a heavy load of gasoline at the start and hence cannot attain full speed; the farther it goes the lighter its load becomes and the higher speed it can make. In such a case the straight-line formula would not be applicable, since the speed of the plane would increase with the distance it had gone. If the straight-line formula were used, it would indicate that it would take just as long to travel the first hundred miles as the last hundred, whereas actually it would take longer than that to travel the first hundred and less than that to travel the final hundred. Only a model which included some value that properly took into account the change in speed with the change in distance could satisfactorily represent this relation.

The Quadratic Equation. Another familiar case in which the rate at which Y increases changes as the value of X increases is that of a weight falling to the ground. Since the attraction of the earth near the earth is practically a constant, it exercises a constant pull on a falling body. Thus, the farther a body falls, the faster it travels. It is just as if, in throwing a ball, a boy did not let go the ball for it to travel by its momentum but was able to keep shoving against it, adding more and more speed to the momentum it already had. Physicists express this relation by saying that the velocity with which an object falls is accelerated at a constant rate. This equation, therefore, is:

$$V = gt$$

where g is a constant measuring the force of gravity, V is velocity in feet per second, and t is time in seconds.

The velocity, or speed, increases with every passing moment, and therefore the distance traveled in each succeeding second is greater than the distance traveled in the previous second.

---

[2] A complete model will also contain a term to recognize departures of individual observations from the continuous line called for by the theory. For example, if accidental errors in recording the train's passage led to fluctuations in the actual distances from those expected, this might be recognized by stating the model $Y = a + bX + e$, where e stood for the accidental errors.

If we assume that the value of g in the equation is already known to be 32, the equation $V = gt$ can then be written $V = 32t$.

We can estimate the distance traversed by a falling body in each successive second by a process like this:

Let us figure that the average speed for each 2 seconds is the same as at the midpoint and then let us estimate the distance traversed in those 2 seconds by multiplying this average speed by the time. Then by adding all the distances together we can get an approximation of the total distance.

First we need to calculate the average speed for each period, using the last equation, $V = 32t$:

End of 1st second, speed = 32(1) = 32 = average speed for 1st two seconds
End of 3d second, speed = 32(3) = 96 = average speed for 2d two seconds
End of 5th second, speed = 32(5) = 160 = average speed for 3d two seconds
End of 7th second, speed = 32(7) = 224 = average speed for 4th two seconds
End of 9th second, speed = 32(9) = 288 = average speed for 5th two seconds

Then we can estimate the distance traveled in each 2-second period, as follows:

Period	Average Speed (feet per second)	Distance in That Period (feet)
1st	32	64
2d	96	192
3d	160	320
4th	224	448
5th	288	576
Estimated total distance		1600

Since the velocity increases at *a uniform rate for each moment of time*, the true average rate of speed for any period will be just half way between the speed at the beginning and at the end.[3]

The average speed during any period of t seconds is therefore $32t/2$. The total distance traversed in the t seconds can therefore be determined by multiplying the average speed, $32t/2$, by the total number of seconds, t. This gives $d = 32(t/2)t$ or $d = 32t^2/2 = 16t^2$.

So far, we have assumed that we know the acceleration, or rate of increase in velocity per second. Suppose instead we had not known it to begin with. How could we have found it out?

If we had used the symbol g to represent this value, we could have carried

---

[3] This would not be true of all types of relations. If, for example, velocity increased at a *changing* rate, the smaller the units taken the more accurate would be the result.

out all the previous calculations, except that we should have used "*g*" where instead we have used "32."

Our last formula then would have been

$$d = g(t^2/2), \quad \text{or} \quad d = (g/2)t^2$$

If we let $g/2 = b$, the equation then would read $d = bt^2$.

We could readily determine the value for b by observing the distance a given body falls in 1 second, in 2 seconds, in 3 seconds, etc., and then working out the probable value for the constant, just as has been done before.

In this case it should be noted that the formula $d = (g/2)t^2$ is derived on the assumption that the attraction of gravity is a constant, tending to increase velocity at a uniform rate per second, or other unit of time. Only if this assumption is correct can the equation be used. The equation is directly based upon this assumption; the reasoning used in deriving the equation also serves to explain what the constants obtained *really represent*. On the basis of this reasoning the equation determined is not a mere empirical expression of the relation between time falling and distance traversed. Instead, it is a fundamental measurement of *why that distance is what it is*, and relates it in a logical manner to the attraction of the earth.

Although it would be quite possible in this particular case to draw a freehand curve expressing the relation between time and distance, it would not be so satisfactory as the mathematical equation. The curve would merely state what the relation was; the equation, in addition, explains *why* it is, in the terms of a particular hypothesis.

The Parabolic Equation. Another physical case in which a definite relationship may be established logically, and then measured statistically, is the firing of a projectile from a gun.

Disregarding the resistance of the air, there are three elements which will determine the height the projectile will have reached at any given instant after it leaves the muzzle of the gun. The simplest of these elements is the height of the muzzle of the gun itself, represented by a in Figure 6.2. All the subsequent changes in elevation will obviously have to be added to that.

The second element is the rate at which the projectile is moving upward at the instant it leaves the muzzle. That is dependent, of course, on the angle at which the gun is elevated and on the muzzle velocity. If the slope were 10 per cent from the horizontal and the muzzle velocity were 1,000 feet per second, the projectile would leave the muzzle moving upward at the rate of 100 feet per second. If there were no resistance of the air, and if there were no force of gravity to pull the projectile off its course, its momentum would carry it on in this direction to infinity, as illustrated by the straight line in the picture. (These directions are, of course, relative to the earth's surface at the moment of firing.) Here b represents the increase in elevation the projectile would attain for each additional second of flight, and a and bt the elevation it would attain if gravity did not influence it.

But gravity is at work too. As we have already seen, as soon as a body is released, the pull of gravity tends to move it downward at ever-increasing speed. Even if it is headed upward as when shot from a gun, the pull of gravity starts tending to pull it down. The diagram illustrates what happens,

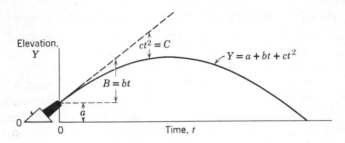

Fig. 6.2. The trajectory of a projectile, illustrating the equation
$$Y = a + bX + cX^2.$$

with C used to represent the distance the body would have fallen if it had no upward velocity. At first the gain in height from its upward momentum is more than enough to offset the tendency to lose height because of the pull of gravity, and the projectile moves upward along the curved course indicated. But finally the loss due to gravity becomes greater than the gain from its original upward momentum and the trajectory gradually turns downward, until the projectile finally comes to rest in the earth or on its target.

The height that the projectile reaches at any moment is the sum of these three components—the original height, the upward course, and the loss by gravity. Its height, then, can be expressed by adding together the three elements.

1. a remains the same, regardless of the time elapsed.

2. B, the height due solely to the original momentum, depends on the time, increasing as the time increases. If we let b represent the initial rate of gain in elevation per second of time, B can then be stated: $B = bt$.

3. Finally, C depends on the time elapsed, and, as we have just seen, varies with the square of time. With the same notation as in our falling-stone problem, but with C substituted for distance fallen, $C = -(g/2)t^2$ $= ct^2$.

Adding these three elements together, we obtain the equation for the height of the projectile at any instant, letting H represent height in feet: $H = a + bt + ct^2$.

It will be seen that this equation is exactly identical in form with the equation for a parabola, $Y = a + bX + cX^2$.

Measurements of the height of the projectile at various given times after firing the charge, made for a given gun, firing the same charge at the same elevation of the gun, would give a series of X and Y values which could be used in computing the constants a, b, and c, even if all were unknown to start with.

If the equation were actually worked out, it would tell much more than merely the graph of the relation. For if the reasoning on which the several different constants were included in the equation was correct, then the equation would furnish a real explanation of why the projectile moved as it did, in terms of the laws of motion and of gravity upon which all such movements depend.

If the projectile were an intercontinental ballistic missile, the equation expressing its flight would be much more complicated, because it would also have to take account of the curvature of the earth over long distances, of the continued acceleration until its fuel had been expended, of the varying resistance offered by the air as it became thinner at higher altitudes, and possibly even of the varying pull of gravity as the missile rose to very high altitudes. And if the missile were of the type that can change its course—that is, steer itself—until its fuel was exhausted, the calculations would have to consider also the elevation and direction needed at that moment, and during its trajectory thereafter, in order to reach its target.

By a similar process of logical analysis, we can work out the type of equation that should properly fit our auto-stopping example, i.e., serve as the *model* for the relation. We know that when a car is not moving, it takes 0 distance to stop. Therefore, when X (speed) $= 0$, Y (distance) should also equal 0. So our curve should have no a constant, in view of this *logical condition*.

Second, it takes some definite period of time after the signal to stop is given for a driver to react, to put his foot on the brake pedal, and for the brakes to start to take hold. During that period of time, the car will continue to move at its original speed—and the higher that speed, the greater will be the distance covered before the brakes are applied. Our equation will therefore need a term of bX, to represent this part of the stopping distance.

Third, once the brakes are applied, they will tend to exert a stopping force which is the product of the friction of the brakes multiplied by the movement of the wheel drums against that friction—and as the wheels slow down, that movement will constantly become less. We may assume that this force of deceleration works in the exact reverse of the way the accelerating force of gravity works—and that the speed of the car falls in a straight line with time, just as the speed of falling increased in a straight line. So the sum of this force, as measured by the distance it takes to stop after the brakes begin to take hold, should vary with the square of the initial speed, just as in the gravity case the distance fallen varied with the square of the time. So we will need a term cX^2 to represent this portion of the stopping distance.

On the basis of this analysis, we arrive at the equation

$$Y = bX + cX^2 + e$$

as the model which, fitted to the observations of the auto-stopping example, should both fit the observed relation, and have a logical explanation of its two constants, b and c.

Reasoning such as this, carried out to much greater lengths, has formed the basis for the scientific "laws" which have been discovered in physics and chemistry and expressed in definite equations. Methods for determining the constants have been devised to serve in determining such relations. But when the same methods are applied to biological, economic, educational, or other relationships in the natural or social sciences, their value is much more limited. Only rarely is there real basis for expecting a particular mathematical relationship such as can be expressed in a given type of equation. In many cases our knowledge of the reasons for the relationship are altogether too limited to enable us to say *why* it exists; and even where we can establish the reasons, they are frequently too complicated or too involved to admit of mathematical treatment. If we express a given relation by a formula, merely on the basis that that formula seems to describe the observed relation satisfactorily, we do not have any greater knowledge of the relation than if we merely drew in a freehand curve. The equation is simply an empirical description of the relation; of and by itself, it offers no clue as to what the relation *means*.

Practical Procedures for Fitting Curves

The equations discussed to this point can all be fitted to the data by relatively elementary arithmetic operations, as will be shown subsequently. There are many other types of more complicated equations which cannot be fitted so readily. These can reproduce curves with recurrent or periodic oscillations, growth curves, and other complicated biological or physical phenomena. Discussion of the use and fitting of such complicated curves lies outside the scope of this book.[4]

The inability of any one equation to represent many simple curves may be illustrated by taking a new example. Table 6.1 shows a series of observations of two variables—the protein content of different samples

Table 6.1

PROTEIN CONTENT AND PROPORTION OF VITREOUS KERNELS FOR EACH
OF A NUMBER OF SAMPLES OF WHEAT*

Sample Number	Protein Content	Proportion of Vitreous Kernels
	Per cent	*Per cent*
1	10.3	6
2	12.2	75
3	14.5	87
4	11.1	55
5	10.9	34
6	18.1	98
7	14.0	91
8	10.8	45
9	11.4	51
10	11.0	17
11	10.2	36
12	17.0	97
13	13.8	74
14	10.1	24
15	14.4	85
16	15.8	96
17	15.6	92
18	15.0	94
19	13.3	84
20	19.0	99

* These values are selected cases, picked so as to show the relationship more clearly. Actually, the correlation is not so high as is shown here.

[4] For examples of such complicated curves and methods of fitting them, see Frederick E. Croxton and Dudley J. Cowden, *Applied General Statistics*, 2d ed., pp. 297–318, Prentice-Hall, Englewood Cliffs, N.J., 1955. Also pp. 540–571, 1st ed., Henry Holt and Co., N.Y., 1939.

of wheat, as determined by chemical analysis, and the proportion of "hard, dark, vitreous kernels" in each sample, as determined by visual examination with the naked eye. The relation here is quite different from the one we have been considering so far. There is no causal connection between these two variables in the sense of one's being caused by the other. Instead, they are merely two different ways of measuring the character of the wheat. It is a short, rapid process, however, to examine the samples by eye and determine the percentage of hard, dark, vitreous

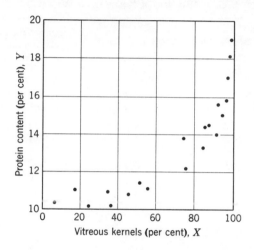

Fig. 6.3. Dot chart showing relation of proportion of vitreous kernels to protein content of wheat.

kernels, whereas it is a long and expensive process to run a chemical test on each lot. For that reason it is of importance to know whether it is possible to estimate the protein content from the percentage of vitreous kernels, and, if so, how closely. So even though the vitreous kernels do not *cause* the differences in protein, we can still regard the proportion of vitreous kernels as the independent variable and the percentage of protein as the dependent variable. That means only that we are going to try to estimate the dependent (protein) from the independent (percentage of vitreous kernels) even though there is no direct cause-and-effect relation present.

The relation between the proportion of vitreous kernels and the per cent of protein may be seen more readily if a dot chart is made, showing the two variables for each of these individual observations. We shall designate the proportion of kernels vitreous as X, and the percentage of protein as Y. In preparing the dot chart, shown in Figure 6.3, we shall therefore plot the X values, or percentage of vitreous kernels, along the horizontal axis and the Y values, the proportion of protein, along the vertical axis.

It is quite obvious from the figure that a straight line could not represent the change in protein with change in vitreous kernels. Some type of curve is necessary. In this case we have no prior logical expectation as to what type of curve would fit. Let us see if the simple parabola is the proper type of curve.

Fitting a Simple Parabola. To represent the relationship between the two variables according to the formula

$$Y = a + bX + cX^2 \tag{6.1}$$

we shall have to determine from the 20 observations the values to assign to the constants a, b, and c, just as before for the straight line we had to determine values for a and b. (Of course the a and b for the parabola will not be the same as the values would be for a straight line fitted to the same data—unless c happens to be zero, which would make the equation for the parabola give a straight line instead.) The values for these constants are determined by constructing and solving the following equations:[5]

$$\left.\begin{array}{l} (\Sigma x^2)b + (\Sigma xu)c = \Sigma xy \\ (\Sigma xu)b + (\Sigma u^2)c = \Sigma uy \end{array}\right\} \tag{6.2}$$

and

$$a = M_y - b(M_x) - c(M_u) \tag{6.3}$$

The values necessary in constructing equations (6.2) and (6.3) are derived as follows:

Use U to represent the X^2 values of equation (6.1).[6]

Then

$$\left.\begin{array}{l} M_x = \dfrac{\Sigma X}{n} \qquad M_u = \dfrac{\Sigma U}{n} \qquad M_y = \dfrac{\Sigma Y}{n} \\[2mm] \Sigma x^2 = \Sigma X^2 - nM_x^2 \\ \Sigma xu = \Sigma XU - nM_xM_u \\ \Sigma u^2 = \Sigma U^2 - nM_u^2 \\ \Sigma xy = \Sigma XY - nM_xM_y \\ \Sigma uy = \Sigma UY - nM_uM_y \end{array}\right\} \tag{6.4}$$

[5] An alternative method is to solve the following three equations simultaneously. The clerical work is about the same in both methods.

$$na + (\Sigma X)b + (\Sigma U)c = \Sigma Y$$
$$(\Sigma X)a + (\Sigma X^2)b + (\Sigma UX)c = \Sigma XY$$
$$(\Sigma U)a + (\Sigma UX)b + (\Sigma U^2)c = \Sigma YU$$

[6] If U is made equal to X^2 divided by some convenient number, say 1,000, the volume of necessary arithmetic can be materially reduced, without affecting the accuracy of the result.

After computing these values, the two equations (6.2) are solved simultaneously to obtain the values for *b* and *c*, and then these values are substituted in equation (6.3) to obtain the value for *a*.

Table 6.2 shows the form of computation in the first step to obtain these values for the data of Table 6.1 (the computations are omitted for the central 16 observations).

Table 6.2

COMPUTATION, FOR WHEAT PROBLEM, OF VALUES NEEDED TO DETERMINE CONSTANTS OF THE SIMPLE PARABOLA

Vitreous Kernels, X	Protein (minus 10),* Y	X^2 and U	XU	U^2	XY	UY
Per cent	*Per cent*					
6	0.3	36	216	1,296	1.8	10.8
75	2.2	5,625	421,875	31,640,625	165.0	12,375.0
...	...	...	...	...	...	...
84	3.3	7,056	592,704	49,787,136	277.2	23,284.8
99	9.0	9,801	970,299	96,059,601	891.0	88,209.0
1,340	68.5	107,566	9,259,238	824,403,226	5,985.6	542,584.6

* To simplify the following calculations, 10.0 has been subtracted from each protein reading.

The values at the foot of the table give the values called for in equations (6.4). Substituting the values as computed for those shown symbolically, the arithmetic appears as follows:

$$M_x = \frac{\Sigma X}{n} = \frac{1,340}{20} = 67$$

$$M_y = \frac{\Sigma Y}{n} = \frac{68.5}{20} = 3.425$$

$$M_u = \frac{\Sigma U}{n} = \frac{107,566}{20} = 5,378.3$$

$$\Sigma X^2 - nM_x^2 = 107,566 - 20(67)^2 = 17,786$$

$$\Sigma XU - nM_x M_u = 9,259,238 - 20(67)(5,378.3) = 2,052,316$$

$$\Sigma U^2 - nM_u^2 = 824,403,226 - 20(5,378.3)^2 = 245,881,008$$

$$\Sigma XY - nM_x M_y = 5,985.6 - 20(67)(3.425) = 1,396.1$$

$$\Sigma UY - nM_u M_y = 542,584.6 - 20(5,378.3)(3.425) = 174,171.05$$

These calculations give the values needed in equations (6.2) which are to be solved simultaneously to obtain the values of b and c. Substituting the values just computed in the equations gives the two equations to be solved as follows:

(A) $(\Sigma x^2)b + (\Sigma xu)c = \Sigma xy$ $17{,}786b + 2{,}052{,}316c = 1{,}396.1$

(B) $(\Sigma xu)b + (\Sigma u^2)c = (\Sigma uy)$ $2{,}052{,}316b + 245{,}881{,}008c = 174{,}171.05$

The simplest way to solve these is by the Doolittle method, as indicated in Appendix 2 pp. 489 to 497.

Solving the equations simultaneously gives $b = -0.0879$, $c = 0.001442$. These values are then substituted in equation (6.3) to obtain the value for a.

$$a = M_y - b(M_x) - c(M_u)$$

$$= 3.425 - (-0.0879)(67) + (0.001442)(5{,}378.3)$$

$$= +1.56$$

With our values for a, b, and c, we can now write out the equation for the parabola, $Y = a + bX + cX^2$, for this particular case as follows:

$$Y = 1.56 - 0.088X + 0.00144X^2$$

Since 10 was subtracted from the percentage of protein before calculating the equation,[7] to estimate the actual percentage 10 must be added back in, making the equation read

(I) $Y = 11.56 - 0.088X + 0.00144X^2$

This then is the equation of the simple parabola which comes nearest to describing the relationships between Y and X. From it the percentage of protein in a given sample of wheat may be estimated from the percentage of hard, dark, vitreous kernels in that sample.

We can see how the estimates are made by working them out for some of the samples. If we take the value of X for the first sample in Table 6.2, and substitute it in equation (I) we obtain an estimated value for Y as follows:

When $X = 6$

$$Y = 11.56 - 0.088(6) + 0.00144(36) = 11.08$$

Substituting each of the values of X in the formula in turn in a similar manner, we obtain estimated values for Y as shown in Table 6.3. So as

---

[7] This does not affect the values obtained for $\Sigma(x^2)$, $\Sigma(xy)$, etc.

to distinguish between the actual values of Y, and the values for Y estimated from X according to the equation of the parabola, we shall designate the latter as Y' values.

Table 6.3

COMPARISON, FOR WHEAT PROBLEM, OF ACTUAL PROTEIN CONTENT
WITH PROTEIN CONTENT ESTIMATED FROM PER CENT OF VITREOUS
KERNELS ON BASIS OF THE SIMPLE PARABOLA

Vitreous Kernels, X	Protein (minus 10), Y	Estimated Protein (minus 10), Y'	Difference Between Actual and Estimated Protein, $Y - Y'$
Per cent	*Per cent*	*Per cent*	
6	0.3	1.08	−0.78
75	2.2	3.06	−0.86
87	4.5	4.80	−0.30
55	1.1	1.08	+0.02
34	0.9	0.23	+0.67
98	8.1	6.79	+1.31
91	4.0	5.50	−1.50
45	0.8	0.52	+0.28
51	1.4	0.83	+0.57
17	1.0	0.48	+0.52
36	0.2	0.26	−0.06
97	7.0	6.60	+0.40
74	3.8	2.95	+0.85
24	0.1	0.28	−0.18
85	4.4	4.51	−0.11
96	5.8	6.41	−0.61
92	5.6	5.68	−0.08
94	5.0	6.04	−1.04
84	3.3	4.35	−1.05
99	9.0	6.99	+2.01

We can plot the actual and the estimated values on a dot chart (Figure 6.4), using dots to represent the values of Y originally observed and crosses to represent the estimated values, Y'. The crosses all lie on a continuous smooth curve, which we can sketch in freehand, as indicated by the dotted line in the figure. To estimate the protein for a sample with a proportion of vitreous kernels not included in our problem, say 65, we can substitute 65 for X in equation (I), and compute it out, or read from our smooth curve the Y value corresponding to an X value of 65. This *graphic interpolation* will not be quite exact, but for many purposes it will be sufficient.

Let us now examine Figure 6.4 and decide whether the formula for the parabola gives a satisfactory "fit" in this case—whether the estimated values do agree fairly well with the actual. The curved line does come closer to agreeing with the actual values than any straight line could, but the shape of the parabolic curve and the general trend of the actual relationship is rather different.

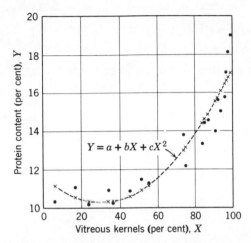

Fig. 6.4. Dot chart showing relation of vitreous kernels to protein content of wheat, and parabolic curve fitted to same.

Apparently the equation of the simple parabola is not adequate to describe this particular relationship. Especially for high proportions of vitreous kernels, the estimates are quite inaccurate. Between 70 and 90 per cent vitreous kernels, the estimates of protein are all too high, with only one exception. For 99 per cent vitreous, the parabola would estimate 17.0 per cent protein, whereas both samples over 97 per cent vitreous kernels had over 18 per cent protein. The failure of this curve to give a satisfactory "fit" is not due to any error in the computations but merely to the fact that this formula cannot give the proper-shaped curve to fit the relationship in this case. The mathematical properties of the equation itself are such that, no matter what constants are used for a, b, and c, it cannot come any closer to describing the true relation.

Fitting a Cubic Parabola. The cubic parabola, type (f) of the equations on page 70, might be tried to see if it would describe this particular relationship more closely.

The equation of the cubic parabola,

$$Y = a + bX + cX^2 + dX^3 \tag{6.5}$$

has four constants a, b, c, and d to be computed. Here again, of course, a, b, and c will be different from those we have computed previously, unless the d value comes out zero. The values b, c, and d are computed by the simultaneous solution of the following three equations:[8]

Use U to represent the X^2 of equation (6.5) and V to represent the X^3.

$$\left. \begin{array}{l} (\Sigma x^2)b + (\Sigma xu)c + (\Sigma xv)d = \Sigma xy \\ (\Sigma xu)b + (\Sigma u^2)c + (\Sigma uv)d = \Sigma uy \\ (\Sigma xv)b + (\Sigma uv)c + (\Sigma v^2)d = \Sigma vy \end{array} \right\} \tag{6.6}$$

The value for a is then computed from the following equation:

$$a = M_y - b(M_x) - c(M_u) - d(M_v) \tag{6.7}$$

The values for Σx^2, Σxu, Σxy, Σu^2, and Σuy are computed as shown previously, equations (6.4). The additional values required in equation (6.6) are computed as follows:

$$\left. \begin{array}{l} M_v = \dfrac{\Sigma V}{n} \\[2ex] \Sigma uv = \Sigma UV - nM_uM_v \\[1ex] \Sigma xv = \Sigma XV - nM_xM_v \\[1ex] \Sigma v^2 = \Sigma V^2 - nM_v^2 \\[1ex] \Sigma vy = \Sigma VY - nM_vM_y \end{array} \right\} \tag{6.8}$$

It should be noted that among the values required to "fit" this cubic parabola, that is, to determine the constants a, b, c, and d, are such values as ΣV^2 and ΣUV. Remembering that $V = X^3$, and $U = X^2$, we need to calculate X^5 and X^6. For $X = 10$, $X^6 = 1,000,000$, so for values of X such as those in Table 6.1, ranging from 6 to 99, it would take a tremendous volume of computation to compute the values required in equations (6.6), (6.7), and (6.8). This may be reduced by letting $U = X^2/100$, and $V = X^3/10,000$. The computation is not shown here in detail. It follows the general form of that given in Table 6.2, and the

---

[8] The alternative method here involves the simultaneous solution of 4 equations, as follows:

$$na + (\Sigma X)b + (\Sigma U)c + (\Sigma V)d = \Sigma Y$$
$$(\Sigma X)a + (\Sigma X^2)b + (\Sigma XU)c + (\Sigma XV)d = \Sigma XY$$
$$(\Sigma U)a + \Sigma(UX)b + (\Sigma U^2)c + (\Sigma UV)d = \Sigma UY$$
$$(\Sigma V)a + (\Sigma VX)b + (\Sigma UV)c + (\Sigma V^2)d = \Sigma VY$$

solution of the equations (6.6), starting in just as shown on pages 178 to 180, may be most conveniently carried through by the methods shown subsequently in Appendix 2.

Even when the cubic parabola is "fitted" to the data given, however, it does not give a satisfactory "fit." Thus Figure 6.5 shows the cubic

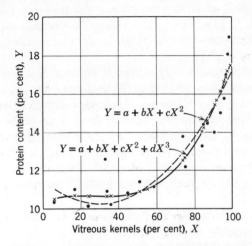

Fig. 6.5. Dot chart, with parabola and cubic parabola.

parabola fitted to the data, worked out as just described. The values found gave the equation

$$Y = 0.35 + 0.0345X - 0.1397(X^2/100) + 0.1788(X^3/10,000)$$

or, clearing of fractions,

$$Y = 0.35 + 0.0345X - 0.0014X^2 + 0.000018X^3$$

Adding in the 10 which was subtracted from Y before making the computations, the equation becomes

$$Y = 10.35 + 0.0345X - 0.0014X^2 + 0.000018X^3$$

In Figure 6.5, the original observations are represented by dots, the estimated values from the cubic parabola are represented by crosses, and the curve of the simple parabola is also shown. A curve has been drawn through the crosses to show the general shape of the cubic parabola.

The last curve comes much closer than the previous curve to describing the relationship which actually exists. Even so, however, it is not entirely satisfactory, for it gives estimates which are still too low at the very highest percentage of vitreous kernels. Except for this portion, and the slight downturn between values of 20 and 40, it seems quite satisfactory.

There are still other types of curves, however, some of which might give better fits than the ones we have tried. For instance the fourth-order parabola,

$$Y = a + bX + cX^2 + dX^3 + eX^4$$

can be fitted by an extension of the methods just described, as can parabolas with even more terms. Those are rarely useful, however, as the greater the number of terms, the greater the tendency becomes for the curve to "wiggle." In addition, the volume of arithmetic required becomes extremely burdensome—the computations for the fourth-order parabolas involving powers of X up to X^8.

Furthermore, there are only a limited number of observations, 20 in all. If a parabola were fitted with 20 constants, for example, it would simply twist and turn so as to pass through every observation. Since it would simply reproduce these 20 observations, it would be of no value at all in indicating the relation which probably holds true in the universe from which the observations in the sample are drawn. (See Chapters 18 and 22 for standard errors of the coefficients of a fitted curve, which indicate its sampling significance, and provide a basis for calculating its confidence limits.)

Fitting Lines or Parabolas to Time Series. In studying time series, it is sometimes desirable to fit a straight line or a curve to the successive observations as a means of determining the long-time trend. The techniques of time-series analysis for individual variables lie outside the scope of this book, and therefore are not given special consideration here. Some special problems involved in correlating two or more time series are considered in Chapter 20.[9] Fitting a mathematical trend to a time series involves regarding the successive months or years as values of the X, or independent, variable. The fact that these values are regularly spaced 1, 2, 3, 4, etc., and that the same succession occurs in many problems, makes possible special methods and special tables, which greatly reduce the labor of fitting the equations. This method of computation, known as *orthogonal polynomials*, should be used in determining lines or parabolic curves for such data.[10]

Fitting a Logarithmic Curve. Some of the other types of curves mentioned on page 70, particularly types (*b*), (*c*), and (*d*), involving logarithms, and type (*e*), using reciprocals, may be fitted with relatively

[9] An excellent discussion of the methods and meaning of time-series analysis for individual series is given by Fredrick C. Mills in his textbook, *Statistical Methods*, 3rd ed., Chapters 10, 11, and 12, Henry Holt, New York, 1955.

[10] For methods of fitting orthogonal polynomials, see Frederick E. Croxton and Dudley J. Cowden, *Applied General Statistics*, 2d ed., pp. 289–90, Prentice-Hall, Englewood Cliffs, N.J., 1955; and R. A. Fisher, *Statistical Methods for Research Workers*, 12th ed., pp. 147–56, Oliver and Boyd, Edinburgh and London, 1954.

little computation. The methods of fitting several of these types may be shown for the present case, even though they may fail to give any better fit than the curves which have already been computed.

The three simple types of logarithmic curves, (*b*), (*c*), and (*d*), may all be fitted by exactly the same method previously used in fitting a straight line, except that the logarithms of *X*, of *Y*, or of both together are employed where otherwise the values of the variables themselves are used. Comparison of the straight-line formula with the logarithmic formula indicates how this is done.

If we use $\bar{Y}$ to represent the logarithms of the *Y* values, and $\bar{X}$ to represent the logarithms of the *X* values, our equations will change as follows:

$$\log Y = a + bX, \text{ to } \bar{Y} = a + bX \tag{b}$$

$$\log Y = a + b \log X, \text{ to } \bar{Y} = a + b\bar{X} \tag{c}$$

$$Y = a + b \log X, \text{ to } Y = a + b\bar{X} \tag{d}$$

In each case it is evident that the new equation is identical in form with the simple straight-line equation,

$$Y = a + bX$$

and the same methods may therefore be used in determining the constants *a* and *b* as were used earlier in equations (5.2) to (5.5).

Some indication as to which one of the three logarithmic formulas will come nearest to fitting a given set of data can be obtained by converting both the *X* and *Y* values to logarithms, variables $\bar{X}$ and $\bar{Y}$, and then making dot charts of $\bar{Y}$ against *X*, of $\bar{Y}$ against $\bar{X}$, and of *Y* against $\bar{X}$. If one chart shows the dots falling in substantially a straight line, the equation corresponding to that chart will give the most satisfactory fit.[11]

The first step in applying any one of the three logarithmic equations to the data of the wheat example is to work out the logarithms and construct the three dot charts, to indicate which formula to use. The form of computation is shown in Table 6.4.

It should be noted that in working out the logarithms nothing can be

---

[11] This is strictly true only if the "goodness of fit" is measured in terms of the logarithms used.

Logarithms may also be used with parabola of higher orders, such as:

$$\text{Log } Y = a + bX + cX^2$$

Such involved curves will not be considered at length in this book, however.

Table 6.4

VARIABLES IN WHEAT PROBLEM AND LOGARITHMS OF VALUES

Protein, Y	Vitreous Kernels, X	Logarithms of Variables*	
		Protein, $\bar{Y}$	Vitreous Kernels, $\bar{X}$
Per cent	Per cent		
10.3	6	1.013	0.778
12.2	75	1.086	1.875
14.5	87	1.161	1.940
11.1	55	1.045	1.740
10.9	34	1.037	1.531
18.1	98	1.258	1.991
14.0	91	1.146	1.959
10.8	45	1.033	1.653
11.4	51	1.057	1.708
11.0	17	1.041	1.230
10.2	36	1.009	1.556
17.0	97	1.230	1.987
13.8	74	1.140	1.869
10.1	24	1.004	1.380
14.4	85	1.158	1.929
15.8	96	1.199	1.982
15.6	92	1.193	1.964
15.0	94	1.176	1.973
13.3	84	1.124	1.924
19.0	99	1.279	1.996

* Logarithms to base 10.

added or subtracted from any of the variables (except for rounding off decimals).[12] In all the previous work the protein had been stated as protein in excess of 10 per cent, but now the original percentage figures are used once more. That is because logarithms deal with *relative* values, and the ratio of 1 to 2 is quite different from that of 11 to 12. All the previous equations have dealt with *absolute* values or differences from the average; and the absolute difference between 1 and 2 is of course just the same as that between 11 and 12.

[12] After the logarithms are once computed, however, they can be "coded" by subtracting a constant or by division, just as other variables have been treated formerly, with the same effect on the final constants obtained.

Figure 6.6 shows the three dot charts in which the three different ways of combining the logarithmic and actual values are shown. None of the three gives a very close linear relation, but the one where $\bar{Y}$ and X are plotted seems to come nearest. The equation

$$\log Y = a + bX, \quad \text{or} \quad \bar{Y} = a + bX$$

will therefore be used.

The values necessary to determine a and b are as follows, using equations (5.2) and (5.3):

$$\Sigma X \bar{Y}, \qquad M_x, \qquad M_{\bar{y}}, \qquad \Sigma X^2$$

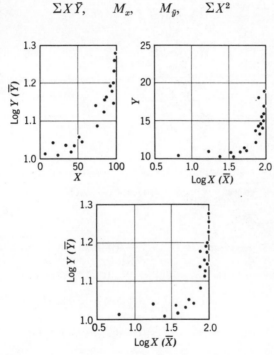

Fig. 6.6. Dot charts illustrating $\log Y = f(X)$; $Y = f(\log X)$; $\log Y = f(\log X)$.

Table 6.5 shows the form of computation of these values from the original values of the two variables, with part of the observations omitted. This computation gives the values necessary to compute a and b.

The averages of X and $\bar{Y}$ of course are:

$$M_x = \frac{\Sigma X}{n} = \frac{1{,}340}{20} = 67.0$$

$$M_{\bar{y}} = \frac{\Sigma \bar{Y}}{n} = \frac{22.389}{20} = 1.11945$$

Table 6.5

COMPUTATION, FOR WHEAT PROBLEM, OF VALUES NEEDED TO
DETERMINE CONSTANTS FOR LOGARITHMIC CURVE

Protein, Y	Vitreous Kernels, X	Logarithms of Y, $\bar{Y}$	Extensions	
			X^2	$X\bar{Y}$
Per cent	*Per cent*			
10.3	6	1.013	36	6.078
12.2	75	1.086	5,625	81.450
14.5	87	1.161	7,569	101.007
...	...	...	...	...
19.0	99	1.279	9,801	126.621
Sums	$\Sigma X = 1{,}340$	$\Sigma \bar{Y} = 22.389$	$\Sigma X^2 = 107{,}566$	$\Sigma X\bar{Y} = 1{,}545.888$

Then, by equation (5.2)

$$b = \frac{\Sigma X\bar{Y} - nM_x M_{\bar{y}}}{\Sigma X^2 - nM_x^2} = \frac{1{,}545.888 - 20(67)(1.11945)}{107{,}566 - 20(67)^2} = 0.002576$$

and by equation (5.3)

$$a = M_{\bar{y}} - b(M_x) = 1.11945 - (0.002576)(67) = 0.9469$$

In terms of the variable, the equation required is therefore

$$\bar{Y} = a + bX = 0.9469 + 0.002576X$$

or

$$\log Y = a + bX = 0.9469 + 0.002576X$$

The percentage of protein can now be estimated from the proportion
of vitreous kernels observed for any sample of wheat, by substituting
the percentage of vitreous kernels (the X values) in this equation and
working it out. Thus for the first example, with 6 per cent of vitreous
kernels, it would work out as follows:

$$\log Y = a + bX = 0.9469 + 0.002576(6) = 0.9624$$

Using a table of logarithms we find that the number corresponding to
the logarithm 0.9624 (that is to say, its antilogarithm) is 9.17. The
estimated proportion of protein is therefore 9.17 per cent.

Similarly if the proportion of vitreous kernels in the second sample,

75, is substituted in the equation, the work to calculate the estimated proportion of protein is:

$$\log Y = a + bX = 0.9469 + 0.002576(75)$$
$$\log Y = 1.1401$$
$$\text{antilog } 1.1401 = 13.81$$

The estimated proportion of protein is therefore 13.81 per cent.

Table 6.6 shows this computation carried through for each of the 20 observations.

Table 6.6

COMPUTATION, FOR WHEAT PROBLEM, OF ESTIMATED PROTEIN
CONTENT FROM PER CENT OF VITREOUS KERNELS ON THE
BASIS OF A LOGARÍTHMIC CURVE

$$(\text{LOG } Y = 0.9469 + 0.00258\ X)$$

Vitreous Kernels, X	Estimated Protein		Actual Protein, Y	Percentage Errors in Estimating Protein Proportion, $100\left(\dfrac{Y}{Y'} - 1.00\right)$
	Estimated logarithm, $\bar{Y}'$	Antilog of Estimate, Y'		
Per cent	*Per cent*		*Per cent*	
6	0.9624	9.2	10.3	+12.0
75	1.1401	13.8	12.2	−11.6
87	1.1710	14.8	14.5	− 2.0
55	1.0888	12.3	11.1	− 9.8
34	1.0345	10.8	10.9	+ 0.9
98	1.1993	15.8	18.1	+14.6
91	1.1813	15.2	14.0	− 7.9
45	1.0628	11.6	10.8	− 6.9
51	1.0783	12.0	11.4	− 5.0
17	0.9907	9.8	11.0	+12.2
36	1.0396	11.0	10.2	− 7.3
97	1.1968	15.7	17.0	+ 8.3
74	1.1375	13.7	13.8	+ 0.7
24	1.0087	10.2	10.1	− 1.0
85	1.1659	14.7	14.4	− 2.0
96	1.1942	15.6	15.8	+ 1.3
92	1.1839	15.3	15.6	+ 2.0
94	1.1890	15.5	15.0	− 3.2
84	1.1633	14.6	13.3	− 8.9
99	1.2019	15.9	19.0	+19.5

It should be noted in this table that errors made in estimating the proportion of protein are stated as relative errors rather than absolute errors. That is done because the thing that is really estimated is the logarithm of the percentages of protein, or $\bar{Y}$, and the errors are really the differences between the actual logarithms and the estimated logarithms. If z is used to stand for the error, in this case z is really in terms of logarithms, that is:

$$z = \log Y - \text{estimated} \log Y, \text{ or } \bar{Y} - \bar{Y}'$$

or in terms of natural numbers:

$$\text{antilog } z = \frac{\text{antilog } \bar{Y}}{\text{antilog} \cdot \bar{Y}'} = \frac{\text{actual } Y}{\text{estimated } Y}$$

Subtracting the constant 1.00 and multiplying by 100 changes this relative figure to the percentage the observed value is above or below the estimate.[13]

Where log Y is taken as the dependent variable, as has been done here, fitting the equation by the methods just shown involves making the sum of squares of the *logarithmic* residuals around the line as small as possible. Instead of minimizing the sum of the *absolute* errors, squared, as heretofore, we now minimize approximately the sum of the *percentage* errors, squared.

In Figures 6.7 and 6.8 the actual proportions of protein, shown as dots, are compared with the estimated values as worked out by the logarithmic relation. In the first of these figures the actual and estimated values are both stated in terms of the logarithms. It is quite apparent here that this equation assumes a straight-line relation between the proportion of vitreous kernels and the logarithms of the proportion of protein; since they were computed by a straight-line equation ($\log Y = a + bX$) the estimated values all lie along the continuous straight line indicated. The next figure, however, compares the actual proportion of protein with the estimated, both stated in actual terms. Here the continuous curve which the logarithms produce in the estimated actual values is clearly shown. The relation between the proportion of vitreous kernels and the percentage of protein, as shown by this curve, does not agree with the actual relation as shown by the original observations even as closely as did the previous curves computed by means of parabolic equations.

Before discussing other ways of expressing the curvilinear relation it might be well to discuss the procedure to determine the constants a and b if either of the other two forms of simple logarithmic equations were used.

---

[13] The reason for making this distinction will be seen later on, when the question of measuring the accuracy of the estimate is taken up.

If the equation $Y = a + b \log X$ is employed, the form $Y = a + b\bar{X}$ is used.

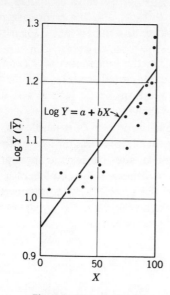
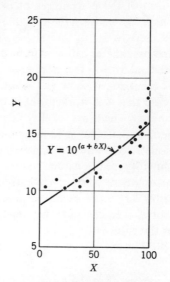

Fig. 6.7. Dot chart showing observations and fitted line for equation $\log Y = a + bX$, in logarithms of Y.

Fig. 6.8. Dot chart showing observations and fitted line for equation $Y = 10^{(a+bX)}$, in natural values of Y.

The values which must be computed are

$$M_y, \qquad M_{\bar{x}}, \qquad \Sigma Y\bar{X}, \qquad \Sigma \bar{X}^2$$

and the constants are determined from the equations

$$b = \frac{\Sigma Y\bar{X} - nM_yM_{\bar{x}}}{\Sigma \bar{X}^2 - nM_{\bar{x}}^2}$$

$$a = M_y - bM_{\bar{x}}$$

Since the equation is in terms of Y itself, the estimated values, computed from the logarithms of X, will be directly in values of Y, and will not have to be converted to the antilogarithms.

If the equation $\log Y = a + b \log X$ is to be fitted, the form $\bar{Y} = a + b\bar{X}$ is used.

The values which have to be computed are:

$$M_{\bar{y}}, \qquad M_{\bar{x}}, \qquad \Sigma \bar{Y}\bar{X}, \qquad \Sigma \bar{X}^2$$

and the constants are determined from the equations

$$b = \frac{\Sigma \bar{Y} \bar{X} - nM_{\bar{y}}M_{\bar{x}}}{\Sigma \bar{X}^2 - nM_{\bar{x}}^2}$$

$$a = M_{\bar{y}} - bM_{\bar{x}}$$

In this case the equation is in terms of $\bar{Y}$, the logarithms of Y, and the estimated values will therefore have to be converted from logarithms into natural numbers to show just what the relationship is, just as was done in the case that was worked out in detail earlier.

No matter which one of the three logarithmic curves is employed, the arithmetic is the same as in determining the simple straight line, with the exception of computing the logarithms and of substituting the appropriate logarithms where the natural values would otherwise be employed.

In cases where other modifications of the straight-line equation, such as type (e), are to be used, the process is to transform the equation to a linear form, then compute the constants just as before.

Thus the type

$$Y = \frac{1}{a + bX}$$

can be converted to the form

$$1/Y = a + bX$$

or, letting $1/Y = Q$,

$$Q = a + bX$$

The computation can then be carried out in the usual way, and after the estimated values of Q, Q', are worked, converted back into Y values by the equation $Y' = 1/Q'$.

Fitting a Conditioned Parabola. Sometimes the logic of a problem requires that one of the constants in an equation be 0. In mathematical language, we impose a "condition" upon the curve of relationship over and above the usual conditions implied in the least-squares method. For example, as discussed earlier, the auto-stopping relation (Table 3.1) should logically be expressed by a curve of the equation

$$Y = bX + cX^2 \tag{6.9}$$

Let us fit that curve, and see how well it represents the data.

To determine the constants b and c, we will let $U = X^2/100$, and $c' = 100c$, and solve the two simultaneous equations:

$$\left. \begin{array}{l} \Sigma X^2 b + \Sigma X U c' = \Sigma(XY) \\ \Sigma X U b + \Sigma U^2 c' = \Sigma(UY) \end{array} \right\} \tag{6.10}$$

The form of computation of the necessary sums is shown in Table 6.7, but the detailed extensions are entered for only 5 of the observations.

Table 6.7

COMPUTATIONS FOR FITTING EQUATION $Y = bX + cX^2$ TO
AUTO-STOPPING DATA

X	Y	X^2	U, $X^2/100$	XU	XY	U^2	UY
5	2	25	0.25	1.25	10	0.0625	0.50
10	8	100	1.00	10.00	80	1.0000	8.00
...	...	...	...	...	...	...	...
35	107	1225	12.25	428.75	3745	150.0625	1310.75
22	35	484	4.84	106.48	770	23.4256	169.40
40	134	1600	16.00	640.00	5360	256.0000	2144.00
Sums		28719		7964.51	66358	2411.8383	19536.51

(These computations can be made more easily from the grouped data of Table 4.1.)

Substituting the necessary sums in equations (6.10), we have

$$28,719b + 7,964.51c' = 66,358$$

$$7,964.51b + 2,411.84c' = 19,536.51$$

Solving the equations simultaneously (one convenient procedure for this is shown on page 173), we obtain the values $b = 0.76192$, and $c' = 5.58420$. c is therefore 0.055842, and our equation for the curve is

$$Y = 0.7619X + 0.05584X^2$$

Calculating the values of Y' for selected values of X, we obtain

X	Y	X	Y
5	5.2	25	53.9
10	13.2	30	73.1
15	24.0	35	95.1
20	37.6	40	119.8

In Figure 6.9 this new curve has been plotted in comparison with the original group averages and their confidence intervals for $P = 0.67$, as shown earlier on Figure 4.3. Throughout almost its entire length, the new curve lies within these confidence intervals of the group averages, and also frequently intersects the irregular line of the original group averages. The curve, logically derived, therefore appears to fit the trend of the original observations far better than any straight line could, and to confirm the logical analysis on which the type of the equation was based. It can therefore be used with reasonable safety for extrapolations beyond the speeds observed in the sample. For example, it would indicate that at 60 miles an hour the distance required for stopping would average 247

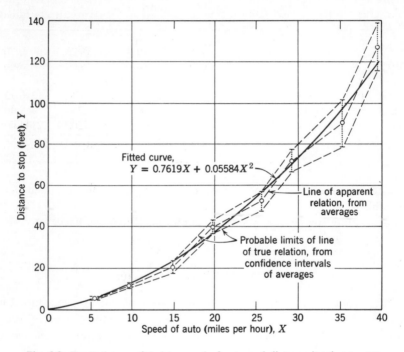

Fig. 6.9. Logical curve fitted to speed of auto and distance it takes to stop, compared to line of group averages and their confidence intervals.

feet, and at 90 miles, 521 feet. Such extrapolations would become highly unreliable, however, if speeds were considered that were so high that additional factors not present at lower speeds, such as wind resistance, became important.

The values we have obtained for our constants b and c and the estimated distances based on them, are of course derived from the relations shown in this single sample of 63 observations. They are still only statistics, and may vary perceptibly from the true parameters in the entire universe. Methods of computing confidence intervals for these values will be presented later (pages 283–287).

Limitations of Equations in Describing Relationships

Up to this point an expression of the relation between the proportion of vitreous kernels and the proportion of protein in each sample has been worked out on the basis of a number of different mathematical formulas. Each different equation has given a different curve. Some, such as the cubic parabola or the logarithmic curve, have given curves coming somewhere near to the relationship shown by the actual observations themselves; others, such as the simple straight line, have entirely failed

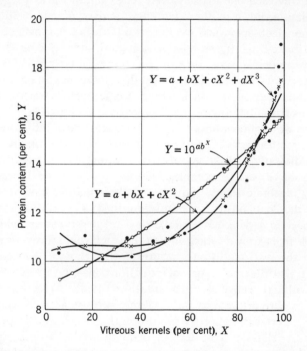

Fig. 6.10. Original observations, and several different types of fitted curves.

to describe the relation. Yet the exact slope or shape of each curve was determined from the same set of observations; the constants of each curve were determined by "fitting" the same data. The diversity in the shape of the different curves is strikingly shown in Figure 6.10, where the several different curves are all drawn on one scale, and the original observations are shown as well. It is quite apparent that the differences in the shapes of the several curves are due solely to the particular form of equation used in computing them. There are certain types of relations which can be accurately represented by each of these equations. When it is "fitted" to data where that type of relation is really present, it can give a curve which accurately represents the central tendency of the data. But when the same equation is fitted to data for which the underlying relationship follows a different function, the resulting curve gives only a distorted representation of the true relation—*it shows the relation only insofar as it is possible to do so within the limits of the particular equation used.*

So far there has been no attempt to show what there is in the "nature" of relations which may make them of the type to be represented accurately by one type of equation or by another. Instead, the purely empirical test of the way each one fits has been relied upon. If, as judged by the eye, the relation shown by the fitted curve *looked like* the relation shown by the original observations, we have said it gave a satisfactory fit; if it has not looked like it, we have said it did not give a satisfactory fit. And in this particular case, none of the computed curves has been really fully satisfactory—we can readily see that there might be some other smooth continuous curve which would come much closer to the actual observations than does any of the curves so far computed.

Of course we might continue the process, using more and more complex equations or other ways of stating the variables,[14] until finally we found one which did satisfactorily describe the relation. Or it might be that the underlying curve was so complex that it could not be represented in elementary algebraic terms. So long as the equation had been derived merely by the "cut-and-try" method described, it would have no logical meaning beyond serving as a simple device for estimating values of the one variable from known values of the other and would throw no particular light upon the real or inherent nature of the relation. Sometimes it is found that two different types of equations may give almost identical estimates within the range of the observations fitted.[15] Which one expresses the "true" nature of the relation? Merely because a given equation *can* reproduce a certain relation is no proof that it really "expresses" the nature of the relation. To establish this, we need a logical explanation which leads to the given equation, which in turn does closely fit the central tendency of the observed data.

If, however, it is not desired to determine what the "real nature" of

[14] A good fit can be obtained in this problem by stating the relation a different way, i.e., by relating the per cent of protein to the logarithm of the *non-vitreous kernels*. The equation used is then

$$Y = a - b \log (1 - X)$$

See Frederick F. Stephan, Alternative statements of percentage data in the fitting of logarithmic curves, *Journal of the American Statistical Association*, Vol. XXVI, pp. 58–61, 1931.

[15] An example of this may be seen in the bulletin by Hugh Killough, What makes the price of oats, *U.S. Department of Agriculture Bulletin* 1351, p. 8, 1925.

the relationship is, but it is merely desired to express it sufficiently well so that values of one variable (such as protein content) can be estimated from known values of another (such as the proportion of vitreous kernels), it does not make any difference what type of equation is used, so long as it represents the observed relationship adequately. As a matter of fact, it is not necessary to have an equation at all. All that is really necessary is a graph of the curve, or a table of values for one variable corresponding to values of another, from which we can construct a graph. Further, by enlarging the chart and making the scale sufficiently detailed, we may read off the estimated values to any degree of accuracy that is desired —much more accurately, as a matter of fact, than our ability to determine the real relation usually justifies, as will be evident later on.

In many cases simply the working expression of the relation may be all that is either needed or desirable. The "true relation" between the variables may be so involved that a very complex mathematical expression would be required to represent it properly. Even simple types of physical relations may require rather complex equations to represent them. In many cases, too, the knowledge of the causes of the relation may be so undeveloped that there is no real basis for expressing the relationship mathematically. The relation between vitreous kernels and percentage of protein would be an example of this type—very complex details of chemical content and physical and biological structure are probably responsible, so complex as to be quite beyond satisfactory reduction to mathematical expression. Yet the original observations undeniably indicate that there is some sort of definite relation. For many practical purposes it may be entirely satisfactory merely to know what the relationship *is*, without bothering at all with what it really means. Even in scientific study that may frequently be satisfactory as a first step, since in many cases it is essential to know what are the facts before trying to work out the reasons *why* they are as they are.

When the expression of the relation is not to be used except as an empirical basis for estimating values of the dependent variable from the independent, a curve can be determined with only a small fraction of the effort required in "fitting" a mathematical equation, yet which fits the data quite as well as any mathematical curve. In such cases the curve may afford quite as satisfactory a description of the relation and a basis for estimating one variable from the other as if elaborate computations had been made. This method is known as *freehand smoothing*.

Expressing a Curvilinear Relation by a Freehand Curve. The process of determining a freehand curve may be simply illustrated. The simplest way to do it would be to plot the original observations on coordinate paper, and then draw a continuous smooth curve through them by eye in such a way as to pass approximately through the center of the observations all along its course. Where the nature of the relation is indicated as closely by the original observations as it is in the wheat problem which we have been discussing, this might yield quite a satisfactory expression of the relation. In other cases, however, the observations might be more widely scattered, and the underlying relation might be more difficult to determine, so that different persons, drawing in the curves freehand, might draw in rather different curves. Some method is therefore needed to give a greater degree of precision to the result, and to insure that the same data would yield substantially the same result even in the hands of different investigators.

This stability of result can be secured by a relatively minor extension of the methods already discussed in the first illustration of a two-variable relationship—the automobile-stopping problem. There it was found that by classifying the observations in appropriate groups, the general nature of the relation could be expressed by an irregular line connecting the several group averages. All that is needed is some method of deriving a continuous smooth curve from that irregular line. The method for smoothing out that irregular line, freehand, is very evident and simple. At the same time, starting with the irregular line of group averages gives a certain stability to the process and insures that different persons would draw in the curve with about the same position and shape.

Applying the process to the wheat problem, the first step is to classify the data into appropriate groups according to the values of the independent variable, the proportion of vitreous kernels, and to determine the average percentage of vitreous kernels and of protein content for the observations falling into each group. The discussion of the automobile problem has shown that, for the differences in averages to be significant, it is necessary for the groups to be large enough so that the averages would not vary erratically from group to group. In some cases a little experimenting might be necessary to determine what this size would be. In the present case, inspection of the dot chart showing the original observations (Figure 6.3) indicates that a class interval of 25 per cent of vitreous kernels will give groups large enough to make the averages of protein content fairly stable from group to group.

The form of computation most convenient for obtaining the group averages, using groups of the size suggested, is shown in Table 6.8.

The averages for the several groups are shown in Figure 6.11, indicated by circles, and original observations are again shown by solid dots. A smooth continuous dashed curve has been drawn through the series of group averages, ignoring the individual observations and following only the general trend shown by the averages. This smooth curve comes quite

Table 6.8

COMPUTATION OF AVERAGES TO USE IN FITTING FREEHAND CURVE, FOR WHEAT-PROTEIN PROBLEM

Vitreous Kernels Below 25 Per Cent		Vitreous Kernels 25–49 Per Cent		Vitreous Kernels 50–74 Per Cent		Vitreous Kernels 75–100 Per Cent	
Vitreous Kernels (per cent)	Protein (per cent)	Vitreous Kernels (per cent)	Protein (per cent)	Vitreous Kernels (per cent)	Protein (per cent)	Vitreous Kernels (per cent)	Protein (per cent)
6	10.3	34	10.9	55	11.1	75	12.2
17	11.0	45	10.8	51	11.4	87	14.5
24	10.1	36	10.2	74	13.8	98	18.1
...	...	...	...	...	...	91	14.0
...	...	...	...	...	...	97	17.0
...	...	...	...	...	...	85	14.4
...	...	...	...	...	...	96	15.8
...	...	...	...	...	...	92	15.6
...	...	...	...	...	...	94	15.0
...	...	...	...	...	...	84	13.3
...	...	...	...	...	...	99	19.0
Totals 47	31.4	115	31.9	180	36.3	998	169.9
No. cases 3	...	3	...	3	...	11	...
Averages 15.67	10.47	38.33	10.63	60.00	12.1	90.73	15.35

near to representing the relation shown by the individual observations through most of its extent; but beyond 95 per cent of vitreous kernels it fails to follow the individual observations—through that portion of the range the protein content rises much faster than is indicated by the average for the whole range from 75 through 100 per cent vitreous kernels.

Because over half of all the observations fall in this upper portion of

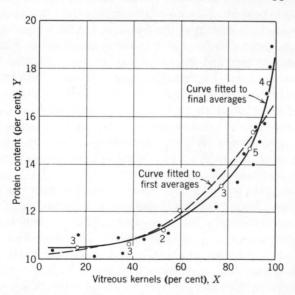

Fig. 6.11. Original observations and averages of protein content, and freehand curve.

the range, it would seem reasonable to classify them into smaller groups so as to give a better basis for determining this portion of the curve. Let us try splitting the observations above 50 into four groups, each with about the same number of observations—say 50 to 69, 70 to 84, 85 to 94, and 95 to 100. The computation of the new averages is shown in Table 6.9.

Table 6.9

COMPUTATION OF SUBAVERAGES FOR LAST GROUPS IN WHEAT PROBLEM, FOR FITTING FREEHAND CURVE

Vitreous Kernels 50–69 Per Cent		Vitreous Kernels 70–84 Per Cent		Vitreous Kernels 85–94 Per Cent		Vitreous Kernels 95–100 Per Cent	
Vitreous Kernels (per cent)	Protein (per cent)	Vitreous Kernels (per cent)	Protein (per cent)	Vitreous Kernels (per cent)	Protein (per cent)	Vitreous Kernels (per cent)	Protein (per cent)
55	11.1	75	12.2	87	14.5	98	18.1
51	11.4	74	13.8	91	14.0	97	17.0
...	...	84	13.3	85	14.4	96	15.8
...	...	...	...	92	15.6	99	19.0
...	...	...	...	94	15.0	...	...
Totals 106	22.5	233	39.3	449	73.5	390	69.9
No. cases 2	...	3	...	5	...	4	...
Averages 53	11.25	77.67	13.1	89.8	14.7	97.5	17.48

These new averages, together with the previous ones for the lower groups, are also plotted in Figure 6.11 and the number of cases that each represents is indicated next to it, to aid in judging what weight to assign to that average. Finally, a smooth continuous curve has been drawn in, to pass as near as possible to the different averages without making illogical twists or turns. As is evident in the figure, it has been possible to draw the line with no point of inflection in it, yet so that it passes quite near to all the group averages and approximately through the middle of the individual observations. Further, the general course of the line is sufficiently well defined by the several group averages so that if it were redrawn, either by the same person or another person, it could have only minor differences from the line actually shown. Making the chart over two or three times, and drawing a separate curve on each trial, then averaging the two or three curves together, is one method of reducing the variation due to individual judgment in drawing the curve.

CAUTIONS IN FREEHAND FITTING. No attempt has been made to have the curve follow all the twists and turns of the irregular line of averages. As was shown previously with the automobile illustration, irregular differences from group to group may be due to chance fluctuations in sampling where the groups are small. Not unless the groups included a very much larger number of cases than these do here would one be justified in bending the curve because of the position of a single group average, and not even then unless there was some logical basis for a curve of that shape. In doubtful cases breaking up a particular group into smaller groups, as was just done in the wheat example, or reclassifying the observations into somewhat different groups, will help to determine whether or not the data positively indicate that an extra inflection is needed. It is also necessary to see if some single observation is responsible for the abnormality; if it is, it is better to disregard it and draw the curve without the extra twist.

In drawing in a freehand curve, it is desirable to place certain logical limitations on the shape of the curve rather than to have it be purely an empirical representation of the data. To do this, it is necessary to decide before the curve is drawn what those limitations should be. The limitations should be based upon a logical analysis of the relation under examination, in the light of all the information available to the investigator. In this case, for example, a consideration of the biological structure of the kernels, of the portions which run high in protein content, and of the appearance and size of those portions might lead one to the following conclusions:

1. An increase in the proportion of vitreous kernels might be associated with no change in the proportion of protein, or with an increase in the proportion, but never with a decrease in the proportion.

2. The relation between vitreous kernels and protein should be a progressive one, consistently changing throughout the range of variation, rather than fluctuating back and forth.

3. The maximum proportion of protein would be found with the largest proportion of vitreous kernels.

These three logical expectations might then be expressed in the following limitations to be placed on the shape of the curve to be drawn:

1. The curve should have no negative slope throughout its length.

2. The curve should have no points of inflection, but should change shape continuously and progressively.

3. The maximum should be reached at the end of the curve.

These three logical limitations are all fulfilled by both the curves shown in Figure 6.11, yet they would exclude other types of curves which might be drawn. For example, they would rule out a curve with a hump or twist in it, or one which sloped down and then up.[16]

In some cases, examination of the data by the method of successive group averages, even after all the tests suggested above, will show the

---

[16] This use of logical analysis in stating the limitations on a freehand curve may be compared with the use of logic in deciding on the type of mathematical equation to employ.

presence of a relation which cannot be expressed within the logical limitations imposed on the shape of the curve. In that case, the reasoning underlying the logical analysis should be reexamined, to see if some step requires restatement and if the limitations themselves should be changed. (For a further discussion of this interaction of induction and deduction, see pages 469–476 of Chapter 26.) For a curve to have real meaning, it must be consistent with a careful logical analysis, no matter whether the curve is obtained mathematically or freehand, or whether the logical limitations are expressed in a mathematical equation or in a set of limitations placed on the shape of the curve drawn by freehand fitting.[17]

INTERPRETING THE FITTED CURVE. The use of the freehand curve in estimating values of the dependent variable, percentage of protein, from known values of the independent variable, proportion of vitreous kernel, may be readily illustrated. Taking the first observation, with 6 per cent of vitreous kernels, and reading off the corresponding proportion of protein from the curve in Figure 6.11, we get 10.4 per cent as the estimated protein content. Similarly for the second observation, 75 per cent vitreous kernels, the curve indicates 12.9 per cent as the proportion of protein. Reading off the estimated protein for each of the 20 observations we get the estimates shown in Table 6.10.

Even though in using the freehand curve we do not have an equation stating the relation between X and Y, we still have a mathematical expression of the relation between them. For we can write

$$Y' = f(X)$$

which simply means that the estimates, or Y' values, are a *function of X*; that is, for every X value there is some corresponding Y' value. Of course, we can find what this corresponding value is only by reading it off the curve; yet that is enough. We have a graphic statement of the functional relation; if we had a definite formula to represent the curve, we would have an *analytical* statement of the relation as well.

Although we have not fitted a definite equation to represent the freehand curve, it is still possible to state the relation shown by the curve other than in graphic form. This can be done by constructing a table showing, for whatever values of the independent variable may be selected, the corresponding estimated values of the dependent variable. Table 6.11 illustrates this method of starting the relation.

---

[17] For a more detailed discussion of the pros and cons of freehand versus mathematical fitting, see W. Malenbaum and J. D. Black, The use of the short-cut graphic method of multiple correlation, *Quarterly Journal of Economics*, Vol. LII, November, 1937, and The use of the short-cut graphic method of multiple correlation: Comment by Louis Bean; Further comment by Mordecai Ezekiel; and Rejoinder and concluding remarks by Malenbaum and Black, *Quarterly Journal of Economics*, February, 1940.

Table 6.10

ACTUAL PER CENT OF PROTEIN AND PROPORTION ESTIMATED ON BASIS OF
FREEHAND CURVE

Proportion of Vitreous Kernels, X	Actual Proportion of Protein, Y	Proportion of Protein Estimated from Vitreous Kernels, $Y' = f(X)$	Difference Between Actual and Estimate, $Y - Y'$
6	10.3	10.4	−0.1
75	12.2	12.9	−0.7
87	14.5	14.5	0
55	11.1	11.4	−0.3
34	10.9	10.7	0.2
98	18.1	17.4	0.7
91	14.0	15.2	−1.2
45	10.8	11.1	−0.3
51	11.4	10.3	1.1
17	11.0	10.5	0.5
36	10.2	10.8	−0.6
97	17.0	17.0	0
74	13.8	12.8	1.0
24	10.1	10.6	−0.5
85	14.4	14.2	0.2
96	15.8	16.7	−0.9
92	15.6	15.5	0.1
94	15.0	15.9	−0.9
84	13.3	14.0	−0.7
99	19.0	18.0	1.0

In the range where the curve is rising most steeply the readings are taken more closely together, to provide for reproducing that portion of the curve more accurately. In addition, no readings are taken beyond the range covered by the original observations.

When to Fit a Mathematical Equation. Mathematical curves will have a distinct advantage over freehand methods when there is some good logical basis for expecting a certain type of relation to hold. When there is a logical basis for using a given formula, the constants of the equation serve as an explanation of the real nature of the relationship. In all other cases the mathematical curve is no more reliable than the freehand curve; the latter may therefore be employed to describe the nature of the relation, and can be determined with much less expenditure of effort. That does not mean that a mathematical curve, based on

Table 6.11

PER CENT OF PROTEIN CORRESPONDING TO VARIOUS PROPORTIONS OF VITREOUS
KERNELS IN SAMPLES OF WHEAT, AS INDICATED BY 20 OBSERVATIONS

Proportion of Vitreous Kernels	Corresponding Proportion of Protein	Proportion of Vitreous Kernels	Corresponding Proportion of Protein
Per cent	*Per cent*	*Per cent*	*Per cent*
10	10.4	70	12.4
20	10.5	80	13.5
30	10.7	90	15.0
40	10.9	95	16.2
50	11.2	99	18.0
60	11.7		

adequate logical analysis, is of no additional value. If it can be shown that such a curve does fit the data, that may verify an hypothesis and so provide a "law" to state the nature of the relationship, which may be of far more value than the mere empirical statement of what the relationship is observed to be.

Where the logical expectations do not lead to a relation which can be formally expressed in a simple equation, they may, as has already been shown, still be sufficient to state a set of limiting conditions to be used in fitting a freehand curve. However, if a mathematical curve is found empirically which fits the data about as well as a freehand curve could, confidence intervals can be calculated by straightforward methods (note Chapter 17). Where good computing equipment is available, the substitution of machine time and clerical labor for the time of the researcher himself may tip the balance in favor of mathematical curves.

A Mathematical Equation Used in an Economic Problem. Economists sometimes find it convenient to classify commodities according to their *elasticities of demand with respect to income*—that is, the ratio between the rate of change in the expenditure made by families for a commodity group as their incomes increase and the corresponding rate of change in their incomes. In general, expenditures for necessities increase less than proportionately to income, and expenditures for luxury goods and services increase more than proportionately to income. Although the elasticity of expenditure with respect to income for a particular commodity group may change from one income level to another, it is often desired to obtain an average elasticity over some specified range of incomes. This is equivalent to assuming that the elasticity is constant over the range in question.

This economic relationship can be expressed in a mathematical equation or model just as readily as the several physical hypotheses which have been discussed, for it makes certain definite assumptions as to the way the two variables (total income and expenditures on a given commodity) are related.

If X is used for income per capita, and E for expenditure for food per capita, the desired elasticity for food as a whole is given by the coefficient b in the equation

$$\log E = a + b \log X$$

The application of this equation may be illustrated by concrete data. Table 6.12 shows data from family consumption studies in 15 selected areas of 13 countries. These have widely varying income levels per person. For each country or area, the average total consumption expenditures per person, and the average expenditures on food, are shown. (Total consumption expenditure is used here rather than total income, as in a number of countries the surveys did not collect data on total income.) All values are adjusted to dollars of 1948 purchasing power, and are shown to 3 significant digits. Each study represents a large number of cases, from 180 consumption units for the smallest (Finland) to 22,705 for the largest sample (Japan).

We now "fit" the equation $\log E = a + b \log X$ to the data by the methods previously discussed for fitting the straight line. The computation of the extensions needed are also shown in Table 6.12. From the totals obtained there, we then compute the statistics for a and b as follows:

$$M_E = \frac{\Sigma \bar{E}}{n} = \frac{31.8060}{15} = 2.1204$$

$$M_X = \frac{\Sigma \bar{X}}{n} = \frac{37.3915}{15} = 2.4928$$

$$\Sigma(\bar{x}^2) = \Sigma(\bar{X}^2) - n(M_{\bar{X}}^2) = 95.80981 - 15(2.4928^2) = 2.59906$$

$$\Sigma(\bar{e}\bar{x}) = \Sigma(\bar{E}\bar{X}) - nM_E M_{\bar{X}} = 81.13743 - 15(2.1204)(2.4928) = 1.85158$$

$$b_{ex} = \frac{\Sigma(ex)}{\Sigma(x^2)} = \frac{1.85158}{2.59906} = 0.7124$$

$$a_{\bar{e}\bar{x}} = M_E - bM_{\bar{X}} = 2.1204 - (0.7124)(2.4928) = 0.3445$$

and then

$$\bar{E} = a + b\bar{X} = 0.3445 + 0.7124\bar{X}$$

and also

$$\log E = 0.3445 + 0.7124 \log X.$$

The size of the constant b, 0.71, indicates that on the average a 1 per cent increase in total consumption expenditure is accompanied by approximately 0.7 per cent increase in expenditures for food—that is, in economic

Table 6.12

EXPENDITURES PER CAPITA ON CONSUMPTION AND ON FOOD,
AND COMPUTATION OF LOGARITHMIC CURVE
$(\log E = a + b \log X)$

| Country or Area | Average Expenditure | | Logarithms of Data | | Extensions | |
	Total, X	Food, E	Total, $\bar{X}$	Food, $\bar{E}$	$\bar{X}^2$	$\bar{X}\bar{E}$
	Dollars	*Dollars*				
India	52.9	33.7	1.7235	1.5276	2.97045	2.63282
Ceylon	76.2	48.1	1.8820	1.6822	3.54192	3.16590
Gold Coast	102.	58.4	2.0086	1.7664	4.03447	3.54799
Japan	143.	69.6	2.1553	1.8426	4.64532	3.97136
Portugal (Porto)	144.	90.3	2.1584	1.9557	4.65869	4.22118
Portugal (Lisbon)	291.	153.	2.4639	2.1847	6.07080	5.38288
Austria	309.	151.	2.4900	2.1790	6.20010	5.42571
Ireland	350.	135.	2.5441	2.1303	6.47244	5.41970
Finland	407.	173.	2.6096	2.2381	6.81001	5.84055
Panama	439.	152.	2.6425	2.1818	6.98281	5.76541
Switzerland	539.	172.	2.7316	2.2355	7.46164	6.10649
Sweden	622.	215.	2.7938	2.3324	7.80532	6.51626
Canada (1948)	919.	247.	2.9633	2.3927	8.78115	7.09029
United States	1295.	409.	3.1123	2.6117	9.68641	8.12839
Canada (large cities, 1953)	1296.	351.	3.1126	2.5453	9.68828	7.92250
Sums			37.3915	31.8060	95.80981	81.13743

L. Goreux, Long-range projections of food consumption, *FAO Monthly Bulletin of Agriculture, Economics, and Statistics*, Vol. 6, No. 6, pp. 1–18, June, 1957.

terms, the "income-elasticity" of food expenditures (with reference to total consumption expenditure) is 0.7.

Studies of the relation of food expenditure to family income within individual countries sometimes suggest that the income elasticity declines as income rises. To test this hypothesis with respect to the data in Table 6.12 we would have to fit a parabola of the form $\log E = a + b \log X + c(\log X^2)$. We can form an opinion about this by plotting both the original observations and the logarithmic straight line just fitted. Figure 6.12 shows this comparison (on the left half of the figure) in terms of the

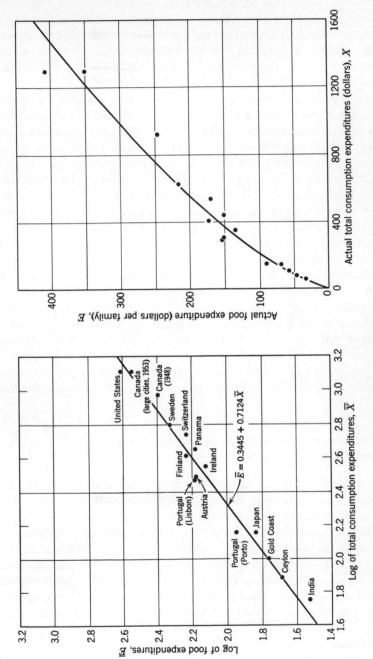

Fig. 6.12. The relation of food expenditure to total consumption expenditure, fitted by a logarithmic straight line, plotted both in logarithms and in natural numbers.

logarithmic values used in the computation and with the logarithmic values of the function which is, of course, a straight line. The straight line seems to fit the original values quite closely and it seems doubtful that a logarithmic parabola would improve the fit. There is a slight indication that income elasticity of food expenditures declines with increasing income.

A comparison may also be made in terms of the original arithmetic values, using the estimated values of the fitted line transformed from logarithms back to natural numbers. The right-hand section of Figure 6.12 shows this comparison. The fitted line is now seen to be a slight curve, with a reasonably close "fit" to the actual observations.

An alternàtive to plotting the logarithms is to plot the variables in natural numbers on double-log paper. Use of the graphic method on double-log paper, or of the values plotted in logarithms, facilitates study of changing elasticity along the demand curve.

The wheat-protein example, on the other hand, illustrated a case where there was no logical basis for the use of any particular equation and where a freehand curve was therefore as satisfactory as any other type and gave a better fit than any of the analytical types which were tried. Many of the problems in the natural and social sciences are of this type, where the relation can be measured even though the specific causes for it cannot be stated mathematically. Only where the relations can be explained on some logical basis which lends itself to mathematical statement is there justification for a large amount of work to "fit" a specific formula; or where it is desired to determine the confidence intervals for the fitted curve.[18]

Limitations in Estimating One Variable from Known Values of Another

The methods shown so far provide a definite technique by which an investigator can determine the way in which the values of one variable differ as the values of another related variable differ. These same operations afford a basis for estimating values of the dependent variable from given values of the independent variable, for cases in addition to those from which the functional relation was determined. Whether such estimated values, for cases not included in the original study, can be expected to agree with the true values if they could be determined, depends upon two groups of considerations: (1) the descriptive value of the curve; and (2) its representative reliability when it comes to applying it to new observations.

[18] For methods of calculating such confidence intervals, see Chapter 17.

These two groups of considerations apply (1) to exactly what a given curve means, with regard solely to the particular cases from which it was determined; and (2) the dependability of the curve with regard both to the ability of those observations to represent the universe (whole group of facts) from which they were drawn and the ability of the curve to represent the true relations existing in that universe. This second group involves an extension of the points which were raised in the first chapter as to the reliability of an average; discussion of these questions will be deferred until Chapters 17 and 19.

Just as an average computed from a sample may differ more or less widely from the true average of the universe from which that sample was drawn, so a regression line or curve determined from a sample may differ more or less widely from the true regression in the universe. The following chapter discusses this problem, and Chapter 17 presents methods of estimating how far the regression line or curve from an individual sample may miss the true regression of the universe, and how the confidence interval for its position may be determined.[19]

The reliability of a curve depends upon the number of observations from which its position was determined and how closely the curve as determined "fits" those observations. Since the number of observations usually differs along the different portions of a curve, it may be much more reliable in its central portions, where the bulk of observations occurs, than in the extreme portions where the number of observations may be much less. This may be especially marked in the case of complex curves fitted by mathematical means, where single extreme observations may have a material effect upon the shape of the end portions. In any event, only those portions of the curve where there are enough observations to make its shape and position definite should be regarded as statistically determined; the end portions, when dependent upon a few observations, should either not be used at all or else stated as very rough indications of the true curve.

It is particularly to be noted that determination of the line or curve of relationship gives no basis for estimating beyond the limits of the values of the independent variable actually observed. No matter whether a formula has been fitted or not, any attempt to make estimates beyond the range of the original data by "extrapolation," i.e., by extending the curve beyond the range of the observed data, gives a result that is not based on the statistical evidence. In case a formula has been used which has a good logical basis, extrapolation may give a result which it is logical

[19] For the straight line, e.g., (5.1), both the *level* (*a*) and *slope* (*b*) may vary between samples drawn from the same universe. The word "position" is used here to designate the combined effects of variations in *a* and *b*.

to expect—but its reasonableness rests on the validity of the logic rather than on a statistical basis. The statistical analysis indicates only what the relations are within the range of the observations which are used in the analysis, and only within the confidence interval for the relation determined.

The "closeness" with which the line or curve fits the original data is another criterion of the reliance which can be placed in it. If the data all fall quite close to the line, that fact inspires more confidence in it than if they differ widely and erratically from it. But there are special statistical measures of just what this "closeness" is, and they will be given separate considerations in the next chapter.

Summary

In some functional relations, the change in the dependent variable with changes in the independent variable cannot be represented by a straight line. Such a relation may be represented by a curve showing the value of the dependent variable for each particular value of the independent variable. Curves may be fitted to given sets of observations either by use of mathematical functions, such as parabolas, logarithmic curves, and hyperbolas, or by various processes of freehand smoothing. When there is a good logical basis for the selection of a particular equation, the equation and the corresponding curve can provide a definite logical measurement of the nature of the relationship. When no such logical basis can be developed, a curve fitted by a definite equation yields only an empirical statement of the relationship and may fail to show the true relation. In such cases a curve fitted freehand by graphic methods, and conforming to logical limitations on its shape, may be even more valuable as a description of the facts of the relationship than a definite equation and corresponding curve selected empirically, but fitting less well.

In any event, estimates of the probable value of the dependent variable cannot be made with any degree of accuracy for values of the independent variable beyond the limits of the cases observed; and can be made most accurately only within the range where a considerable number of observations is available. It may be possible to extrapolate the curve if its equation is based on a logical analysis of the relation as well as on the cases observed; but in that case the logical analysis, and not the statistical examination, must bear the responsibility for the validity of the procedure.

Note 6.1. When an equation is used with the dependent variable stated as a logarithm, as types (*b*) and (*c*) on page 70, the further assumption is involved that the errors to be minimized vary proportionately with the size of the dependent variable.

The standard error of estimate also must be stated as a percentage of the value estimated, rather than as a natural number. For an example of a problem where the range of error increases with the size of the dependent variable, and where a logarithmic equation would therefore be justified, see Figure 8.2, page 142.

REFERENCES

Mills, Frederick C., The measurement of correlation and the problem of estimation, *Quart. Pub., Amer. Stat. Assoc.*, Vol. XIX pp. 273–300, September, 1924.

Ezekiel, Mordecai, A method of handling curvilinear correlation for any number of variables, *Quart. Pub. Amer. Stat. Assoc.*, Vol. XIX pp. 431–453. December, 1924.

CHAPTER 7

Measuring accuracy of estimate and degree of correlation

The methods developed up to this point may be used to estimate the values of one variable when the values of another are known or given. They also furnish an explicit statement of the average difference or change in the values of the estimated or dependent variable for each particular difference or change in the value of the known or independent variable. But that is not enough. In addition it is frequently desirable to answer three queries: (1) How closely can values of the dependent variable be estimated from the values of the independent variable? (2) How *important* is the relation of the dependent variable to the independent variable? (3) How far are the regression curve and these relations, as shown by the particular sample, likely to depart from the true values for the universe from which the sample was drawn? Special statistical devices, termed (1) the *standard error of estimate* and (2) the *coefficient* and *index of correlation*, have been developed to meet the need indicated by the first two questions. Error formulas and knowledge of the distributions of these coefficients, and standard errors for the regression line or curve, provide approximate answers for the third, under the assumption that certain conditions of sampling are met.

The Closeness of Estimate—Standard Error of Estimate

Attention has previously been called to the fact that when some dependent variable, such as the distance required for an automobile to stop after the brake is applied or the protein content in wheat samples, is estimated from another variable, such as the speed at which the car is moving or the proportion of vitreous kernels in the sample, the estimated values in many cases will not be the same as the values of the dependent variable that were originally observed. These differences are obviously due to *residual* causes; that is, to variations in the dependent variable which were unrelated to changes in the particular independent variable used in the analysis. For that reason the differences between the estimated values and the actual values are termed residual differences or, more simply, *residuals*.

For Linear Relations. The meaning of the residuals and their use in determining the standard error of estimate and the coefficient and index of correlation can best be understood if illustrated by a concrete case. Such an illustration is given in Table 7.1. Each of the 22 observations relates to a *subregion* of several counties drawn at random from a larger number of subregions in the North Central States. One variable (X) measures the degree of industrialization in terms of the per cent of all employed persons in the subregion as of 1940 who were engaged in manufacturing. The other variable (Y) measures the effect of migration during 1940–1950 upon the population of the subregion. During this decade employment in manufacturing expanded rapidly and many people migrated from rural areas to industrial centers.

In Table 7.1 the observations have been fitted by a straight line to estimate net migration on the basis of manufacturing employment. The estimated net migration figures, Y', and the residuals, z, or differences between the estimate and the actual, are also shown.

The residuals vary from $+13.33$ to -14.79. If we wish to say how large they are on the average, we can ignore the plus and minus signs and compute the average deviation. For the 22 residuals in Table 7.1, the average deviation is 4.08, and the standard deviation is 5.60. If these residuals are grouped in a frequency distribution, they fall as shown in Table 7.2.

The standard deviation of z is different from the standard deviations previously computed. Instead of showing the standard deviation of net migration from its mean (that is, s_y), it shows the standard deviation around a changing quantity, depending on the per cent of employment in manufacturing. The s_z is thus the standard deviation around the fitted line of relation for the sample we have analyzed, and may be indicated graphically on a dot chart as a certain distance above and below the fitted line (note Figure 8.1, in the next chapter).

The standard deviation of the residuals is 5.60, so we should expect two-thirds of the residuals to come between $+5.60$ and -5.60. Of the 22 cases, 16 came within this range of the line, or 73 per cent of all the cases. Similarly, only 5 per cent of the cases would be expected to fall outside the range $\pm 2s_z$, or below -11.20 or above $+11.20$. Actually two observations, or 9 per cent of the cases, fall outside this range. These

Table 7.1

EMPLOYMENT IN MANUFACTURING, 1940, NET MIGRATION, 1940–1950, AND
NET MIGRATION ESTIMATED FROM EMPLOYMENT IN MANUFACTURING,
IN SELECTED SUBREGIONS

Employment in Manufacturing, X	Net Migration, Y	Estimated Net Migration* Y'	Excess of Actual over Estimate z
Per cent†	*Per cent‡*	*Per cent §*	*Per cent §*
28.1	−6.9	1.41	−8.31
5.1	−29.2	−14.41	−14.79
5.9	−7.5	−13.86	6.36
29.4	7.9	2.30	5.60
41.9	8.3	10.90	−2.60
18.7	−5.8	−5.06	−0.74
9.3	−10.5	−11.52	1.02
18.7	1.0	−5.06	6.06
24.1	1.9	−1.35	3.25
17.5	−9.3	−5.88	−3.42
18.0	−3.9	−5.54	1.64
9.0	−12.3	−11.73	−0.57
8.7	−16.1	−11.94	−4.16
14.4	−13.0	−8.02	−4.98
4.7	−13.3	−14.69	1.39
7.6	−7.9	−12.69	4.79
4.1	−15.6	−15.10	−0.50
1.7	−19.9	−16.75	−3.15
2.7	−14.9	−16.06	1.16
10.3	2.5	−10.83	13.33
3.3	−17.3	−15.65	−1.65
4.0	−14.9	−15.17	0.27

Source of data: Paul J. Jehlik and Ray E. Wakeley, Population change and net migration in the North Central States, 1940–50, *Iowa Agricultural Experiment Station, Research Bulletin* 430, July, 1955.

* Computed by regression formula $Y = -17.918 + 0.6877X$.

† Per cent of all employed persons in the subregion as of 1940 who were engaged in manufacturing.

‡ Per cent change in population of the subregion through migration, 1940–1950.

§ Same units of measure as for Y.

Table 7.2

FREQUENCY DISTRIBUTION OF RESIDUALS IN ESTIMATING NET MIGRATION

Residual	Number of Times Occurring	Residual	Number of Times Occurring
−14.99 to −10.00	1	0 to 4.99	7
−9.99 to −5.00	1	5.00 to 9.99	3
−4.99 to 0	9	10.00 to 14.99	1

are sufficiently close to the expected proportions for a normal distribution with this limited number of observations.

The symbol S is used to denote the standard error of estimate. $S_{y \cdot x}$ indicates the standard error for estimates of Y made from a linear relation to X by the equation $Y = a + bX$. Similarly, $S_{y \cdot f(x)}$ would indicate the standard error for estimates of Y made on the basis of a freehand curve relation to X, as indicated by the equation $Y = f(X)$.

The standard error of estimate is therefore defined by the two equations:

$$S_{y \cdot x}^2 = s_z^2 = \frac{\Sigma z^2}{n} \qquad (7.1)$$

$$S_{y \cdot f(x)}^2 = s_{z''}^2 = \frac{\Sigma (z'')^2}{n} \qquad (7.2)$$

The standard error of estimate in estimating net migration from employment in manufacturing, by the linear equation, is therefore 5.60 per cent.

For Curvilinear Relations. Where a curvilinear regression is represented by a fitted algebraic equation, as with the cubic parabola fitted to the wheat-protein example in the preceding chapter, the standard error of estimate can be calculated from the residuals between the actual values of Y and the estimated values, Y'', based on the fitted equation. (Methods of calculating $s_{z''}$ without actually computing each individual value of Y'' and z'' are given later, in Chapters 12 and 13.)

The calculation of the standard error of estimate for a freehand regression curve may be illustrated by the migration data. From a freehand curve, fitted by methods already described, estimates of Y from the relation $Y = f(X)$ were obtained, as shown in Table 7.3.

The standard deviation of the new residuals is 5.49. This is then the standard error of estimate for estimates based on the curve.

Table 7.3

EMPLOYMENT IN MANUFACTURING, NET MIGRATION, AND NET MIGRATION
ESTIMATED FROM EMPLOYMENT IN MANUFACTURING, BY FREEHAND CURVE

Employment in Manufacturing, X	Net Migration, Y	Estimated Net Migration, Y''	Excess of Actual over Estimate, z''
Per cent	*Per cent*	*Per cent*	*Per cent*
28.1	−6.9	1.9	−8.8
5.1	−29.2	−14.8	−14.4
5.9	−7.5	−14.0	6.5
29.4	7.9	2.4	5.5
41.9	8.3	7.2	1.1
18.7	−5.8	−3.2	−2.6
9.3	−10.5	−10.5	0
18.7	1.0	−3.2	4.2
24.1	1.9	0	1.9
17.5	−9.3	−4.2	−5.1
18.0	−3.9	−3.9	0
9.0	−12.3	−11.0	−1.3
8.7	−16.1	−11.3	−4.8
14.4	−13.0	−6.4	−6.6
4.7	−13.3	−15.3	2.0
7.6	−7.9	−12.2	4.3
4.1	−15.6	−15.7	0.1
1.7	−19.9	−19.3	−0.6
2.7	−14.9	−17.6	2.7
10.3	2.5	−9.8	12.3
3.3	−17.3	−17.0	−0.3
4.0	−14.9	−16.0	1.1

The standard error of estimate of 5.49 from the curve, compared with that of 5.60 from the straight line, indicates that in both cases the net migration from a subregion can be estimated, for the cases included in the sample, from the per cent of employment in manufacturing with a standard deviation of about $5\frac{1}{2}$ percentage points. It appears at this stage that the estimates made on the basis of the curvilinear relation are only a little more reliable than those based on the linear relation.

Where the same set of conditions prevails as those under which the original data were selected and only the independent variable is known, it may be desired to estimate the probable value of the dependent variable from the known value of the independent. Thus if the 1940 percentages of employment in other subregions are known, it may be desired to estimate the probable extent of net migration during 1940–1950. To actually measure net migration for these other subregions would require detailed and laborious calculations. Or in a case where yield of cotton with various applications of irrigation water has been determined (note the example in the next chapter) it may be desired to estimate the most probable yield on other fields, solely from the amount of water applied. In case the estimates were to be made for new observations taken from the same universe—for example, on the same soil type, in the same area, and for the same year—as were the previous samples, a knowledge of the standard deviation of the residuals for original samples gives a basis for judging how closely the new estimates are likely to approximate the true, but unknown, yields for the new observations. Similarly, in the net migration case it is evident that the errors of estimate will not often be greater than 11.20 per cent and usually will be less than 5.60 per cent.

Because the standard deviation of the residuals may thus serve as a basis for indicating the closeness with which new estimated values may be expected to approximate the true but unknown values, it has been named the *standard error of estimate*.[1]

The standard error of estimate can be used to indicate the probable reliability of a series of estimates of the values of the dependent variable for new observations when only the values of the independent variable are known, but only where it is definitely known that the new cases are drawn at random from exactly the same universe—the same set of conditions—as were the observations from which the relation was determined. In case they do not represent exactly the same conditions—as if, for example, they represent a different period of time[2]—then the standard error of estimate has meaning only with respect to the scatter of the residuals around the regression line *for the cases used in determining the relationship*. It measures (when adjusted) what the differences probably would have been in the universe from which the observations came but does not give more than a clue or a possible indication as to what the differences may be when the same relations are applied to data obtained under new or different conditions.

Adjustment of Standard Error of Estimate for the Number of Observations. The standard deviations of a series of samples drawn from any stable universe will vary from one to another, owing to statistical

---

[1] Chapter 19 gives more refined measures of the accuracy with which estimates may be made for individual new observations.

[2] See Chapters 2 and 17 for other conditions assumed before error formulas apply exactly.

fluctuations. The same is true for the standard error of estimate computed for a fitted line. The standard deviations and standard errors of estimate not only vary but on the average also are slightly smaller than would be obtained from an extremely large sample from the same universe. Because of this tendency of the standard error of estimate from the sample to understate the standard error in the universe, an adjustment is necessary before it can be used other than for the sample. An unbiased estimate of the value of the square of the standard error of estimate for the entire universe may be calculated from the standard error of estimate for the sample by the use of the following equations:

$$\bar{S}_{y\cdot x}^2 = \frac{ns_z^2}{n-2} = \frac{nS_{y\cdot x}^2}{n-2} \tag{7.3}$$

hence

$$\bar{S}_{y\cdot x}^2 = \frac{\Sigma(z^2)}{n-2} = s_z^2\left(\frac{n}{n-2}\right) \tag{7.4}$$

And for curvilinear functions

$$\bar{S}_{y\cdot f(x)}^2 = \frac{ns_{z''}^2}{n-m} = \frac{nS_{y\cdot f(x)}^2}{n-m} \tag{7.5}$$

hence

$$\bar{S}_{y\cdot f(x)}^2 = \frac{\Sigma(z''^2)}{n-m} = s_{z''}^2\left(\frac{n}{n-m}\right) \tag{7.6}$$

In these equations, $\bar{S}_{y\cdot x}^2$ is used to indicate the estimated squared standard error of estimate for the universe, just as $\bar{s}$ was used (in Chapter 2) to indicate the estimated standard deviation in the universe from which the sample was drawn.

In equations (7.3) to (7.6), n stands for the number of observations. In all four equations, m stands for the number of constants in the regression equation, such as a, b, and c. In the case of a parabola of the second order (type a, Chapter 6), m would be 3; for a cubic parabola (type f), it would be 4. Where a freehand curve has been used, it is necessary to estimate how many constants would be needed to represent the curve mathematically. (See pages 70–72 for examples of the constants needed to represent various shapes of curves.)

The standard error of estimate in estimating net migration by the linear equation, after the standard deviation of the residuals is adjusted by equation (7.3), works out to be:

$$\bar{S}_{y\cdot x}^2 = \frac{ns_z^2}{n-2}$$

$$= \frac{22(5.60)^2}{22-2} = 34.50$$

$$\bar{S}_{y\cdot x} = 5.87$$

The new value indicates that the errors in estimating net migration from employment in manufacturing, when the estimate is made for new observations drawn at random from the same universe, will run slightly larger than was indicated by the residuals for the cases included in the study, as tabulated in Table 7.1.

When the standard deviation for the curvilinear function is calculated by equation (7.5), a different result appears. If it is assumed that the regression curve used could have been represented mathematically by an equation with three constants (such as a parabola) then the correction works out to be:

$$\bar{S}^2_{y \cdot f(x)} = \frac{n s^2_{z''}}{n - m}$$

$$= \frac{22(5.49)^2}{22 - 3} = 34.90$$

$$\bar{S}_{y \cdot f(x)} = 5.91$$

The adjusted standard error of estimate for the curvilinear relation, 5.91, is slightly larger than that for the linear equation, 5.87. This indicates that when estimates are made for new observations from the same universe, the straight line is likely to give fully as reliable results as is the regression curve. Not unless the adjusted standard error for the curve is materially smaller than for the straight line can the curvilinear regression be expected to improve the accuracy of estimate.[3]

Units of Statement for Standard Error of Estimate. The standard error of estimate is necessarily stated in exactly the same kind of units as the original dependent variable. Where the dependent variable is stated in feet, as in the automobile problem, the standard error of estimate will be in feet; where it is in percentage points, as in the wheat problem, the standard error will be in percentage points; and where it is in logarithms, as in Table 6.11, the standard error will be in logarithms. Thus in a case like that shown there, the standard error might be the logarithm 0.038. That would mean that the logarithm of the estimates is likely to agree with the logarithm of the true values to within ±0.038, two-thirds of

---

[3] The values of $\bar{S}_{y \cdot x}$ are subject to errors of sampling, just as the values of s_y are subject to errors of sampling. Accordingly, the values of $\bar{S}_{y \cdot x}$ must be regarded only as estimates of the true values, σ_z, which prevail in the universe from which the sample is drawn. Also, it must be remembered that the adjustment, m, for the number of degrees of freedom removed, is only an approximate adjustment in the case of a freehand curve, and that this introduces a further limitation to the accuracy of $\bar{S}_{y \cdot f(x)}$. Even for relations estimated from an algebraic equation, the accuracy of estimates for new observations will vary somewhat from one observation to another, depending on how unusual is the value of the independent variable. See Chapter 19 for fuller discussion.

the time. With an estimated logarithm of 1.00, the logarithm of the true value would then be between 0.962 and 1.038, two-thirds of the time. In terms of antilogarithms, this gives values of 9.16 and 10.91, or between 9.1 per cent above and 8.4 per cent below the value 10.

The *standard error of estimate* is thus computed from the standard deviation of the residuals for the cases on which the relation is based. It indicates the closeness with which values of the dependent variable may be estimated from values of the independent variable. Its exact interpretation differs with the particular units in which the values of the dependent variable are expressed.

The Relative Importance of the Relationship—Correlation

In certain problems it might be found that every bit of variation in one variable could be explained, or accounted for, by associated differences in the value of an accompanying variable. Thus all the variation in the volume of a cube can be explained by the corresponding difference in the length of one side. No other variable is needed to account for the volume of the cube. If we know what the length of the side is, we can compute accurately what the volume will be. All the variation in volume can therefore be said to be explained, or accounted for, by the known relation to the length of the side.

In most problems with which the statistician has to deal, however, all the variation cannot be explained by the relation to another variable, and residual variation is left over. As has just been pointed out, this residual variation can be measured and used as an indication of the errors of estimate.

It is obvious that if no relation has been found, the independent variable considered does not explain any of the observed variation in the dependent variable, and so none of the variation can be explained as due to, or associated with, the independent variable. If, as in the case of the cube, the estimates all agree exactly with the actual values, there are no residual elements, and the variation is perfectly explained. But between these two extremes lie the cases of partial explanation, where a portion of the variation can be explained by the independent variable considered, and a portion cannot. In the automobile case, part of the variation in stopping distance, but not all, was associated with the speed; in the wheat case, part of the variation in protein content, but not all, could be estimated from variations in the proportion of vitreous kernels; and in the migration case, part of the variation in net migration, but not all, could be accounted for by variations in manufacturing employment. In many problems it is of interest to determine what proportion of the variation in the dependent variable can be explained by the particular independent variable considered, according to the relation observed.

Measurement of the relative importance of the relation between two variables calls for a different type of statistical constant than the standard error of estimate. The standard error of estimate simply indicates the size of the residuals without regard to the amount of variation in the dependent variable as first observed. If the standard error of estimate for a cotton-yield problem, for example, were 50 pounds, that would be the standard error no matter whether the yield of cotton in the original cases varied only between 200 and 400 pounds or between 50 and 1,200. If the yields varied only between 200 and 400 pounds, and the standard error was 50, practically all the variation in the original yields would still be left in the residuals; whereas if the yields varied between 200 and 1,200 and the standard error was 50, only a very small portion of the original variation would be left in the residuals. Yet the standard error of estimate would be of the same size in both cases.

What is needed to indicate the relative importance of the relationship is some measure that shows what *proportion* of the original variation has been accounted for. The regression line separates each original value of the dependent variable into two parts, an estimated value (Y') and a residual (z). If the original variation in Y is measured by its standard deviation squared (its "variance" as defined on page 10), and the unexplained variation by s_z^2, then the difference $s_y^2 - s_z^2$ is a logical measure of the amount of variation accounted for by the regression line.

As it turns out, this difference is exactly equal to the variance of the Y' values estimated from the regression line. Thus, in the migration example, $s_y^2 = 81.23$, $s_z^2 = 31.36$, and $s_{y'}^2$, calculated from the Y' values in Table 7.1, is 49.91. If we determine how large $s_{y'}^2$ is compared to the original variance, we get $s_{y'}^2/s_y^2 = 49.91/81.23$, or 0.614. This is the proportion of variation in Y accounted for by X according to the mathematically fitted straight line relationship. The square root of this proportion, $s_{y'}/s_y$, is termed the *coefficient of correlation*.

Linear Relations—Coefficients of Correlation and Determination. The symbol r is used to represent the coefficient of correlation where the relationship between the two variables is found or assumed to be a straight line. When values of Y are estimated from values of X according to the straight-line equation, the coefficient of correlation is indicated by the notation r_{yx}, which is read "the coefficient of correlation between Y and X."

The coefficient of correlation may therefore be defined

$$r_{yx} = \frac{s_{y'}}{s_y} \qquad (7.7)$$

In this particular case, $r_{yx} = 7.06/9.01 = 0.784$.

This formula gives values of r identical with those given by the more usual formula, equation (8.3), presented in the next chapter, as can be proved by simple algebra.[4]

Curvilinear Relations—Index of Correlation. In case the relation has been determined as a curvilinear function instead of a straight line, the ratio $s_{y''}/s_y$ is termed the *index of correlation*, and is represented by the symbol i_{yx}.

The index of correlation may therefore be approximately defined as

$$i_{yx} = \frac{s_{y''}}{s_y} \tag{7.8}$$

Computing the index of correlation for the migration case, $s_{y''}/s_y = 7.15/9.01 = 0.794$. From this figure, it would appear that the correlation is slightly higher for the curve than for the straight line.

Characteristics of the Measures of Correlation. It should be noted that in the case of straight-line relations, if the line has a positive slope, so that as X increases the values of Y' (the estimated values of Y) increase, the correlation is said to be *positive*, and a plus sign is affixed to the correlation coefficient. Similarly, if the line has a negative slope, so that as the values of X (the independent variable) are larger, the values of Y' (the estimated values for the dependent variable) become smaller, the correlation is said to be *negative*, and a minus sign is affixed to the correlation coefficient. The coefficient of correlation thus takes the same sign as the constant b of the corresponding linear equation. In the case of the correlation index, the slope may be positive in one portion and negative in another, so no sign is used, and reference to the curve is necessary to indicate the nature of the relationship.

In a case where the observed relation explains *all* the variation in the dependent variable, the estimated values will be identical with the actual values. The standard deviation of Y' will therefore be exactly as large as the standard deviation of Y, and the ratio $s_{y'}/s_y$ will equal 1.0. This is termed *perfect correlation*, and is indicated when $i = 1.0$, or when $r = +1.0$ or -1.0.

At the other extreme of no relation, no variation can be accounted for by the particular independent variable considered, and the estimated values Y' are therefore all the same, being merely the average of Y. In that case the standard deviation of the estimated values is zero, and the ratio $s_{y'}/s_y = 0/s_y = 0$. The case of complete absence of correlation, therefore, is indicated by values of zero for either r or i.

---

[4] The correlation observed in a sample is designated r, and ρ (Greek rho) is used to represent the true correlation in the universe from which the sample was drawn.

The possible values of the coefficient of correlation therefore range from 0 to $+1.0$ or to -1.0; whereas the values for the index of correlation range from 0 to 1.0. Since most problems with which the investigator has to deal involve cases that are intermediate, where there is some but not perfect correlation, it is these intermediate cases which are of most importance. The precise significance of different values of r and i will be considered next.

The correlation coefficient was originally defined in terms of the special situation in which Y and X each followed a normal distribution (see Chapter 1, Figure 1.1), and the universe of all possible paired values of Y and X formed what is called a "bivariate normal distribution." If a completely random sample were drawn from such a universe, the coefficient r calculated for that sample could be regarded as an estimate of the true correlation, ρ, existing in the universe. Similarly, the square of the sample coefficient of correlation could be regarded as an estimate of the proportion of variation in Y associated with variations in X *in the universe*.

Precisely the same arithmetic is involved in calculating r for pairs of observations in which the values of X have been chosen in some non-random fashion, such as in a controlled experiment. But the value of r is now strongly influenced by the way in which the X values are selected, being high if only extremely large and extremely small values of X are chosen and low if the chosen values of X are concentrated in a narrow range. (See fuller explanation in Chapter 17.) In such cases r is little more than a description of one aspect of the particular set of observations under study. It will be fairly stable from one experiment to another only if the values of X are selected in the same way in every experiment. In controlled experiments, therefore, some statisticians do not even bother to calculate r, but content themselves with the line of relationship, the standard error of estimate, and measures of the accuracy with which the level and slope of the regression line have been estimated.

There are other situations in which the values of X are not controlled by the investigator but in which they can hardly be regarded as random drawings from a definite, stable universe. This may often be the case in economic time series; the correlation between a given series and, say, consumer income will typically be higher in a period characterized by wide fluctuations in consumer income than in one during which income is relatively stable. As in the experimental case above, r is not likely to be stable from period to period unless the economic system happens to generate the same values of income as before or at least about the same total amount of variation in income.

Finally, the observations on X and Y may be drawn at random from some definite, stable universe that does not follow the bivariate normal distribution. In some such cases it may happen that certain functions of X and Y (such as their logarithms or reciprocals) do form a bivariate normal distribution; if so, the coefficient r based on appropriately transformed sample values may be regarded as an estimate of the true correlation ρ, in the universe of transformed values. There are other cases in which the original values of X and Y, or transformed ones, are distributed in something other than "normal" fashion; in these cases, r may still be a rough estimate of correlation in the universe, but we can no longer be sure that certain formulas appropriate to the normal distribution still apply.

In contrast, the interpretation of the regression coefficient b_{yx} is the same regardless of whether the X values are drawn at random or are subjected to purposeful selection. For this reason many statisticians now place primary emphasis upon *regression* and stress *correlation* only when both X and Y values are drawn at random from a universe approximating the bivariate normal form. In the latter case the regression of X upon Y *may* under certain conditions be just as meaningful as that of Y upon X; this is clearly not true if the values of X have been set by the researcher.

Where both X and Y are assumed to be built up of simple elements of equal variability, all of which are present in Y but some of which are lacking in X, it can be proved mathematically that r^2 measures the proportion of all the elements in Y which are also present in X. For that reason in cases where the dependent variable is known to be causally related to the independent variable, r^2 may be called the *coefficient of determination*. It may be said to measure the percentage to which the variance in Y is determined by X, since it measures that proportion of all the elements of variance in Y which are also present in X.[5] The coefficient of determination, d_{xy}, may be defined by the equation

$$d_{xy} = r^2_{xy} \tag{7.9}$$

Where some elements are present in each variable which occur in the other, the coefficient of determination is the product of these joint proportions. That is, if $\frac{2}{3}$ of the elements in X are the same as $\frac{2}{3}$ of the elements in Y, then the coefficient of determination will be equal to $\frac{4}{9}$.

Although the coefficient of correlation was the earliest measure used, it can be seen that it may be misinterpreted. Thus if half the variance in Y were directly due to X, the coefficient of correlation would be 0.707 $(= \sqrt{\frac{1}{2}})$. Since the coefficient of determination is the most direct and

---

[5] See Note 1, Appendix 3.

unequivocal way of stating the proportion of the variance in the dependent factor which is associated with the independent factor, it should be used in preference to the correlation coefficient.

Where curvilinear relations have been used in determining the relationship, the term *index of determination* will be used to denote the value of i^2, thus retaining the same relation to the index of correlation that the coefficient of determination bears to r, the coefficient of correlation. The index of determination, $d_{y \cdot f(x)}$ may be defined

$$d_{y \cdot f(x)} = i_{yx}^2 \qquad (7.10)$$

When an expression is used such as "Forty per cent of the variance in yield is due to differences in rainfall," it will be understood that it is either the coefficient or the index of determination which is being stated.

RELATION OF THE MEASURES OF CORRELATION TO THE TWO REGRESSION LINES. Attention has been called in several previous chapters to the fact that two regression lines can be fitted to any set of observations. These are denoted by the two coefficients b_{yx} and b_{xy} in the two equations

$$Y = a_{yx} + b_{yx}X$$

and

$$X = a_{xy} + b_{xy}Y$$

Although there are these two regression lines, there is only a single coefficient of correlation for any one set of observations. In fact, the coefficient of correlation has certain definite relations to the two lines. It indicates how closely the two lines approach one another. The higher the correlation, the closer the two lines come together; the lower the correlation, the farther they diverge. In perfect correlation ($r = \pm 1$) the two lines coincide. When there is no correlation ($r = 0$) the two lines will be at right angles to one another.

This relationship is so exact that the value of the correlation coefficient can be computed from the slopes of the two lines according to the equation

$$r_{yx} = \sqrt{b_{yx}b_{xy}} \qquad (7.11)$$

It follows from this equation that when $r = 1$, $b_{yx} = 1/b_{xy}$, and therefore the two regression lines will coincide.[6]

---

[6] This property of the two lines can be used to estimate graphically the closeness of correlation. When the two variables, X and Y, are stated in terms of standard deviation units, X/s_x and Y/s_y, by dividing each observation by the standard deviation of the series, the coefficient of correlation will then be a precise mathematical function of the angle between the two lines. By stating the variables in this way, plotting them on a dot chart, and drawing in the two lines graphically, a fairly close approximation to the coefficient can be obtained.

Although there can be only a single coefficient of correlation for a single set of observations, there can be two *indexes* of correlation. This follows from the fact that the curve which expresses the relation of Y to X,

$$Y = f(X)$$

may be a curve of quite a different type from that which expresses the relation of X to Y, which we can designate

$$X = \phi(Y)$$

Accordingly, the index of correlation i_{yx}, which measures the closeness of correlation according to the first curve, may be quite different from the index of correlation i_{xy}, which measures the closeness according to the second curve. Only in the special case where all the observations lie precisely along the curve, so that $i = 1$, will the two indexes have the same value. In that case it will also hold true that the curves $Y = f(X)$ and $X = \phi(Y)$ will be identical with the coordinates reversed.

There is only one correlation coefficient r, however. It measures the correlation according to both regression lines. Since $r = r_{yx} = r_{xy}$, the subscript notations can be used interchangeably.

ADJUSTMENTS FOR NUMBER OF OBSERVATIONS. Just as the standard error of estimate from the sample tends to be biased downward, as compared to the value that is most likely to prevail in the universe, so the correlation coefficient or index from a small sample is likely to be biased upward. This will be important, however, only in those special cases where there is some valid basis for using the sample to make generalizations concerning the probable closeness of correlation in the universe. Adjustments to use in such cases, and cautions in their use, are presented in the second portion of Chapter 17.

The Sampling Significance of the Regression Line or Curve and of the Measures of Correlation. Chapter 2 showed how a series of samples drawn from the same universe would yield varying estimates of the true average in that universe. It also presented methods of estimating how far the average from a single sample might miss the true average in the universe. In exactly the same way, if regression lines or curves are determined for a series of samples from the same universe, they will vary among themselves. Similarly, the coefficients or indexes of correlation and the standard errors of estimate will vary from sample to sample. Standard errors of some of these measures are available. These measures of reliability are more complicated, both in computation and in interpretation, than the standard error of an average. Accordingly, their presentation is deferred to Chapter 17. In addition, the special problem of the reliability of an individual estimate for an individual new observation, from the results shown by a sample, is treated in Chapter 19. The methods given in the present chapter and in Chapter 8 are sufficient for determining the correlation and regression *as shown in the individual sample.* Before a student or research worker uses the results of the sample to draw more general conclusions as to the relations which are likely to hold true in other samples or in the universe as a whole, or before he makes estimates for new observations, he should master these later chapters and should apply the checks and limitations set forth there in stating his general conclusions or in making his estimates.

Summary

This chapter has pointed out that the closeness of relation between two variables may be measured either by the absolute closeness with which values of one may be estimated from known values of the other or on the basis of the proportion of the variance in one which can be explained by, or estimated from, the accompanying values of the other. The accuracy of estimate is measured by the standard error of estimate, which indicates the reliability of values of the dependent variable estimated from observed values of the independent variable.

The relative closeness of the relation is best measured by the coefficient of determination, in the case of linear relationship, or by the index of determination, in the case of curvilinear relationship. These measures show the proportion of the variance in the dependent variable which is associated with differences in the other variable. In the case of variables causally related, they measure the proportion of the variance in one which can be said to be "caused by" variations in the other.

CHAPTER 8

Practical methods for working out two-variable correlation and regression problems

Terms to Be Used. The preceding discussion has developed the means by which values of one variable may be estimated from the values of another, according to the functional relation shown in a set of paired observations. Simple correlation involves only the means for making such estimates, and for measuring how closely those estimates conform to, and account for, the original variation in the variable which is being estimated, for the given set of observations.

The *regression line* is used, in statistical terminology, to designate the straight line used to estimate one variable from another by means of the equation

$$Y = a + bX$$

This equation is termed the *linear regression* equation; and the coefficient *b*, which shows how many units (or fractional parts) *Y* changes for each unit change in *X*, is termed the *coefficient* of *regression*.

Where a curvilinear function has been determined, either by the use of an equation or by graphic methods, the corresponding curve is similarly designated as the *regression curve*. Either the mathematical equation or, if none has been computed, the expression

$$Y = f(X)$$

where the symbol $f(X)$ stands for the relation shown by the graphic curve, is termed the *regression equation*.

The coefficient of correlation and the index of correlation have both been defined as the ratio of the standard deviation of the estimated values of *Y* to the standard deviation of the actual values, whereas the standard error of estimate has been defined as the standard deviation of the residuals

134

from the estimates so made. In linear relations, however, the coefficient of correlation and the standard error of estimate can both be computed directly from the same values that were employed in computing the constants of the regression equation, and from the standard deviation of the dependent variable, s_y. This will be illustrated by the practical example which follows.

Working Out a Linear Regression. As was illustrated in Chapter 5, the values for a and b of the regression equation can be determined for any two variables, X and Y, between which it may be desired to determine the relation, by working out the values, M_x, M_y, ΣX^2 and $\Sigma(XY)$, and then substituting them in the appropriate equations. To calculate the coefficient of correlation, r_{xy}, and the standard error of estimate, S_{yx}, it is necessary to compute in addition only the value ΣY^2 and substitute

Table 8.1

COMPUTING THE VALUES NEEDED TO DETERMINE LINEAR REGRESSION AND CORRELATION COEFFICIENTS

Irrigation Water Applied per Acre,* X	Yield of Pima Cotton per Acre,* Y	X^2	XY	Y^2
Feet	*Units of 10 pounds*			
1.8	26	3.24	46.8	676
1.9	37	3.61	70.3	1,369
2.5	45	6.25	112.5	2,025
1.4	16	1.96	22.4	256
1.3	9	1.69	11.7	81
2.1	44	4.41	92.4	1,936
2.3	38	5.29	87.4	1,444
1.5	28	2.25	42.0	784
1.5	23	2.25	34.5	529
1.2	18	1.44	21.6	324
1.3	22	1.69	28.6	484
1.8	18	3.24	32.4	324
3.5	40	12.25	140.0	1,600
3.5	65	12.25	227.5	4,225
Total 27.6	429	61.82	970.1	16,057
Mean 1.97	30.64			

* From James C. Muir and G. E. P. Smith, The use and duty of water in the Salt River Valley, *Agricultural Experiment Station Bulletin* 120, 1927. All the plots were on the same type of soil, Maricopa sandy loam.

it in appropriate formulas. The data given in Table 8.1 illustrate the necessary operations.

The computations shown in this table—squaring both X and Y, calculating the product XY, summing both X, Y, and the three columns of extensions, and dividing the first two sums by the number of cases to give the mean of X and Y—provide all the basic data necessary.[1] The values a and b for the regression equation may next be computed by substituting these extensions in equations (5.2) and (5.3).

$$b_{yx} = \frac{\Sigma(XY) - nM_xM_y}{\Sigma(X^2) - n(M_x)^2} = \frac{970.1 - 14(1.97)(30.64)}{61.82 - 14(1.97^2)}$$

$$= \frac{125.0488}{7.4874} = 16.701$$

$$a = M_y - bM_x = 30.64 - 16.701(1.97) = -2.26$$

The *regression line*, $Y = a + bX$, therefore is for this case

$$Y = -2.26 + 16.70X$$

The unadjusted coefficient of correlation, r_{xy}, may now be computed from the following new formula:

$$r_{xy} = \frac{\Sigma(XY) - nM_xM_y}{\sqrt{[\Sigma(X^2) - nM_x^2][\Sigma(Y^2) - nM_y^2]}} \tag{8.1}$$

$$= \frac{970.1 - 14(1.97)(30.64)}{\sqrt{[61.82 - 14(1.97)^2][16,057 - 14(30.64)^2]}} = 0.847$$

It should be noticed that the numerator of this fraction is the same as that in the equation for b and that half of the denominator is the same, except that it is under the radical sign.

Comparison of equations (5.2) and (8.1) with equation (1.5) for the standard deviation

$$s_x = \sqrt{\frac{\Sigma(X^2)}{n} - M_x^2}$$

shows that they may be written more simply

$$b_{yx} = \frac{\Sigma(XY) - nM_xM_y}{ns_x^2} \quad \text{or} \quad = \frac{\Sigma(xy)}{ns_x^2} \tag{8.2}$$

$$r_{xy} = \frac{\Sigma(XY) - nM_xM_y}{ns_xs_y} = \frac{\Sigma(xy)}{ns_xs_y} \tag{8.3}$$

[1] Where the number of cases to be handled is large, various short cuts may be used to reduce the volume of computation required in computing the sums of extensions ΣX^2, ΣXY, and ΣY^2. See pages 455–460 of the 2d edition.

The second form, in each case, uses the notation $\Sigma(xy)$ for $\Sigma(XY) - n(M_x M_y)$ as discussed in Chapter 5.[2] The forms shown in equations (5.2), (5.3) and (8.1), however, are the ones ordinarily used in actual computation and should be kept clearly in mind.

In Chapter 7, we noted that the unadjusted standard error of estimate satisfied the relation

$$S^2_{y \cdot x} = s^2_z = s^2_y - s^2_{y'}$$

and that

$$r^2_{yx} = \frac{s^2_{y'}}{s^2_y}$$

It follows that

$$S^2_{y \cdot x} = s^2_y(1 - r^2_{yx})$$

The adjusted standard error of estimate squared (equation 7.3) is equal to

$$S^2_{y \cdot x} \frac{n}{n - 2}$$

The adjusted standard error can be calculated conveniently by means of the following equation:

$$\bar{S}_{y \cdot x} = \sqrt{\frac{\Sigma(Y^2) - n(M_y)^2}{n - 2} (1 - r^2_{xy})} \tag{8.4}$$

$$= \sqrt{\frac{16,057 - 14(30.64^2)}{14 - 2} [1 - (0.847)^2]}$$

$$= \sqrt{68.5912} = 8.28.$$

As noted earlier, though $r_{xy} = r_{yx}$, b_{xy} is *not* the same as b_{yx}. The former regression, showing the change in X for each unit change in Y (that is, regarding the dependent factor as the independent factor instead), is obtained by modifying equation (5.2) to the following form:[3]

$$b_{xy} = \frac{\Sigma(XY) - nM_x M_y}{\Sigma(Y^2) - n(M_y)^2}$$

The new regression coefficient, b_{xy}, shows the average change in water applied with each additional unit (10 pounds) of cotton harvested. With the quantity of water subject to human control, as in this case, this relation

---

[2] The value of $\Sigma(xy)$ is sometimes called the *product moment, and* $\Sigma(xy)/n$ is called the *covariance.*

[3] When the correlation is perfect, so that $r_{xy} = 1$, the two regression coefficients will have the definite relation $b_{yx} = 1/b_{xy}$. Under these conditions the regression lines will be identical, no matter which variable is regarded as the independent variable and which as the dependent.

would have little meaning. However, if it is desired to chart it on Figure 8.1 along with the other regression line, it can be charted according to the linear regression equation

$$X = a_{xy} + b_{xy}Y$$

The value of the new a can be computed by restating equation (5.3) in the form

$$a_{xy} = M_x - b_{xy}M_y$$

Equation (8.4) completes the computation of all the values needed[4] except the coefficient of determination d_{xy}, which is simply r_{xy}^2. That is:

$$d_{xy} = r_{xy}^2 = (0.847)^2 = 0.717$$

Interpreting the Measures of Linear Regression. The next step is to examine the values of the several statistics which have been computed from the sample and see what they mean.

The coefficient of regression of Y on X, $b_{yx} = 16.70$, shows that on the average the acre yield of cotton increases 16.7 10-pound units, or 167 pounds, for each additional acre-foot of water applied. The constant a shows that with no water applied, a yield of -2.26 10-pound units, -22.6 pounds, or less than no cotton at all, might be expected. Since these results are based on observations extending from 1.2 acre-feet of water to 3.5, the relations shown by the regression line do not necessarily hold beyond those limits, and it is not certain what the yield would be when no water is applied. Extrapolating the regression line to that point has no meaning, by itself.

The regression equation

$$Y = -2.26 + 16.7(X)$$

or

$$\text{Yield} = -22.6 + 167 \text{ (feet of water)}$$

then gives the yields of cotton estimated as most likely to be obtained from the quantity of water applied within the limits of 1.2 to 3.5 feet. Figure 8.1 shows how these estimated values, along the regression line, compare with the actual yields observed.

The standard error of estimate, 8.28 10-pound units or 82.8 pounds, shows that the (adjusted) standard deviation of the differences between the actual and the estimated values is 82.8 pounds of cotton. Two lines have been drawn in Figure 8.1, at 82.8 pounds above and below the regression line. It will be seen that of the 14 cases, 9 fell between these two lines, or in the zone within one standard error on either side of the regression line.

---

[4] Except also the calculation of measures of reliability, as explained in Chapter 17.

The coefficient of correlation, $r_{xy} = 0.85$, and the coefficient of determination, $d_{xy} = 0.72$, show that about 72 per cent of the variance in the yield of this crop in this area, on the farms from which these records were obtained, could be accounted for by the differences in the quantity of water used in irrigation. Since this leaves only 28 per cent of the variance to be accounted for by all other factors, it would appear that the quantity

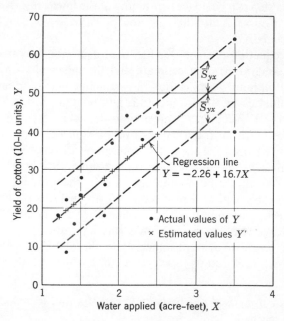

Fig. 8.1. Relation of yield of cotton to irrigation water applied; estimated yields from a linear regression and zone of probable yields indicated by the adjusted standard error of estimate.

of water applied (or other factors associated with it) was the most important factor which was associated with the yield of cotton on these farms and on this type of soil.[5]

The fact that 72 per cent of the variance in yield can be explained by corresponding differences in the quantity of water applied does not in itself mean that the differences in irrigation *caused* the differences in yield. For example, it might be possible that the quantity of water applied was regulated to conform to the fertility of the land and that the differences in yield were really due in part to the differences in fertility. The statistical measure merely tells how closely the variance in one variable was associated with the variance in the other; whether that association is due to, or can

[5] Methods of adjusting these percentages for estimated bias are presented in Chapter 17.

be taken as evidence of, cause-and-effect relation is another matter, and is outside the scope of the statistical analysis. (For more extended discussion of this point, see Chapters 25 and 26.)

Working Out a Curvilinear Regression. The next step is to consider whether the straight line is adequate to describe the way that the yield increases as more water is applied, or whether a curve had better be employed. (This step can be taken before any of the linear results are worked out, and, if a curve is decided on, the previous work can be skipped entirely, if desired.)

Where the equation of the curve has been determined by mathematical means, the standard error of estimate and the index of correlation may be computed without working out the estimates and residuals for each of the individual cases. These methods will be described in Chapter 12.

Before fitting the curve, we must consider what type of curve it is logical to expect. In most agricultural production problems, diminishing returns are experienced.[6] That is, the application of successive increments of fertilizer or other productive aid on the same areas will be expected to produce a smaller and smaller increase in the product. Also, it is known that if too much of some factors are applied, the result may be to produce a decline in output. The decline after the point of maximum output is reached may be gradual, or it may be sudden, owing to a toxic effect of too much of one substance upon the plant or animal. These considerations would lead us to expect a curve with the following characteristics:

1. It should rise steeply at first, and then less and less sharply until a maximum is reached.

2. It might show a decline after the maximum is reached, either gradual or sharp.

3. It would have only the single point of maximum yield.

These are the conditions we shall apply in fitting the curve.

A mathematical curve can be derived expressing these logical conditions. The curve which has been found most suitable to express them, however, has a quite complicated mathematical form, and cannot be fitted to the data by least-squares or other simple arithmetic methods. We shall therefore use a freehand fit, keeping in mind the conditions stated.

Examining Figure 8.1 more closely, we see that, in the range up to 1.8 acre-feet of water, the actual yields lie below the regression line 4 times, and above 4 times; in the range from 1.9 to 3 acre-feet, the actual yields lie above in all 4 observations; and above 3 acre-feet the 1 yield

---

[6] William J. Spillman, *The Law of Diminishing Returns*, World Book Co., Yonkers-on-the-Hudson, New York, and Chicago, 1924.

below the line is much farther below than is the 1 above. These facts suggest that a curve convex from above, giving lower estimated yields than the straight line for the lowest and highest applications of water and higher estimated yields for the intermediate applications, would more accurately represent the relations in this case. The facts also agree with the logical conditions stated. (The number of observations is too small to serve as an accurate indication of the shape of the curve, but it will serve at least as a simple illustration of the way the whole problem may be worked through.)

The next step is to group the observations according to the value of X (the quantity of water) and average both X and Y, water and yield (Table 8.2). In view of this small number of observations, rather large groups are taken; were more cases available, the groups might be made narrower.

Table 8.2

COMPUTATION OF GROUP AVERAGES TO INDICATE REGRESSION CURVE—
COTTON EXAMPLE

X (water) 1–1.4		X (water) 1.5–1.9		X (water) 2.0–2.9		X (water) 3.0–3.9	
X	Y	X	Y	X	Y	X	Y
1.4	16	1.8	26	2.5	45	3.5	40
1.3	9	1.9	37	2.1	44	3.5	65
1.2	18	1.5	28	2.3	38		
1.3	22	1.5	23				
		1.8	18				
Sums 5.2	65	8.5	132	6.9	127	7.0	105
Means 1.3	16.25	1.7	26.4	2.3	42.33	3.5	52.5

These averages are then plotted, as shown in Figure 8.2, an irregular line dotted in connecting them, and as smooth a curve as possible which fulfills the stated conditions drawn in freehand through the averages and the broken line, just as discussed in Chapter 6. This then gives the regression curve. It is seen to fit the data well, and yet to fulfill the logical conditions stated. The point of maximum yield, however, apparently lies beyond the limit of the observations.

Next the estimated yields for each different application of water are read off from this curve, and the differences between the actual and the estimated yields are determined. These residuals are then squared to determine their standard deviation, the Y values are also squared so as to determine their standard deviation, and so give the basis for measuring the amount of correlation. The sum of the Y'' values is slightly smaller than the sum of the Y values, and the mean of the z'' values is therefore not exactly zero, but 0.264. That indicates that the curve shown in Figure 8.2 should be shifted up 0.264 unit, or 2.64 pounds, to make the

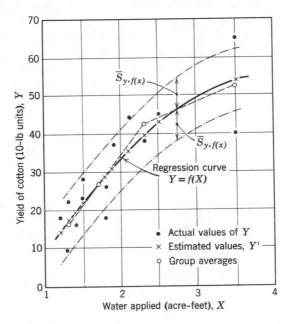

Fig. 8.2. Relation of yield of cotton to irrigation water applied; estimated yields from a curvilinear regression; and zone of probable yields as indicated by the adjusted standard error of estimate.

estimated and actual averages agree.[7] Representing this curve by $f(X)$, the regression equation may therefore be written:

$$Y = k + f(X)$$

$$Y = 2.64 + f(X)$$

[7] In problems with many observations, the sum of the Y values and of the Y'' values may be determined separately for the several different portions of the curve, to see if its position should be shifted in one portion and not in another. This process can be carried too far, however, for if the divisions are made too small the effect will be to make the curve pass through each successive group average, without smoothing out the irregularities into a continuous function.

The values at the foot of Table 8.3 now give the sums necessary to measure the closeness of the correlation. First the standard deviations of Y and z'' are computed, using the formula

$$s_y = \sqrt{\frac{\Sigma Y^2 - n(M_y^2)}{n}} = 14.44$$

$$s_{z''} = \sqrt{\frac{\Sigma(z'')^2 - n(M_{z''}^2)}{n}} = \sqrt{\frac{718.51 - 14(0.264^2)}{14}} = 7.16$$

Table 8.3

COMPUTATION OF RESIDUALS AND STANDARD DEVIATION FOR CURVILINEAR REGRESSION—COTTON EXAMPLE

Water per Acre, X	Yield, Y	Yield Estimated from X, Y''	$Y - Y''$, z''	$(z'')^2$	Y^2
	(10-pound units)	(10-pound units)			
1.8	26	29.0	−3.0	9.00	676
1.9	37	31.0	6.0	36.00	1,369
2.5	45	42.8	2.2	4.84	2,025
1.4	16	19.2	−3.2	10.24	256
1.3	9	16.8	−7.8	60.84	81
2.1	44	35.2	8.8	77.44	1,936
2.3	38	39.5	−1.5	2.25	1,444
1.5	28	21.9	6.1	37.21	784
1.5	23	21.9	1.1	1.21	529
1.2	18	14.2	3.8	14.44	324
1.3	22	16.8	5.2	27.04	484
1.8	18	29.0	−11.0	121.00	324
3.5	40	54.0	−14.0	196.00	1,600
3.5	65	54.0	11.0	121.00	4,225
Sums	429	425.3	+3.7	718.51	16,057

Then, by equation (7.6),

$$\bar{S}_{y \cdot f(x)}^2 = s_{z''}^2 \left(\frac{n}{n - m}\right) = (7.16^2)\left(\frac{14}{14 - 3}\right) = 65.23$$

$$\bar{S}_{y \cdot f(x)} = 8.07$$

Here 3 is used for the value of m, since it is judged that a parabolic equation of type (*a*), Chapter 6, with 3 constants, would be adequate to reproduce the freehand curve.

The standard error of estimate for the graphic regression curve is thus 8.07 10-pound units, or 80.7 pounds. This is 2.1 pounds smaller than the corresponding value in the case of the linear correlation, indicating how much more closely the curve fits the data than does the straight line, even after allowing for its greater flexibility. In Figure 8.2 two dotted lines have been drawn in, each 80.7 pounds away from the regression curve, indicating the zone of estimate within which approximately two-thirds of the cases fall (10 out of 14 in this instance) and within which two-thirds of the actual yields may be expected to fall if new estimates of yield are made from the water applied for additional cases drawn from the same universe. (Note also the discussion, in Chapters 17 and 19, of more exact methods of calculating confidence intervals for individual estimates.)

The index of correlation may be computed from the formula

$$i_{yx}^2 = 1 - \frac{s_{z''}^2}{s_y^2} \tag{8.5}$$

which is equivalent to Equation (7.8). In this case it works out to be

$$i_{yx}^2 = 1 - \frac{(7.16)^2}{(14.44)^2} = 1 - 0.2459 = 0.7541$$

$$i_{yx} = 0.868$$

Since the index of determination is simply i_{yx}^2, it is 75.4 per cent. The index of determination, 75.4 per cent, compares with the coefficient of determination of 71.7 per cent. Apparently taking the curvilinear nature of the relations into account, has increased the proportion of the variance in yield accounted for by differences in water application by 4 per cent of the total variance.[8] (Only the measures of determination can be directly compared in this way. If the coefficient of correlation, 0.847, were subtracted from the index of correlation, 0.868, that would give an incorrect idea of the importance of taking account of the curvilinear nature of the relation.)

Interpreting the Measures of Curvilinear Regression. The index of determination and the accompanying standard error of estimate have

---

[8] See Chapter 23 for tests as to whether this difference is large enough to be significant, and see Chapter 17 for corrections for number of constants involved. By using the curve without these adjustments, the difference between r^2 and i^2 exaggerates the real gain, especially in small samples.

been interpreted for the curve in much the same manner as were the coefficient of determination and the standard error of estimate for the straight line. In the case of the regression curve itself, however, a somewhat different method of presentation may be best, since a mathematical equation expressing the relation has not been computed.

The regression curve just worked out for the cotton problem, for example, may be presented either as a curve showing graphically the yield to be expected for various applications of water, as is illustrated in Figure 8.2, or as a table showing the same thing, as in Table 8.4. In both instances the constant which has been determined from the average of z'' is added to the values read from the curve in Figure 8.2, $f(X)$, so as to give the final estimates which would be made by taking into account this slight shift in the position of the curve.

Table 8.4

YIELD OF PIMA COTTON, WITH DIFFERENT APPLICATIONS OF IRRIGATION WATER, ON MARICOPA SANDY LOAM SOILS IN THE SALT RIVER VALLEY, ARIZONA, IN 1913, 1914, AND 1915

Irrigation Water Applied	Average Yield of Cotton Lint
Acre feet	*Pounds per acre*
1.25	156
1.50	222
1.75	283
2.00	335
2.25	385
2.50	431

In preparing the table, the relation is shown only for that range of water application within which the bulk of the observations falls. Similarly, only this range should be shown by the solid line in the chart; a dotted line might be used to indicate the relations beyond that up to the extremes observed. Neither the regression line nor curve should, ordinarily, be carried beyond the limits of the observations on which it was based. Also, before general conclusions are drawn as to the application of the results to cases other than those included in the sample (as, in this instance, to other fields in the same area), the standard errors set forth in Chapter 17 and 19 should be calculated and confidence intervals based upon them should be included in the interpretation, for curves mathematically fitted.

Summary

This chapter has illustrated the way in which regression analysis may be applied to a specific problem, the manner in which linear and curvilinear regressions may be determined most simply, and the way in which they may be interpreted. In addition, the simplest manner of computing the standard error of estimate and the coefficient and the index of correlation have been illustrated, and their significance has been briefly discussed.

CHAPTER 9

Three measures of correlation and regression—the meaning and use for each

So many different statistical coefficients have been introduced in the discussion of correlation that there may be some confusion as to the meaning and use of the different statistics. Particularly in the linear situation, there are three statistics which summarize nearly all that a regression analysis reveals.

First, the standard error of estimate indicates how nearly the estimated values agree with the values actually observed for the variable being estimated. This coefficient is stated in the same units as the original dependent variable, and its size can be compared directly with those values.

Second, the coefficient of determination (r^2) shows what proportion of the variance in the values of the dependent variable can be explained by, or estimated from, the concomitant variation in the values of the independent variable.[1] Since this coefficient is a ratio, it is a "pure number"; that is, it is an arbitrary mathematical measure, whose values fall within a certain limited range, and it can be compared only with other statistics like itself, derived from similar problems.

Third, the coefficient of regression measures the slope of the regression line; that is, it shows the average number of units increase or decrease in the dependent variable which occur with each increase of a specified unit in the independent variable. Its exact size thus depends not only on the relation between the variables but also on the units in which each is stated. It can be reduced to another form, however, by stating each of the variables in units of its own individual standard deviation. In this

---

[1] These statements are all subject to the error limitations set forth later, in Chapters 17 and 19.

form it has been termed β or the *beta* coefficient.[2] The relation between beta and the coefficient of regression may be indicated by stating the regression equation in both ways:

$$Y = a + b_{yx}X$$

$$\frac{Y}{s_y} = a' + \beta_{yx}\left(\frac{X}{s_x}\right)$$

$$\beta_{yx} = b_{yx}\left(\frac{s_x}{s_y}\right) \tag{9.1}$$

$$a' = \frac{M_y}{s_y} + \beta_{yx}\frac{M_x}{s_x}$$

Stated in this way, β for the cotton-yield problem is 0.845. That is, for each increase of one standard deviation (0.73 acre-foot of water) in X, the yield of cotton increased 0.845 of one standard deviation. Since the standard deviation of Y was 144.3 pounds, that is equal to 121.9 pounds of cotton for each 0.73 acre-foot of water. This is at the rate of 167 pounds of cotton for each foot of water, which is the same thing as was shown by the coefficient of regression. However, for comparisons between problems where the standard deviations are much different, the beta coefficient may have value. It is evident that in simple correlation the value of beta is the same as that of r.

Relation of the Different Coefficients to Each Other. Even though each of the three coefficients measures certain aspects of the relation between variables, it does not follow that all three coefficients will vary together, or that a problem which shows a high coefficient of determination will also show a high regression coefficient or a low standard error of estimate. That is because they measure different aspects of the relation.

If either of the standard deviations involved has been artificially modified by selection of the sample, as in the "regression model" of Chapter 17, then the beta coefficient will be of less significance, just as will the correlation coefficient.

The particular usefulness of each of the three different groups of correlation measures is illustrated in Figure 9.1, which shows three sets of simple relationships, with hypothetical data.

Here the regression coefficient is smaller in (*a*) than (*b*). In (*a*) an additional inch of rain causes an average increase of 2.5 bushels in yield,

[2] See Truman Kelley, *Statistical Method*, p. 282, Macmillan, New York, 1924; and George W. Snedecor, *Statistical Methods*, 5th ed., p. 416, Iowa State College Press, Ames, 1956. Snedecor uses the term "standard regression coefficient" for β.

as compared with an increase of 3.1 bushels in (*b*). But in case (*a*), a considerable part of the variation in yield is apparently due to rainfall, as shown by the high correlation ($r = 0.83$) and the small size of the standard error of estimate (2.2 bushels); whereas in case (*b*), factors other than rainfall apparently cause most of the differences in yield, as indicated by the lower correlation ($r = 0.71$) and the larger standard

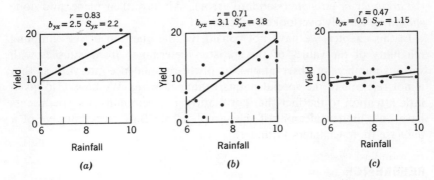

Fig. 9.1. Hypothetical sets of data, illustrating three types of correlation coefficients.

error of estimate (3.8 bushels). In terms of determination apparently about 69 per cent of the differences in yield are related to differences in rainfall in the first case, and only about 50 per cent in the second.

In comparison with (*a*) and (*b*), case (*c*) has much less variable yields, ranging only from about 8 bushels to 12 bushels, compared with a range of 8 to 21 in case (*a*) and 0 to 20 in case (*b*). Only a small part (22 per cent) of the variation in yields is associated with rainfall differences, as indicated by the low correlation (0.47). An increase of 1 inch in rainfall apparently causes only 0.5 bushel increase in yield. Yet in spite of this low relation, it is possible to estimate yields more accurately, given the rainfall, in this case than in either of the other two, as is shown by the standard error of estimate of 1.15 bushels as compared to 2.2 bushels for (*a*) and 3.1 for (*b*). The original variation in yields is so slight in case (*c*) that even the small relation shown to rainfall is enough to make it possible to estimate yields more accurately than in either of the other cases.[3]

These three cases illustrate the relative place of each of the three types of correlation measure. Case (*b*) shows the greatest change in yield for a given change in rainfall (the regression measure); case (*a*) shows the

[3] The standard deviation of Y is so small in case (*c*) (only about 1.3) that Y could be estimated from M_y in this case more closely than from the regression line in either of the other cases. A similar case could be constructed with $s_y = 2.2$ or more, and r very low, which would still have S_{yx} smaller than in (*a*) or (*b*).

highest proportion of differences in yields accounted for by rainfall (the correlation or determination 'measure); and case (*c*) shows the greatest accuracy of estimate (the error of estimate measure). Which of these measures should have most attention in a particular investigation depends upon the phase of the investigation which is most important: the *amount* of change (regression); the *proportionate* importance (correlation); or the *accuracy of estimate* (standard error). All have their place, and none should be entirely overlooked or ignored.

In this chapter, we have not considered the question of the sampling reliability of the values of the statistics determined from samples with various numbers of observations, or of their confidence intervals, as that is not relevant to the special points examined here. We have also given little attention to the fact that correlation and determination coefficients have only limited meaning if the sample is not fully representative of a universe (note Chapters 17 and 18).

REFERENCE

Ezekiel, Mordecai, Meaning and significance of correlation coefficients, *Amer. Econ. Rev.*, Vol. XIX, No. 2, pp. 246–50, June, 1929.

Multiple
Linear Regressions

CHAPTER 10

Determining
multiple linear regressions:
(1) by successive elimination

The Problem of Multiple Relations

The relations studied up to this point have all been of the type where the differences in one variable were considered as due to, or associated with, the differences in one other variable. But in many types of problems the differences in one variable may be due to a number of other variables, all acting at the same time. Thus the differences in the yield of corn from year to year are the combined result of differences in rainfall, temperature, winds, and sunshine, month by month or even week by week through the growing season. The premiums or discounts at which different lots of wheat sell on the same day vary with the protein content, the weight per bushel, the amount of dockage or foreign matter, and the moisture content. The speed with which a motorist will react to a dangerous situation may vary with his keenness of sight, his speed of nervous reaction, his intelligence, his familiarity with such situations, and his freshness or fatigue. The price at which sugar sells at wholesale may depend upon the production of that season, the carryover from the previous season, the general level of prices, and the prosperity of consumers. The weight of a child will vary with its age, height, and sex. The volume of a given weight of gas varies with the temperature and the barometric pressure.

The physicist and the biologist use laboratory methods to deal with problems of compound or multiple relationship. Under laboratory conditions all the variables except the one whose effect is being studied can be held constant, and the effect determined of differences in the one remaining varying factor upon the dependent variable, while effects of differences in the other variables are thus eliminated. In the case of a gas, for example, the temperature may be held constant while the volume at different barometric pressures is determined experimentally, and then the pressure held constant while the volume at different temperatures is determined. For many of the problems with which the statistician has to deal, however, such laboratory controls cannot be used. Rainfall and temperature and sunshine vary constantly, and only their combined effect upon crop yields can be noted. The astronomer or meteorologist can observe events in the heavens or the atmosphere, but cannot control them (though seeding clouds to make rain is a step in that direction). Economic conditions are constantly shifting, and only the total result of all the factors in the existing situation can be measured at any time. And so on through many other types of multiple relations similar to those mentioned—the statistician has to deal with facts arising from the complex world about him, and frequently has but little opportunity to utilize laboratory checks or artificial controls.

Theoretical Example.   Where a dependent variable is influenced not only by a single independent variable, as in the relation of Y to X, but also by two or more independent variables, we can represent the relation symbolically by the equation

$$X_1 = a + b_2 X_2 + b_3 X_3 + \ldots b_m X_m \qquad (10.1)$$

Here X_1 represents the dependent variable, and $X_2, X_3, \ldots X_m$ represent the several independent variables.

The meaning of the several constants in this equation and the way in which it may be interpreted geometrically can be shown by making up a simple example.

Let us assume that in a new irrigation project the farms are all alike in quality of land and kinds of buildings and that the price at which each one is sold to the settlers is computed as follows:

Buildings, \$1,000 per farm.
Irrigated land, \$100 per acre.
Range (non-irrigated) land, \$20 per acre.

Using X_1 to represent the selling price per farm in dollars, X_2 to represent the number of acres of irrigated land in each farm, and X_3 to represent the number of acres of range land, we can state the method of computing the selling price in the single equation

$$X_1 = 1,000 + 100 X_2 + 20 X_3$$

The relations stated in this equation may be represented graphically as shown in Figure 10.1. The representation is broken up into halves. The upper half shows the relation of farm value to irrigated land for farms that have no range land; the lower half shows the relation of farm value to range land for farms that have no irrigated land. This figure is constructed exactly the same as was Figure 5.2. Thus in the upper section of

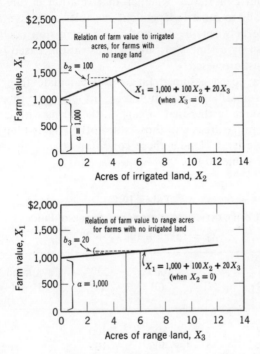

Fig. 10.1. Graph of the function $X_1 = 1,000 + 100X_2 + 20X_3$.

Figure 10.1, each increase of 1 unit in X_2, as, for example, from 3 to 4, adds $100 to the farm value. Similarly, in the lower section, each change of 1 unit in X_3, as, for example, from 5 to 6, adds $20 to the farm value. In each case, as for zero acres, the line begins with the value of a, 1,000, to cover the value of the buildings.

Equation (10.1) is called the *multiple regression equation*. The term *multiple* is added to indicate that it explains X_1 in terms of two or more independent variables, $X_2, X_3 \ldots X_m$. The coefficients b_2 and b_3 are termed *net regression coefficients*. The term *net* is added to indicate that they show the relation of X_1 to X_2 and X_3, respectively, excluding, or *net of*, the associated influences of the other independent variable or variables. In contradistinction, the regression coefficient b_{yx} of equation (5.1),

$$Y = a + b_{yx}X$$

may be termed the *gross regression coefficient*. The term *gross* is added here to indicate that it shows the apparent, or gross, relation between Y and X without considering whether that relation is due to X alone, or partly or wholly to other independent variables associated with X.

The difference between the net and gross regression coefficients may be further shown by a simple arithmetic illustration, based on the farm-value formula just discussed.

Let us take a dozen assumed farms and calculate from the pricing equation what their selling prices should be. In setting up these illustrative farms, let us assume further that in general the farms with large irrigated areas had small range areas and those with little irrigated land had larger amounts of range land. Under these conditions the computation works out as shown in Table 10.1.

Table 10.1

COMPUTATION OF ESTIMATED SELLING PRICE

$$X_1 = 1,000 + 100X_2 + 20X_3$$

Observation Number	X_2 (1)	X_3 (2)	$100X_2$ (3)	$20X_3$ (4)	Calculated Values of X_1 (3) + (4) + 1,000
1	8	5	800	100	1,900
2	4	5	400	100	1,500
3	3	10	300	200	1,500
4	7	8	700	160	1,860
5	7	10	700	200	1,900
6	8	15	800	300	2,100
7	6	12	600	240	1,840
8	1	15	100	300	1,400
9	4	17	400	340	1,740
10	2	22	200	440	1,640
11	4	20	400	400	1,800
12	5	13	500	260	1,760

The apparent relation of the values of X_1, as just computed, to X_2 and X_3 may be shown by preparing dot charts of the X_1 to X_2 relation and the X_1 to X_3 relation. These dot charts are shown in Figure 10.2.

Examining this figure, we find that X_1 is fairly closely related to X_2 but that it has no definite relationship to X_3. We could calculate the regression lines for each of the two relationships shown. The regression coefficient, b_{12}, for the first comparison, would show the average change in X_1 with unit changes in X_2. The regression coefficient, b_{13}, for the second comparison, would show the average change in X_1 with unit changes in

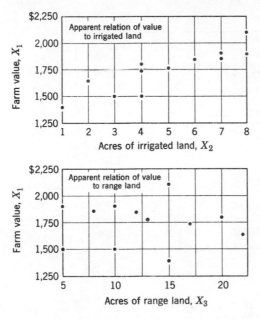

Fig. 10.2. The apparent relation of farm value to acres of irrigated land and to range land reveals little of the underlying net relationship.

X_3. The latter coefficient would come very close to zero, to judge visually from the chart. Both these would be *gross regression coefficients*, measuring only the apparent relation between X_1 and each of the other variables. We know in this case that the values of X_1 are completely determined by the values of X_2 and X_3. If we could hold constant, or eliminate, the true effect of X_2 on X_1, we should find that the relation of the corrected values of X_1 to X_3 was just as close as to X_2. In spite of the fact that the gross regression, b_{13}, appears to be zero, the net regression, b_3, is really 20.

By using the known net regression of X_1 on X_2, we can correct the X_1 values to eliminate that part of their variation which is due to X_2 and then relate the remaining fluctuation to X_3. Let us do that by subtracting $b_2 X_2$ from X_1. This process is shown in Table 10.2.

Table 10.2

CORRECTION OF COMPUTED X_1 FOR CONTRIBUTION OF X_2

Observation Number (1)	X_2 (2)	X_3 (3)	X_1 (4)	$b_2 X_2$ $(100 X_2)$ (5)	$X_1 - b_2 X_2$ (6)
1	8	5	1,900	800	1,100
2	4	5	1,500	400	1,100
3	3	10	1,500	300	1,200
4	7	8	1,860	700	1,160
5	7	10	1,900	700	1,200
6	8	15	2,100	800	1,300
7	6	12	1,840	600	1,240
8	1	15	1,400	100	1,300
9	4	17	1,740	400	1,340
10	2	22	1,640	200	1,440
11	4	20	1,800	400	1,400
12	5	13	1,760	500	1,260

We can now plot the values of X_1 corrected for X_2, or $X_1 - b_2 X_2$ as shown in the sixth column, against the X_3 value, as shown in the third column. The resulting dot chart is shown in Figure 10.3.

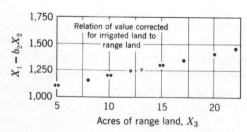

Fig. 10.3. After the net influence of irrigated land has been removed, the underlying relation of farm value to acres of range land is very clear.

This figure now shows the underlying relation between X_1 and X_3, with all the dots falling exactly on one straight line. If we now draw in the regression line and calculate its slope, we shall find it is exactly the same as the line for b_3, which was illustrated in the lower section of Figure 10.1. Figure 10.3 illustrates the *net* regression of X_1 on X_3, as contrasted to the *gross* regression which was represented by the lower section of Figure 10.2. If X_1 were similarly corrected for X_3 and the values $X_1 - b_3 X_3$ were plotted against X_2, the net regression of X_1 on X_2 would similarly be shown. (This step is left for the student to perform.)

If we had not known the underlying relationships as given in this case to start with, but merely had the series of observations of X_1, X_2, and X_3 shown in Table 10.1 and Figure 10.2 would it be possible to work out from those observations the underlying, or *net*, relationships? That is the problem which next will be explored. This time we shall use a series where we do not know the relationship, and see how we can proceed to to work it out. Also, as in most practical cases, we shall use an example where all the causes of variation are not known and where we must deal with independent variables which explain only a part of the variation in the dependent variable.

Illustrative Example. The problem of multiple relations is illustrated by the data in Table 10.3. These represent 20 farms in one area, with varying crop acreages, dairy cows, and incomes.[1] To determine from these records what income might be expected, on the average and under the same conditions, with a given size of farm and with a given number of cows, it is necessary to estimate the effects of differences in the number of acres and also of differences in the number of cows on income.

From these data it would seem that both the size of the farm and the size of the dairy herd influence farm income, to judge from dot charts showing the relation of income to acres (Figure 10.4) and of income to number of cows (Figure 10.5). It appears from these charts that there may be a slight tendency for the farms with the larger acreage in crops to have larger incomes and a rather marked tendency for the farms with the larger number of cows to have larger incomes.

ANALYSIS BY SIMPLE AVERAGES NOT ADEQUATE. The simple comparison alone, however, is not sufficient to tell exactly how incomes change with acres and with number of cows. That is because there is a marked relation between the size of the farms and the number of cows, as is illustrated in

[1] The dollar income levels reflect conditions on small to medium-sized dairy farms in Wisconsin in the late 1920's. In the 1950's, income from farms with these sizes and numbers of cows would have been roughly twice as large. The regression coefficients expressing the net effects of acres and cows would also be about twice as large.

Table 10.3

ACRES, NUMBER OF COWS, AND INCOMES, FOR 20 FARMS

Record No.	Size of Farm	Size of Dairy	Income
	Number of acres	*Number of cows*	*Dollars per year*
1	60	18	960
2	220	0	830
3	180	14	1,260
4	80	6	610
5	120	1	590
6	100	9	900
7	170	6	820
8	110	12	880
9	160	7	860
10	230	2	760
11	70	17	1,020
12	120	15	1,080
13	240	7	960
14	160	0	700
15	90	12	800
16	110	16	1,130
17	220	2	760
18	110	6	740
19	160	12	980
20	80	15	800

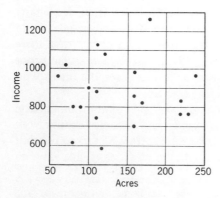 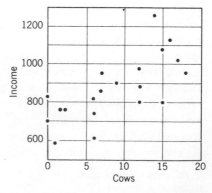

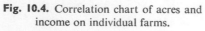

Fig. 10.4. Correlation chart of acres and income on individual farms.

Fig. 10.5. Correlation chart of number of cows and income on individual farms.

Figure 10.6. There is a definite tendency in this case for the larger farms to have smaller dairy herds. As a result, the difference in incomes in Figure 10.4, which appeared to be directly associated with differences in acreages, may reflect in part the differences in the sizes of the dairy herds on the farms with different acreages in crops. If we make groups of farms

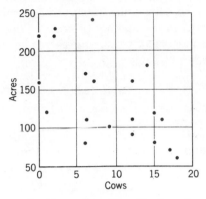

Fig. 10.6. Correlation chart of number of cows and number of acres on individual farms.

of 50 to 99 acres, 100 to 150 acres, and so on, and average the acres, cows, and incomes for each group, as is shown in Table 10.4, we find a marked difference in the number of cows from group to group, as well as in the number of acres and in the incomes.

Table 10.4

AVERAGE NUMBER OF COWS AND INCOME, FOR FARMS OF DIFFERENT SIZES

Size Group (acres)	Number of Farms	Average Size	Average Size of Dairy	Average Income
		Number of acres	*Number of cows*	*Number of dollars*
50–99	5	76	13.6	838
100–149	6	112	9.8	887
150–199	5	166	7.8	924
200–249	4	228	2.8	828

The farms of 50 to 99 acres, with an average size of 76 acres, have incomes which average $838; the farms of 150 to 199 acres, with an average size of 166 acres, show incomes which average $924. Does this difference in income reflect the difference in size? Before this can be definitely answered we must consider that the two groups also differ in the average number of cows, with 13.6 in the first group and only 7.8 in the second. So far, there is nothing to indicate whether the difference in income is associated with the difference in the size of the farms or in the number of cows; we have shown that both vary from group to group, and that is all.

If, on the other hand, we should attempt to determine how far income varied with differences in the number of cows by classifying the records with respect to the number of cows, and averaging incomes, we should secure the result shown in Table 10.5.

Table 10.5

AVERAGE ACRES AND INCOME, FOR FARMS WITH DIFFERENT NUMBERS OF COWS

Size of Herd (cows)	Number of Farms	Average Size of Dairy	Average Size of Farms	Average Income
		Number of cows	*Number of acres*	*Number of dollars*
Under 5	5	1.0	190	728
5–9	6	6.8	143	815
10–14	4	12.5	135	980
15 and over	5	16.2	88	998

Even though the income is higher on the farms with more cows, Table 10.5 does not indicate how much of that can be credited to the cows and how much to other factors. It is evident from the table that as the number of cows goes up, the number of acres goes down; are the differences in income associated with changes in number of cows, in number of acres, or in part with both?

ELIMINATING THE APPROXIMATE INFLUENCE OF ONE VARIABLE. What we need to know is how far income varies with size of farm, for farms with the same number of cows; and how far income varies with the number of cows, for farms of the same number of acres. One way of determining this would be to adjust the income on each farm to eliminate the differences due to (or associated with) the number of cows, and then compare the adjusted incomes with the size of the farm to determine the effect of size on income. To start this process, the effect of the number of cows upon incomes is needed. We can secure an approximate measure of this by determining the straight-line equation for estimating incomes from cows—approximate only, since the differences in the size of the farms are ignored at this point.

Determining the straight-line relation according to Chapter 5, we find that the apparent relation between cows and income is given by the equation:

$$\text{Income} = 694 + 20.11 \times \text{number of cows}$$

According to this equation, farms with no cows averaged about 694 income, and these incomes increased 20.11 for each cow added, on the average. Knowing this relation, we can adjust the incomes on the several farms by deducting that part of the income which would be assumed to be due to the cows, according to this average relation.

Table 10.6 illustrates the process of adjusting the incomes to a no-cow basis, by subtracting this approximate effect of cows on incomes. The next step is to see what the relation is between the acres in the farm and

Table 10.6

ADJUSTING FARM INCOMES FOR DIFFERENCES IN NUMBER OF COWS

Size of Farm	Size of Dairy	Income	Income Assumed Due to Cows	Income Adjusted to No-Cow Basis
Number of acres	*Number of cows*	*Number of dollars*	*Number of dollars*	*Number of dollars*
60	18	960	362	598
220	0	830	0	830
180	14	1,260	282	978
80	6	610	121	489
120	1	590	20	570
100	9	900	181	719
170	6	820	121	699
110	12	880	241	639
160	7	860	141	719
230	2	760	40	720
70	17	1,020	342	678
120	15	1,080	302	778
240	7	960	141	819
160	0	700	0	700
90	12	800	241	559
110	16	1,130	322	808
220	2	760	40	720
110	6	740	121	619
160	12	980	241	739
80	15	800	302	498

these adjusted incomes. Plotting both on a dot chart, Figure 10.7, shows this relation graphically. Comparing this figure with Figure 10.4, where the relation between the acres and the unadjusted incomes were plotted, we see that the relation is much closer and more definite for the adjusted incomes than for the unadjusted incomes. This is only natural; now that the marked relation of number of cows to income has been removed, even if only approximately, the underlying relation of size to income can be more clearly seen.

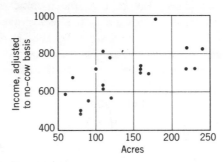

Fig. 10.7. Relation of income, adjusted for number of cows, to number of acres.

It is evident from Figure 10.7 that size has a more marked effect upon income than was apparent in Figure 10.4, where the effect of cows was mixed in also. As was pointed out earlier, the fact that cows and acres were correlated meant that the effects of differences in cows were mixed in with the effects of differences in acres. Now that the effect of cows has been at least roughly removed, the change in incomes with changes in acres can be more accurately determined.

Fitting straight lines to the relations shown in Figures 10.4 and 10.7, to determine the average change in income with changes in acres, we obtain regression equations as follows:

$$\text{Income} = 868.74 + 0.0234 \times \text{number of acres}$$

Income, effect of cows removed $= 508.51 + 1.33 \times$ number of acres

(For these early calculations, these values might be rounded off by using only, say, 3 significant digits in the values for a, b_2, and b_3.)

It is evident that the calculation of the effect of acres upon income without making some allowance for the effect of the correlated variable, number of cows, in this case would have seriously underestimated the effect of acres upon income. Such a determination would have shown only 0.02 increase in income with each acre increase in size, whereas the later determination shows 1.33 increase instead.

The relation now shown between income (in dollars) and acres illustrates the extent to which one variable may really influence a second, even though its influence is concealed by the presence of a third variable. From Figure 10.4, which indicates that there is practically no correlation between acres and income, one might conclude that differences in income were not at all associated with differences in acreage; yet when the variation in income associated with cows is removed, even by the rough method shown, a very definite relation of income to size is found. For that reason one cannot conclude that, because two variables have no correlation, they are not associated with each other; the lack of correlation may be due to the compensating influence of one or more other variables, concealing the real relation.

ELIMINATING THE APPROXIMATE INFLUENCE OF BOTH VARIABLES. We now have two equations, one showing the effect of cows upon income and the other the effect of acres:

(A) Income = 694 + 20.11 × number of cows
(B) Income, effect of cows removed,
   = 508.51 + 1.33 × number of acres

These two equations can be combined into a single equation by taking that part of the first one which shows the increase in income for each cow and adding it to the second one. This gives an equation which includes allowances for both factors, as follows:

(C) Income = 508.51 + 1.33 × number of acres
   + 20.11 × number of cows

The last equation gives a basis for indicating the effect of both acres and cows on income and for computing the income that might be expected, on the average, with a farm of a given size and with a given number of cows. For example, for a farm of 120 acres and 15 cows, the expected income would work out as follows:

Income = 508.51 + 1.33(120) + 20.11(15)
   = 508.51 + 159.60 + 301.65 = 970

If 5 cows were added, making it 120 acres and 20 cows, the estimated income would be:

Income = 508.51 + 1.33(120) + 20.11(20)
   = 1,070

Equation (C) can be used as illustrated to work out what income might be expected for each farm. The estimated income can then be compared with the actual income and the difference, if any, determined.

As given in Table 10.7, the estimated incomes still vary somewhat from the actual. This is just another way of saying that all the differences in income cannot be accounted for by the effect of differences in acres and in cows, according to the relations summarized in equation (C). This failure of the estimated values to agree exactly with the original values is seen graphically in Figure 10.7 by the fact that all the dots do not lie

Table 10.7

ACTUAL INCOME AND INCOME ESTIMATED FROM NUMBER OF ACRES AND COWS

Acres	Cows	Estimate for Acres, 1.33 × acres (A)	Estimate for Cows, 20.11 × cows (C)	Estimated Income, (A) + (C) +508.51	Actual Income	Actual Income Minus Estimated Income
60	18	80	362	950.5	960	9.5
220	0	293	0	801.5	830	28.5
180	14	239	282	1,029.5	1,260	230.5
80	6	106	121	735.5	610	−125.5
120	1	160	20	688.5	590	−98.5
100	9	133	181	822.5	900	77.5
170	6	226	121	855.5	820	−35.5
110	12	146	241	895.5	880	−15.5
160	7	213	141	862.5	860	−2.5
230	2	306	40	854.5	760	−94.5
70	17	93	342	943.5	1,020	76.5
120	15	160	302	970.5	1,080	109.5
240	7	319	141	968.5	960	−8.5
160	0	213	0	721.5	700	−21.5
90	12	120	241	869.5	800	−69.5
110	16	146	322	976.5	1,130	153.5
220	2	293	40	841.5	760	−81.5
110	6	146	121	775.5	740	−35.5
160	12	213	241	962.5	980	17.5
80	15	106	302	916.5	800	−116.5

exactly along the regression line. Subtracting the estimated values from the actual values gives the residual differences of the actual income above or below the income estimated from the two factors, acres and cows.

CORRECTING RESULTS BY SUCCESSIVE ELIMINATION. It may now be recalled that, even though the incomes were adjusted to eliminate the effects of cows upon income before determining the relation between income and acres, the determination of the relation between income and cows was·made without making any allowance for the concurrent effect of acres. Since we now have an approximate measure of the effect of acres determined while eliminating to some extent the effect of cows, we can use that new measure, equation (B), to adjust the incomes for the effect of the acres and then get a more accurate measure of the true effect of cows alone upon incomes. This process is shown in Table 10.8. Here

Table 10.8

ADJUSTING FARM INCOMES FOR DIFFERENCES IN NUMBER OF ACRES

Size of Farm	Size of Dairy	Income	Income Estimated for Acres, with no Cows	Income with Effects of Acreage Differences Eliminated*
Number of acres	*Number of cows*	*Dollars*	*Number of dollars*	*Number of dollars*
60	18	960	588	372
220	0	830	801	29
180	14	1,260	748	512
80	6	610	615	−5
120	1	590	669	−79
100	9	900	642	258
170	6	820	735	85
110	12	880	655	225
160	7	860	722	138
230	2	760	815	−55
70	17	1,020	602	418
120	15	1,080	669	411
240	7	960	828	132
160	0	700	722	−22
90	12	800	629	171
110	16	1,130	655	475
220	2	760	802	−42
110	6	740	655	85
160	12	980	722	258
80	15	800	615	185

* Where the actual income is below that expected for a farm of that size with no cows, the deficit is indicated by the minus sign.

estimates of income are worked out by equation (B) on the basis of acres, showing what the incomes might be expected to average if all the farms had no cows. The difference between these estimates and the actual incomes may then be considered to be the part due to cows alone, while eliminating the effect of differences in the numbers of acres. On the first farm, for example, equation (B) indicates that with no cows the income for 60 acres should be 588. Subtracting this from the 960 actually received leaves 372 as the income apparently accompanying the 18 cows.

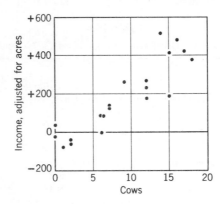

Fig. 10.8. Relation of income, adjusted for number of acres, to number of cows.

The adjusted incomes may then be plotted on a dot chart with the number of cows as the other variable, as shown in Figure 10.8. Comparing this figure with Figure 10.5, where the number of cows was plotted against income without first making any adjustment in the original incomes, we easily see how much closer the relation is after making the adjustment. Further, it is evident that cows have a greater effect upon income than was indicated by the earlier comparison. Computing the straight-line relationship for Figure 10.8 gives the following equation:

(D) Income, adjusted to constant acres,
$$= -68.77 + 27.88 \times \text{number of cows}$$

By this last computation [equation (D)], each increase of one cow is accompanied by an average increase in income of 27.88, whereas according to the earlier comparison [equation (A)], the increase was only 20.11. The second value is larger than the first, again showing the necessity of making due allowance for the effect of one factor before the true value of the other can be properly measured.

Now that we have a new measure of the effect of cows, we might go on to adjust incomes for cows by this new measure and then get a revised value for the effect of acres upon incomes on a no-cow basis, in place of the relation shown in equation (B). This possibility of further correction will be referred to later. But before that we will make some experiments with the new equation (D).

We now have equations for the relation of incomes, adjusted for the other factors, to the remaining factors. These two equations, (B) and (D), are:

(B) Income, effect of cows removed,

$$= 508.51 + 1.33 \times \text{number of acres}$$

(D) Income, adjusted to constant acres,

$$= -68.77 + 27.88 \times \text{number of cows}$$

These two equations may be combined to give a revised equation to indicate the effect of both cows and acres upon incomes:

(E) Income $= 439.74 + 1.33 \times$ number of acres

$$+ 27.88 \times \text{number of cows}$$

Equation (E) is exactly the same as the previous equation (C) except that the revised effect of cows is included, and the constant term has also been changed owing to changing the allowance for cows.

In exactly the same way that equation (C) could be used to work out the estimated income for any given combination of cows and acres, equation (E) can be used also. Thus for 120 acres and 15 cows, it would give

Estimated income $= 439.7 + 1.33(120) + 27.88(15)$

$$= 439.7 + 159.6 + 418.2 = 1{,}018$$

The result, 1,018, is 48 higher than the 970 worked out by equation (C). This higher estimate is due to the fact that equation (E) makes a larger allowance for the effect of each cow, and 15 is more than the average number of cows. If less than the average number of cows were used, equation (E) would give a lower estimate than equation (C).

Working out the estimated incomes for each of the original observations according to equation (E), we obtain results as shown in Table 10.9.

Comparing the residuals, or differences between the actual and estimated income, obtained by means of this new equation with those obtained using the equation in its first form (shown in Table 10.7), we see that in more than half the cases they are smaller with the revised form. A more definite comparison can be made by computing the standard deviation of the residuals in each case. The standard deviation of the residuals shown in Table 10.7, using equation (C), is 90.29, whereas the standard deviation of the residuals shown in Table 10.9, using equation (E), is but 78.70. It is apparent from this that the revised equation, determined after the effects of the other variable had been more closely allowed for, gives more accurate estimates of income than does the original equation in which the effects of the other variable had not been so fully eliminated.

Table 10.9

ACTUAL INCOME AND INCOME ESTIMATED FROM NUMBER OF ACRES AND NUMBER OF COWS, REVISED RELATIONS

		Computation of Estimated Income		Estimated Income, (A) + (B) +439.7	Actual Income	Actual Income Minus Estimated Income
Acres	Cows	Estimate for Acres, 1.33 × acres (A)	Estimate for Cows, 27.88 × cows (B)			
60	18	80	502	1,021.7	960	−61.7
220	0	293	0	732.7	830	97.3
180	14	239	390	1,068.7	1,260	191.3
80	6	106	167	712.7	610	−102.7
120	1	160	28	627.7	590	−37.7
100	9	133	251	823.7	900	76.3
170	6	226	167	832.7	820	−12.7
110	12	146	335	920.7	880	−40.7
160	7	213	195	847.7	860	12.3
230	2	306	56	801.7	760	−41.7
70	17	93	474	1,006.7	1,020	13.3
120	15	160	418	1,017.7	1,080	62.3
240	7	319	195	953.7	960	6.3
160	0	213	0	652.7	700	47.3
90	12	120	335	894.7	800	−94.7
110	16	146	446	1,031.7	1,130	98.3
220	2	293	56	788.7	760	−28.7
110	6	146	167	752.7	740	−12.7
160	12	213	335	987.7	980	−7.7
80	15	106	418	963.7	800	−163.7

It was suggested previously that the last corrected values for the relation of cows to income gave a new basis for correcting income so as to measure more accurately the relation of acres to income. This in turn would give a new basis for measuring the effect of cows, and so on, until a final stable value had been reached. So long as a new correction would result in a further change in the computed effect of either variable, the new values would give a better basis for estimating income than did the previous values. Only when the point was reached where no further change need be made in the effect of either variable could it be said that the relation of each variable to income had been quite correctly measured while allowing for the influence of the other factor, and that might involve a large number of successive corrections.

This method of allowing for the effect of other factors so as to determine the true relation of each one to the dependent factor (as income, in this case), by first correcting for one, and then for another, is known as the *method of successive elimination.* This method can be used where there are three or more independent factors related to (or accompanying variations in) a dependent (or resultant) factor, just as it was used here for two factors, except that then the dependent needs to be corrected in turn to eliminate the effects of all the other independent factors except the particular one whose effect is being measured. But although it is possible to measure the relations by this method, it would be a very slow and laborious process. A shorter mathematical method which gives the same result by more direct processes is available instead. This method, known as the method of multiple regression, is presented in Chapter 11.

Summary

This chapter has shown that when two related factors both affect a third factor it is difficult to measure the extent to which the dependent variable is associated with one independent factor without making allowances for its relation to the other independent variable. Allowing for this duplication by eliminating the effects of each factor in turn (successive elimination) can gradually determine the net association with each, but the method is long and laborious.

REFERENCE

Foote, Richard J., The mathematical basis for the Bean Method of graphic multiple correlation, *Jour. Amer. Stat. Assoc.*, Vol. 48, pp. 778–88, 1953.

CHAPTER II

Determining multiple regressions: (2) by fitting a linear regression equation

In equation (E) in Chapter 10 it was shown that an equation could be arrived at to express the average relation between income, acres, and cows, as follows:

Income $= 439.74 + 1.33 \times$ number of acres $+ 27.88 \times$ number of cows

If we designate the three series of variable quantities, income, acres, and cows, by the symbol X with different subscripts, using X_1 to represent dollars of income, X_2 to represent number of acres, and X_3 to represent the number of cows, we can rewrite the equation in the form

$$X_1 = 439.74 + 1.33X_2 + 27.88X_3$$

If now we use the symbol a to represent the constant quantity 439.74; b_2 to represent 1.33, the amount which X_1 increases for each increase of one unit in X_2 (one acre); and b_3 to represent 27.88, the amount which X_1 increases for each increase of one unit in X_3 (one cow); the equation appears as

$$X_1 = a + b_2X_2 + b_3X_3 \tag{11.1}$$

Comparing this equation with the regression equation for the straight-line relation between two variables

$$Y = a + bX$$

we see that the two equations are just alike, except for the difference in the symbols used to represent the different variables and for our having added the expression for an additional variable. In equation (11.1), X_1, the variable which is being estimated, is termed the *dependent* variable, since its estimated value depends upon those of the other variable or variables; and X_2 and X_3 are termed *independent* variables, since their values are taken just as observed, independent of any of the conditions of the problem. Since there is more than one independent variable concerned, the equation is said to be a multiple estimating equation, or a *multiple linear regression equation*.

Chapter 10 showed that the values of the constants a, b_2, and b_3, could be worked out by a cut-and-try method which gradually approached nearer and nearer to the right values. For any particular criterion of "rightness" only one set of values for these constants can be exactly right. If the criterion of "rightness" is taken as that which will make the standard deviation of the residuals, when income is estimated from the other two variables, as small as possible, the values of a, b_2, and b_3 which will give this result can be determined by a direct mathematical process, known as the method of *linear multiple regression*.

Determining a Regression Equation for Two Independent Variables. The best values for a, b_2, and b_3 in the multiple regression equation (11.1) can be worked out by an extension of the same process used in working out the values for the estimating equation when only one independent variable was considered. Just as before, the value of the b constants will be determined first from equation (11.2) and then the a values will be worked out from them:[1]

$$\left. \begin{array}{l} \Sigma(x_2^2)b_2 + \Sigma(x_2x_3)b_3 = \Sigma(x_1x_2) \\ \Sigma(x_2x_3)b_2 + \Sigma(x_3^2)b_3 = \Sigma(x_1x_3) \end{array} \right\} \tag{11.2}$$

$$a = M_1 - b_2M_2 - b_3M_3 \tag{11.3}$$

Here, just as in Chapter 5, the symbol M represents the mean value of each variable, and the subscript indicates the particular variable.

Similarly, the symbols $\Sigma(x_2x_3)$, $\Sigma(x_1x_2)$, and $\Sigma(x_1x_3)$ represent the sums of the products of the variables, corrected to adjust them to deviations from the mean; that is, $\Sigma(x_1x_2) = \Sigma[(X_1 - M_1)(X_2 - M_2)]$. Likewise the symbols $\Sigma(x_2^2)$, etc., represent the sums of the squares of the variables also adjusted to deviations from the mean.

Using the two basic formulas

$$\Sigma(x_1x_2) = \Sigma(X_1X_2) - nM_1M_2 \tag{5.4}$$

and

$$\Sigma(x_2^2) = \Sigma(X_2^2) - n(M_2^2)$$

---

[1] These are the *normal equations* for two independent variables, corresponding to the normal equations for one independent variable given on p. 62, in the footnote.

the other values shown in equations (11.2) may be worked out as follows:

$$\Sigma(x_1 x_3) = \Sigma(X_1 X_3) - n M_1 M_3$$
$$\Sigma(x_2 x_3) = \Sigma(X_2 X_3) - n M_2 M_3$$
$$\Sigma(x_3^2) = \Sigma(X_3^2) - n(M_3^2)$$

COMPUTING THE EXTENSIONS. Inspection of these equations shows that there are eight arithmetic values which must be computed from the original data to work out the values to substitute in equations (11.2) and (11.3). These are ΣX_1, ΣX_2, ΣX_3, $\Sigma(X_2^2)$, $\Sigma(X_3^2)$, $\Sigma(X_1 X_2)$, $\Sigma(X_1 X_3)$, and $\Sigma(X_2 X_3)$. The work of computing these values for the farm-income data originally presented in Table 10.3 is shown in Table 11.1. [The value

Table 11.1

COMPUTATION OF VALUES TO DETERMINE MULTIPLE REGRESSION EQUATION
TO ESTIMATE ONE VARIABLE FROM TWO OTHERS

Acres,* X_2 (1)	Cows, X_3 (2)	Dollars Income,* X_1 (3)	$X_2{}^2$ (4)	$X_2 X_3$ (5)	$X_1 X_2$ (6)	$X_3{}^2$ (7)	$X_1 X_3$ (8)	$X_1{}^2$ (9)
6	18	96	36	108	576	324	1,728	9,216
22	0	83	484	0	1,826	0	0	6,889
18	14	126	324	252	2,268	196	1,764	15,876
8	6	61	64	48	488	36	366	3,721
12	1	59	144	12	708	1	59	3,481
10	9	90	100	90	900	81	810	8,100
17	6	82	289	102	1,394	36	492	6,724
11	12	88	121	132	968	144	1,056	7,744
16	7	86	256	112	1,376	49	602	7,396
23	2	76	529	46	1,748	4	152	5,776
7	17	102	49	119	714	289	1,734	10,404
12	15	108	144	180	1,296	225	1,620	11,664
24	7	96	576	168	2,304	49	672	9,216
16	0	70	256	0	1,120	0	0	4,900
9	12	80	81	108	720	144	960	6,400
11	16	113	121	176	1,243	256	1,808	12,769
22	2	76	484	44	1,672	4	152	5,776
11	6	74	121	66	814	36	444	5,476
16	12	98	256	192	1,568	144	1,176	9,604
8	15	80	64	120	640	225	1,200	6,400
Sums 279	177	1,744	4,499	2,075	24,343	2,243	16,795	157,532
Means 13.95	8.85	87.2						
Adjustment item			3,892.05	2,469.15	24,328.80	1,566.45	15,434.40	152,076.80
Adjusted sums			606.95	−394.15	14.20	676.55	1,360.60	5,455.20

* In these computations, X_2 and X_1 have been divided by 10.

$\Sigma(X_1^2)$ is not needed in solving equations (11.2) or (11.3); but, as it will be needed later, it is also worked out here for convenience in calculation.][2]

After we have multiplied all the extensions shown in this table, and added each of the columns, our next step is to compute the values M_2, M_3, and M_1, by dividing the sums of each of the first three columns by

[2] Alternative methods of making the computations and solving the equations, and of controlling the computations to prevent error, are shown in Appendix 2, pp. 489 to 520.

the number of cases. The adjustment item for each of the products is then computed and entered below the value from which it is to be subtracted. Thus the value below the sum of the fourth column, $\Sigma(X_2^2)$, is its adjustment, $n(M_2^2)$. This is equal to $20(13.95)^2$, or 3,892.05, which is the value entered. Similarly, the value below the sum of the fifth column, $\Sigma(X_2X_3)$, is its adjustment, $n(M_2M_3)$, or $20(8.85)(13.95)$, which equals 2,469.15. All the other adjustments are similarly worked out and entered. Then subtracting each adjustment from the value above it gives the values all ready for equations (11.2). Thus the value at the foot of column 4 is the value for $\Sigma(x_2^2)$; and so on. When these values are substituted in the appropriate spaces of equation (11.2), they become

(I) $\quad \Sigma(x_2^2)b_2 + \Sigma(x_2x_3)b_3 = \Sigma x_1x_2$ $\left.\begin{array}{l} \\ \\ \end{array}\right\} = \left\{ \begin{array}{l} 606.95b_2 - 394.15b_3 = 14.20 \\ -394.15b_2 + 676.55b_3 = 1360.60 \end{array}\right.$

(II) $\quad \Sigma(x_2x_3)b_2 + \Sigma(x_3^2)b_3 = \Sigma x_1x_3$

SOLVING THE EQUATIONS. The next step is to solve the two algebraic equations simultaneously to determine the values for b_2 and b_3.

One simple way to carry this through is by the Doolittle method. The first equation is divided by the coefficient of b_2, with the sign changed, giving the first derived equation (I′):

(I) $\qquad\qquad 606.95b_2 - 394.15b_3 = 14.20$

(I′) $\qquad\qquad -b_2 + 0.64939b_3 = -0.02340$

Then equation (II) is entered, and under it is written equation (I) multiplied by the coefficient of b_3 in equation (I′), 0.64939. The sum of these two equations is then taken, eliminating the values in b_2:

(II) $\qquad -394.15b_2 + 676.55b_3 = 1360.60$

(0.64939) (I) $\qquad +394.15b_2 - 255.96b_3 = 9.22$

(ΣII) $\qquad\qquad\qquad 420.59b_3 = 1369.82$

(II′) $\qquad\qquad\qquad\qquad b_3 = 3.25690$

As indicated above, this step gives the value of b_3. This is then substituted in equation (I′) and the value of b_2 determined:

$$-b_2 + 0.64939(3.25690) = -0.02340$$
$$b_2 = 0.02340 + 2.11500 = 2.13840$$

The values of b_2 and b_3 being thus obtained, the next step is to substitute them, together with the other values required, in equation (11.3) to work out the value for a:

$$a = M_1 - b_2M_2 - b_3M_3$$
$$= 87.2 - (2.1384)(13.95) - (3.2569)(8.85)$$
$$= 87.2 - 29.83 - 28.82 = 28.55$$

ESTIMATING X_1 FROM X_2 AND X_3. Having computed the values for a, b_2, and b_3, we can now write out our regression equation (11.1), with the *best* values, as determined by the mathematical calculation:

$$\left(\frac{X_1}{10}\right) = 28.55 + 2.1384\left(\frac{X_2}{10}\right) + 3.2569X_3$$

$$X_1 = 285.5 + 2.1384X_2 + 32.569X_3$$

Comparing this equation with the last one obtained in Chapter 10, (page 167), we see that the least-squares calculation has changed the 1.33 allowed for the effect of each acre (b_2) to 2.14, and increased the 27.88 allowed for the effect of each cow (b_3) to 32.57. Just what effect this has on the accuracy of the equation as a basis for estimating income

Table 11.2

ACTUAL INCOME AND INCOME ESTIMATED FROM NUMBER OF ACRES AND COWS, ON BASIS OF MATHEMATICALLY DETERMINED RELATIONS

Acres, X_2	Cows, X_3	Estimated for Acres, b_2X_2	Estimated for Cows, b_3X_3	Constant, a	Estimated Income, X_1'	Actual Income, X_1	Actual Minus Estimated Income, $X_1 - X_1'$ z
60	18	128	586	286	1,000	960	−40
220	0	470	· · ·	286	756	830	74
180	14	385	456	286	1,127	1,260	133
80	6	171	195	286	652	610	−42
120	1	257	33	286	576	590	14
100	9	214	293	286	793	900	107
170	6	364	195	286	845	820	−25
110	12	235	391	286	912	880	−32
160	7	342	228	286	856	860	4
230	2	492	65	286	843	760	−83
70	17	150	554	286	990	1,020	30
120	15	257	489	286	1,032	1,080	48
240	7	513	228	286	1,027	960	−67
160	0	342	· · ·	286	628	700	72
90	12	192	391	286	869	800	−69
110	16	235	521	286	1,042	1,130	88
220	2	470	65	286	821	760	−61
110	6	235	195	286	716	740	24
160	12	342	391	286	1,019	980	−39
80	15	171	489	286	946	800	−146

from cows and acres may be judged by working out an estimated income for each of the 20 cases according to these last results, and then comparing the estimated values with the original values, just as was done before with the equations worked out by the approximation method. The necessary computation is shown in Table 11.2.

The operations that have been performed in this table may be mathematically stated as follows:

First, an estimated value of income, X_1, has been worked out by substituting in equation (11.1) the values for X_2 and X_3 given by each successive observation. Using the symbol X_1' to represent this estimated value of X_1 it may be defined

$$X_1' = a + b_2 X_2 + b_3 X_3 \tag{11.4}$$

Each estimated income has next been subtracted from the corresponding actual income. With the symbol z used to represent the *residual*, the amount by which the actual value exceeds or falls below the estimated value, it may be defined

$$z = X_1 - X_1' \tag{11.5}$$

The residual z has exactly the same meaning when the estimated values of the dependent variable are based upon two or more variables, using multiple regression, as it had previously when the estimate was based on a single variable, with simple regression.

The accuracy of the last estimating equation, derived by an exact mathematical process, can now be compared with the accuracy of previous equations, obtained by a cut-and-try process. Computing the standard deviation of the residuals shown in this last table and comparing it with the standard deviations of the residuals worked out in Tables (10.7) and (10.9), we find the comparison to be:

Standard deviations of residuals using various straight-line equations:

First approximation equation, $s_z = 90.29$

Second approximation equation, $s_z = 78.70$

Mathematically determined equation, $s_z = 70.48$

The equation determined mathematically gives closer estimates of the actual incomes from which it was derived than do either of the two previous equations. This will always hold true. The mathematically determined equation gives once and for all the estimates of X_1 which will make s_z the smallest that can be obtained, assuming linear relations. The best that could be done by the approximation method would be to obtain the same conclusions as would be obtained by the other method. The successive steps in Chapter 10 have shown how difficult it is to do this when the several independent variables are correlated with each other, and so tend to vary with one another. The mathematical method for determining the estimating equation, as illustrated in this chapter (or some alternative form of computation involving the same principle), has therefore been practically universally adopted as the standard way of determining the precise way in which one variable is related to, or may be estimated from, two or more variables related among themselves, if only straight-line relations are to be assumed.

NOMENCLATURE IN MULTIPLE LINEAR REGRESSION. When the constants of the estimating equation are determined by the exact mathematical process, the equation is called a *multiple regression equation*, and the constants b_2 and b_3, which show, in this case, the average increase in income (X_1) associated with unit increases in acres (X_2), and cows (X_3) are termed *net regression coefficients*. The constant b_2 is termed *the net regression of X_1 on X_2, holding X_3 constant*, and b_3 is termed *the net regression of X_1 on X_3, holding X_2 constant*. All that that means for b_2, for example, is "the average change observed in X_1 with unit changes in X_2, determined while simultaneously eliminating from X_1 any variation accompanying (hence temporarily assumed due to) changes in X_3."[3]

In order that the mathematical notation for the net regression coefficients may show quite clearly which independent variables were held constant when a particular coefficient was determined, the subscripts under the b are sometimes more elaborate, showing first the dependent variable, then the independent variable whose effect is stated, then a period followed by the independent variables which were held constant in the process. Thus the b_2 we have been using would be written $b_{12.3}$. The whole regression equation would appear

$$X_1 = a_{1.23} + b_{12.3}X_2 + b_{13.2}X_3 \qquad (11.6)$$

This notation serves to distinguish these net regression coefficients from those which would be obtained if additional independent variables were included. Thus if a third independent variable, say X_4, were also considered, the equation would read

$$X_1 = a_{1.234} + b_{12.34}X_2 + b_{13.24}X_3 + b_{14.23}X_4 \qquad (11.7)$$

For still another variable it would be

$$X_1 = a_{1.2345} + b_{12.345}X_2 + b_{13.245}X_3 + b_{14.235}X_4 + b_{15.234}X_5 \quad (11.8)$$

The notation for a is changed as well as for each of the b's; $a_{1.234}$ will probably be a different value from $a_{1.23}$, just as $b_{12.34}$ is likely to be somewhat different from $b_{12.3}$.

[3] The term *partial regression coefficient* is used by some authors in place of *net regression coefficient*.

Determining a Regression Equation for Three Independent Variables.
Solely to illustrate the method, we may take the number of men on each
of these 20 farms as given in Table 11.3 and work out an estimating
equation considering men as well as acres and cows. (In practice, 20
observations are often too few to determine, with a satisfactory degree of
reliability, the net relations of one variable to three independent variables.)

Table 11.3

COMPUTATION OF ADDITIONAL VALUES TO DETERMINE MULTIPLE REGRESSION
EQUATION, ADDING A THIRD INDEPENDENT FACTOR

Item Number	Acres,* X_2	Cows, X_3	Men, X_4	Dollars Income,* X_1	X_2X_4	X_3X_4	X_1X_4	X_4^2
1	6	18	2	96	12	36	192	4
2	22	0	3	83	66	0	249	9
3	18	14	4	126	72	56	504	16
4	8	6	1	61	8	6	61	1
5	12	1	1	59	12	1	59	1
6	10	9	1	90	10	9	90	1
7	17	6	3	82	51	18	246	9
8	11	12	2	88	22	24	176	4
9	16	7	2	86	32	14	172	4
10	23	2	3	76	69	6	228	9
11	7	17	2	102	14	34	204	4
12	12	15	3	108	36	45	324	9
13	24	7	4	96	96	28	384	16
14	16	0	2	70	32	0	140	4
15	9	12	1	80	9	12	80	1
16	11	16	3	113	33	48	339	9
17	22	2	2	76	44	4	152	4
18	11	6	1	74	11	6	74	1
19	16	12	2	98	32	24	196	4
20	8	15	2	80	16	30	160	4
Sums	279	177	44	1,744	677	401	4,030	114.00
Means	13.95	8.85	2.2	87.2				
Adjustment items					613.80	389.40	3836.80	96.80
Adjusted sums					63.20	11.60	193.20	17.20

* Coded by dividing by 10.

With the number of men designated as X_4, the unknown constants to be determined are those given in equation (11.7); $a_{1.234}$, $b_{12.34}$, $b_{13.24}$, and $b_{14.23}$. They can be obtained by the solution of the following set of equations.

$$\left. \begin{aligned} \Sigma(x_2^2)b_{12.34} \; + \Sigma(x_2x_3)b_{13.24} + \Sigma(x_2x_4)b_{14.23} = \Sigma(x_1x_2) \\ \Sigma(x_2x_3)b_{12.34} + \Sigma(x_3^2)b_{13.24} \; + \Sigma(x_3x_4)b_{14.23} = \Sigma(x_1x_3) \\ \Sigma(x_2x_4)b_{12.34} + \Sigma(x_3x_4)b_{13.24} + \Sigma(x_4^2)b_{14.23} \; = \Sigma(x_1x_4) \end{aligned} \right\} \quad (11.9)$$

$$. \, a_{1.234} = M_1 - b_{12.34}M_2 - b_{13.24}M_3 - b_{14.23}M_4 \qquad (11.10)$$

COMPUTING THE EXTENSIONS. All except 4 of the arithmetic values for equations (11.9) which need to be calculated from the original data have been worked out previously. Only the values which involve X_4, and its mean, are additional. The new values needed are therefore M_4, $\Sigma(x_1x_4)$, $\Sigma(x_2x_4)$, $\Sigma(x_3x_4)$, and $\Sigma(x_4^2)$. The computation of these values is shown in Table 11.3.

All the calculations, including correcting for the means at the end, are carried out just as in Table 11.1. The figures at the foot of each column provide the remaining values necessary to write out equations (11.9) in full. For convenience in writing these equations, we shall again use the abridged notation of b_2 for $b_{12.34}$, b_3 for $b_{13.24}$, etc., remembering, however, that b_2 here is a different constant from b_2 previously.

$$\left. \begin{aligned} &\text{(I)} \quad \Sigma(x_2^2)b_2 + \Sigma(x_2x_3)b_3 \\ &\qquad + \Sigma(x_2x_4)b_4 = \Sigma(x_1x_2) \\[4pt] &\text{(II)} \; \Sigma(x_2x_3)b_2 + \Sigma(x_3^2)b_3 \\ &\qquad + \Sigma(x_3x_4)b_4 = \Sigma(x_1x_3) \\[4pt] &\text{(III)} \; \Sigma(x_2x_4)b_2 + \Sigma(x_3x_4)b_3 \\ &\qquad + \Sigma(x_4^2)b_4 \; = \Sigma(x_1x_4) \end{aligned} \right\} = \left\{ \begin{aligned} &606.95b_2 - 394.15b_3 \\ &\qquad + 63.20b_4 = 14.20 \\[4pt] &-394.15b_2 + 676.55b_3 \\ &\qquad + 11.60b_4 = 1360.60 \\[4pt] &63.20b_2 + 11.60b_3 \\ &\qquad + 17.20b_4 = 193.20 \end{aligned} \right.$$

SOLVING THE EQUATIONS. The three equations are now to be solved simultaneously to determine the values for b_2, b_3, and b_4. This can be done by the usual algebraic processes, but the peculiar symmetrical character of the equations, which the attentive reader has probably already noticed, makes it possible to use a much shorter method. Since the saving in clerical labor by the use of this method is quite significant, it will be shown in full.

The first step is to set down the first equation (I) and divide it by the coefficient of the first term, Σx_2^2, *with the sign changed*, or -606.95 in this case. The resulting derived equation (I′) is set down just below it:

(I) $\qquad\qquad 606.95b_2 - 394.15b_3 + 63.20b_4 = 14.20$

(I′) $\qquad\qquad\quad -b_2 + 0.64939b_3 - 0.10413b_4 = -0.02340$

The next step is to set down the second equation (II). The first equation (I) is then multiplied by the coefficient of the *second term* in the derived equation (I′), which is +0.64939 in this case, and the products set down just below equation (II). These two equations are added, giving the sum equation (Σ_2), which cancels out the first term, as shown below. The sum equation is then divided by the coefficient of its first term, with the sign changed, giving the second derived equation (II′). The second portion of the work now appears as follows:

$$\text{(II)} \qquad -394.15b_2 + 676.55b_3 + 11.60b_4 \;=\; 1360.60$$

$$\text{(0.64939) (I)} \qquad 394.15b_2 - 255.96b_3 + 41.04b_4 \;=\; 9.22$$

$$(\Sigma_2) \qquad\qquad 420.59b_3 + 52.64b_4 \;=\; 1369.82$$

$$(\text{II}′) \qquad\qquad -b_3 - 0.12516b_4 \;=\; -3.25690$$

The final step in the process of elimination is to write down equation (III), multiply the first equation (I) by the coefficient of the *third* term of the first derived equation (I′), which is −0.10413 in this case, and set the products down below equation (III); multiply the sum equation (Σ_2) by the corresponding coefficient (the second term) from the second derived equation (II′), −0.12516; and set these products down below the previous equation. Equation (III) and the two new equations are then added, giving an equation (Σ_3), from which values in both b_2 and b_3 have been eliminated. This equation is then divided by the coefficient of its first term, with the sign changed, −4.03 in this case, and the resulting new derived equation entered as equation (III′). (A method of checking each step in these computations is shown in Appendix 2, p. 493.) All the computations to this point are:

$$\text{(I)} \qquad 606.95b_2 - 394.15b_3 + 63.20b_4 \;=\; 14.20$$

$$\text{(I}′) \qquad -b_2 + 0.64939b_3 - 0.10413b_4 \;=\; -0.02340$$

$$\text{(II)} \qquad -394.15b_2 + 676.55b_3 + 11.60b_4 \;=\; 1360.60$$

$$\text{(0.64939) (I)} \qquad 394.15b_2 - 255.96b_3 + 41.04b_4 \;=\; 9.22$$

$$(\Sigma_2) \qquad\qquad 420.59b_3 + 52.64b_4 \;=\; 1369.82$$

$$(\text{II}′) \qquad\qquad -b_3 - 0.12516b_4 \;=\; -3.25690$$

$$\text{(III)} \qquad 63.20b_2 + 11.60b_3 + 17.20b_4 \;=\; 193.20$$

$$\text{(−0.10413) (I)} \qquad -63.20b_2 + 41.04b_3 - 6.58b_4 \;=\; -1.48$$

$$\text{(−0.12516) } (\Sigma_2) \qquad -52.64b_3 - 6.59b_4 \;=\; -171.45$$

$$(\Sigma_3) \qquad\qquad\qquad 4.03b_4 \;=\; 20.27$$

$$(\text{III}′) \qquad\qquad\qquad -b_4 \;=\; -5.02978$$

It is now very easy to compute the values of b_2, b_3, and b_4, from the three derived equations. From equation (III'), $b_4 = 5.02978$.

Substituting this value in equation (II'), which may be transposed to read

$$b_3 = 3.25690 - 0.12516b_4$$

we find

$$b_3 = 3.25690 - (0.12516)(5.02978)$$

$$= 3.25690 - 0.62953 = 2.62737$$

Then, transposing equation (I'), we find

$$b_2 = 0.02340 + 0.64939b_3 - 0.10413b_4,$$

and substituting the values for b_3 and b_4,

$$b_2 = 0.02340 + (1.70619) - (0.52375),$$

we find

$$b_2 = 1.20584$$

The values of b_2, b_3, and b_4, just computed, may next be verified by substituting them in the last equation (III). *Equations* (I) *or* (II) *should not be used for this verification, since they will not provide a complete check.* Equation (III), $63.20b_2 + 11.60b_3 + 17.20b_4 = 193.20$, which becomes, when the newly calculated values are substituted,

$$(63.20)(1.20584) + (11.60)(2.62737) + (17.20)(5.02978) = 193.20$$

This works out to $76.21 + 30.48 + 86.51 = 193.20$, or $193.20 = 193.20$. This proves the accuracy of all the previous work.

The work just summarized is all that is needed to solve these three simultaneous equations. In view of the way the terms cancel out during the second and subsequent steps of the process, the work can be still further simplified by omitting all entries to the left of the solid line which has been drawn in through the last full set of computations.

Having calculated the values of the three b's, we can calculate a very readily.

$$a = M_1 - b_2M_2 - b_3M_3 - b_4M_4$$

$$= 87.2 - (1.20584)(13.95) - (2.62737)(8.85) - (5.02978)(2.20)$$

$$= 36.06$$

The regression equation for the three variables is therefore

$$\left(\frac{X_1}{10}\right) = 36.06 + 1.20584\left(\frac{X_2}{10}\right) + 2.62737X_3 + 5.02978X_4$$

If we clear the fractions, the equation becomes

$$X_1 = 360.60 + 1.20584X_2 + 26.2737X_3 + 50.2978X_4$$

Using this equation, we may work out values of X_1 and of z just as we did previously. (This will be left as an exercise for the student. Is s_z for the new estimates larger or smaller than for the previous estimates? Why should it be?)

INTERPRETING NET REGRESSION COEFFICIENTS. It should be noted that though the value of 1.20584 for $b_{12.34}$, just determined, compares with the value of 2.13840, for $b_{12.3}$, determined previously, they do not measure exactly the same thing. The coefficient $b_{12.34}$ shows the average increase in income for each acre increase in size of farm, with both the number of *cows* and the number of *men* remaining unchanged. The coefficient $b_{12.3}$ shows the average increase in income for each increase of one acre in size, with the number of *cows* remaining unchanged, but without making any allowance for differences in the number of men. Apparently a considerable portion of the differences in income which on the earlier analysis would have been ascribed to the additional acreage is shown by this more complete analysis really to have been associated with the larger labor force on the greater acreages, rather than to the greater acreages themselves. This result illustrates one property of net regression coefficients in common with all other regression results. They ascribe to any particular independent variable not only the variation in the dependent variable which is directly due to that independent variable but also the variation which is due to such other independent variables correlated with it as have not been separately considered in the study. In the same way that acres, taken alone, included part of the effect due to cows, the effect of acres eliminating cows still included part of the effect due to men; and even the effect of acres holding constant the effect of both cows and men may still include variation due to other variables correlated with them, such, for example, as fertility of the land. These considerations illustrate the extreme care which is necessary in examination of the data and the theoretical analysis of the problem before deciding on the variables to be correlated, and the caution which must be employed in interpreting the results.

Determining the Regression Equation for Any Number of Independent Variables. The same mathematical principle which has been used to determine the constants for regression equations involving one, two, or three independent variables can be extended to problems involving any number of variables it may be desired to employ.

For four independent variables the equations are:

$$
\left.
\begin{aligned}
\Sigma(x_2^2)b_{12.345} &+ \Sigma(x_2x_3)b_{13.245} + \Sigma(x_2x_4)b_{14.235} \\
&+ \Sigma(x_2x_5)b_{15.234} = \Sigma(x_1x_2) \\[4pt]
\Sigma(x_2x_3)b_{12.345} &+ \Sigma(x_3^2)b_{13.245} + \Sigma(x_3x_4)b_{14.235} \\
&+ \Sigma(x_3x_5)b_{15.234} = \Sigma(x_1x_3) \\[4pt]
\Sigma(x_2x_4)b_{12.345} &+ \Sigma(x_3x_4)b_{13.245} + \Sigma(x_4^2)_{14.235} \\
&+ \Sigma(x_4x_5)b_{15.234} = \Sigma(x_1x_4) \\[4pt]
\Sigma(x_2x_5)b_{12.345} &+ \Sigma(x_3x_5)b_{13.245} + \Sigma(x_4x_5)b_{14.235} \\
&+ \Sigma(x_5^2)b_{15.234} = \Sigma(x_1x_5)
\end{aligned}
\right\}
\qquad (11.11)
$$

$$
a_{1.2345} = M_1 - b_{12.345}M_2 - b_{13.245}M_3 - b_{14.235}M_4 - b_{15.234}M_5 \qquad (11.12)
$$

When this set of equations is compared with equations (11.9) for three independent variables, it is evident that adding the additional variable, X_5, has made it necessary to add the additional equation, in which X_5 appears in each of the product terms, and also to add an additional term to each of the previous equations, the additional term including a product summation [such as $\Sigma(x_2x_5)$ and $\Sigma(x_3x_5)$] in which X_5 appears, and also the net regression coefficient $b_{15.234}$. The equation to compute a has also been extended by adding the term $-b_{15.234}M_5$. In the same way the equations to be solved to determine the constants for any number of variables can be built up, if it is remembered that for each variable added a new term must be added to each of the previous equations and a new equation must be added, each term added including the new variable in some way.

The products which must be computed for any given set of variables, and the equations which will need to be solved, may be worked out readily by the use of the following scheme:

Write out the required regression equation (in terms of deviations from the mean), as, for example, for six variables:

$$
b_2x_2 + b_3x_3 + b_4x_4 + b_5x_5 + b_6x_6 = x_1
$$

Multiply each term by the coefficient of the first unknown (that is, by x_2) and sum. This gives the first of the required equations:

$$
\Sigma(x_2^2)b_2 + \Sigma(x_2x_3)b_3 + \Sigma(x_2x_4)b_4 + \Sigma(x_2x_5)b_5 + \Sigma(x_2x_6)b_6 = \Sigma(x_2x_1)
$$

Then multiply by the coefficient of the second unknown (x_3) and sum. The second equation is, therefore,

$$
\Sigma(x_2x_3)b_2 + \Sigma(x_3^2)b_3 + \Sigma(x_3x_4)b_4 + \Sigma(x_3x_5)b_5 + \Sigma(x_3x_6)b_6 = \Sigma(x_3x_1)
$$

The same process is carried out for the coefficient of each unknown in turn, giving five equations to be solved simultaneously to determine the values for the five unknowns. Setting up these equations may be reduced to a tabular form, as in Table 11.4.

The variables to be considered are listed at the head of columns from the left to right, ending with the dependent variable at the right. Then the independent variables are entered down the beginning of the lines

Table 11.4

FORM FOR WORKING OUT THE EQUATIONS TO DERIVE
NET REGRESSION CONSTANTS

Independent Variables	Independent Variables (in Deviations from Means)							Dependent Variable,
	x_2	x_3	x_4	x_5	x_6	x_7	x_8	x_1
x_2	$\Sigma(x_2^2)b_2$	$\Sigma(x_2x_3)b_3$	$\Sigma(x_2x_4)b_4$					$= \Sigma(x_1x_2)$
x_3	$\Sigma(x_2x_3)b_2$	$\Sigma(x_3^2)b_3$	$\Sigma(x_3x_4)b_4$					$= \Sigma(x_1x_3)$
x_4	$\Sigma(x_2x_4)b_2$	$\Sigma(x_3x_4)b_3$	$\Sigma(x_4^2)b_4$					$= \Sigma(x_1x_4)$
x_5	$\Sigma(x_2x_5)b_2$	$\Sigma(x_3x_5)b_3$	$\Sigma(x_4x_5)b_4$					$= \Sigma(x_1x_5)$
x_6	$\Sigma(x_2x_6)b_2$	$\Sigma(x_3x_6)b_3$	$\Sigma(x_4x_6)b_4$					$= \Sigma(x_1x_6)$
x_7	$\Sigma(x_2x_7)b_2$	$\Sigma(x_3x_7)b_3$	$\Sigma(x_4x_7)b_4$					$= \Sigma(x_1x_7)$
x_8	$\Sigma(x_2x_8)b_2$	$\Sigma(x_3x_8)b_3$	$\Sigma(x_4x_8)b_4$					$= \Sigma(x_1x_8)$

at the left in the same order. The cells of the table are then filled by multiplying the variable at the head of the column by the variable at the end of the line. These products indicate the values to be computed [by equations (5.2) and (6.4)], to give the arithmetic values for the equations. The b terms represent, of course, the net regression coefficients for the particular number of variables concerned; that is, b_2 would be $b_{12.3}$ for two independent variables, $b_{12.34}$ from three independent variables, and so on. The illustration is carried out to seven independent variables, but the scheme can be extended to as many as it is desired to consider.

The equation to compute a is simply the value of the mean of the dependent variable, minus the product of the mean of each independent variable multiplied by the coefficient for the net regression of the dependent variable on that independent variable.

As a matter of practical procedure, it is seldom that a problem is so complicated or that enough observations are available so that significant results for each variable will be obtained using ten or more variables; and, ordinarily, analyses involving not more than five variables are all that will yield stable results. Various methods for simplifying the necessary calculations in carrying through a problem involving a large number of observations are presented in Appendix 2.

Interpreting the Multiple Regression Equation. The same limitations apply in interpreting regression coefficients worked out with the effect of one or more variables held constant as when only two variables are considered. Thus for the data shown in Table 11.3: there were no observations with more than 18 cows, or 4 men, and none below 60 acres or above 240 acres. For that reason, there is no basis for using the regression equation to estimate income beyond those limits. Furthermore, for the extreme ranges where only a few observations were available—for example, less than 80 acres—the relations could not be expected to hold as well as where there were more observations upon which to base the conclusions. In Chapter 17 a more definite basis for determining the probable accuracy of such estimates is discussed, together with ways of working out the confidence intervals for each constant which appears in the regression equation. For the present the caution may be restated, that the results may be expected to hold true only within the range covered by the bulk of the observations upon which they were based.[4]

The meaning of the regression equation

$$X_1 = 360.60 + 1.21X_2 + 26.27X_3 + 50.30X_4$$

may be made clearer, in publishing correlation results, by working out the estimated values for a representative variety of conditions. Such a statement of the conclusions covered by the previous regression equation would be made as in Table 11.5.

Table 11.5

AVERAGE INCOME ON FARMS WITH VARYING NUMBERS OF ACRES, COWS, AND MEN

(As indicated by regression analysis)

Labor Force	Income on Farms with 100 Acres			Income on Farms with 160 Acres		
	0 Cows	8 Cows	16 Cows	0 Cows	8 Cows	16 Cows
Men	*Dollars*	*Dollars*	*Dollars*	*Dollars*	*Dollars*	*Dollars*
1	532	742	952	*	*	*
2	*	792	1,003	655	865	*
3	*	*	1,053	705	915	1,125

* Omitted because of absence of observations representing this combination of factors.

[4] Even within the limits of the range of observations of each variable taken separately, there may be combinations of values of independent variables which are not represented by the data, either exactly or even approximately. Estimates for such combinations will have less reliability than for those combinations which are represented. For a fuller discussion of this source of unreliability, see Chapter 19.

It should be noted in Table 11.5 that, according to these results, increasing the number of men from 1 to 2, or from 2 to 3, will add $50 to income, no matter whether the farm has 100 acres and 8 cows, or 160 acres and 16 cows. Similarly, adding 8 more cows is indicated as having the same effect on income, no matter how large the farm is or how many men are employed. But that this conclusion has been reached is no proof that it is really true of the universe represented by the original data. Instead, such a conclusion is inherent in the linear equation (11.6, 11.7, or 11.8) which has been used. That equation necessarily assumes that an increase of one unit in any one independent variable will always be accompanied by an equal change in the dependent variable. Only insofar as the actual facts agree with that assumption can they be represented by a linear equation. Subsequent chapters (particularly 14, 16, and 21) take up methods of analysis which may be employed when this type of relation is not true, and the linear equation is therefore unable to express the facts adequately.

Net regression coefficients, computed from a sample, may vary more or less widely from the true values for the universe from which that sample is drawn. Tests to indicate the reliability of such sample results are given in Chapter 17. They should always be calculated and considered before generalizing from such sample results.

Use of Card Tabulators or Electronic Computers to Perform the Operations. If the samples are large, calculating the extensions, as shown in Tables 11.1 and 11.3, involves a good deal of hand computing even with a calculating machine. Card tabulators can speed up this process especially if they are equipped with automatic multiplying devices.[5] In the latter case, after the equations have been solved, the tabulators can also calculate the estimated values of X_1 for each observation, and compute the residuals z, as shown in Table 11.2. Electronic calculators can perform all these operations, and also can solve the equations for any number of variables within the scope of the machines, by developing a suitable set of instructions to the machines.[6]

[5] D. H. W. Allan and R. F. Attridge, The application of an IBM calculating punch to solve multiple regression problems, *Proceedings Seventh Annual Conference of the American Society for Quality Control*, pp. 521–33, 1953.

[6] J. A. C. Brown, H. S. Houthakker, and S. J. Prais, Electronic computation in economic statistics, *Journal of the American Statistical Association*, Vol. 48, pp. 414–28, September, 1953.

K. D. Tocher, Application of automatic computers to statistics, *Automatic Digital Computer*, pp. 166–78, Her Majesty's Stationery Office, London, 1954.

Gordon Spencer, Statistics and automatic computers, *Computers and Automation*, Vol. 4, No. 1, pp. 6–7, January, 1955.

D. A. Quarles, Jr., Operating notes, D F EOd (program designed to handle multiple

(footnote continued on page 186)

Since many research workers will not have such equipment available to them, full suggestions on ways to control errors and simplify the procedures in doing the work by hand are given in Appendix 2, and are summarized in Chapter 13 for one of the most efficient computing forms. These checking devices can also be used with mechanical or electronic equipment, to insure against any false functioning.

The availability of automatic or electronic equipment may lead the investigator to try out a large number of different combinations of independent variables, or to use solutions involving a very large number of variables simultaneously, in order to obtain the best fit, without regard to the logic of the equations employed. This may mislead him into obtaining results of little reliability (see page 436), or of making analyses without really studying and understanding the various series with which he is dealing. This temptation is thus a drawback to the ease of computation provided by modern computing machines.

Summary

This chapter has presented mathematical methods for determining the constants of a multiple linear regression equation, so that changes in one variable may be estimated from changes in two or more independent variables. Equations so determined afford a more exact basis for making such estimates than do linear equations obtained by any other method.

regression and correlation analysis problems on machine 701), IBM N. Y. Data Processing Center, *701 Program Library*, August, 1955.

———, Description of the printed output of the IBM 701 Multiple Regression and Correlation Analysis Program, E-03, IBM N. Y. Data Processing Center, *701 Program Library*, August, 1955.

S. J. Prais and H. S. Houthakker, The analysis of family budgets. *D.A.E. Monograph* 4, Cambridge University Press, 1955.

J. Aitchison and J. A. C. Brown, The lognormal distribution. *D.A.E. Monograph* 5, Chapter 13, Cambridge University Press, 1957.

F. S. Beckman and D. A. Quarles, Jr., Multiple regression and correlation analysis on the IBM type 701 and type 704 electronic data processing machines, *The American Statistican*, Vol. 10, No. 1, pp. 6–9, February, 1956.

———, Multiple Regression and Correlation Analysis, IBM N. Y. Scientific Computing Center, NY MR 1, 1/11, and Machine Operating Notes, 1/7, November, 1956.

———, Multiple Regression and Correlation Analysis, IBM SBC N. Y. Data Processing Center, *704 Program Library*, NY MR 2, April, 1957.

E. J. Laurie and R. W. Heald, *Data Processing Bibliography*, San Jose State College, 1958.

Malcolm R. Fisher, A sector model—the poultry industry of the U. S., *Econometrica*, Vol. 26, No. 1, pp. 37–66, January, 1958. (This is a large-scale multiple regression investigation, using electronic computation.)

Furthermore, the multiple regression equation serves to sum up all the evidence of a large number of observations in a single statement which expresses in condensed form the extent to which differences in the dependent variable tend to be associated with differences in each of the other variables.

REFERENCES

Tolley, H. R., and Mordecai Ezekiel, A method of handling multiple correlation problems, *Quart. Pub. Amer. Stat. Assoc.*, pp. 994–1003, No. 144, Vol. XVIII, December, 1923.

Ezekiel, Mordecai, The assumptions implied in the multiple regression equation, *Jour. Amer. Stat. Assoc.*, No. 151, Vol. XX, pp. 405–8, September, 1925.

Wallace, H. A., and George W. Snedecor, Correlation and machine calculation, *Iowa State College Bul.* 35, 1925.

Friedman, Joan, and Richard J. Foote, Computational methods for handling systems of simultaneous equations, U.S. Dept. of Agr., Agr. Mktg. Serv., *Agriculture Handbook No. 94*, 109 pp., November, 1955.

Williams, E. J., *Regression Analysis*, John Wiley & Sons, Inc., 1959.

Acton, Forman S., *Analysis of Straight-Line Data*, John Wiley & Sons, Inc., 1959.

CHAPTER 12

Measuring accuracy of estimate and degree of correlation for multiple linear regressions

Standard Error of Estimate. After working out equations by which values of one variable may be estimated from those for two or more independent variables, it is frequently desirable to have some measure of how closely such estimates agree with the actual values and of how closely the variation in the dependent variable is associated with the variation in the several independent variables. Attention has been called in the preceding chapters to the computation of the residuals, z, when the value of a variable is estimated from that of several others. Where the estimate is based on several independent variables the standard deviation of these residuals serves as a measure of the closeness with which the original values may be estimated or reproduced just as well as where the estimate is based on a single variable. Continuing the same terminology as before, this standard deviation is still called the "standard error of estimate." Thus for the regression equation for estimating income from known numbers of acres, cows, and men, the standard error of estimate is designated $S_{1.234}$. The subscripts $_{1.234}$ indicate that that is the standard error for variable X_1 when estimated from the independent variables X_2, X_3, and X_4.

Where the size of the sample is small in proportion to the number of variables involved, the standard deviation of the residuals for the cases included in the sample tends to have a downward bias. That is, it tends to be smaller than the standard error which would be observed if the same constant were computed from large samples drawn from the same universe.

For that reason it is necessary to adjust the square of the observed standard deviation of the residuals, s_z, before it will give an unbiased estimate of the square of the value of the standard error of estimate for estimates made for new observations drawn from the same universe. This adjustment is:

$$S^2_{1.234} = \frac{ns^2_{z1.234}}{n - m} \qquad (12.1)$$

where n = number of sets of observations in the sample,

m = number of constants in the regression equation, including a and the b's.

(Where the adjusted value for $S^2_{1.234}$ exceeds the value of s^2_1, the latter value should be used for the standard error.)

The standard errors for the equations obtained when one, two, and three independent variables were considered in the farm-income study in Chapter 11 may be summarized as follows:

Independent Variables	Observed s_z	n	m	Adjusted Standard Error
X_2	165.15*	20	2	$\bar{S}_{1.2} = 165.15$
X_2, X_3	70.48	20	3	$\bar{S}_{1.23} = 76.45$
X_2, X_3, X_4	66.77	20	4	$\bar{S}_{1.234} = 74.65$

* This value has not been shown previously. It is calculated from the data of Chapter 11.

(In this case the correlation between X_1 and X_2 is practically zero, so $s_z = s_1$. Under the rule given above, $\bar{S}_{1.2} = \bar{s}_1$.) The values tabulated in the last column illustrate the increase in the reliability of estimate as additional variables are taken into account.

So far, the standard errors of estimate (except for simple or two-variable regression) have been determined by actually working out all the estimated values, subtracting to get the individual residuals, z, and then determining their standard deviation. For linear multiple regression equations, however, a much simpler process can be used. To compute the standard deviation of the residuals by this process, all that is required in addition to the values which have been used in computing the b's is the value, $\Sigma(x_1^2)$. The formula is as follows:

$$S^2_{1.234\ldots m} = \frac{\left\{ \begin{array}{c} \Sigma(x_1^2) - [b_{12.34\ldots m}(\Sigma x_1 x_2) + b_{13.24\ldots m}(\Sigma x_1 x_3) \\ + \ldots + b_{1m.23\ldots(m-1)}(\Sigma x_1 x_m)] \end{array} \right\}}{n - m} \qquad (12.2)$$

Substituting the values for the regression equation computed with two independent variables, p. 172 and 174, the equation becomes

$$S^2_{1.23} = \frac{\Sigma(x_1^2) - [b_{12.3}(\Sigma x_1 x_2) + b_{13.2}(\Sigma x_1 x_3)]}{n - 3}$$

In terms of coded values for X_1,

$$\frac{S^2_{1.23}}{10^2} = \frac{5{,}455.20 - (2.1384)(14.20) - (3.2569)(1{,}360.60)}{20 - 3}$$

$$\frac{S_{1.23}}{10} = \sqrt{\frac{993.50}{17}} = 7.645$$

$$S_{1.23} = 76.45$$

The result is seen to be identical with the value computed (after adjustment) by the lengthy process illustrated in Table 11.2, of working out all the individual estimates, computing their standard deviation, and then adjusting by equation (12.1).

Multiple Correlation. The standard error of estimate for a multiple regression equation, just as with simple regression, measures the *closeness* with which the estimated values agree with the original values. The standard error, however, offers no measure of the *proportion* of the variation in the dependent factor which can be explained by, or is associated with, variation in the independent factor or factors. For example, in one area the farm income might be twice as variable as in another. If two or three independent factors such as those discussed came as near accounting for all the variation in incomes in one area as in the other, the standard errors of estimate would be the same in both cases. There was originally more variance in income in the one case than in the other; therefore with the same amount left unaccounted for the independent factors would have been associated with a larger proportion of the original variance, in the case where it was largest to begin with, and would have been relatively more important in that case. In simple regression, the *relative* importance of the independent factors was measured by the ratio of the standard deviation of the estimated values to the standard deviation of the actual values, and the name *coefficient of correlation* was given to this ratio. In exactly similar manner, when the estimates are based on several variables, instead of on one, the relative importance of all those variables combined may be measured by dividing the standard deviation of the estimated values by that of the original values. This ratio is named the *coefficient of multiple correlation*, since it measures the combined importance of the several independent factors as a means of explaining the differences in the dependent factor. It is a useful measure when the sample is constructed on the "correlation model" (see Chapter 17).

Using X_1' to designate the estimates of X_1 made from variables X_2, X_3, and X_4, and $R_{1.234}$ to represent the unadjusted *coefficient of multiple correlation*, the coefficient may be defined:

$$X_1' = a_{1.234} + b_{12.34}X_2 + b_{13.24}X_3 + b_{14.23}X_4 \tag{12.3}$$

$$R_{1.234} = \frac{s_{x'}}{s_1} \tag{12.4}$$

The same short formula which has been shown for computing the standard error of estimate may be employed to facilitate the computation of the coefficient of multiple correlation, using only values already involved in equation (12.2). The equation for computing the coefficient of correlation by this method is:[1]

$$R^2_{1.234\ldots m} = \frac{\left\{ \begin{array}{c} b_{12.34\ldots m}(\Sigma x_1 x_2) + b_{13.24\ldots m}(\Sigma x_1 x_3) \\ + \ldots + b_{1m.23\ldots(m-1)}(\Sigma x_1 x_m) \end{array} \right\}}{\Sigma(x_1^2)} \tag{12.5}$$

The square of the coefficient of multiple correlation, R^2, may be termed the *coefficient of multiple determination*.

The same relations hold between the coefficient of multiple correlation and the unadjusted standard error of estimate in the case of multiple correlation as in the case of simple correlation. For that reason, one of these measures may be computed from the other, whichever is determined first, according to the following equations:

$$R^2_{1.234\ldots m} = 1 - \left(\frac{S^2_{1.234\ldots m}}{s_1^2} \right) \tag{12.6}$$

$$S^2_{1.234\ldots m} = s_1^2(1 - R^2_{1.234\ldots m}) \tag{12.7}$$

Using equation (12.6) to compute the values of R from the values of S, the multiple coefficients for the three regression equations previously worked out may be stated in the following different ways:

Dependent Variable, Income	Independent Variable(s)			Standard Error of Estimate	Coefficient of Multiple Correlation	Coefficient of Multiple Determination
	Acres	*Cows*	*Men*			
X_1	X_2			165.15	0.008*	0
X_1	X_2	X_3		70.48	0.904	0.818
X_1	X_2	X_3	X_4	66.77	0.915	0.837

* Simple correlation coefficient, r_{12}.

[1] This may be computed conveniently by following the form shown on pp. 496 or 512.

It is evident that the correlation increases as the standard error decreases. Here the residual variation in each case is being compared with the same original standard deviation, so that that necessarily follows. Where different studies are being compared, however, such as two samples with widely different original deviations in the dependent variable, the standard error of estimate would not necessarily decrease as the correlation increased, since the former is an *absolute* measure whereas the latter is a *relative* measure.

If such a statement is to be made as "75 per cent of the variance in income was associated with (or related to) variances in numbers of acres farmed, of cows milked, and men hired," it is more accurate to use the coefficient of multiple determination than to use the coefficient of multiple correlation. The latter would overstate the case. This principle holds true both for simple correlation (r) and multiple correlation (R): the square of the coefficient indicates the proportion of the variance in the dependent variable which has been mathematically accounted for; whereas, $1-$ the square of the coefficient indicates the proportion which has not been accounted for.[2]

The coefficient of multiple correlation, $R_{1.234\ldots m}$, may also be defined as the simple correlation between the actual X_1 values and the X' values estimated from the several independent factors.

For convenient methods of calculating the various measures discussed in this chapter, see Appendix 2, pages 489–520.

Where a parabola or other algebraic equation has been fitted as a regression curve by least squares, the index of correlation may be computed by equation (12.5), and the standard error of estimate by equation (12.2) or (12.7), using the several terms in the fitted equation, X, X^2, etc., in place of X_2, X_3, etc.

Measuring the Separate Effect of Individual Variables. In addition to the measures of the importance of all of the independent variables combined, it is sometimes desirable to have measures of the importance of each of the individual variables taken separately, while simultaneously allowing for the variation associated with remaining independent variables. There are two different types of these measures: *the coefficient of partial correlation* and the *beta coefficient*.

PARTIAL CORRELATION. Coefficients of *partial correlation* measure the correlation between the dependent factor and each of the several independent factors, while eliminating any (linear) tendency of the remaining independent factors to obscure the relation. Thus in the problem where income was correlated with numbers of acres, cows, and men, the partial correlation of income with acres, while holding constant cows

---

[2] See Note 2, Appendix 3.

and men, indicates what the average correlation would probably be between acres and income in samples of farms in which all the farms in each sample had the same number of cows and the same number of men.

If an average of this series of correlations was calculated,[3] it would correspond to the partial correlation of income with acres, while holding cows and men constant ($r_{12.34}$). A similar interpretation can be made for the other two partial correlation coefficients. Even in problems (such as the present one) where the number of observations is not sufficient to permit of many such subgroups being formed, the partial correlation coefficient indicates about what such an average correlation in selected subgroups would be, if computed from a larger sample drawn from the same universe.

Any group of independent variables may serve to explain some, but not all, of the variation in a dependent variable. If an additional independent variable is added, it may also account for part of the variation left unexplained by the factors previously considered. The coefficient of partial correlation may be defined as a measure of the extent to which that part of the variation in the dependent variable which was *not* explained by the other independent factors can be explained by the addition of the new factor. For example, in the farm-income problem, considering only acres and cows, the correlation was $R_{1.23} = 0.904$. When acres, cows, and men were considered, the correlation was $R_{1.234} = 0.915$. Squaring both values shows that, whereas the two variables explain 81.8 per cent of the variance in income, the three variables explain 83.7 per cent. Whereas 18.3 per cent of the variance is left to be explained when the two variables are considered, only 16.3 per cent is left to be explained when three are considered. Adding the additional variable has increased the variance which can be explained by the difference between these two figures, or 2.0 per cent (18.3–16.3). If the importance of this increase is determined by comparing it with the variance left unexplained before the new variable was added, we find that 2.0/18.3, or 10.93 per cent of the variance left unexplained by acres and cows, has now been found to have been associated with differences in numbers of men. Taking its square root gives the coefficient of partial correlation, 0.33.

The coefficient is designated $r_{14.23}$, since it shows the partial correlation between X_1 and X_4, after X_2 and X_3 have been taken into account. As is indicated in the discussion, its square may be computed by the formula[4]

$$r_{14.23}^2 = \frac{(1 - R_{1.23}^2) - (1 - R_{1.234}^2)}{1 - R_{1.23}^2}$$

[3] The calculation of the average of a series of correlation coefficients would involve the use of Fisher's z-transformation.

[4] This is different from the formula customarily given. See Note 3, Appendix 3.

For purposes of computation, this formula may be simplified to

$$r_{14.23}^2 = 1 - \frac{1 - R_{1.234}^2}{1 - R_{1.23}^2} \tag{12.8}$$

If it is desired to compute coefficients of partial correlation for the other independent variables, acres and cows, the corresponding formulas are

$$r_{13.24}^2 = 1 - \frac{1 - R_{1.234}^2}{1 - R_{1.24}^2}$$

$$r_{12.34}^2 = 1 - \frac{1 - R_{1.234}^2}{1 - R_{1.34}^2}$$

In each case, the coefficient should be given the same sign as the corresponding net regression coefficient. That is, if $b_{13.24}$ is negative, $r_{13.24}$ will also be negative.

It should be noticed that, although the numerator of the fraction is the same in each case, the denominator is different. This is a peculiarity of coefficients of partial correlation—they measure the importance of each of the several variables by determining how much it reduces the variation *after all the other variables except it are taken into account.*

If we work out the new multiple correlations necessary,[5] $R_{1.24}$ and $R_{1.34}$, and substitute them in the equations given just above, the whole set of coefficients of partial correlation and partial determination for the farm-income problem works out as in the following equations and in Table 12.1:

$$r_{13.24}^2 = 1 - \frac{1 - 0.837}{1 - 0.633} = 0.556$$

$$r_{12.34}^2 = 1 - \frac{1 - 0.837}{1 - 0.804} = 0.168$$

[5] The two new coefficients of multiple correlation are obtained by rearranging the arithmetic values previously computed so as to give the necessary regression coefficients, and then determining the value of R by equation (12.5). The two new sets of equations are:

To determine $R_{1.24}$,

$$(\Sigma x_2^2)b_{12.4} + (\Sigma x_2 x_4)b_{14.2} = (\Sigma x_1 x_2)$$

$$(\Sigma x_2 x_4)b_{12.4} + (\Sigma x_4^2)b_{14.2} = (\Sigma x_1 x_4)$$

Similarly, for $R_{1.34}$,

$$(\Sigma x_3^2)b_{13.4} + (\Sigma x_3 x_4)b_{14.3} = (\Sigma x_1 x_3)$$

$$(\Sigma x_3 x_4)b_{13.4} + (\Sigma x_4^2)b_{14.3} = (\Sigma x_1 x_4)$$

For a method of computation which facilitates the calculation of the partial r's, as well as of the earlier values, see Chapter 13, and Appendix 2, pages 503–506.

When income was correlated with acres alone, there was virtually no correlation ($r_{12} = 0.01$). Yet the partial correlation of income with acres, while holding constant the variation associated with cows and men, has just been seen to be 0.41. Although this is not high, it is certainly more than no correlation. Furthermore, even though the correlation of income with cows alone is 0.71, the correlation with both acres and cows is 0.90.

Table 12.1

RELATIVE IMPORTANCE OF INDIVIDUAL FACTORS AFFECTING INCOME,
AS INDICATED BY COEFFICIENTS OF PARTIAL CORRELATION

Factors Already Considered	Factor Added	Coefficient of Partial Correlation, $r_{12.34}$, etc.	Reduction in Unexplained Variance, $r^2_{12.34}$, etc.
Cows (X_3), men (X_4)	Acres (X_2)	0.41	0.168
Acres (X_2), men (X_4)	Cows (X_3)	0.75	0.556
Acres (X_2), cows (X_3)	Men (X_4)	0.33	0.109

On the surface of the data there appears to be no relation between acres and income, since the positive relation of acres to income is hidden. Acres are negatively correlated with cows to a sufficient extent so that the decreased income with decreased number of cows offsets the increases with more acres. Only when the number of cows is allowed for can the influence of acres be seen.

It is evident that a mere surface examination of a set of data cannot reveal which independent factors are important and which are unimportant. A variable which shows no correlation with the dependent variable may yet show significant correlation after the relation to other variables has been allowed for.

Investigators sometimes think they are doing "research" when they study the relation of a given variable, say the price of a commodity, to a number of other factors, discard all those factors that show no correlation with price, and select for further study by multiple correlation the factors that show the highest simple correlation with the price. As the preceding discussion shows, that procedure may result in discarding factors which would show a truly important relation to price after the effect of other associated factors had been allowed for. A careful, logical examination of the problem, the selection of the factors to be considered on the basis of these qualitative considerations, and then preliminary examination of all the intercorrelations among the selected independent factors, will provide more trustworthy results. (See Chapter 26 for a more detailed discussion of the places of qualitative and quantitative analysis in such studies.)

The test whether a given independent variable may really be related to the dependent variable, even if it shows no apparent correlation, is whether that independent variable is correlated with other independent variables, which in turn are correlated with the dependent. Thus in the example just discussed, although acres showed no correlation with income, they did show significant correlation with cows. If acres had had no correlation with either income, cows, or men, it would have been impossible for acres to have correlation with income even after the relation to cows and men was allowed for.

BETA COEFFICIENTS. The importance of individual variables may also be compared by their net regression coefficients. The size of the regression coefficients, however, varies with the units in which each variable is stated. They may be made more comparable by expressing each variable in terms of its own standard deviation, using the beta coefficients mentioned in Chapter 9. In terms of betas, the regression equation for four variables would be

$$\frac{X_1}{s_1} = \beta_{12.34} \frac{X_2}{s_2} + \beta_{13.24} \frac{X_3}{s_3} + \beta_{14.23} \frac{X_4}{s_4} + a'$$

Hence the partial betas may be defined

$$\beta_{12.34} = b_{12.34} \frac{s_2}{s_1} \tag{12.9}$$

For the problem we have been considering, the betas may be calculated very readily:

$$\beta_{12.34} = b_{12.34} \frac{s_2}{s_1} = 1.2058 \left(\frac{5.51}{16.52} \right) = 0.402$$

$$\beta_{13.24} = b_{13.24} \frac{s_3}{s_1} = 2.6274 \left(\frac{5.82}{16.52} \right) = 0.926$$

$$\beta_{14.23} = b_{14.23} \frac{s_4}{s_1} = 5.0298 \left(\frac{0.927}{16.52} \right) = 0.282$$

If the relative importance of each of the different factors, as judged by the two different types of individual measurement, is compared, the relations are as shown in Table 12.2.

Table 12.2

RELATIVE IMPORTANCE OF INDIVIDUAL FACTORS AFFECTING INCOME,
AS INDICATED BY TWO DIFFERENT COEFFICIENTS

Independent Factor	Factors Held Constant	Coefficients of Partial Correlation $(r_{12.34})$	Beta Coefficients $\beta_{12.34}$
Acres (X_2)	Cows (X_3), men (X_4)	0.41	0.402
Cows (X_3)	Acres (X_2), men (X_4)	0.75	0.926
Men (X_4)	Acres (X_3), cows (X_3)	0.33	0.282

It is evident from this comparison that, although the exact values differ for the two sets of measures, the rank of the three variables in order of importance is the same and the relative sizes are comparable. This does not always hold true, owing to the mathematical differences in the meaning of the two sets.

Reliability of Results from a Sample. All the coefficients presented in this chapter are subject to fluctuations of sampling just as are simpler coefficients. Chapter 17 discusses the extent of these fluctuations with various sizes of samples and gives methods of estimating confidence intervals when the sample statistics are used to draw inferences as to the probable values of the parameters in the universe. The comments made previously about the effects of purposeful selection of values of the independent variables upon simple and multiple coefficients of correlation also apply to coefficients of partial correlation and to beta coefficients. If only extremely low and extremely high values of X_2 are selected for inclusion in the sample, the coefficients $r_{12.34}$ and $\beta_{12.34}$ will tend to be high; if the values of X_2 are confined to a narrow range, they will tend to be low. A similar, though less obvious, situation may arise when no special efforts are made to select values of X_2, but when it happens that X_2 is correlated with another independent variable the values of which *are* purposefully selected. Thus, if the partial correlation coefficient is to be regarded as an estimate of a universe parameter, it is important to keep clearly in mind the nature of the universe to which the inferences are expected to apply, as related to the universe represented in the sample.

If the conditions of the sample are such that sample values can be used as bases for estimating the parameters in the universe (i.e., if the sample is of the "correlation model" type), the adjustments for degrees of freedom explained in Chapter 17, as well as the error formulas, may become important.

Summary

This chapter has shown that the accuracy of a regression equation for estimating one variable from two or more others may be measured by the standard error of estimate. The extent to which variation in the dependent variable is associated with the variation in the several independent variables may be measured by the coefficient of multiple correlation, or, with respect to variance, by the coefficient of multiple determination. The closeness of association between the dependent and each of the independent variables after the effects of the other variables have first been removed may (1) be measured by the coefficient of partial correlation, or (2) be indicated by the beta coefficients, which reduce the net regression coefficients to a comparable basis; but both are of limited value except in the *correlation model*, as discussed in Chapter 17.

NOTE. In computing the index of correlation where a regression curve has been determined by using a transformed value of the dependent variable (as with the use of log Y instead of Y on pages 93–96), the index of correlation must be computed from the transformed values instead of the natural values. That is,

$$i_{yx} = \sqrt{1 - \frac{(s_{\hat{z}}'')^2}{(s_{\hat{y}})^2}} = \frac{s_{\hat{y}}''}{s_{\hat{y}}}$$

where $\bar{z} = \bar{Y} - \bar{Y}''$.

CHAPTER 13

Practical methods for working multivariable linear correlation and regression problems

Here we will take a new illustration of a multivariable correlation problem, and work through the calculations of the net regression coefficients, the coefficients of multiple and partial correlation, and the standard errors for these statistics. The meaning of such standard errors is discussed in more detail in Chapters 17 and 19.

For this exercise, we will use data on the per capita consumption of beef in the United States, and on three variables logically related to it—retail price of beef, deflated for changes in price levels; disposable income per capita, similarly deflated; and the consumption of pork per capita. The hypothesis assumed in selecting these data is that consumption of beef per person will be affected by real prices of beef, real income available for expenditure, and consumption of the other main meat, pork. The figures, thus adjusted, are given in the first five columns of Table 13.1.

The final column in the table, headed check sum, is calculated by making a total of all the other values for each observation, from X_1 through X_4.

The next step is to add all the columns, including the Σ_0, and to divide by n, the number of observations (20 in this case), and enter the average. The results are shown at the foot of the table. All the calculations to this point are now verified by the two equations:

$$\Sigma X_1 + \Sigma X_2 + \Sigma X_3 + \Sigma X_4 = \Sigma(\Sigma_0) \tag{13.1}$$

and

$$M_1 + M_2 + M_3 + M_4 = M_0 \tag{13.2}$$

That is, in each of these last two lines, the sum of the entries in the four columns from X_1 to X_4 will equal the sum of the Σ_0 column.

Table 13.1

FACTORS RELATED TO BEEF CONSUMPTION, U.S.A.

Year	Beef Consumption per Capita, X_1	Beef Price at Retail, Deflated,* X_2	Disposable Income per Capita, Deflated,* X_3	Pork Consumption per Capita, X_4	Check Sum, Σ_0
	Pounds	*Cents*	*Dollars*	*Pounds*	
1922	59.1	23.1	452	65.7	599.9
1923	59.6	23.6	505	74.2	662.4
1924	59.5	24.1	499	74.0	656.6
1925	59.5	24.5	507	66.8	657.8
1926	60.3	24.8	515	64.1	664.2
1927	54.5	26.5	520	67.7	668.7
1928	48.7	30.5	533	70.9	683.1
1929	49.7	32.0	556	69.6	707.3
1930	48.9	30.3	506	67.0	652.2
1931	48.6	27.6	474	68.4	618.6
1932	46.7	25.5	400	70.7	542.9
1933†	51.5	23.3	394	69.6	538.4
1934†	55.9	24.4	430	63.1	573.4
1935†	52.9	31.1	468	48.4	600.4
1936†	58.1	28.9	522	55.1	664.1
1937	55.2	31.7	537	55.8	679.7
1938	54.4	28.5	502	58.2	643.1
1939	54.7	29.7	542	64.7	691.1
1940	54.9	29.5	575	73.5	732.9
1941	60.9	30.0	663	68.4	822.3
Σ	1093.6	549.6	10,100	1315.9	13059.1
Means	54.680000	27.480000	505.0000	65.795000	652.955000

* Divided by Consumer Price Index, with 1935–39 = 1.00.

† Excludes quantities of beef diverted from normal market channels under emergency Government programs.

The next step is to make the extensions and sums for all the variables and the check sum, as follows:

$$\Sigma(X_1^2) + \Sigma(X_1X_2) + \Sigma(X_1X_3) + \Sigma(X_1X_4) = \Sigma(X_1\Sigma_0)$$
$$\Sigma(X_2^2) \quad + \Sigma(X_2X_3) + \Sigma(X_2X_4) = \Sigma(X_2\Sigma_0)$$
$$\Sigma(X_3^2) \quad + \Sigma(X_3X_4) = \Sigma(X_3\Sigma_0)$$
$$\Sigma(X_4^2) \quad = \Sigma(X_4\Sigma_0)$$

The accuracy of the extensions for each variable are controlled by the identities

$$\Sigma(X_1^2) \quad + \Sigma(X_1X_2) + \Sigma(X_1X_3) + \Sigma(X_1X_4) = \Sigma(X_1\Sigma_0) \quad (13.3)$$
$$\Sigma(X_1X_2) + \Sigma(X_2^2) \quad + \Sigma(X_2X_3) + \Sigma(X_2X_4) = \Sigma(X_2\Sigma_0)$$

etc.

The sums are then adjusted to departures from the means of each variable, and the normal equations, together with certain additional terms to provide the partial correlation coefficients and the standard errors, are set up and solved. The method of computing the solutions and the constants is given in detail in Appendix 2, pp. 507–520. This method, based on matrix algebra, provides all the needed values more rapidly than does the Doolittle method used elsewhere in this book. The new method, too, can be learned and followed by any intelligent student or statistical clerk, without advanced mathematical training.

The solution of these equations, as shown in Appendix 2, gives the following statistics:

1. Regression equation:

$$X_1 = 90.814 - 1.850X_2 + 0.0832X_3 - 0.415X_4$$
$$\quad\quad\quad (0.146) \quad\quad (.0069) \quad\quad (0.054)$$

2. Standard error of estimate:

$$S_{1.234} = 1.371$$

3. Coefficient of multiple correlation:

$$R_{1.234} = 0.980$$

4. Coefficients of partial correlation:

$$r_{12.34} = -0.954$$

$$r_{13.24} = 0.950$$

$$r_{14.23} = -0.887$$

And computing the standard deviations of the four variables, the β coefficients can then be calculated as follows:

$$\beta_{12.34} = -1.27 \quad\quad \beta_{13.24} = 1.130 \quad\quad \beta_{14.23} = -0.631.$$

Interpretation of the Results. The regression equation shows that during the 20-year period, beef consumption was significantly related to all three variables, beef prices, income, and pork consumption. Since the computed net regression coefficients are all more than 7 times their own standard errors, the chance of getting values as large as this from a universe in which X_1 is uncorrelated with the other variables is practically zero—even if we take into account the fact that after allowing for the 4 constants in the equation, we have only 16 degrees of freedom to consider in consulting tables such as Table 2.3. If ± 2 times the standard error is taken as a reasonable confidence interval, there is a high probability ($P = 0.93$) that the coefficients of X_2 and X_3 are within 16 percent, and that of X_4 within 26 percent, of the corresponding universe values. (See Chapter 20 for a discussion of sample and universe relationships in time series.)

The regression coefficients as they stand indicate that per capita beef consumption tended to decline 1.85 pounds for every increase of 1 cent in beef prices, to rise 0.83 pound for every advance of 10 dollars in per capita income; and to decline about 0.4 pound for every increase of 1 pound in pork consumption. (These increases of pork consumption presumably were usually accompanied by corresponding decreases in pork prices—since that was not considered separately the effect here is the net result of changes in both pork consumption and associated pork prices.) Altogether, the standard error of estimate indicates that the errors in estimating beef consumption from the three other factors, for the period studied, had a standard deviation of only 1.37 pounds. For estimating the average errors expected in using this equation to make estimates for other observations drawn from the same universe, however, we must compute $\bar{S}_{1.234}$. With $n = 20$ and $m = 4$, this comes out, by equation 12.1, to 1.53 pounds, only slightly larger.

To really measure the average elasticity of demand with respect to beef price and income, we should have used logarithmic values for our variables instead of arithmetic. We can, however, make a rough approximation by estimating the response for changes at the mean values, as in Table 13.2.

Table 13.2

Independent Variable	Mean Value	1% Change from Mean	Net Regression	Corresponding Change in Consumption, X $(2) \times (3)$	Change Divided by Mean Beef Consumption $(4) \div 54.68$
	(1)	(2)	(3)	(4)	(5)
Beef price, X_2	27.48	0.2748	−1.85	−0.5084	−0.930%
Disposable income, X_3	505	5.05	0.0832	0.4202	0.768%
Pork consumption, X_4	65.795	0.658	−0.415	−0.2731	−0.499%

At their means, the elasticity of demand for beef with respect to price is thus —0.93, or almost unity; the elasticity with respect to income is 0.77, and the cross elasticity with respect to pork consumption is —0.50.

(For an exercise in the new method of computation, students may wish to rework this problem using logarithmic values of the four variables. This will also serve to show how much the approximate economic conclusions just presented are changed by using the logarithmic transformations, which logically are better suited for measuring these relations.)

Finally, the coefficient of multiple correlation, 0.980, indicates that the three variables together explained 96 per cent (R^2) of all the variance in annual beef consumption observed during this 20-year period. If this is interpreted from Figure 17.2, as explained in Chapter 17, there is only 1 chance out of 20 that so high a multiple correlation could have been obtained from a sample of this size, unless the true correlation in the universe was at least 0.95. (For other considerations in applying such an interpretation to time series, see Chapter 20). Since the variables are logically consistent, as related to the independent variable, the results are thus highly significant as measuring the main factors explaining the fluctuations in beef consumption.

Multiple Curvilinear Regressions

CHAPTER 14

Determining multiple curvilinear regressions by algebraic and graphic methods

The discussion of multiple regression to this point has been limited to linear relationships—where the change in the dependent variable accompanying unit changes in each independent variable was assumed to be of exactly the same amount, no matter how large or how small the independent variable became. Thus in the farm income example, it was assumed that each additional cow would be accompanied by the same increase in income, no matter whether it was the first, the tenth, or the thirtieth. Similarly, each additional acre in crops or each additional man employed was assumed to be accompanied by an identical contribution to the income, no matter how large or how small the business already was. It is quite evident that such an analysis makes no provision for there being an otpimum size of operation for given circumstances; in this particular case, it assumes that there is no such thing as the principle of diminishing returns. Such an analysis might therefore fail entirely to reveal the proper size of productive unit, or the number of each of the several elements to be employed to yield maximum returns.

In many other types of problems for which multiple regression analysis might be used, limitation of the analysis to linear relations would seriously restrict its value or prevent its use altogether. In dealing with the effect of weather upon crop yields, several variable weather factors are usually concerned. There may be an optimum point for growth, with respect to both temperature and precipitation, with values either above or below the optimum tending to produce lower yields. Linear regressions are obviously unfitted to express such relations. In problems such as these, and many others which might be enumerated, determination of the curvilinear relation between independent and dependent variable, while simultaneously eliminating the effect of other factors which also affect the dependent variable, is the most important feature in the investigation.

The problem in its simplest outlines may be stated as follows: Given a series of paired observations of the values of a dependent variable X_1 and two or more independent variables X_2, X_3, X_4, etc., required to find the change in X_1 accompanying the changes in X_2, X_3, and X_4, in turn, while holding the remaining independent factors constant, so that for any given values of X_2, X_3, and X_4, etc., values may be estimated for X_1, according to the regression equation

$$X_1 = a' + f_2(X_2) + f_3(X_3) + f_4(X_4) + \cdots \tag{14.1}$$

The expression $f_2(X_2)$ is used here simply as a perfectly general term meaning any regular change in X_1 with given changes in X_2, whether describable by a straight line or a curve. The equation is read "X_1 is a *function* of X_2 plus a *function* of X_3," etc.

The several partial (or "net") regression curves may be determined either by the use of definite mathematical expressions, one for each independent variable, with the constants all determined simultaneously just as in linear multiple correlation; or by a method known as "successive graphic approximation," which involves no prior assumptions as to the shapes of the curves.

Multiple Regression Curves Mathematically Determined

In using definite mathematical functions, it is necessary to express the curvilinear relations by simple mathematical curves of some type, so that the constants for the curves may be determined by methods similar to those already presented. If simple parabolas were used, involving only the first and second powers of each independent variable, equation (14.1) could be expressed

$$X_1 = a + b_2 X_2 + b_{2'}(X_2^2) + b_3 X_3 + b_{3'}(X_3^2) + b_4 X_4 + b_{4'}(X_4^2) \tag{14.2}$$

In practical research we must distinguish between two situations. If we know, on the basis of established theory or of previous studies involving the same variables, the mathematical form of the net relationship between the dependent and each independent variable, we can proceed at once to fit the appropriate multiple curvilinear regression function by the method of least squares. The only unknowns in this situation are the numerical values of the constants of the prescribed function.

If we do not know the forms of the net regression curves, we have a more complex version of the problem in Chapter 6 in which we determined *from the data* a freehand curve expressing the relationship between wheat protein and the proportion of vitreous kernels. Unless we were willing to compute a large number of alternative mathematical functions, being guided as to goodness of fit only by the respective standard errors of estimate, we would be obliged first to examine our data by graphic means to determine the approximate forms of the net relationships implicit in them.

The bulk of this chapter will deal with the second situation. In the next few pages, however, we shall outline computational methods for the situation in which the mathematical forms of the net regression curves are known.

Determining the Curves by Least Squares.   The process of determining net regression curves by the use of a definite mathematical equation may be illustrated for the following data:

X_2	X_3	X_1	X_2	X_3	X_1
1	3	8	0	2	7
2	2	10	4	5	9
4	7	8	3	3	10
9	8	9	1	2	9
5	5	10	6	5	10
2	3	9	1	2	9
2	2	9	2	2	10
7	14	7	4	14	6
9	8	9	1	2	9
2	4	8	10	11	8

Let us assume (on the basis of theory or of other studies of these variables) that the net regressions of X_1 on both X_2 and X_3 may be appropriately represented by parabolas. Accordingly, the multiple regression equation is of the form

$$X_1 = a + b_2 X_2 + b_2'(X_2^2) + b_3 X_3 + b_3'(X_3^2)$$

The arithmetic required to determine the five constants can be reduced by "coding" the squared values. Let $U = X_2^2/10$, and $V = X_3^2/10$. The regression equation then may be written

$$X_1 = a + b_2 X_2 + b_u U + b_3 X_3 + b_v V$$

The normal equations to determine the constants are next obtained in exactly the same manner as described in Chapter 11 for a multiple regression involving four independent variables. The resulting normal equations are:

$$(\Sigma x_2^2)b_2 + (\Sigma x_2 u)b_u + (\Sigma x_2 x_3)b_3 + (\Sigma x_2 v)b_v = \Sigma x_1 x_2$$

$$(\Sigma x_2 u)b_2 + (\Sigma u^2)b_u + \Sigma(x_3 u)b_3 + (\Sigma uv)b_v = \Sigma x_1 u$$

$$(\Sigma x_2 x_3)b_2 + (\Sigma ux_3)b_u + \Sigma(x_3^2)b_3 + (\Sigma x_3 v)b_v = \Sigma x_1 x_3$$

$$(\Sigma x_2 v)b_2 + (\Sigma uv)b_u + \Sigma(x_3 v)b_3 + (\Sigma v^2)b_v = \Sigma x_1 v$$

Carrying out the required computations, the equations are found to be:

$$171.750b_2 + 170.625b_u + 165.000b_3 + 207.600b_v = -2.50$$

$$170.625b_2 + 181.165b_u + 153.540b_3 + 192.316b_v = -5.31$$

$$165.000b_2 + 153.540b_u + 295.200b_3 + 441.480b_v = -50.80$$

$$207.600b_2 + 192.316b_u + 441.480b_3 + 696.072b_v = -86.52$$

The $(\Sigma x_1^2) = 24.20$.

Solving these equations by the usual method, and computing a from equation (11.12) by restating it

$$a_{1.2u3v} = M_1 - b_2 M_2 - b_u M_u - b_3 M_3 - b_v Mv$$

we find the regression equation to be

$$X_1 = 9.411 + 1.2709X_2 - 0.7337U - 0.9957X_3 + 0.3309V$$

The net regressions of X_1 on X_2 and X_3 are now shown by the two parabolic equations:

$$X_1 = 5.596 + 1.2709X_2 - 0.07337X_2^2$$

and

$$X_1 = 12.515 - 0.9957X_3 + 0.03309X_3^2$$

The graph of these two curves is shown in Figure 14.1.

The multiple correlation of X_1 with X_2, U, X_3, and V is 0.968. This is the *index* of multiple correlation of X_1 with X_2 and X_3, according to the parabolic regressions. The standard error of estimate, adjusted by equation (12.2) for the five degrees of freedom used up in determining the five constants, is found to be 0.319.

It should be noted that where net curvilinear regressions are found by this method, the number of constants assumed in the regression equation is definitely known, and there can be no question as to the exact correction to apply to the standard error of estimate.

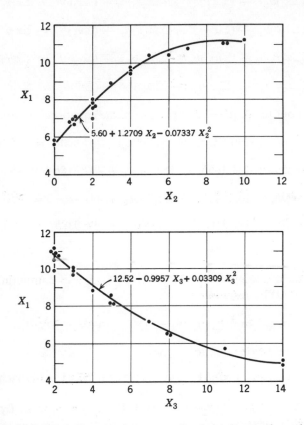

Fig. 14.1. Parabolic regression curves, fitted simultaneously, and net residuals.

Testing the Fit of the Curves. If there were any doubt as to the appropriateness of the parabolic relationships in this case, we could check the goodness of fit of each net regression curve much as we appraised the fit of the curves in Chapter 6 which involved only one independent variable. To make this check, after the regression equation is determined for the particular curves selected, estimated values of X_1 are calculated from the equation. The residual differences between X_1 and these estimated values are then computed. These residuals are then plotted as departures from the mathematical net regression curves. Carrying this out for the problem illustrated, we obtain results as follow:

X_2	X_3	X_1	X_1'	z	X_2	X_3	X_1	X_1'	z
1	3	8	7.9	0.1	0	2	7	6.7	0.
2	2	10	9.8	0.2	4	5	9	9.2	−0.2
4	7	8	8.0	0.0	3	3	10	9.9	0.1
9	8	9	9.1	−0.1	1	2	9	8.8	0.2
5	5	10	9.8	0.2	6	5	10	10.2	−0.2
2	3	9	9.0	0.0	1	2	9	8.8	0.2
2	2	9	9.8	−0.8	2	2	10	9.8	0.2
7	14	7	7.2	−0.2	4	14	6	5.9	0.1
9	8	9	9.1	−0.1	1	2	9	8.8	0.2
2	4	8	8.2	−0.2	10	11	8	7.8	0.2

The residuals obtained above are plotted as departures from the parabolic net regressions, as shown in Figure 14.1. It is evident in this case that the parabolic regressions represent the relations quite well, with the departures in general evenly distributed on both sides of each curve throughout their length.

Any other type of mathematical function, the parameters of which can be expressed in the first degree, can be used to determine net regressions by the method of least squares. Besides the third, fourth, or higher powers of X_2, such transformations as $10/X_2^2$, $100/X_2^3$, $\log X_2$, and $1/\log X_2$ may be employed as independent variables, either in place of the previous independent variables or as an addition to the simple statement of them.

The simple parabola is not very flexible and will often fail to represent adequately the relationships existing in sets of data. On the other hand, situations in which theory or previous experiments point unequivocally to higher order parabolas as expressing the true relationships are relatively infrequent.

If the more flexible cubic parabola is employed, involving the first, second, and third powers of each independent variable, the equation becomes

$$X_1 = a + b_2(X_2) + b_{2'}(X_2^2) + b_{2''}(X_2^3) + b_3(X_3) + b_{3'}(X_3^2)$$
$$+ b_{3''}(X_3^3) + b_4(X_4) + b_{4'}(X_4^2) + b_{4''}(X_4^3) \quad (14.3)$$

This last equation for three independent variables involves ten constants, and the clerical labor of dealing with the squared and cubed values is large (unless they are coded). The curves corresponding to the three functions in equation (14.3) are:

$$f_2(X_2) = b_2 X_2 + b_{2'}(X_2^2) + b_{2''}(X_2^3)$$
$$f_3(X_3) = b_3 X_3 + b_{3'}(X_3^2) + b_{3''}(X_3^3)$$
$$f_4(X_4) = b_4 X_4 + b_{4'}(X_4^2) + b_{4''}(X_4^3)$$

Whether or not these curves will actually be a good fit to the true functions can rarely be told beforehand.

Very frequently we find ourselves in the second situation referred to above, and must determine the net regression curves by examining the data at hand. A dot chart of the observations on X_1 and X_2 *may* suggest some definite mathematical function. However, the true net relationship may be obscured by the effects on X_1 of other independent variables. These may be linearly intercorrelated with X_2; we have seen in earlier chapters how intercorrelation can mask the true net relationships even in linear regression. If X_1 is related in an unknown curvilinear fashion to X_3 and other variables it may be quite impossible to choose an appropriate mathematical equation for the net regression of X_1 on X_2 simply by looking at the dot chart for these two variables.

In such a situation we need a new method to guide our exploration of the data for the underlying curvilinear relationships. Such a method, called the method of successive approximations, is presented below.

Multiple Regression Curves by Successive Approximations

The general method of determining partial regression curves by the successive approximation method may be outlined as follows:

The conditions to be imposed on the shape of each curve, in view of the logical nature of the relations, are first thought through and stated. This procedure, for each curve, is similar to that described on pages 103 to 108 of Chapter 6.

The linear partial regressions are computed next. Then the dependent variable is adjusted for the deviations from the means of all independent variables except one, and a correlation chart, or dot chart, is constructed between these adjusted values and that independent variable. This provides the basis for drawing in the first approximation curve for the net regression of the dependent variable on that independent variable, within the limitations of the conditions stated. The dependent variable is then adjusted for all except the next independent variable, the adjusted values plotted against the values of that variable, and the first approximation curve determined with respect to that variable. This process is carried out for each independent variable in turn, yielding a complete set of first approximations to the net regression curves. These curves are then used as a basis for adjusting the dependent factor for the approximate curvilinear effect of all independent variables except one, leaving out each in turn; and second approximation curves are determined by plotting these adjusted values against the values of each independent variable in turn. New adjustments are made from these curves, and the process is continued until no further change in the several regression curves is indicated.

The process of determining net curvilinear regressions by the successive graphic approximation method may be illustrated by the data shown in Table 14.1. These data show, for a period of 38 years, the average rainfall during June, July, and August, for nine weather stations scattered through the Corn Belt. This precipitation has been designated as variable

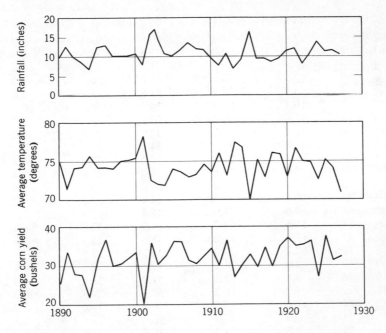

Fig. 14.2. Rainfall, temperature, and corn yields in the Corn Belt, 1890 to 1927.

X_3. The average temperature during the same months, at the same stations, has been designated as X_4. The average yield of corn per acre, in the six leading Corn Belt states, is shown as X_1—the variable whose fluctuations are to be explained, so far as possible, by the other factors.

It is evident from the table that there was an upward trend in corn yield during this period, although there was not a similar trend in rainfall or temperature. Plotting each one of the three factors, X_3, X_4, and X_1 as shown in Figure 14.2, we notice, however, that there have been marked though irregular long-time cycles in rainfall and temperature during the period. To a certain extent the swings in yields have agreed with the high point of the rainfall cycles. It is not safe, therefore, to fit a long-time

Table 14.1

YIELD OF CORN, RAINFALL, AND TEMPERATURE IN SIX LEADING STATES; AND
YIELD ESTIMATED BY LINEAR REGRESSIONS ON THREE FACTORS*

Year	Time, X_2	Rainfall, X_3	Temperature, X_4	Yield, X_1	Estimated yield, X_1'	Difference, $X_1 - X_1'$ z
		inches	degrees	bushels		
1890	0	9.6	74.8	24.5	28.4	−3.9
1891	1	12.9	71.5	33.7	31.6	2.1
1892	2	9.9	74.2	27.9	29.1	−1.2
1893	3	8.7	74.3	27.5	28.5	−1.0
1894	4	6.8	75.8	21.7	27.0	−5.3
1895	5	12.5	74.1	31.9	30.9	1.0
1896	6	13.0	74.1	36.8	31.4	5.4
1897	7	10.1	74.0	29.9	30.0	−0.1
1898	8	10.1	75.0	30.2	29.7	0.5
1899	9	10.1	75.2	32.0	29.8	2.2
1900	10	10.8	75.7	34.0	30.1	3.9
1901	11	7.8	78.4	19.4	27.5	−8.1
1902	12	16.2	72.6	36.0	34.6	1.4
1903	13	14.1	72.0	30.2	33.8	−3.6
1904	14	10.6	71.9	32.4	32.1	0.3
1905	15	10.0	74.0	36.4	31.1	5.3
1906	16	11.5	73.7	36.9	32.2	4.7
1907	17	13.6	73.0	31.5	33.7	−2.2
1908	18	12.1	73.3	30.5	32.9	−2.4
1909	19	12.0	74.6	32.3	32.5	−0.2
1910	20	9.3	73.6	34.9	31.6	3.3
1911	21	7.7	76.2	30.1	29.8	0.3
1912	22	11.0	73.2	36.9	33.0	3.9
1913	23	6.9	77.6	26.8	29.1	−2.3
1914	24	9.5	76.9	30.5	31.0	−0.5
1915	25	16.5	69.9	33.3	37.7	−4.4
1916	26	9.3	75.3	29.7	31.8	−2.1
1917	27	9.4	72.8	35.0	33.0	2.0
1918	28	8.7	76.2	29.9	31.4	−1.5
1919	29	9.5	76.0	35.2	32.1	3.1
1920	30	11.6	72.9	38.3	34.6	3.7
1921	31	12.1	76.9	35.2	33.4	1.8
1922	32	8.0	75.0	35.5	32.1	3.4
1923	33	10.7	74.8	36.7	33.8	2.9
1924	34	13.9	72.6	26.8	36.5	−9.7
1925	35	11.3	75.3	38.0	34.2	3.8
1926	36	11.6	74.1	31.7	35.0	−3.3
1927	37	10.4	71.0	32.6	35.7	−3.1

* Data from E. G. Misner, Studies of the relation of weather to the production and price of farm products, I. Corn, Mimeographed publication, Cornell University, March, 1928. The six states are Iowa, Illinois, Nebraska, Missouri, Indiana, and Ohio.

trend to yield and to assume that in removing that trend we are merely taking out the effects of such factors as better varieties, improved methods of tillage, or concentration of acreage in the more fertile sections. Since there is some association between rainfall and time, at least over considerable periods, in eliminating all the variation associated with time we might be eliminating a part of the variation which really reflected differences in rainfall. Accordingly we may make time itself one of the factors in the multiple regression and ascribe to time only that part of the long-time change in yields which is not associated with differences in rainfall or in temperature. Each year, numbered from 0 up, is therefore included as one of the factors in the multiple regression[1] and is designated as variable X_2.

Before starting the statistical process, we must state the conditions to be observed in fitting a curve to each function. For rainfall, the considerations are quite similar to those discussed in Chapter 8 for irrigation water applied, so we shall use the same conditions as stated there (page 140).

For temperature, the range of possible relations might be wider. There may be certain temperatures to which the plant does not respond and then certain higher temperatures which produce a marked response. Again, if the temperature is too high, a marked reduction in yield might be produced.[2] These considerations lead to the following conditions for the temperature curve:

1. It might rise not at all or slowly in the lower range, then more steeply, then taper off until a maximum is reached.

2. It might decline after the maximum, gradually or sharply, but would have only one maximum.

3. It might have one point of inflection at a temperature below that which has the maximum favorable effect on yield.

With respect to the third curve, that for trend, there is no *a priori* reason to expect any given shape during the period concerned, except that there be no sudden changes from year to year. Accordingly, the only condition imposed is that the trend have a smooth, gradual change, with no sharp inflections.

[1] Note the parallel treatment of changes in time as an independent factor in R. A. Fisher, *Statistical Methods for Research Workers*, 12th ed., pp. 206–8, Oliver and Boyd, Edinburgh and London, 1954.

[2] More elaborate investigations, experimental and statistical, have shown that the effect of both temperature and rainfall vary at different times of the season, and especially at certain critical times in the growth of the plant, such as at tasseling. Also, the particular combination of moisture and heat may be important. These possibilities will be referred to subsequently, in connection with more refined and elaborate methods of analysis.

As a preliminary step before starting to determine the net regression curves, we may examine the apparent relation of yield to rainfall, before the other factors (temperature and time) are taken into account.

The apparent relation between rainfall (X_3) and yield (X_1) is indicated in Figure 14.3, by a dot chart of the relation, with the average yield indicated for each group of years of similar rainfall. The broken line connecting these averages indicates that there is a marked curvilinear relation, a one-inch increase in rainfall being associated with a larger increase in yield when the basic level of rainfall is low than when the basic level of rainfall is high. Fitting a straight regression line to these two variables, the relation is found to be

$$X_1 = 23.55 + 0.776X_3$$

This line is accordingly drawn in on the chart, cutting across the curve indicated by the line of group averages.

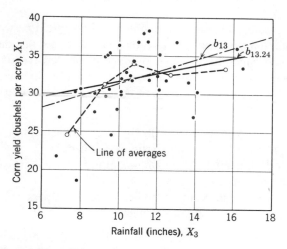

Fig. 14.3. Apparent relation of corn yields to rainfall (with simple and net regression lines).

Although Figure 14.3 shows yields to be definitely associated with differences in rainfall, it must be noted that rainfall is significantly correlated with X_4, temperature, the correlation being $r_{34} = -0.67$, and is also slightly correlated with time. To some extent, then, the changes in yield shown in the figure to be associated with differences in rainfall may really be due to concomitant differences in the other two factors. The extent to which these other two factors may have influenced the relations can be judged by determining the multiple correlation of X_1 with all three factors, and then noting how the slope of the regression of X_1 on X_3 alone

(b_{13}), which has just been shown plotted in the figure, compares with the slope of the net regression of X_1 on X_3 $(b_{13.24})$ determined while simultaneously holding constant the linear effects of X_2 and X_4. The first step toward determining the net regression curve, therefore, is to determine the multiple regression equation and the coefficient of multiple correlation, according to the methods outlined in Chapters 12 and 13.

The regression equation works out to be

$$X_1 = 53.505 + 0.146X_2 + 0.537X_3 - 0.405X_4$$

and the multiple correlation, $R_{1.234}$, is 0.55.[3]

This result shows that when the net linear influence of trend and of temperature is allowed for, yield increases on the average only 0.54 bushel for each increase of 1 inch in rainfall, whereas, before these other factors were taken into account, yield appeared to increase 0.78 bushel with each additional inch of rainfall. The difference between the simple regression and the net regression may be shown by plotting the latter as well in Figure 14.3.[4] It is then quite apparent how different are the relations as shown by the two lines.

Considering the effect of the other factors reduces the slope of the linear regression of X_1 on X_3 by nearly one-third. If other factors have so much effect on the average linear relation, they may have an even greater effect on the shape of the curve. The net regression line in Figure 14.3 shows the average change in the values of X_1 with different values of X_3, after the differences in X_2 and X_4 are taken into account. The average yield for different groups according to rainfall, connected by the broken line, shows definitely that the simple regression line is but a poor indication of the underlying relation between X_1 and X_3. The

[3] Using units of time in years, rainfall in tenths of inches, temperature in tenths of degrees, and corn yields in tenths of bushels, we find the normal equations for the data of Table 14.1 to be:

$$4,569.50b_{12.34} + \qquad 248.00b_{13.24} - \qquad 8.50b_{14.23} = \quad 6,813.00$$
$$248.00b_{12.34} + 18,989.06b_{13.24} - 10,279.41b_{14.23} = \ 14,726.97$$
$$-8.50b_{12.34} - 10,279.41b_{13.24} + 12,408.86b_{14.23} = -8,442.64$$

$ns_1^2 = 70,455.03$; $s_1 = 43.0$, or 4.3 bushels.

[4] The net regression line, showing the change in yield with changes in rainfall while holding constant time and temperature, may be computed from the multiple regression equation by substituting the average values for time and for temperature for X_2 and X_4, and then working out the new constant. For the data given in Table 14.1, the averages are:

$$M_2 = 18.500; \quad M_3 = 10.784; \quad M_4 = 74.276; \quad M_1 = 31.916$$

If we substitute the means of X_2 and X_4 for their values in the multiple regression equation, that equation becomes:

$$X_1 = 53.505 + (0.146)(18.500) + 0.537X_3 - (0.405)(74.276) = 26.124 + 0.537X_3$$

The net regression line in Figure 14.3 is therefore drawn in from this last equation.

net (or partial) regression line may be an equally poor indication of the relation with the other factors held constant. What is needed is some way of seeing the differences in the *individual* values of X_1 for different values of X_3, after the variation due to X_2 and X_4 has been eliminated. It is impossible to do this entirely, for we have as yet no measure of the *curvilinear* relation of X_1 to X_2 or X_3. But we do have our net regression coefficients, which measure the linear regression of X_1 on these other factors, and by using them we can eliminate from X_1 that part of its variation associated with the linear effects of X_2 and X_4, and then see if that gives us any clearer picture of the curvilinear relation between X_1 and X_3.

Determining the "First Approximation" Net Regression Curves. Having determined the linear multiple regression equation, we next calculate the estimated value of X_1 for each one of the 38 observations, by substituting the corresponding values of X_2, X_3, and X_4 in the equation. Each of the estimated values (X_1') is then subtracted from the actual value (X_1), giving the residual values (z), as also shown in Table 14.1.

The next step is to construct a scatter diagram to show the relation between variations in X_3 and the variation in X_1 after that associated with X_2 and X_4 has been eliminated. To do that, the net regression line for X_1 on X_3 is plotted on Figure 14.4, just as it had been on Figure 14.3.[5]

The residuals for each observation, from Table 14.1, are then plotted on the chart, with their X_3 value for abscissa and with the value of z as ordinate *from the net regression line as zero base.* For the first observation, $X_3 = 9.6$ and $z = -3.9$. The ordinate of the point on the net regression line corresponding to $X_3 = 9.6$ is 31.3, and the dot for this observation is correspondingly plotted 3.9 lower than that, at 27.4. For the second observation, $X_3 = 12.9$ and $z = +2.1$. The ordinate of the point on the regression line corresponding to $X_3 = 12.9$ is 33.1; so the dot for this observation is plotted at 33.1 + 2.1, or 35.2. After the corresponding operation has been carried out for all the observations, the figure appears as shown in Figure 14.4.[6]

---

[5] To plot the line, all that is necessary is to take the equation of the line to be used (see previous footnote)

$$X_1 = 26.124 + 0.537X_3$$

and substitute any two convenient values for X_3, say 6 and 16.

For $X_3 = 6$, $X_1 = 26.124 + (0.537)(6) = 29.35$
For $X_3 = 16$, $X_1 = 26.124 + (0.537)(16) = 34.71$

With these two sets of coordinates, the line is then drawn in with a straight edge through the points indicated.

[6] The simplest way of plotting the individual observations is to use a scale, which can be slid along the regression line as zero. The values of z are then plotted directly as vertical deviations from the points on the regression line corresponding to the particular values of the independent variable considered, as X_3 in the present case.

If Figure 14.4 is compared with Figure 14.3, it is readily seen that the scatter of the dots has been reduced. This will always be true when the other variables show any significant relation to the dependent factor; that is, when $R_{1.234}$ exceeds r_{13}. The scatter is reduced because that part of the variation in X_1 which can be expressed as net linear functions of X_2 and X_4 has now been eliminated.[7]

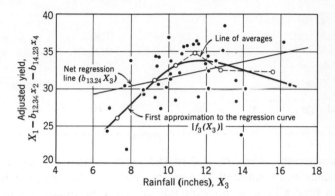

Fig. 14.4. Rainfall and yield of corn adjusted to average temperature and year, and first approximation curve fitted to the averages.

Consideration of Figure 14.4 can be facilitated by computing the means of the ordinates corresponding to the values of X_3 falling within convenient intervals. These can be obtained by simply averaging together the z values for each selected group of values of X_3 and plotting those

---

[7] This can readily be proved. Each point on the net regression line was obtained by the formula:

(A) $$X_1 = a_{1.234} + b_{12.34}M_2 + b_{13.24}X_3 + b_{14.23}M_4$$

To these values have been added the residuals, z. These residuals equal $X_1 - X_1'$, and therefore for each observation are equal to

(B) $$X_1 - a_{1.234} - b_{12.34}X_2 - b_{13.24}X_3 - b_{14.23}X_4$$

The ordinate of each dot in Figure 14.4 is the ordinate of the regression line plus z, and is therefore equal to the sum of the two equations, (A) and (B). If we use π to represent these ordinates, they are therefore equal to

$$\pi = a_{1.234} + b_{12.34}M_2 + b_{13.24}X_3 + b_{14.23}M_4 + X_1 - a_{1.234} - b_{12.34}X_2$$
$$- b_{13.24}X_3 - b_{14.23}X_4$$
$$\pi = X_1 - b_{12.34}(X_2 - M_2) - b_{14.23}(X_4 - M_4)$$
$$\pi = X_1 - b_{12.34}x_2 - b_{14.23}x_4$$

The adjusted values shown on Figure 14.4 are therefore simply the values of X_1 less net linear corrections for deviations in X_2 and X_4 from their mean values.

averages as deviations from the regression line, just as the individual deviations were plotted previously. The necessary averages are as shown in Table 14.2.

Table 14.2

Average Values of z, for Corresponding X_3 Values

X_3 Values	Number of Cases	Average of X_3	Average of z
Under 8.0	4	7.30	−3.85
8.0– 9.9	10	9.19	+0.16
10.0–10.9	8	10.35	+1.49
11.0–11.9	5	11.40	+2.56
12.0–13.9	8	12.76	−0.52
14.0 and over	3	15.60	−2.20

These averages, when plotted in the same manner as the individual observations and connected by a broken line, give the irregular line also shown in Figure 14.4. Comparing this line with the similar one in Figure 14.3, we see that though the lines are in general similar there are some marked differences. The average for the second group ($X_3 = 8.0$–9.9) is now above the straight net regression line, whereas previously it was below it. Likewise the average for $X_3 = 14$ and over is now slightly below the average for $X_3 = 12.0$–13.9, whereas before it was a little above it. Also, the difference between the first two averages is not so large as it appeared before. Apparently part of the previous deviations reflected other independent factors.

It is quite evident that a regression curve is indicated, rising sharply to a maximum yield between 10 and 12 inches of rain, then declining gradually for higher rainfalls. Such a curve is accordingly drawn in freehand, passing as near to the several group averages as is consistent with a continuous smooth curve, and yet conforming to the limiting conditions as to its shape. This curve is the first approximation to the curvilinear function.

$$X_1 = f_3(X_3)$$

which was required to be determined while simultaneously taking into account the curvilinear effects of X_2 and X_4 on X_1. It is only a first approximation because it has been determined while allowing for only the net *linear* effects of the other two variables. If their *curvilinear* effects were determined and allowed for, that might change somewhat the shape of this curve.

The next step is to determine similar first approximations to the curvilinear relation between X_1 and X_2, and between X_1 and X_4, with the net linear effects of the other variables eliminated just as has been done for X_3. It is not necessary to plot the apparent relation between X_1 and X_2 or X_1 and X_4. This was done in the case of X_3 (Figure 14.3) solely to illustrate the difference between taking the apparent relations and taking

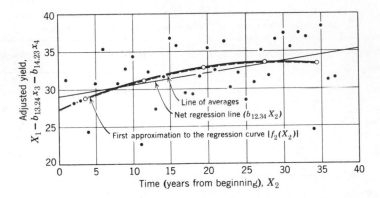

Fig. 14.5. Time and yield of corn adjusted to average temperature and rainfall, and first approximation curve fitted to the averages.

the net relations after the linear influence of the other factors had been allowed for (Figure 14.4). Instead, we may proceed at once to examine the net relation of X_1 to X_2. Figure 14.5 shows this step. This figure is constructed exactly as was Figure 14.4, by the following steps: (1) Plot the net regression line.[8] (2) Plot in the individual residuals, z, as deviations from that line.[9] (3) Average the residuals grouped according to X_2,

[8] The regression equation, for mean values of X_3 and X_4, becomes

$$X_1 = 53.505 + 0.146X_2 + 0.537(M_3) - 0.405(M_4)$$

$$= 53.505 + 0.146X_2 + (0.537)(10.784) - (0.405)(74.276)$$

$$= 29.214 + 0.146X_2$$

This equation is then the equation to which the net regression line in Figure 14.5 is drawn. Substituting the values $X_2 = 0$ and $X_2 = 20$ in the equation, values for X_1 of 29.214 and 32.13 are obtained, giving the coordinate points for drawing in the line.

[9] For the first observation, $X_2 = 0$ and $z = -3.9$. The point on the regression line corresponding to $X_2 = 0$ has an ordinate of 29.2. The dot for this observation is accordingly plotted at $29.2 - 3.9$, or 25.3. For the next observation, $X_2 = 1$ and $z = 2.1$. The corresponding ordinate on the regression line is 29.4, so the dot is plotted at $29.4 + 2.1$, or 31.5. The dot for each observation is plotted in turn in the same way, with a sliding graphic scale to place the dots above or below the regression line.

plot the group averages, and connect them by a broken line. (4) Draw in a smooth curve through the line of averages, if a curve is indicated, conforming to the limiting conditions stated for this curve.

After the first two steps have been carried out, just as described for Figure 14.4, grouping and averaging the residuals with respect to X_2 give the averages shown in Table 14.3.

Table 14.3

AVERAGE VALUES OF z FOR CORRESPONDING X_2 VALUES

X_3 Values	Number of Cases	Average of X_2	Average of z
0– 7	8	3.5	−0.38
8–15	8	11.5	+0.24
16–23	8	19.5	+0.64
24–31	8	27.5	+0.26
32–37	6	34.5	−1.00

The average residuals shown in the table are then plotted above and below the regression line in Figure 14.5 and connected by a broken line. This line of averages indicates that corn yield (for years of similar rainfall and temperature) rose rapidly during the earlier years, then more and more gradually, until during the last ten years it tended to remain about on the same level. A smooth continuous curve is therefore drawn through the averages, completing step (4) and giving the first approximation to the curvilinear net regression of X_1 on X_2, $f_2(X_2)$.

The same operations are then carried out for X_4, as shown in Figure 14.6. After drawing in the net regression line,[10] and plotting in the individual observations,[11] we group the residuals on X_4 and average, with the results shown in Table 14.4.

[10] The net regression line for X_1 and X_4 may be determined by an alternative method to that used before. On such charts as Figures 14.4, 14.5, or 14.6, the net regression line will always pass through the mean of the two variables. For Figure 14.6, therefore, X_1 will have its mean value, 31.92, when X_4 has its mean value, 74.28. From the net regression coefficient, $b_{14.23}$, it is evident that each unit increase in X_4 is accompanied by −0.405 unit increase in X_1. If X_4 is increased from 74.28 to 78.28, or 4 units, X_1 will change by (−0.405)(4), or −1.62. For $X_4 = 78.28$, X_1 will therefore be 31.92 − 1.62, or 30.30. This gives the two sets of points necessary to locate the line; when $X_4 = 74.28$, $X_1 = 31.92$; and when $X_4 = 78.28$, $X_1 = 30.30$.

[11] The individual residuals are plotted in the same way as indicated in the other two cases; the residual −3.9 for $X_4 = 74.8$ is plotted 3.9 units below the corresponding point on the regression line, and similarly for the other observations.

Table 14.4

AVERAGE VALUES OF z FOR CORRESPONDING X_4 VALUES

X_4 Values	Number of Cases	Average of X_4	Average of z
Under 72.0	4	71.08	−1.28
72.0–72.9	5	72.58	−1.24
73.0–73.9	5	73.36	+1.46
74.0–74.9	10	74.30	+0.49
75.0–75.9	7	75.33	+0.91
76.0–76.9	5	76.44	+0.64
77.0 and over	2	78.00	−5.20
76.0 and over	7	76.89	−1.03

The last group, on the first grouping, has but two cases, so the last two groups are combined, giving the averages shown in the last line. The fact that both the items above 77 degrees are low, also evident in Figure 14.6, would give a little more reliability to the average based on only two items; but it is generally unsafe to give such an extreme bend to the end of a regression curve as this would call for, on the basis of so few observations. The larger grouping will therefore be used in this case,

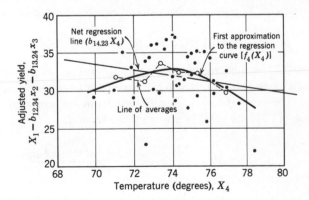

Fig. 14.6. Temperature and yield of corn adjusted to average rainfall and year, and first approximation curve fitted to the averages.

leaving the subsequent approximations to determine whether the more extreme bend is justified.

The line of averages in Figure 14.6 indicates that yields may tend to rise as temperature increases up to between 73 and 75 degrees, and then to fall as the temperature goes still higher. A smooth curve is therefore drawn in, averaging out the irregularities shown in the broken line of the group averages and conforming to the limiting conditions stated on page 213. It does not make much difference if these first approximation curves are not drawn in in exactly the right position or shape, as the subsequent operations will tend to correct them to the proper shape if the original one is incorrect. It is for that reason that fairly accurate results can be secured by this graphic process, even though the true shape of the curves is not known at the beginning.

ESTIMATING X_1 FROM THE FIRST APPROXIMATION CURVES. We have now arrived at first approximations to the net regression curves for X_1, against each of the three factors. It must be remembered that in making the adjustments on X_1 to arrive at these curves, only the net *linear* effects of the other independent variables have been eliminated. Now that we have at least an approximate measure of the curvilinear relations of X_1 to the independent variables, making adjustments to eliminate these approximate curvilinear effects may enable us to determine more accurately the true curvilinear relation to each variable.

The first step in the next stage of the process is to work out estimated values of X_1 based on the curvilinear relations. To do this we may designate the relation between X_1 and X_2 shown by the curve in Figure 14.5 as $f_2'(X_2)$; the relation between X_1 and X_3 shown in Figure 14.4 as $f_3'(X_3)$; and the relation between X_1 and X_4 shown in Figure 14.6 as $f_4'(X_4)$. The estimates of X_1 may then be worked out by the regression equation

$$X_1'' = a_{1.234}' + f_2'(X_2) + f_3'(X_3) + f_4'(X_4) \tag{14.4}$$

The symbol X_1'' is used to designate this second set of estimates, just as X_1' was used to designate the first set, worked out from the linear regression equation. The constant $a_{1.234}'$ is different from the constant $a_{12.34}$ used in equation (11.7); its value is given by the formula

$$a_{1.234}' = M_1 - \frac{\Sigma[f_2'(X_2) + f_3'(X_3) + f_4'(X_4)]}{n} \tag{14.5}$$

To work out $a_{12.34}'$ according to equation (14.5), it is first necessary to work out the value $f_2'(X_2) + f_3'(X_3) + f_4'(X_4)$ for each observation. For the first observation, for example, $X_2 = 0$, $X_3 = 9.6$, and $X_4 = 74.8$.

From $f_2'(X_2)$, given in Figure 14.5, the curve reading (or ordinate) corresponding to a value of 0 for X_2 is 27.3. For $f_3'(X_3)$, Figure 14.4, the ordinate of the curve corresponding to $X_3 = 9.6$ is 31.7. For $f_4'(X_4)$, Figure 14.6, the curve ordinate corresponding to $X_4 = 74.8$ is 32.5. The value $[f_2'(X_2) + f_3'(X_3) + f_4'(X_4)]$ for the first observation is therefore $[27.3 + 31.7 + 32.5]$, or 91.5. The sum of these values for each observation is the value required in equation (14.5).

Before continuing the process of reading each value from the charts for the remaining observations, it should be noted that, since many observations of each variable have the same values, the same point would be read from each chart many times. The process of working out the computations can be much simplified by reading each required value from each chart once and for all, and recording it so that it can be used each time. Since each chart indicates each individual observation for each independent variable, only those points for which there are observations need be recorded. Carrying out this process, we may record the functional relations as shown in Tables 14.5, 14.6, and 14.7, which show the readings from Figures 14.5, 14.4, and 14.6, respectively.[12]

Table 14.5

VALUES OF X_1 CORRESPONDING TO GIVEN VALUES OF X_2, FROM THE FIRST APPROXIMATION CURVE

X_2	$f_2'(X_2)$	X_2	$f_2'(X_2)$	X_2	$f_2'(X_2)$	X_2	$f_2'(X_2)$
0	27.3	10	30.8	20	32.8	29	33.4
1	27.8	11	31.0	21	33.0	30	33.5
2	28.2	12	31.3	22	33.1	31	33.5
3	28.6	13	31.5	23	33.1	32	33.5
4	29.0	14	31.7	24	33.2	33	33.5
5	29.4	15	31.9	25	33.2	34	33.5
6	29.7	16	32.1	26	33.3	35	33.5
7	30.0	17	32.3	27	33.3	36	33.5
8	30.3	18	32.5	28	33.4	37	33.5
9	30.6	19	32.6				

[12] In entering these values it is not worth while reading further than the first decimal, for the line is not drawn more accurately than to within 0.1 or 0.2. The accuracy depends, of course, on the scale; but it is not worth using very large charts to secure spuriously high accuracy, when the standard error of any particular point on the curve is probably several units and when the curve is only a first approximation, subject to subsequent modification.

Table 14.6

VALUES OF X_1 CORRESPONDING TO GIVEN VALUES OF X_3, FROM THE FIRST
APPROXIMATION CURVE

X_3	$f_3'(X_3)$	X_3	$f_3'(X_3)$	X_3	$f_3'(X_3)$	X_3	$f_3'(X_3)$
6.8	24.6	9.5	31.5	10.8	33.4	12.9	33.3
6.9	25.0	9.6	31.7	11.0	33.5	13.0	33.2
7.7	27.1	9.9	32.4	11.3	33.6	13.6	32.9
7.8	27.4	10.0	32.5	11.5	33.7	13.9	32.7
8.0	27.9	10.1	32.6	11.6	33.7	14.1	32.5
8.7	29.7	10.4	33.1	12.0	33.7	16.2	31.0
9.3	31.0	10.6	33.3	12.1	33.6	16.5	30.8
9.4	31.2	10.7	33.4	12.5	33.5		

Table 14.7

VALUES OF X_1 CORRESPONDING TO GIVEN VALUES OF X_4, FROM THE FIRST
APPROXIMATION CURVE

X_4	$f_4'(X_4)$	X_4	$f_4'(X_4)$	X_4	$f_4'(X_4)$	X_4	$f_4'(X_4)$
69.9	30.2	73.0	32.5	74.2	32.8	75.7	31.6
71.0	31.0	73.2	32.6	74.3	32.7	75.8	31.5
71.5	31.4	73.3	32.6	74.6	32.6	76.0	31.3
71.9	31.7	73.6	32.7	74.8	32.5	76.2	31.0
72.0	31.8	73.7	32.7	75.0	32.3	76.9	30.1
72.6	32.2	74.0	32.8	75.2	32.1	77.6	29.0
72.8	32.3	74.1	32.8	75.3	32.0	78.4	27.6
72.9	32.4						

The values to determine $a_{1.234}'$ may now be worked out in orderly
manner, as shown in Table 14.8, in the fourth to the seventh columns.

This computation gives us the sum of the respective functional values
for the 38 observations. Substituting this sum and the number of observations in equation (14.5), we find the required constant to be

$$a_{1.234}' = 31.916 - \frac{3{,}621.9}{38} = -63.397$$

Since the functional values for our regression equation are expressed only
to one decimal point, we shall use -63.4 for $a_{1.234}'$, which will result in
the estimated values being 0.003 unit too low, on the average.

Table 14.8

COMPUTATION OF FUNCTIONAL VALUES CORRESPONDING TO INDEPENDENT VARIABLES, OF THE ESTIMATED VALUE OF X_1, AND THE NEW RESIDUAL, FOR EACH OBSERVATION

X_2	X_3	X_4	$f_2'(X_2)$	$f_3'(X_3)$	$f_4'(X_4)$	$f_2'(X_2)$ $+f_3'(X_3)$ $+f_4'(X_4)$	$\Sigma(f)+a'$ $=X_1''$	X_1	X_1-X_1'' z''
(1)	(2)	(3)	(4)	(5)	(6)	(7)	(8)	(9)	(10)
0	9.6	74.8	27.3	31.7	32.5	91.5	28.1	24.5	−3.6
1	12.9	71.5	27.8	33.3	31.4	92.5	29.1	33.7	4.6
2	9.9	74.2	28.2	32.4	32.8	93.4	30.0	27.9	−2.1
3	8.7	74.3	28.6	29.7	32.7	91.0	27.6	27.5	−0.1
4	6.8	75.8	29.0	24.6	31.5	85.1	21.7	21.7	0
5	12.5	74.1	29.4	33.5	32.8	95.7	32.3	31.9	−0.4
6	13.0	74.1	29.7	33.2	32.8	95.7	32.3	36.8	4.5
7	10.1	74.0	30.0	32.6	32.8	95.4	32.0	29.9	−2.1
8	10.1	75.0	30.3	32.6	32.3	95.2	31.8	30.2	−1.6
9	10.1	75.2	30.6	32.6	32.1	95.3	31.9	32.0	0.1
10	10.8	75.7	30.8	33.4	31.6	95.8	32.4	34.0	1.6
11	7.8	78.4	31.0	27.4	27.6	86.0	22.6	19.4	−3.2
12	16.2	72.6	31.3	31.0	32.2	94.5	31.1	36.0	4.9
13	14.1	72.0	31.5	32.5	31.8	95.8	32.4	30.2	−2.2
14	10.6	71.9	31.7	33.3	31.7	96.7	33.3	32.4	−0.9
15	10.0	74.0	31.9	32.5	32.8	97.2	33.8	36.4	2.6
16	11.5	73.7	32.1	33.7	32.7	98.5	35.1	36.9	1.8
17	13.6	73.0	32.3	32.9	32.5	97.7	34.3	31.5	−2.8
18	12.1	73.3	32.5	33.6	32.6	98.7	35.3	30.5	−4.8
19	12.0	74.6	32.6	33.7	32.6	98.9	35.5	32.3	−3.2
20	9.3	73.6	32.8	31.0	32.7	96.5	33.1	34.9	1.8
21	7.7	76.2	33.0	27.1	31.0	91.1	27.7	30.1	2.4
22	11.0	73.2	33.1	33.5	32.6	99.2	35.8	36.9	1.1
23	6.9	77.6	33.1	25.0	29.0	87.1	23.7	26.8	3.1
24	9.5	76.9	33.2	31.5	30.1	94.8	31.4	30.5	−0.9
25	16.5	69.9	33.2	30.8	30.2	94.2	30.8	33.3	2.5
26	9.3	75.3	33.3	31.0	32.0	96.3	32.9	29.7	−3.2
27	9.4	72.8	33.3	31.2	32.3	96.8	33.4	35.0	1.6
28	8.7	76.2	33.4	29.7	31.0	94.1	30.7	29.9	−0.8
29	9.5	76.0	33.4	31.5	31.3	96.2	32.8	35.2	2.4
30	11.6	72.9	33.5	33.7	32.4	99.6	36.2	38.3	2.1
31	12.1	76.9	33.5	33.6	30.1	97.2	33.8	35.2	1.4
32	8.0	75.0	33.5	27.9	32.3	93.7	30.3	35.5	5.2
33	10.7	74.8	33.5	33.4	32.5	99.4	36.0	36.7	0.7
34	13.9	72.6	33.5	32.7	32.2	98.4	35.0	26.8	−8.2
35	11.3	75.3	33.5	33.6	32.0	99.1	35.7	38.0	2.3
36	11.6	74.1	33.5	33.7	32.8	100.0	36.6	31.7	−4.9
37	10.4	71.0	33.5	33.1	31.0	97.6	34.2	32.6	−1.6
Totals			1,208.4	1,204.2	1,209.3	3,621.9			

It is now possible to complete the process of computing X_1'', the estimated value of X_1, using the first approximation curves, according to equation (14.4), and the constant which has just been computed. When equations (14.4) and (14.5) are compared, it is evident that, except for the constant term, X_1'' is equal to the values that have just been computed in the seventh column of Table 14.8. Accordingly, all that is necessary is to subtract 63.4 from each of those values. This step is shown also in Table 14.8, in the eighth column.

The column headed X_1'' shows the estimated values obtained by this process. The next step is to see whether the new estimates come any nearer to reproducing the observed values of X_1 than did the first set of estimates, based on the linear regression equation. We therefore compute a new set of residuals, z'', by subtracting the new estimates from the actual values of X_1. This step, also, is shown in Table 14.8.

$$z'' = X_1 - X_1'' \tag{14.6}$$

If the individual residuals shown are compared with the residuals obtained by the linear regression, as computed in Table 14.1, it will be seen that in general the new residuals are smaller than the previous ones, though the reverse is true in many cases. There are 23 cases in which the new residual is smaller, and 15 in which it is larger than the original residual. A more accurate comparison can be obtained by comparing the adjusted standard deviations of the residuals for the two sets. If we assume that the curves involving X_3 and X_4 each use up three degrees of freedom while the relationship to X_2 requires four, the multiple curvilinear function as a whole involves eleven degrees of freedom, compared with four degrees for the multiple linear regression. For the linear correlation, the adjusted standard deviation of the residuals was 3.8 bushels, whereas the adjusted standard deviation of the new residuals is 3.5 bushels. Apparently the new estimates do come nearer to the observed values, on the average, than did the first set of estimates.

Determining the Second Approximation Net Regression Curves. The regression curves used in constructing the estimate X_1'' were only the first approximations to the true curvilinear relations, since they were determined by eliminating only the linear effects of the other independent factors. Now that the residuals obtained by the use of the first approximation curves have been computed, however, we can determine whether any change in the shape of the several curves is necessary.

To do this we construct Figure 14.7 by drawing in the regression curve from Figure 14.5, using the same scale as before. Use of Table 14.5 makes it easier to reproduce the curve. Next we plot each of the last residuals as a deviation just as before, except that now the residuals are plotted as deviations from the regression curve, instead of from the regression line, at the point corresponding to the independent variable X_2. Thus the first observation, with $X_2 = 0$, has $z'' = -3.6$. The point on the curve corresponding to $X_2 = 0$ is 27.3; so the dot has for ordinate 27.3 − 3.6, or 23.7. The values for the next observation are $X_2 = 1$ and $z'' = 4.6$. The corresponding value of $f_2'(X_2)$ is 27.8, so the ordinate

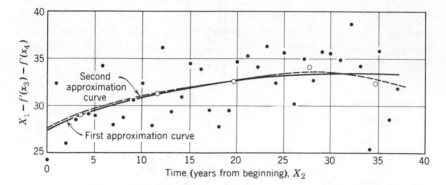

Fig. 14.7. Time, and yield of corn adjusted to average temperature and rainfall on basis of first approximation curves, and second approximation to $f_3(X_2)$.

for the dot is 27.8 + 4.6, or 32.4. The coordinates for this dot are therefore 1 and 32.4. The remaining observations are plotted in the same manner, shortening the process by scaling the value for z'' directly above or below from the corresponding point on the regression curve.

With the dots all plotted, it is evident that the scatter is too great to indicate definitely changes which may be needed in the curve, if any, simply from the dots alone. Accordingly the residuals are averaged in groups, employing the same grouping as before (Table 14.3), which eliminates the need of averaging the corresponding X_2 values over again. The new averages work out as shown in Table 14.9.

Table 14.9

AVERAGE VALUES OF z'', FOR CORRESPONDING X_2 VALUES

X_2 Values	Number of Cases	Average of X_2	Average of z''
0– 7	8	3.5	+0.10
8–15	8	11.5	+0.16
16–23	8	19.5	−0.08
24–31	8	27.5	+0.64
32–37	6	34.5	−1.08

The averages are next plotted as deviations from the first approximation curve. They indicate that a slight raise in the lower part of the curve may be needed, and a downward bend toward the end. It appears that now that the influence of rainfall and temperature on yield have been more accurately allowed for, the upward trend with time is slightly less than it seemed before in the early years; and the trend seems to have turned downward toward the end of the series—the exact year or extent of the turn is indeterminate. A new curve is therefore drawn in in Figure 14.7, and, as it happens, a smooth, continuous curve can be drawn exactly through each of the first three group averages, but not having the extreme bend indicated by the last two group averages.[13]

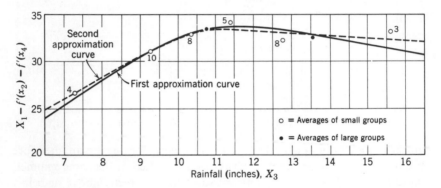

Fig. 14.8. Rainfall, and yield of corn adjusted to average temperature and time on the basis of first approximation curves, and second approximation to $f_3(X_3)$

The same process may now be applied to X_3, to see if any change need be made in the first regression curve for the change in X_1 with changes in that variable. This process is carried out as shown in Figure 14.8, the first approximation curve being drawn in just as before, using the data given in Table 14.6.

Instead of plotting the individual residuals for each observation, as was just done with respect to X_2, we may proceed at once to compute the average residuals for each of the groups of values of X_3, since it is sufficiently apparent from Figure 14.7 that the scatter of the individual observations is still too great to serve as a guide in correcting the first approximation curves. Averaging the residuals gives the averages shown in Table 14.10.

---

[13] As Figure 14.7 shows, it is the coincidence of 3 low years out of the last 4 in the series that provides the basis for this downturn. This is really too small a sample to provide a firm basis for changing a curve.

Table 14.10

AVERAGE VALUES OF z'', FOR CORRESPONDING X_3 VALUES

X_3 Values	Number of Cases	Average of X_3	Average of z''	Average of X_3	Average of z''
Under 8.0	4	7.30	+0.58		
8.0– 9.9	10	9.19	+0.03		
10.0–10.9	8	10.35	−0.15 ⎫		
11.0–11.9	5	11.40	+0.48 ⎭	10.75	+0.09
12.0–13.9	8	12.76	−1.11 ⎫		
14.0 and over	3	15.60	+1.73 ⎭	13.53	−0.34

Again the averages are somewhat irregular when plotted, so the last four groups are reduced to two, and the new averages plotted and indicated separately. The number of observations represented by each of the first set of averages is indicated next to it, so that averages based on a small number of observations will not be given undue weight in drawing in the curve. It might be desirable in some cases, also, to try regrouping the cases into different groups—say from 8.5–9.4, 9.5–10.4, etc.—and see if that would change at all the indications as to the shifts needed in the first curve. Working that out in this case, the changes needed are still found to be about the same as shown by the group averages in Figure 14.8,

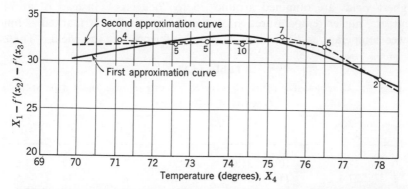

Fig. 14.9. Temperature, and yield of corn adjusted to average rainfall and time on the basis of first approximation curves, and second approximation to $f_4(X_4)$.

though somewhat less regular, owing to the smaller size of groups. A new curve is then drawn in freehand, as indicated by the group averages, rising somewhat higher than formerly at both ends, and not rising quite so high in the central portion as before.

Turning to the relation between X_1 and X_4, the first approximation curve for $f_4'(X_4)$ is reproduced in Figure 14.9, using the values given in

Table 14.7. The next step is to average the values of z'' for corresponding values of X_4. Using the same groupings used in Table 14.4, we arrive at the averages shown in Table 14.11.

Plotting these new averages, and connecting them by a broken line, we see that the relation of yield to temperature may be quite different

Table 14.11

AVERAGE VALUES OF z'', FOR CORRESPONDING X_4 VALUES

X_4 Values	Number of Cases	Average of X_4	Average of z''
Under 72.0	4	71.08	+1.15
72.0–72.9	5	72.58	−0.36
73.0–73.9	5	73.36	−0.58
74.0–74.9	10	74.30	−0.86
75.0–75.9	7	75.33	+0.63
76.0–76.9	5	76.44	+0.90
77.0 and over	2	78.00	−0.05
76.0 and over	7	76.89	+0.63

from the way it appeared on the first approximation. Apparently the highest yields are obtained around 75 to 76 degrees, instead of at 74 degrees; higher temperatures appear to reduce the yield markedly, but lower temperatures have only a slight influence on the yield. These

Table 14.12

VALUES OF X_1 CORRESPONDING TO GIVEN VALUES OF X_2, FROM THE SECOND APPROXIMATION CURVE

X_2	$f_2''(X_2)$	X_2	$f_2''(X_2)$	X_2	$f_2''(X_2)$	X_2	$f_2''(X_2)$
0	27.4	10	31.0	20	32.7	29	33.6
1	27.9	11	31.2	21	33.0	30	33.5
2	28.4	12	31.4	22	33.2	31	33.4
3	28.8	13	31.6	23	33.3	32	33.2
4	29.2	14	31.8	24	33.4	33	33.0
5	29.5	15	32.0	25	33.5	34	32.8
6	29.8	16	32.1	26	33.6	35	32.6
7	30.2	17	32.3	27	33.7	36	32.4
8	30.4	18	32.5	28	33.7	37	32.2
9	30.7	19	32.6				

indications are all within the theoretical limitations on the shape of the curve, as stated on page 213. The new curve, drawn in freehand so as to pass as nearly through these new averages as possible and still maintain a smooth continuous shape, with only a single maximum, expresses these relations.

ESTIMATING X_1 FROM THE SECOND APPROXIMATION CURVES. Now that the second approximation curves have been determined for each variable, we can proceed to estimate values of X_1 on the basis of the revised curves, to see whether the new curves enable us to estimate X_1 any more accurately than the first set of curves did. To facilitate the process we first construct Tables 14.12, 14.13, and 14.14 for $f_2''(X_2)$, $f_3''(X_3)$, and $f_4''(X_4)$, showing the readings for the functions from the revised curves.

Table 14.13

VALUES OF X_1 CORRESPONDING TO GIVEN VALUES OF X_3, FROM THE SECOND APPROXIMATION CURVE

X_3	$f_3''(X_3)$	X_3	$f_3''(X_3)$	X_3	$f_3''(X_3)$	X_3	$f_3''(X_3)$
6.8	25.5	9.5	31.5	10.8	33.3	12.9	33.0
6.9	25.7	9.6	31.7	11.0	33.4	13.0	33.0
7.7	27.5	9.9	32.2	11.3	33.4	13.6	32.8
7.8	27.8	10.0	32.3	11.5	33.3	13.9	32.7
8.0	28.2	10.1	32.5	11.6	33.3	14.1	32.7
8.7	29.9	10.4	32.9	12.0	33.2	16.2	32.2
9.3	31.1	10.6	33.1	12.1	33.2	16.5	32.1
9.4	31.3	10.7	33.2	12.5	33.1		

Table 14.14

VALUES OF X_1 CORRESPONDING TO GIVEN VALUES OF X_4, FROM THE SECOND APPROXIMATION CURVE

X_4	$f_4''(X_4)$	X_4	$f_4''(X_4)$	X_4	$f_4''(X_4)$	X_4	$f_4''(X_4)$
69.9	31.6	73.0	32.0	74.2	32.2	75.7	32.2
71.0	31.7	73.2	32.0	74.3	32.2	75.8	32.2
71.5	31.8	73.3	32.0	74.6	32.2	76.0	32.1
71.9	31.8	73.6	32.1	74.8	32.3	76.2	32.0
72.0	31.8	73.7	32.1	75.0	32.3	76.9	30.7
72.6	31.9	74.0	32.2	75.2	32.3	77.6	29.1
72.8	32.0	74.1	32.2	75.3	32.3	78.4	27.3
72.9	32.0						

To simplify the calculations, 20 is subtracted from each of the functional values in making subsequent entries. The computations to determine the estimated values are then carried out as shown in detail in Table 14.15,

Table 14.15

COMPUTATION OF FUNCTIONAL VALUES, FROM THE SECOND APPROXIMATION CURVES, CORRESPONDING TO INDEPENDENT VARIABLES FOR EACH OBSERVATION, AND COMPUTATION OF ESTIMATED VALUE FOR X_1 AND OF NEW RESIDUALS

Independent Variables			Corresponding Functional Values*			$f_2''(X_2)$ $+ f_3''(X_3)$ $+ f_4''(X_4)$	$(7) - a$ $= X_1'''$	Dependent Variable, X_1	$X_1 - X_1'''$ z'''
X_2 (1)	X_3 (2)	X_4 (3)	$f_2''(X_2)$ (4)	$f_3''(X_3)$ (5)	$f_4''(X_4)$ (6)	(7)	(8)	(9)	(10)
0	9.6	74.8	7.4	11.7	12.3	31.4	28.0	24.5	−3.5
1	12.9	71.5	7.9	13.0	11.8	32.7	29.3	33.7	4.4
2	9.9	74.2	8.4	12.2	12.2	32.8	29.4	27.9	−1.5
3	8.7	74.3	8.8	9.9	12.2	30.9	27.5	27.5	0
4	6.8	75.8	9.2	5.5	12.2	26.9	23.5	21.7	−1.8
5	12.5	74.1	9.5	13.1	12.2	34.8	31.4	31.9	0.5
6	13.0	74.1	9.8	13.0	12.2	35.0	31.6	36.8	5.2
7	10.1	74.0	10.2	12.5	12.2	34.9	31.5	29.9	−1.6
8	10.1	75.0	10.4	12.5	12.3	35.2	31.8	30.2	−1.6
9	10.1	75.2	10.7	12.5	12.3	35.5	32.1	32.0	−0.1
10	10.8	75.7	11.0	13.3	12.2	36.5	33.1	34.0	0.9
11	7.8	78.4	11.2	7.8	7.3	26.3	22.9	19.4	−3.5
12	16.2	72.6	11.4	12.2	11.9	35.5	32.1	36.0	3.9
13	14.1	72.0	11.6	12.7	11.8	36.1	32.7	30.2	−2.5
14	10.6	71.9	11.8	13.1	11.8	36.7	33.3	32.4	−0.9
15	10.0	74.0	12.0	12.3	12.2	36.5	33.1	36.4	3.3
16	11.5	73.7	12.1	13.3	12.1	37.5	34.1	36.9	2.8
17	13.6	73.0	12.3	12.8	12.0	37.1	33.7	31.5	−2.2
18	12.1	73.3	12.5	13.2	12.0	37.7	34.3	30.5	−3.8
19	12.0	74.6	12.6	13.2	12.2	38.0	34.6	32.3	−2.3
20	9.3	73.6	12.7	11.1	12.1	35.9	32.5	34.9	2.4
21	7.7	76.2	13.0	7.5	12.0	32.5	29.1	30.1	1.0
22	11.0	73.2	13.2	13.4	12.0	38.6	35.2	36.9	1.7
23	6.9	77.6	13.3	5.7	9.1	28.1	24.7	26.8	2.1
24	9.5	76.9	13.4	11.5	10.7	35.6	32.2	30.5	−1.7
25	16.5	69.9	13.5	12.1	11.6	37.2	33.8	33.3	−0.5
26	9.3	75.3	13.6	11.1	12.3	37.0	33.6	29.7	−3.9
27	9.4	72.8	13.7	11.3	12.0	37.0	33.6	35.0	1.4
28	8.7	76.2	13.7	9.9	12.0	35.6	32.2	29.9	−2.3
29	9.5	76.0	13.6	11.5	12.1	37.2	33.8	35.2	1.4
30	11.6	72.9	13.5	13.3	12.0	38.8	35.4	38.3	2.9
31	12.1	76.9	13.4	13.2	10.7	37.3	33.9	35.2	1.3
32	8.0	75.0	13.2	8.2	12.3	33.7	30.3	35.5	5.2
33	10.7	74.8	13.0	13.2	12.3	38.5	35.1	36.7	1.6
34	13.9	72.6	12.8	12.7	11.9	37.4	34.0	26.8	−7.2
35	11.3	75.3	12.6	13.4	12.3	38.3	34.9	38.0	3.1
36	11.6	74.1	12.4	13.3	12.2	37.9	34.5	31.7	−2.8
37	10.4	71.0	12.2	12.9	11.7	36.8	33.4	32.6	−0.8
Totals			447.6	445.1	448.7	1,341.4			

* Less 20.0 for each functional reading.

just as for Table 14.8. In practical computation these entries, for the second approximation curves, would be made on the same sheet as were the entries in Table 14.8 for the first approximation curves, thus eliminating the work of entering the values of X_1, X_2, X_3, and X_4 over again.

Table 14.15 is worked out just as was Table 14.8. Thus the data for the first observation show values of 0, 9.6, and 74.8 for X_2, X_3, and X_4, respectively. Looking up the corresponding values in Tables 14.12, 14.13, and 14.14 gives values of 27.4, 31.7, and 32.3, for the three functional values. Subtracting 20 from each value, to reduce the subsequent clerical work, we enter 7.4, 11.7, and 12.3, in the functional columns. The three functional values are then added, and the sum entered in the seventh column. The entries for the functional readings are completed as shown, and the sum computed for each observation. Then the average of the seventh column is determined, giving the value 35.30. As the average of X_1 is 31.916, the value of the new constant, $a''_{1.234}$, is found by equation (14.5) to be

$$a''_{1.234} = 31.916 - 35.300$$
$$= -3.384$$

Accordingly, 3.4 is subtracted from each of the values in column 7 to give the estimated value of X_1, X'''_1, which is then entered in the eighth column.

The final step in computing the table is to subtract each of the estimated values, X'''_1, from the actual value X_1, giving the residuals z''', which appear in the last column.

Comparing the new residuals, z''', with the previous ones, z'', given in Table 14.9, we find that their size has been increased in just about as many cases as it has been decreased. But when we compute the standard deviation of the new residuals, we find that the adjusted standard deviation of z''' is 3.3 bushels, or slightly smaller than the adjusted standard deviation of z'', 3.5 bushels, assuming that the new curves also use a total of eleven degrees of freedom.

Correcting the Curves by Further Successive Approximations. The process could be carried through one or more additional approximations by repeating the steps shown. Thus, the last residuals, z''', when averaged and plotted with respect to the second set of approximation curves, would indicate whether any further modifications were needed in the curves; if any were needed, new readings would be made from the new curves, new estimates of X_1 obtained from them, and another set of residuals determined.

The number of successive approximations used in any given case would depend upon several considerations. In Chapter 10 it was noted that, with linear relations, repeated applications of the method of successive elimination would approach more and more closely to the net regression lines that would be obtained if a multiple linear regression equation were fitted to the same data by least squares. The successive elimination method could never improve upon the least squares norm. As long as the standard deviation of the residuals continued to decline, we would know that the successive steps were bringing us closer to the least squares fit.

The method of successive approximations would have a similar norm or limit for curvilinear relations only if we held the mathematical form of each net regression curve exactly the same during the entire process. With freehand methods we often change the shapes of the net regression curves somewhat from one step to the next, and we may sometimes make offsetting changes in the shapes of two net regression curves that leave the standard deviation of the residuals unchanged. In most practical situations, however, one can arrive fairly soon at an approximation which conforms to the hypothetical limitations placed upon the curves and also approaches the minimum standard deviation of residuals for the general types of curves chosen. As there is no definite criterion of best fit in freehand curvilinear regression, it may be advisable to average the curve readings from two or three successive approximations after the standard error of estimate has approached a stable minimum value to somewhat reduce arbitrary elements in the final positions of the curves. This lack of definiteness has sometimes been regarded as a serious weakness of freehand methods. But it is equally possible to choose two or more mathematical functions that will yield almost identical standard errors of estimate when fitted to the same data by the method of least squares. The mathematical net regression curves will differ at least slightly from one function to another, and the selection of any one function as "best" will be arbitrary in much the same sense as will the selection of final curves in the freehand case.

Stating the Final Conclusions. After the final shape of the several net regression curves has been determined either by graphic or algebraic processes, it still remains to state those curves in such a manner that their meaning is perfectly clear. The several functions may be stated to show the value of the dependent factor associated with given values of the particular independent factor when values of other independent factors are held at their means. There are two alternative ways of stating the associated values: (1) as actual values; and (2) as deviations from the mean values.

To state the associated values as actual values, we may use the following procedure for graphic curves:

First, the mean of all the values read from the final curve is determined. For $f_2(X_2)$, this mean may be designated $M_{f(2)}$. The values from the curve are read off for selected intervals of X_2. Then the estimated values of X_1 for each of these values of X_2 (with values of X_3, X_4, etc., at their means) are determined by subtracting the mean of the curve readings from each of these actual readings and adding to the result the mean of X_1. That is, if we use $X_1 = F_2(X_2)$ to designate these values of X_1, estimated from the net curvilinear relation to X_2, we can define them by the equation

$$X_1' = F_2(X_2) = f_2(X_2) - M_{f(2)} + M_1 \tag{14.7}$$

If, however, the expected values of X_1 for given values of X_2 are to be stated merely as deviations from the mean values, those deviations may be determined by subtracting from each curve reading the mean of all the curve readings. If we use $F_2(x_2)$ to designate these expected deviations from the mean values, we may define them by the equation

$$x_1' = F_2(x_2) = f_2(X_2) - M_{f(2)} \tag{14.8}$$

It is evident, from equations (14.7) and (14.8), that

$$F_2(X_2) = F_2(x_2) + M_1$$

In the actual statement of the results of a regression study, it is frequently desirable to state the relation of the dependent factor to the most important independent factor according to equation (14.7), and to state the relation for the remaining independent factors according to equation (14.8). When that is done, the estimated values of X_1, based on all the independent factors, may be readily computed by taking the estimate from the most important factor, and then adding to or subtracting from that the adjustments to take account of the departures of other factors from their means. Using X_1' to designate this final estimate of the value X_1, and taking X_3 as the most important factor, we make the estimate by the equation

$$X_1' = F_2(x_2) + F_3(X_3) + F_4(x_4) + \ldots + F_k(x_k) \tag{14.9}$$

The process of working out these final statements of the net curvilinear regression lines may be illustrated by the data of the corn-yield problem. Since the rainfall (X_3) was apparently the most important factor, that may be taken as the one for which the regression is to be stated according to equation (14.7). If we regard the second approximation curve shown in Figure 14.8 and Table 14.13 as the final curve, then Table 14.15 gives the readings from this curve for each of the individual observations.

The mean of the readings of $f_3(X_3)$ is next computed from the values of Table 14.15. The sum of the 38 $f''(X_3)$ readings is 445.1, so

$$M_{f(X_3)} = \frac{445.1}{38} = 11.71$$

The mean value of X_1 is $M_1 = 31.92$. From equation (14.7),

$$F_3(X_3) = f_3(X_3) - M_{f(3)} + M_1$$

which is

$$F_3(X_3) = f_3(X_3) - 11.71 + 31.92$$

$$= f_3(X_3) + 20.21$$

All that is necessary, therefore, is to add the new constant, 20.2, to the values read from the curve. This process is shown in Table 14.16.

Table 14.16

COMPUTATION OF AVERAGE YIELD OF CORN WITH VARYING RAINFALL, HOLDING TREND IN YIELD AND INFLUENCE OF TEMPERATURE CONSTANT

Inches of Rainfall, X_3	Readings from Final Curve,* $f_3''(X_3)$	Constant, $M_1 - M_{f_3}$	Average Yield, $F_3(X_3)$
7	6.0	20.2	26.2
8	8.2	20.2	28.4
9	10.5	20.2	30.7
10	12.3	20.2	32.5
11	13.4	20.2	33.6
12	13.2	20.2	33.4
13	13.0	20.2	33.2
14	12.7	20.2	32.9
15	12.5	20.2	32.7
16	12.3	20.2	32.5

* Curve readings minus 20, just as entered in Table 14.15.

The computation for $F_4(x_4)$ follows the same form as that for $F_3(X_3)$, save that equation (14.8) is used instead, hence the mean of X_1 is not involved. First the mean of all the readings for $f_4(X_4)$, as shown in Table 14.15, is computed, giving the value of 11.81. The values for $F_4(x_4)$ are therefore given by the equation

$$F_4(x_4) = f_4''(X_4) - M_{f(X_4)}$$

$$= f_4''(X_4) - 11.81$$

These values are worked out in Table 14.17.

The net correction in the estimated yield to allow for the influence of trend can be obtained by carrying through a similar computation for

Table 14.17

COMPUTATION OF DEVIATION OF CORN YIELDS FROM YIELDS OTHERWISE EXPECTED, BECAUSE OF DIFFERENCES IN TEMPERATURE FOR SEASON

Average Temperature, X_4	Readings from Final Curve,* $f_4''(X_4)$	Constant, M_{f_4}	Correction to Expected Yield, $F_4(x_4)$
70.0	11.6	−11.8	−0.2
71.0	11.7	−11.8	−0.1
72.0	11.8	−11.8	0
73.0	12.0	−11.8	0.2
74.0	12.2	−11.8	0.4
75.0	12.3	−11.8	0.5
76.0	12.1	−11.8	0.3
77.0	10.5	−11.8	−1.3
78.0	8.3	−11.8	−3.5

* Curve readings minus 20, just as entered in Table 14.15.

$F_2(x_2)$. The readings for $f_2''(X_2)$ sum to 447.6, so $M_{f(2)} = 11.78$. The values of $F_2(x_2)$ are then given by the equation

$$F_2(x_2) = f_2''(X_2) - 11.78$$

This computation is carried out in Table 14.18.

The conclusions of the study can then be stated as shown in the last column of each of the last three tables, free from all the previous details.

Table 14.18

COMPUTATION OF DEVIATION OF CORN YIELDS FROM THOSE OTHERWISE EXPECTED, BECAUSE OF NET TREND IN YIELDS

Number of Year, X_2	Date	Readings from Final Curve,* $f_2''(X_2)$	Constant, M_{f_2}	Correction to Expected Yield, $F_2(x_2)$
0	1890	7.4	−11.8	−4.4
5	1895	9.5	−11.8	−2.3
10	1900	11.0	−11.8	−0.8
15	1905	12.0	−11.8	0.2
20	1910	12.7	−11.8	0.9
25	1915	13.5	−11.8	1.7
30	1920	13.5	−11.8	1.7
35	1925	12.6	−11.8	0.8

* Curve readings minus 20.

The relations for each of the variables can also be combined to show the expected or estimated yield for various combinations of the independent factors. Thus for the present case, it might be desired to combine the findings into a table showing the expected or probable yield for any given combination of rainfall and temperature, with the 1927 trend of yield. These values can be obtained by taking the trend correction for 1927, plus 0.4 (the 12.2 under $f_2''(X_2)$ in Table 14.15 minus the constant 11.8 in Table 14.18), and combining it with the estimated influence of various quantities of rain and degrees of temperature. These estimates would then be defined by equation (14.9):

$$X_1' = F_2(x_2) + F_3(X_3) + F_4(x_4)$$
$$= 0.4 + F_3(X_3) + F_4(x_4)$$

Combining the readings for $F_3(X_3)$ from Table 14.16 with those for $F_4(x_4)$ from Table 14.17, and adding in the correction for $F_2(x_2)$ as just stated, we obtain estimated yields as shown in Table 14.19.[14]

Table 14.19

ESTIMATED YIELD OF CORN, IN BUSHELS PER ACRE, WITH VARYING RAINFALL
AND TEMPERATURE CONDITIONS, FOR 1927

Inches of Rainfall*	Average Temperature†				
	70°	72°	74°	76°	78°
7	‡	‡	27.0	26.9	23.1
9	30.9	31.1	31.5	31.4	‡
11	33.8	34.0	34.4	34.3	‡
13	33.4	33.6	34.0	‡	‡
15	32.9	33.1	‡	‡	‡

* Total for June, July, and August; average for nine Corn Belt stations.
† Average for June, July, and August, at same nine stations.
‡ This combination of factors was not represented in the observations analyzed.

In preparing a table such as Table 14.19, we should not enter values for combinations of the several factors which were not represented in the data on which the relations were based. Examination of a dot chart of the relation between rainfall and temperature, for the data included in the analysis, shows that no combinations of rainfall below 9 inches and temperature below 74 degrees appeared in the record, and no cases of temperature above 78 degrees with rainfall above 9 inches occurred. Accordingly, these combinations, and other combinations which were

[14] Table 14.19 may be compared with the results secured by cross-classifying and averaging the same data, by the methods of Chapter 23.

not represented, are left blank in the table, as shown. (A more exact method for measuring the representativeness of the relations is referred to in Chapter 19, on page 324.)

By combining a table such as Table 14.19 with a statement of the extent to which yields averaged higher or lower than those shown at different times through the period, all the conclusions from the study can be presented in simple form, easy to understand.

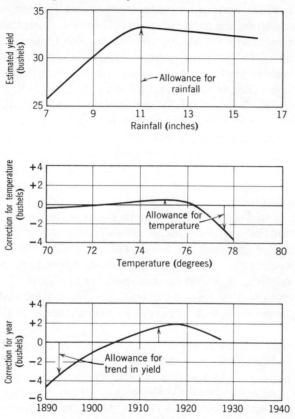

Fig. 14.10. Relation of yield of corn to rainfall, temperature, and time.

The final results of curvilinear correlation studies, after being simplified to the form shown in Tables 14.16 to 14.18, or in Table 14.19, may also be expressed graphically for final publication. Thus all three relations might be combined into a single figure, such as Figure 14.10, to present in relatively simple form the final conclusions reached by the statistical analysis.[15]

[15] A three-dimensional chart illustrating Table 14.19 is shown on page 349.

It might be noted at this point that Table 14.19 is much more than merely a table of average yields for various rainfall and temperature groups. There were only 38 observations to begin with, and only 14 of those were under 74 degrees temperature. If these 14 observations had been grouped according to year and rainfall, and the average yield determined for each class, only the roughest sort of groups could have been made, and even then the averages would have had but little reliability. As the result of the correlation study, however, all 38 observations have been drawn on to determine the relations. The table shows the yield most likely to be received with any of 16 different combinations of rainfall and temperature, for the trend in 1927. Other estimates could be shown for a large number of other combinations. Furthermore, it is known that estimates made from such tables agreed with the actual yields to within 2.8 bushels in about two-thirds of the original cases. The reliability of these estimated yields is thus greater than it would be for any average of a few cases alone. This example illustrates the ability of regression analysis both to bring out of a series of observations relations which are not observable on the surface, and to provide a basis for estimating the probable effect on the dependent factor of new combinations of the independent factors.

It should be noted that none of the three net regression curves in this example could be approximated at all satisfactorily by straight lines. The constants of the multiple linear regression equation on page 215 show only whether the net slope of the relationships were preponderantly positive or negative. It would be quite possible to obtain a linear partial regression coefficient of zero if the true curvilinear relationship were a second-degree parabola with its maximum near the mean of the observed data.

In the present example, the final net regression curves are not drastically different from the simple regression curves based on the original lines of averages (Figures 14.3, 14.4, and 14.5). The regression of yield on temperature showed the most substantial alteration, but judging from the scatter of residuals about the final curve (Figure 14.11), it was the least reliably determined from a sampling standpoint. In other examples more striking changes might be found, similar to the differences noted between simple and partial linear regressions in Chapters 10 through 13.

Limitations on the Use of the Results. The results of the corn-yield analysis apply only to the same area from which the data were drawn and to the period which they covered. Thus they provide no basis for estimating corn yields in other sections, and their use in estimating yields in other periods—as in subsequent years—is attended by increasing risk due especially to the necessity of extrapolating the trend regression.

Although this may give fair results for a year or two, it may tend to become increasingly inexact. For example, it may be that the trend of yield did not really turn downward about 1920, but only flattened out—additional years of observations are needed to tell which is correct.

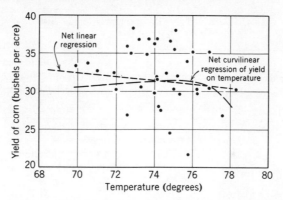

Fig. 14.11. Comparison of apparent relation of corn yields to temperature with net relation after eliminating influence of rainfall and of trend in yield.

Sufficient time has elapsed since the publication of the first edition, which included the preceding paragraph, to afford a check on the stability of the corn-yield relationships over the years. From the standpoint of stability, these relationships are vulnerable to the peculiar hazards attending time series which are subject in part to human control. These hazards will be further discussed in Chapter 20. They are also subject to sampling variability, which has been ignored in our treatment of multiple regression analysis up to this point but which will be discussed in Chapters 17 through 19.

Table 14.20 presents a continuation for 1928–1956 of the same series given for earlier years in Table 14.1. The time variable has, however, been renumbered, starting with 1928 as the first year. The yields as estimated from the 1890–1927 relationships were obtained as in Table 14.15; the trend, X_2, was extrapolated assuming a continued decline of 0.2 bushel a year, as indicated in Figure 14.10 for the period 1920–1927.

The basic data in Table 14.20 may be used as an exercise, starting with a simple plotting of rainfall, temperature, and yield as in Figure 14.2. It will be noted from such a figure that the average levels of the rainfall and temperature series are approximately the same as in the earlier period and that, as before, there is a fairly high negative correlation between rainfall and temperature. These two variables may apparently be regarded as "sampled" from the same statistical universe in both periods. However, from about 1939 on corn yields were substantially and continuously higher than would have been expected on the basis of 1890–1927 conditions. It is also clear that, except for temperature effects in 1934 and 1936, "time" accounts for a much greater proportion of the 1928–1956 variation

Table 14.20

YIELD OF CORN, RAINFALL, AND TEMPERATURE IN SIX LEADING STATES, 1928–1956

Year	Time, X_2	Rainfall, X_3	Temperature, X_4	Yield, X_1	Yield Estimated from 1890–1927 Relationships, X_1	Difference, $X_1 - X_1'$ z
		Inches	*Degrees*	*Bushels*		
1928	1	15.1	72.8	33.4	33.1	0.3
1929	2	10.6	73.4	31.5	33.7	−2.2
1930	3	6.4	76.4	25.8	24.3	1.5
1931	4	10.4	76.9	32.7	31.7	1.0
1932	5	13.5	76.0	35.4	32.7	2.7
1933	6	7.2	77.3	29.4	24.0	5.4
1934	7	7.5	80.0	18.9	17.4	1.5
1935	8	9.6	76.2	31.7	30.6	1.1
1936	9	4.9	80.0	18.5	10.6	7.9
1937	10	10.1	76.6	36.4	30.6	5.8
1938	11	12.9	76.2	36.5	31.3	5.2
1939	12	12.4	75.6	41.3	31.9	9.4
1940	13	9.1	75.4	38.1	28.9	9.2
1941	14	9.8	75.8	42.8	30.0	12.8
1942	15	13.4	74.4	49.1	30.9	18.2
1943	16	11.9	76.6	44.1	30.1	14.0
1944	17	11.0	75.5	42.2	31.2	11.0
1945	18	10.2	72.4	41.2	29.6	11.6
1946	19	12.3	73.8	47.8	30.4	17.4
1947	20	12.2	75.2	32.2	30.3	1.9
1948	21	11.9	74.7	53.8	30.0	23.8
1949	22	12.8	76.3	45.8	29.1	16.7
1950	23	14.3	71.4	46.5	28.6	17.9
1951	24	14.9	72.9	43.7	28.5	15.2
1952	25	10.8	77.1	52.5	27.5	25.0
1953	26	7.7	77.2	47.0	21.5	25.5
1954	27	12.2	77.4	46.2	26.3	19.9
1955	28	9.4	76.3	46.6	25.8	20.8
1956	29	11.5	75.5	52.9	28.7	24.2

Source: Computed from June, July, and August records for nine weather stations in Corn Belt states. Stations averaged include Kansas City, St. Louis, Toledo, Omaha, Peoria, Cincinnati, Topeka, Indianapolis, and the Iowa state average, as in the original study.

in corn yields than do rainfall and temperature combined. This is clearly indicated by the strong trend in the residuals in Table 14.20; these residuals evidently result from factors *other than* temperature and rainfall.

In view of this circumstance it seems desirable in analyzing the new observations to start out with the relation between corn yield and time. We have knowledge in addition to the yield series itself to provide some support in this analysis.

First, we know that hybrid seed corn became commercially important beginning in the middle 1930's. Controlled experiments in the various Corn Belt states have indicated that hybrid corn yields at least 20 per cent more bushels per acre than the open-pollinated varieties used during 1890–1927. From 1934 to date we have annual estimates of the acreage of corn planted with hybrid seed in the leading states (see Table 14.21). Starting from 1 per cent of total corn acreage in 1934, hybrids were planted on 94 per cent of the acreage in 1945. By the latter year, the effect of hybrid seed on corn yields in the 6 states must have been approximately 6 bushels per acre (94 per cent times 20 per cent times 33.2 bushels, the average yield during 1923–1927).

Table 14.21

USE OF HYBRID SEED CORN AND COMMERCIAL FERTILIZER,
SELECTED YEARS, 1930–1955

Year	Percentage of Corn Acreage Planted with Hybrid Seed, 6 States	Commercial Fertilizers Used in the East North Central and West North Central Regions
	Per cent	*Plant nutrients, thousands of tons*
1930	—	185
1935	2.6	162
1940	60.8	252
1945	94.5	598
1950	98.3	1,255
1955	99.2	2,246

Second, starting about 1940 the use of commercial fertilizer on corn, previously unimportant in the 6 states, increased very rapidly. Figures for an area which includes the 6 states are shown in Table 14.21; the rate of change in fertilizer use in the 6 states would be very similar. Experimental studies of the response of corn yields to fertilizer do not give us as accurate a figure as we have for the effects of hybrid seed. However, scattered studies suggest that the yield increase in the 6 states resulting from greater use of fertilizer was on the order of 10 bushels per acre between the 1930's and the mid-1950's.

Third, we have the analysis of the effects of rainfall and temperature upon corn yields presented earlier in the chapter. This gives us a basis for adjusting trend yields, particularly in the drouth years 1934 and 1936, for the probable effects of rainfall and temperature, thereby refining our initial approximation to the appropriate freehand trend.

The averages of corn yields, rainfall, and temperature for successive periods are given in Table 14.22. If we examine the net regression curves

Table 14.22

AVERAGES OF CORN YIELDS, RAINFALL, AND TEMPERATURE,
SELECTED PERIODS, 1928–1956

Period	Number of Cases	Average Yield of Corn	Average Rainfall	Average Temperature
1928–1932	5	31.76	11.2	75.1
1933–1938	6	28.57	8.7	77.7
1939–1944	6	42.93	11.3	75.6
1945–1950	6	44.55	12.3	74.0
1951–1956	6	48.15	11.1	76.1

in Figure 14.10 it appears that the average of rainfall and temperature for all periods other than 1933–1938 would normally be associated with corn yields varying over a range of less than a bushel. In other words, these averages fall on portions of the net regression curves that are nearly horizontal. But during 1933–1938 both rainfall and temperature were strongly adverse to good corn yields—their combined effect would be to reduce yields about 7 bushels relative to those for the other periods, judging from the net regression curves for the earlier period. Furthermore, neither hybrid seed nor commercial fertilizer could have had a great effect upon the average yield during 1933–1938; it seems probable that the earlier relationships were still applicable during that time.

The average yield for 1933–1938 may therefore be raised about 7 bushels above the actual average before drawing in a first approximation to the trend. It should be noted that this adjustment could not have been made without the knowledge gained from the 1890–1927 analysis, or related knowledge based on test-plot records from experimental farms in the 6 states over considerable periods.

With these preliminaries, the rest of the analysis may be left as an exercise for the student. Although the major trend effects have been allowed for, it is possible that hybrid seed, commercial fertilizer, and other cultural practices may also have affected the net relations between yield, temperature, and rainfall. On the graphic level, the student may investigate this possibility for himself, by making a separate analysis for the period 1928–1956, perhaps starting with a first-approximation trend and then determining the other two regressions and the final net trend, by successive approximations. The resulting net curves could then be compared with those for 1890–1927.

Another possibility is to assume that the yield of corn is influenced by weather conditions as a *per cent* of the normal yield from improving technology. This could be explored by fitting an analysis to the entire period using the equation

$$\log X_1 = a + f_2(X_2) + f_3(X_3) + f_4(X_4)$$

(with the symbols for the variables having the same meaning as earlier in this chapter) and then relating the values of $f_2(X_2) + z'''$ for each observation to the changing use of fertilizer and of hybrid seed (so far as data were available). This last step would show how much of the net trend could be explained by these two variables and correlated ones, and how much reflected other independent improvements in technology.

Reliability and Use of Regression Curves

The regression curves show the net relation between the dependent variable and each independent variable, with the net variation associated with the other independent variables held constant, for the particular observations included in the sample. If another sample were drawn from the same universe, and similar net regression curves were determined, they would vary somewhat from the curves determined from the first sample. (It should be noted that the 1890–1927 relationships gave good estimates of corn yields for some 8 years after the end of that period, suggesting that the 1928–1935 observations were drawn from essentially the same universe—before the effects of hybrid seed and commercial fertilizer became important.) The lower the multiple correlation in the universe, or the smaller the sample, the larger would be this variation between successive samples. Methods have been developed for estimating the proportion of such samples which will give regression results falling within given ranges of the true regressions prevailing in the universe. (See Chapter 17, pages 290–293.) In publishing regression results, as shown in Tables 14.16 to 14.19, or in presenting charts of the regression results, such as Figure 14.10, the reliability range of the regressions should be indicated, as shown subsequently. Even if the regressions (as in the example here) are determined from a time series, and so are based upon *all* the evidence for that portion of the constantly evolving universe, the reliability limits may still be used as a preliminary indication of possible significance, in view of the closeness with which the relations can be determined. (For a more extended discussion of the meaning of sampling errors with respect to time, see Chapter 20.)

Where net regression curves are fitted as algebraic equations of the types illustrated in this chapter [equations (14.1) to (14.3)], or to other equations with linear coefficients, the coefficients for the equations can be computed by electronic calculators just as readily as they can for linear equations, as discussed in Chapter 11. The machines can perform the necessary calculations for the extensions and solve the equations so fast, once a method of "programming" the successive operations has been worked out, that it is possible to try a variety of alternative types of equations for the curves and to see empirically which ones give the best fit. It is also possible to compute the standard error of the coefficients for each term in the equations at the same time, and then by dropping out the terms which show coefficients of nonsignificant size compared with their standard errors and solving the equations with these terms omitted, to fit algebraic functions all of whose terms show significant values. (This may, however, affect the meaning of the calculated standard errors —see Chapter 17, page 295.

There are many problems, however, where initial exploration by the successive approximation method is very valuable for deciding what the net regression curves present are really like, even if subsequently they are fitted by algebraic equations suitable to represent those types. Also, even where there is some specific hypothesis as to the type of curve or curves that should be present, it may be useful to examine the data, unconditioned by any specific type of assumed equation, to see if the apparent net relations do conform at all to those logically expected, or if the facts are strongly in contrast with the expectations. With problems involving a relatively small number of observations, this exploratory fitting can readily be done by the short-cut method described in Chapter 16. Where the samples are large, however, or where the number of variables which needs to be considered is large (say five or more) the longer but more exact successive approximation method of Chapter 14 may yield better results.

Use of Electronic Calculating Machines in Computing Multiple Curvilinear Regressions

For such cases, either card-tabulation machines equipped with automatic multipliers, or electronic calculators could be used to greatly accelerate the successive approximation process. This would involve the following steps.

1. Computation of the extensions, and then (for electronic calculators), solution of the linear equations.

2. Computation of the values of X_1' and z for each observation, and entering them on the tape or card for each.

3. Sorting the observations into groups for each independent variable in turn, and computing the average values of the independent variable and z for each of these groups.

4. With these values computed, charts such as Figures 14.3, 14.4, and 14.5, would be made by hand, and first approximation curves fitted to each, if indicated. With card tabulators, values of X'' and of z'' could then be calculated by hand, and entered on the cards; with electronic calculators and large samples, the machines could be "instructed" concerning each net regression first-approximation function, and they could then calculate the values of the several functions corresponding to the value of each independent observation for each variable. In either case, the card tabulator or electronic machines could then sum them to obtain X_1', subtract from X_1 to obtain z'', and calculate the $s_{z''}^2$.

5. With these new values of z'', the entire process could be repeated as under 4, and carried through as many successive approximations as was found necessary to achieve stability in all the curves and reduce $\bar{S}_1$ to a minimum consistent with logical restrictions on the shape of the regression curves. At each step the s_z would be calculated by the machines, and adjusted by hand to $\bar{S}_1$ to determine the progress made, if any, in getting a better fitting set of net regression curves.

By this machine-aided process, the successive approximation operation could be carried out quite rapidly (assuming the machines to be available to work with the investigator immediately as soon as each hand operation was completed), so that in a relatively short time a set of graphic curves could be fitted to a relatively large sample for a number of independent variables, with the investigator having as close and intimate a touch with the process, and as much awareness of the kind of curves being obtained, as in the full hand-operated process described in this chapter.

An alternative method would be to take a random sample of, say, 40 observations from the large sample and determine the approximate shape of the net regression curves from it. Two or three alternative algebraic expressions of these shapes could then be fitted by machine processes to the full set of observations, and the best fit selected for the final expression.

Summary

In this chapter methods of determining curvilinear multiple regressions have been discussed. These show the extent to which changes in the dependent variable are associated with changes in each particular independent variable, while simultaneously removing that part of the variation in the dependent variable which is associated (linearly or curvilinearly) with other independent variables. A method of determining the curves by successive graphic approximations is presented step by step. Since this method does not involve making definite assumptions as to the final shape of the curves, it is to be preferred, at least for exploratory studies, to more mathematical methods, presented earlier in this chapter, unless there is a logical basis for the choice of specific functions. Methods of simplifying the conclusions for popular statement are illustrated, and the universe to which they are applicable is briefly considered.

REFERENCE

Ezekiel, Mordecai, A method of handling curvilinear correlation for any number of variables, *Quart. Pub. Amer. Stat. Assoc.*, Vol. XIX, No. 148, pp. 431–453, December, 1924.

Note 14.1. In applying the process described in the middle paragraph on page 246, terms (or independent variables) whose values are insignificant compared to their standard errors should be dropped out one at a time, starting with the one which shows the least significance, and then the whole set of equations solved again with that one omitted (note pages 520–522). This is necessary because two terms (or independent variables) which are highly intercorrelated may show no significance when both are included, yet if one of them is dropped out the remaining one may then show a significant regression. Note also discussion of the underlying logic of this situation beginning in the last paragraph on page 473.

CHAPTER 15

Measuring accuracy of estimate and degree of correlation for curvilinear multiple regressions

In presenting linear multiple regression methods it was observed that coefficients could be computed to show (1) how closely estimated values of the dependent variable, based on the linear regression equation, could be expected to agree with the actual values; and (2) what proportion of the total observed variance in the dependent factor could be explained or accounted for by its relation to the independent factors considered. These coefficients were, respectively, the standard error of estimate and the coefficient of multiple correlation. Exactly parallel coefficients can be computed to show the importance of the relationship for curvilinear multiple regression, employing curvilinear net regressions such as those discussed in Chapter 14. The term *standard error of estimate* is again used to indicate the measure of the probable accuracy of estimated values of the dependent factor. In measuring the proportion of variance explained we will follow the usage in simple curvilinear regression, and use the term index to denote the fact that curvilinear regressions have been employed. The proportion of variance accounted for is therefore shown by the *index of multiple determination*, which is the square of the *index of multiple correlation*.

Standard Error of Estimate. Values of X_1 may be estimated from $X_2, X_3, \ldots, X_k$ by a multiple curvilinear regression equation of the type

$$X_{1.f(2,3,\ldots k)} = a + f_2(X_2) + f_3(X_3) + \ldots + f_k(X_k) \qquad (15.1)$$

where the net regression functions f specify curves fitted simultaneously either as mathematical equations or as graphic curves determined by a successive approximation process. In either case, when estimated values,

X_1', are determined from equation (15.1), for the observations included in the sample, and the residuals z are determined

$$z_{1.f(2,3,\ldots,k)} = X_1 - X_1' \tag{15.2}$$

the standard error of estimate is then defined as the s_z, so that

$$S_{1,f(2,3,\ldots,k)} = s_z \tag{15.3}$$

The standard error of estimate may be used to indicate the closeness with which new values of the dependent variable drawn from the same universe may be expected to agree with corresponding estimated values based upon the observed relationship with the independent variables.

Where the *net regression curves* are determined as *algebraic equations*, as in the three-variable exercise presented on pages 206 to 208, it is not necessary to go through the actual process of working out the X_1' and z values for each observation. Instead, the standard error can be computed directly from the values of s_1 and of R calculated by the usual equations. In the example mentioned, the regression equation used was

$$X_1 = a + b_2 X_2 + b_2' X_2^2 + b_3 X_3 + b_3' X_3^2 \tag{15.4}$$

and the values computed were $s_1 = 1.1$, and $R_{1.(2,2^2,3,3^2)} = 0.968$. The standard error of estimate for such mathematically determined curves may then be calculated in the usual way by the equation

$$S_{1.f(2,3,\ldots,k)}^2 = s_1^2(1 - R_{1.2,2',3,3',\ldots,k}^2) \tag{15.5}$$

which for this case $\quad = (1.1)^2(1 - 0.968^2) = 0.0762$

$$S_{1.f(2,3)} = 0.28$$

Similarly, by equation (12.1),

$$\bar{S}_{1.f(2,3,\ldots,k)}^2 = S_{1.f(2,3,\ldots,k)}^2 \frac{n}{n-m} \tag{15.6}$$

For this example,

$$\bar{S}_{1.f(2,3)}^2 = (0.0762)\tfrac{20}{15} = 0.1016$$
$$\bar{S} = 0.32$$

Where *graphic curves* are used for the *net regressions*, s_z has to be computed by carrying out the operations of working out X_1' and z for each observation. Thus in the problem worked through in Chapter 16 for steel costs, the standard deviation of the final z, adjusted by equation (15.6), is \$3.86 per ton. This is therefore the standard error of estimate in calculating steel costs for additional observations from the fitted net regression curves. This compares with the unadjusted S of \$2.88. Because of the larger size of m, the adjustment for $n - m$ is even more important for curvilinear regressions than for linear ones.

Index of Multiple Correlation. This coefficient has a meaning exactly corresponding to that of the coefficient of multiple correlation, but applies to curvilinear net regressions instead of linear. Both measure the simple correlation of the original values of X_1 (for the cases in the sample) with the estimated values X_1', calculated according to a regression equation including two or more independent variables. The *coefficient* of multiple correlation $R_{1.23\ldots m}$ is used when the estimates are based on linear net regressions for each of the independent variables $X_2, X_3, \ldots, X_m$. The *index* is used when the estimates are based on curvilinear net regressions for one or more of the independent variables, and is designated $I_{1.23\ldots k}$.

If the *curvilinear net regressions* are determined as *algebraic equations*, as in the first example mentioned above, the index of multiple correlation $I_{1.23\ldots k}$ is the same as the coefficient of multiple correlation $R_{1.(2,2',3,3',\ldots,k)}$ for all the transformations of all the variables included in the regression equation.

For the problem used as the example, where equation 15.4 was the regression equation used, $R_{(1.2,2^2,3,3^2)} = 0.968$. This is, accordingly, the value of $I_{1.23}$ for that problem.

Where *net regression curves* are fitted by a *graphic process*, however, the estimated values of X_1 and X_1' must be worked out for each observation, the residuals z calculated, and their standard deviation worked through. The multiple correlation index is then calculated from these values by the equation

$$I^2_{1.23\ldots k} = 1 - \frac{s_z^2}{s_1^2} = 1 - \frac{S^2_{1.f(2,3,\ldots,k)}}{s_1^2} \tag{15.7}$$

For the steel-cost example, this becomes

$$I^2_{1.234} = 1 - \frac{2.88^2}{7.19^2} = 0.83956$$

$$I_{1.234} = 0.916$$

Indexes of multiple correlation can be interpreted in much the same way as other correlation coefficients or indexes, as measuring what part of the variation in X_1 can be explained by its relation to X_2, X_3, etc. For the simplest interpretation the squares of the values should be used, which will then be called the *index of multiple determination*, following the usage of linear correlation. This measures the per cent of the variance in X_1 which can be explained by the net curvilinear relations to X_2, X_3, etc. This will be designated $d_{1.f(2,3,\ldots k)}$, and is defined as

$$d_{1.f(2,3,\ldots,k)} = I^2_{1.23\ldots k} \tag{15.8}$$

For the two problems we have used as illustrations, the first case, with mathematically determined net regressions, had $I = 0.968$, and the second, the steel-cost case with graphic regressions, had $I = 0.916$. The corresponding d values, 0.937 and 0.840, show that 94 per cent and 84 per cent, respectively, of the total variance in X_1 was explained in the two cases by the curvilinear relations determined.

Measuring the Net Importance of Individual Factors, according to Their Curvilinear Regressions. The importance of each of the factors could be measured by an *index of net curvilinear correlation* $i_{12.34\ldots k}$, which would be defined in a way parallel to that of the coefficient of net or partial correlation $r_{12.34\ldots m}$. Where curves are fitted as algebraic equations, such as equation (15.4), such a net correlation index might be defined as

$$i_{12.3\ldots k} = r_{1(2,2^2)\cdot(3,3^2\ldots k,k^2)} \tag{15.9}$$

It might be possible to work out a way of computing this value algebraically, by a process similar to that used in footnotes 5 to 8 in Chapter 14. This has not yet been done, however, and it seems likely it would involve some arbitrary assumptions in combining the variances due to cross-products of X_2 and X_3.

A more direct method of calculating $i_{12.3}$ can be used by some additional computation. This method is applicable to all net regression curves, whether fitted algebraically or graphically. This is done by correlating the curve readings for each independent variable for each observation with the original value of the dependent variable X_1, so as to obtain new linear partial regression coefficients between X_1 and these transformations for each independent variable. The equation to be used may be written:

$$X_1 = a' + b_{12'.3'4'}[f_2(X_2)] + b_{13'.2'4'}[f_3(X_3)] + b_{14'.2'3'}[f_4(X_4)] \tag{15.10}$$

When that is done, and the partial correlation coefficients such as $r_{12'.3'4'}$, $r_{13'.2'4'}$, etc., are computed, these values can be taken as *indexes of partial correlation*, with

$$i_{12.34} = r_{12'.3'4'} \tag{15.11}$$

This process may also change slightly the exact shape or slope of the net regression curves, and raise slightly the size of the index of multiple correlation. Standard errors may also be computed for the b values of equation (15.10), which will serve to judge the significance of each variable according to the transformations provided by the functions f_2, etc.

Standard Errors and Confidence Intervals of the Sample Statistics. The values of S and I from a sample drawn from a real universe will tend to vary from the true values of the corresponding parameters in the universe, and the shape of the net regression curves from the sample will also tend to vary from the true functions in the universe. In cases where the sample is based on a correlation model, methods are available for making inferences as to the true values in the universe, based on confidence intervals. These are discussed in Chapter 17.

Summary

For curvilinear multiple regression equations it is possible to obtain standard errors of estimate, indexes of multiple correlation and determination, and indexes of partial correlation, which serve the same purpose that the comparable coefficients serve for linear multiple regressions.

REFERENCES

Ezekiel, Mordecai, The application of the theory of error to multiple and curvilinear correlation, *Proc. Amer. Stat. Assoc.*, pp. 99–104, Vol. XXIV, March, 1929.

———, A first approximation to the sampling reliability of multiple correlation curves obtained by successive graphic approximations, *Annals of Math. Stat.*, Vol. 1, September, 1930.

Mills, Frederick C., *Statistical Methods*, 3d edition, pp. 580–601. Henry Holt, New York, 1955.

(Note: This reference treats the issues for simple curvilinear regression only. The use of the analysis of variance shown here, to determine the significance of a fitted simple curvilinear regression with attention to the degrees of freedom used up, may also be extended to multiple curvilinear regression.)

CHAPTER 16

Short-cut graphic method of determining net regression lines and curves

In problems where the correlation is fairly high, the number of variables is not too large, and the number of observations is not over 50 to 100 cases, net regression lines and curves may be determined by a combination of inspection and graphic approximation which takes only a fraction of the time required by the methods previously presented in detail. This graphic method is very speedy, and in the hands of a careful worker can yield results almost as accurate as those obtained by the longer methods previously set forth. The short-cut method was invented by Louis H. Bean.[1]

The general basis of the short-cut method is to select, by inspection, several individual observations for which the values of one or more independent variables are constant, and then note the changes in the dependent variable for given changes in the remaining independent variable. This process is repeated for additional groups of observations for which the other independent variable or variables are constant (or practically so) but at a different level than for the first group. The relation between the dependent variable and the remaining independent variable, as indicated by a series of such groups, approaches the *net* regression line or curve, since the cases have been selected so as largely to eliminate the variation associated with other independent variables. A first approximation line or curve is then drawn in by eye, and the residuals from this curve, measured graphically, are used to determine the regression for the next variable, cases again being selected so as to eliminate the influence of other independent variables. The final fit of the several lines or curves is tested by the same successive approximation process employed in

[1] L. H. Bean, A simplified method of graphic curvilinear correlation, *Journal of the American Statistical Association*, Vol. XXIV, pp. 386–397, December, 1929.

Chapters 10 and 14, or by a shorter graphic equivalent of it. Since the initial lines or curves approach much more closely to the final net regressions, and since graphic transfers of residuals are substituted for reading all the curves and calculating X_1' and z arithmetically, the process is much shorter, and fewer steps are required. However, as mentioned earlier, the standard error of estimate converges to its minimum value more rapidly than the net regression curves converge to their final slope and shape.

The Short-Cut Method Applied to Linear Net Regressions. The short-cut method can be used to determine either linear net regressions or curvilinear ones. The process of determining linear net regressions by this method was illustrated on pages 269 to 276 of the second edition, but is omitted here to make room for other more important materials. In that illustration, with the approximations carried through two stages, an adjusted standard error of estimate of 76.09 was obtained, as compared with 74.65 for the least-squares solution, and an R^2 of 0.830, compared with the exact value of 0.837. The net regression coefficients also differed only slightly from their exact values. While successive approximations would ordinarily not be used for determining linear net regressions, their use does have the advantage that if curvilinear regressions are actually present, that fact will become apparent during the process.

The Short-Cut Method Applied to Curvilinear Net Regressions. The greatest usefulness of the short-cut method is in determining net curvilinear regressions. Since the method of successive graphic approximations presented in Chapter 14 also depends on the convergence of successive approximate curves, the short-cut method secures results which are just as reliable, at a great saving of time.

The procedure will be illustrated by a problem of four variables. The same method may be applied to larger or smaller problems equally well, up to the limit of the number of observations which can be kept separate on the chart paper.

The data to be considered are given in Table 16.1.

Data for 1938 and 1939 were also available when this study was made, but were disregarded until the analysis was completed, and then were used for checking the results.

LOGICAL RELATION OF THE VARIABLES. These data are from a study of the relation of volume of steel output to cost per ton. The qualitative examination of the problem (see discussion in publication cited in the footnote to Table 16.1) indicated that changes in wage rates might be expected to have a relative, or multiplying, effect upon the cost for a given output, so that the relation might best be examined in terms of:

$$\log X_1 = f_2(X_2) + f_3(X_3)$$

Also, the qualitative examination revealed that major changes in technical methods of production, especially the beginning of the substitution of continuous-strip mills for hand mills, had taken place during the period under consideration, and that these improvements in technology might need to be included, either directly as a labor-efficiency factor, or indirectly as a trend factor.

Table 16.1

DATA FOR SHORT-CUT METHOD OF DETERMINING REGRESSION CURVES*

Year, X_4	Cost per Ton of Finished Steel, X_1	Proportion of Capacity Operated, X_2	Average Hourly Earnings, X_3
	Dollars per ton	*Per cent*	*Cents per hour*
1920	72.3	88.3	77.5
1921	78.5	47.5	60.2
1922	57.9	71.3	58.5
1923	63.0	88.3	67.0
1924	63.7	69.0	70.8
1925	62.9	78.4	70.3
1926	60.3	88.0	70.8
1927	59.6	78.9	71.3
1928	55.2	83.4	71.8
1929	51.5	89.2	72.5
1930	58.6	65.6	73.2
1931	65.6	38.0	70.8
1932	81.4	18.3	61.0
1933	65.0	28.7	59.0
1934	64.6	31.2	70.0
1935	65.4	38.8	73.0
1936	61.1	59.3	74.0
1937	65.6	71.2	86.0

* The data are calculated from regular published reports of the U. S. Steel Corporation. See Kathryn H. Wylie and Mordecai Ezekiel, The cost curve for steel production, *Journal of Political Economy*, Vol. XLVIII, pp. 777–821, Dec., 1940.

To simplify this illustrative presentation, the data will be used in absolute values, instead of in logarithms. The charts will be examined for indications of multiplying relationship, however, since (as is shown in detail on pp. 273–275) this graphic method can also be used to spot the presence of such non-additive relations.

CONDITIONS ON THE CURVES TO BE DRAWN. Before proceeding to the statistical steps in the examination of these data, the types of curves logically expected and the resulting conditions to be placed upon the shapes of the curves to be obtained must also be considered. Without going into the underlying technical reasons (presented more fully in the original study), let us assume that the following conditions will be imposed:

On the net relation of cost to capacity:

1. The curve may fall, at a declining rate, until a minimum is reached, and may then increase gradually after that minimum is passed. No points of inflection are expected.

On the net relation of cost to wages:

2. The curve will rise steadily, possibly at an increasing rate with higher wages, but otherwise will be fairly uniform—that is, will be either a straight line or a shallow curve concave from above. There should be no inflections.

On the net relation of cost to the time elements (efficiency, etc.):

.3. The curve will tend to decline, perhaps slowly at first and then more and more rapidly as new techniques are introduced. There might also be irregular changes reflecting the changes in general price level (and in various purchased materials and services other than labor) during the period under examination, especially in the early 1920's and after 1929. (This trend factor lumps together labor efficiency, price levels, and perhaps other factors, each of which might be given separate consideration in a more elaborate investigation.)

PRELIMINARY EXAMINATION OF INTER-RELATIONSHIPS AMONG THE INDEPENDENT VARIABLES. As before, the inter-relationships of the several independent variables (including time for the trend factor) must be examined before the short-cut approximations can be begun. These are presented in Figure 16.1, the years being used to designate the observations. After the dots were located, the successive years were connected by a light line, making it possible to consider the relations of X_4 (time) to X_3 and X_2, as well as of X_2 to X_3, all on this one chart. (This same method could be used even in non-time-series data by first classifying the data on the ascending values of one independent variable. Successive observations, by number, would then indicate increasing values for that variable.)

Examining first the location of the dots in Figure 16.1, without regard to their sequence, a moderate intercorrelation between wages (X_3) and rate of operations (X_2) is evident. No low values of X_2 are found, except together with low values of X_3. In the higher ranges of X_2 the values of X_3 fan out more, varying from quite low to quite high. Apparently there is enough independence in the occurrence of the two variables to permit of fairly good separation of their effects.

When examined with regard to time, however, the intercorrelation between X_2 and X_3 is considerably higher. The low wages at high output all occurred in one period—1921 to 1923. The marked positive correlation of wages and operations from 1930 to 1937 is also a correlation with

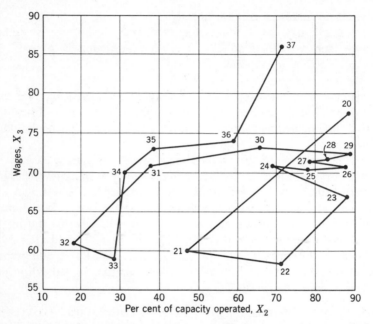

Fig. 16.1. Wages and per cent of capacity operated, with successive observations connected to indicate shift in the X_2X_3 relationship with time.

time, both generally declining from 1930 to 1933, and both rising from 1933 to 1937. Since this was the period when technological changes were greatest, it may be difficult to disentangle the time or trend elements here, reflecting these technological changes, from the effects of the associated advances in output and in wages. We shall have to be on guard for this as we proceed with the analysis.

Looking for groups of observations which hold the other factor constant, we note on Figure 16.1 that there were a considerable number of years when wages[2] fell between 70 and 75 cents per hour. The observations

---

[2] "Wage rates per hour" is quite a different thing from "average earnings per hour employed," since the latter is a weighted figure reflecting all changes in the composition of the labor force. The latter is the figure used here (note Table 16.1), since an average wage-rate figure was not available. For brevity, however, the term "wages" will be used here to describe the data, even though that is not the technically correct designation.

for these years may be used to hold wages substantially constant, while the data are examined for the apparent effects of operation rate and time.

DETERMINATION OF FIRST APPROXIMATION CURVE FOR FIRST INDE-PENDENT VARIABLE. The observations for the years with wages of 70 to 75 cents are accordingly plotted on Figure 16.2 with per cent of capacity operated (X_2) as the abscissa and cost per ton (X_1) as the ordinate.[3]

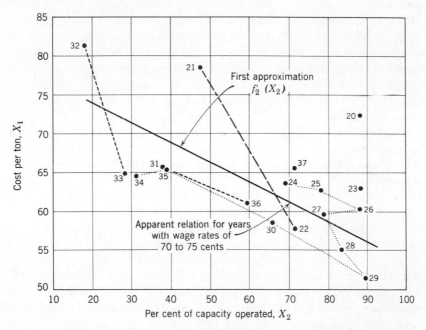

Fig. 16.2. Cost per ton and per cent of capacity operated, and first approximation to $f_2(X_2)$.

After the dots are plotted, successive observations (when they occur in this group) are connected by a "drift line" of short dashes. This enables us to examine the relation of cost to operation rate and time while holding wages constant.

These observations indicate at once a marked negative correlation between operation rate and cost. The data from 1924 to 1929 suggest a rapid fall in cost for a given rate, especially from 1927 to 1929. Apparently there was some further decline from 1931 to 1934, but the data for 1935 to 1936 fall almost precisely on the same line as those for 1930 to 1931. (However, examination of Figure 16.1 shows that wages were

[3] Great care should be exercised in plotting these values, as their exact location becomes the basis for all the successive graphic transfers. Chart paper of adequate size to separate the dots should be used.

slightly higher in this latter period, which might obscure the trend factor at this point.) No curve is indicated as yet. Accordingly, a line is drawn in lightly, as indicated, to show the relation of cost to operation rate for these observations, with the trend factor also considered.[4]

The observations for years of very low wage rates—1921, 1922, 1932, and 1933—are next plotted, and consecutive years again connected by a line with long dashes for the first two, and short dashes for the latter. Both show exaggerated drops in costs with increases in output. Only 1933 shows a cost lower than might be expected from the observations previously plotted. If 1932 were also to show a cost below the usual relation, the regression curve would have to swing up sharply, so as to pass above it. The high value for 1921 may be ignored for the moment, as possibly reflecting the high price levels at the end of the inflation period after World War I.

The two years of high wages—1920 and 1937—and the one remaining year of moderately low wages, 1923, are next plotted. The dot for 1937 falls above the other observations, and that for 1920 much higher still, apparently confirming the unusual (trend?) factors affecting the position of the 1921 observation. Similarly 1923 is fairly high, despite its moderate wage rate, as compared to subsequent years.

The indications as to the effects of wage rates, to this point, sum up as follows: 1920 to 1923 all show relatively high costs (with the exception of 1922). Apparently trend elements outweighed the effects (if any) of the low wages in 1921 and 1923. With low wage rates, 1933 shows quite a low cost for the low rate of output, whereas 1932, with somewhat higher wage rates, shows a much higher cost. Apparently the fall in output to near zero increases cost per unit very greatly. On the basis of these considerations, a curve could be drawn in as the first approximation, extending the previous line but bending it up to pass well above 1932, with its low wage rate. With only one or two observations to support that bend at this stage, it seems best to be more conservative until the other factors have been more definitely allowed for, and until the evidence for a curve (if any) is more clearly established (even though a curve of declining costs was expected.)

Accordingly the straight line previously drawn in lightly is extended and used as the first approximation toward the net regression, $f_2'(X_2)$. (If a curve had been clearly indicated by the examination of the data as

---

[4] By drawing this line parallel to the lines connecting successive years, all trend is eliminated except the one-year change. If the line were tilted slightly more steeply than the line connecting successive years, that would provide an approximate correction for the year-to-year change, also. With the uncertainty of trend effects after 1931, however, that was not done here, but was left for subsequent approximations to clarify.

described above, it would have been drawn in at this point, thus starting the successive approximations from a curve instead of from a straight line.)

DETERMINATION OF FIRST APPROXIMATION CURVE FOR SECOND INDEPENDENT VARIABLE. The next step is to examine the relation of costs, as now approximately corrected for the relation to operation rate by $f_2'(X_2)$, to wages and time. Accordingly, the vertical departures of the dots on Figure 16.2 from the line of $f_2'(X_2)$ are scaled off, and are plotted in Figure 16.3.[5] The departures are plotted as ordinates above and below

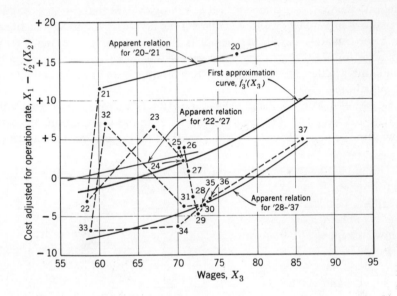

Fig. 16.3. Wages and cost per ton adjusted to average operation rate on the basis of the first approximation, and first approximation to $f_3(X_3)$.

the zero line, with the values of X_3, wages, as abscissas. If the fourth variable, X_4, were not a time series, or not arranged in order, it would be necessary to group these observations according to its value, also, as was done in plotting Figure 16.2. Since the numbers of the successive years indicate the successive values of X_4, that is not necessary. After the dots are all plotted, the successive years are connected by a light dotted line, to aid in separating the trend influences from that of wages.

If the dotted line to the successive years is followed, it is apparent that there was a general downward trend in the adjusted costs. The years

[5] The job of making these readings and transfers can be made swifter and more accurate by using the technique outlined on pages 526 to 530.

1920 and 1921 appear on one level, the years 1922 to 1927 on a lower level, and the years from 1928 on (with the exception of 1932) on a still lower level. In each of these groups of years there is a positive relation between adjusted costs and wages, as indicated by the light lines drawn through each group. Only the last group has any indication of a curve. Even there, the curve depends entirely on the position of the two extreme observations, one at each end. Here, however, the lower portion of this curve parallels, almost exactly, the lines indicating the apparent positions for the two other groups, which in turn lie mainly on the left half of the lower group of observations. Furthermore, the shape of the curve— shallowly concave—is consistent with that logically expected. Accordingly, a shallow curve passing through the center of the observations is drawn in, approximately paralleling the drift lines and curve representing the relations for the three groups. The succeeding successive approximations will show whether this curve is justified or whether a straight line should be substituted.

DETERMINATION OF FIRST APPROXIMATION CURVE FOR THIRD INDEPENDENT VARIABLE. The next step is to examine the relation of costs, now approximately adjusted for both wages and operation rate, to time.

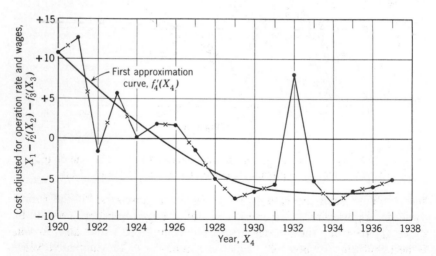

Fig. 16.4. Time and cost per ton adjusted to average operation rate and wages, on the basis of the first approximation curves, and first approximation to $f_4(X_4)$.

Accordingly, the vertical departures of the dots on Figure 16.3 from the curve $f_3'(X_3)$ are scaled off, and are plotted in Figure 16.4. Again the departures are plotted as ordinates, with this time the values of X_4 as abscissas. Since this is the last independent variable to be considered, it is not necessary to group the observations with respect to any other variable but all can be plotted and examined as a whole. Figure 16.4 shows the resulting chart. Connecting the successive years makes it easier to study the type of trend present.[6]

Except for the single wide departure in 1932, Figure 16.4 indicates a definite downward trend from the beginning, tapering off about 1930 and running flat or gradually rising thereafter. Taking midpoints between each pair of observations (indicated by the crosses) helps to locate the approximate level of this trend. The one extreme departure, 1932, is disregarded in the process. Its position in Figure 16.2 at the extreme end of the line, meant that its adjustment for X_2 was in doubt. A smooth curve is then drawn in, declining to about 1930, and running flat thereafter. The rising trend indicated by the observations for 1936 and 1937 is left for subsequent approximations to confirm. In general it is unwise to give an extra "twist" to a regression curve simply on the evidence of one or two observations.

DETERMINATION OF SECOND APPROXIMATION CURVE FOR FIRST INDEPENDENT VARIABLE. We now have determined first approximations to the net regression lines or curves of X_1 on X_2, X_3, and X_4. The departures of the dots on Figure 16.4 from the regression line $f_4'(X_4)$ are the residuals, z'', from this first set of curves. The remaining steps involve the graphic transfer of these residuals to each curve in turn, the correction of each curve on the basis of the fit of the new residuals, and in turn the transfer of the newly corrected residuals to the next curve, and so on until no further change is indicated in any of the curves. Ordinarily the residuals from Figure 16.4 would be plotted back on the original curve for X_2, Figure 16.2. To show the process clearly, however, the dots and the first approximation curve for $f_2'(X_2)$ from Figure 16.2 are reproduced again as Figure 16.5.

The vertical departures of the dots on Figure 16.4 from the approximation curve, $f_4'(X_4)$, are then plotted on Figure 16.5 as departures above and below the regression line, $f_2'(X_2)$, with the corresponding values of X_2 as abscissas. To prevent confusion with the original values shown as solid dots, the corrected values are indicated as hollow dots.

It is at once apparent, on inspection of Figure 16.5, after the corrected values are all plotted in, that the new values show much less scatter than the original values. Closer inspection reveals that every one of the adjusted

---

[6] If joint functions are suspected (see Chapter 21) the data might again be grouped for values of X_2 and X_3, in plotting Figure 16.4. If these groups showed varying relations to X_4, even after the approximate relations to X_2 and X_3 had now been eliminated, that would indicate the presence of a joint relation. Note Figure 16.8, and the discussion on pages 273 to 275 of this chapter.

observations below 60 per cent of capacity falls *above* the first approximation line, with a single exception. In the range from 60 per cent to 80 per cent, 3 cases fall below the first approximation line (2 widely) and 3 slightly above, indicating in this range that the new line should be lower than before. The 5 observations above 80 per cent fall 2 below, 2 about the same distance above, and 1 right on the line, indicating that the position of the line here is about correct. These departures confirm

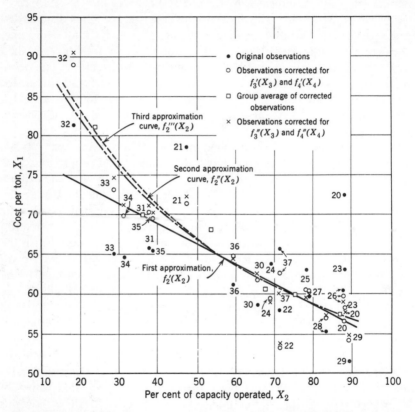

Fig. 16.5. Per cent of capacity operated, and cost per ton unadjusted and adjusted to average values of other variables, and second and third approximations to $f_2(X_2)$.

the suggestion previously given by the 1932 value in Figure 16.2 that the regression should be a curve, concave from above. This accords, also, with the logical conditions originally imposed on this relation. Accordingly such a curve is drawn in freehand, passing as near as possible through the averages of the adjusted values in each successive group. (To facilitate drawing the curve, the average of the residuals in successive ranges of

10 to 15 units of X_2 are estimated graphically and drawn in as hollow squares.)

DETERMINATION OF SECOND APPROXIMATION CURVE FOR SECOND INDE-PENDENT VARIABLE. The vertical departures of the adjusted values (the hollow dots) above or below the second approximation curve, $f_2''(X_2)$, are next scaled off graphically and plotted as ordinates from the values of the $f_3'(X_3)$ curve as zero, with the corresponding X_3 values as abscissas. This is

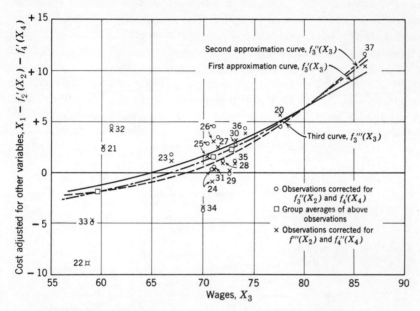

Fig. 16.6. Wages, and cost per ton adjusted to average values of all other variables, and second and third approximations to $f_3(X_3)$.

generally done on the original X_1X_3 chart (Figure 16.3). For clarity, however, the curve of Figure 16.3 is here reproduced on Figure 16.6, and the departures from Figure 16.5 are transferred to this new chart. The 4 observations around 60 for X_3 average definitely below the line; both the next group up to 72.5 and the next group 72.5 up to 75 average slightly below, whereas the single observation above 85 falls above the line. These averages are indicated by squares on Figure 16.6[7]. The single high observation at the end alone would not be enough to indicate a change in the curve, but it is consistent with the group averages, which indicate the need for a slightly steeper curve than the original one.

[7] These averages have been estimated graphically, by the technique explained on page 530.

Accordingly this new curve is drawn in, approximately through the group averages, but still conforming to the conditions stated on p. 257. To this point none of the relations, as indicated by the data, has differed sufficiently from the shapes logically expected to require any reconsideration of the logical analysis from which the conditions limiting the shapes to be drawn were derived.

DETERMINATION OF SECOND APPROXIMATION CURVE FOR THIRD INDEPENDENT VARIABLE. The same process is used in determining the second approximation for the next variable. The vertical departures of the dots on Figure 16.6 above or below the second approximation curve, $f_3''(X_3)$, shown as a dashed line, are scaled off and plotted as departures from the $f_4'(X_4)$ curve, with the corresponding X_4 values as abscissas. Again a new chart is prepared, Figure 16.7, with $f_4'(X_4)$ reproduced, although the original chart, Figure 16.4, is still clear enough so that these new values could readily have been plotted upon it. Again, as the observations are equally spaced in time, a continuous light line is drawn in, connecting the successive observations.

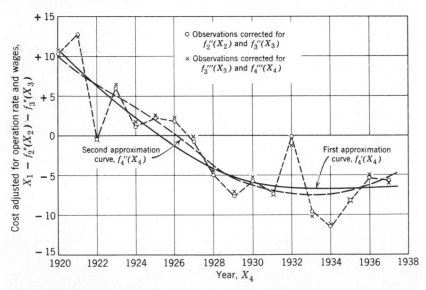

Fig. 16.7. Time, and cost per ton adjusted to average operation rate and wages on basis of second approximation curves, and second approximation to $f_4(X_4)$.

If the curve were any ordinary function—anything except a trend allowance for a number of unrepresented factors—there would be little evidence, from the dots in Figure 16.7, for any further change in the fitted curve. Since it is a trend allowance, however, and was expected to be irregular on logical grounds (note the conditions stated on page 251), more flexibility may be in order. Comparing Figure 16.7 with Figure 16.4, we see that the observations have been changed only slightly by the further adjustments for $f_2(X_2)$ and $f_3(X_3)$. The individual observations on both charts show a pronounced fall from 1920 to 1924, a flattening out then for three or four years, then another fall to 1929. Between 1923 and 1927, Figure 16.7 shows that 4 out of 5 observations fall above the $f_4'(X_4)$ line, whereas, between 1928 and 1935, 6 out of the 8 observations fall below the line. These departures indicate that some changes in the first curve are justified. It is apparent that these changes would not be inconsistent with the possible composite effects of price-level changes and a general upward trend in production efficiency, with a corresponding downward trend in production costs relative to wage rates and other factors. The sharp fall from 1920 to 1924, however, largely reflects the two high observations for 1920 and 1921, offset somewhat by a very low observation in 1922. Accordingly, the trend may be interpreted as moderately downward from 1920 to 1926, more sharply downward to about 1929, then gradually tapering off to a low about 1933 or 1934, and rising gradually thereafter. A more flexible trend is therefore drawn in according to these general changes but not following single observations to the extremes of their departures.[8]

DETERMINATION OF THIRD APPROXIMATION CURVES. The same process as before is now repeated, plotting the departures from $f_4''(X_4)$ around the $f_2''(X_2)$ curve, with X_2 values as abscissas. This time the new departures shown on Figure 16.7 are plotted back on the previous chart, Figure 16.5. Crosses are used for the new departures, to distinguish them from the previous values shown as hollow dots. To prevent confusing the chart, the observation (year) number is not shown with the cross, except where there are two or more observations with about the same X_2 value.

Examining the location of these new crosses on Figure 16.5, we notice that, for every observation with a value below 50 for X_2, the cross is one to one and one-half units (of X_1) higher than the corresponding dot. For values of X_2 above 50, however, the crosses fall alternately above and below the corresponding dots, with the averages of the crosses hitting just about the curve. This pattern indicates that the $f_2''(X_2)$ curve should be raised somewhat below 50, to be still steeper. Accordingly, a new curve is drawn in, changed as indicated, to pass as near as possible through the group averages of the crosses (as graphically estimated) and yet conform with the logical limitations on its shape.

---

[8] Only in rare instances would a curve with this much flexibility be justified. In this particular case its use is in line both with the theoretical analysis and the resulting conditions imposed on the shape of the curve.

The vertical departures of the crosses from the new curve, $f_2'''(X_2)$, are then carried forward to Figure 16.6, as departures from $f_3''(X_3)$. Again crosses are used to represent the new values.

Inspection of Figure 16.6, after the crosses are inserted, discloses a different situation from that in the previous chart. In the left portion of Figure 16.6, for values of X_3 below 65, the crosses fall very close to the corresponding dots, with no change for the average. In the right-hand portion, for values of X_3 above 75, the crosses also fall above and below the corresponding dot. Between 65 and 75, however, a number of the crosses fall a considerable distance below the corresponding dot, so that out of the twelve observations in this range, six crosses fall slightly above the f'' line and six fall a considerable distance below. This pattern indicates that the f'' curve should be made more sharply concave, without changing the elevation of either end. A new curve is therefore drawn in to correct this, through the group averages of the crosses. (To prevent confusion, these averages are not shown on Figure 16.6.) The sharp lift in the last portion of this curve is dependent only upon the two observations, 1920 and 1937. However, the shape of this part of the curve is consistent with the logical limitations and with the other observations. Except for these two observations, a straight line would fit the crosses almost as well as the curve. The evidence for the existence of a curve, or for its exact shape, is thus very uncertain, as the data are distributed here.[9]

If the f''' curves are compared with the f'' curves on both Figure 16.7 and Figure 16.6, it is evident that we have determined the shape of these curves about as well as we can with the data at hand. Even with the material change in the trend by using the much more flexible curve of $f_4''(X_4)$, the differences between the f'' curves and the f''' curves for X_2 and X_3 are insignificant. However, to complete the process we carry the final residuals, the departures of the crosses on Figure 16.6 from the $f_3'''(X_3)$ curve, over to Figure 16.7, as departures from the trend line $f_4''(X_4)$.

There is little improvement in the average closeness of the crosses to the trend line, $f_4''(X_4)$, as a result of the slight changes in f_2 and f_3. The general characteristics of the trend, as fitted by the previous flexible curve, remain the same. From 1923 to 1930, every cross falls slightly above the corresponding dot, suggesting the possibility of a slightly better fit if the trend was raised a little in this portion. The single high value in 1932 continues to stand out, alone and unexplained. It seems hard to justify it on any trend basis. We could eliminate the wide departure for 1932 by

---

[9] See page 291 of Chapter 17 for the sampling reliability of the portion of a curve determined by such extreme observations, where the theory of random sampling may be properly applied.

twisting the lower end of $f_2(X_2)$ up sharply to pass through this single observation. In the absence of confirmatory evidence from another such low year for percentage of capacity operated, this would be a risky assumption.

Although it would be possible to modify the trend further, as suggested in the preceding paragraph, it seems best to let it stand unchanged. In view of the slight changes in the f_2 and f_3 curves in the last approximation, we end the successive approximation process at this point, feeling we have carried the process about to the point of diminishing returns in increased accuracy.

It should be noted, in Figures 16.5, 16.6, and 16.7, that the final curves at the end of the approximation process differ significantly from the first approximations only in the case of $f_2(X_2)$. Almost the same flexible trend of $f_4''(X_4)$ could have been drawn in the first approximation on Figure 16.4. The closeness with which $f_3'(X_3)$, $f_4'(X_4)$, and $f_2''(X_2)$ approximate the final curves is an indication of the great power of the graphic method in making a rapid approach to the underlying relations. The routine of comparing selected observations for which the values of the other independent variables are constant, or almost so, and judging the net relations from these selected comparisons provides a much closer initial approximation to the final curves than does the initial assumption of linear net regressions, used as the starting point in the successive approximation process presented in Chapter 14.

(For an exercise, the student might take the example which has just been analyzed and determine the net regression curves by the method of Chapter 14, using the same limitations on the shape of the curves as used here. That will enable him to compare the relative speed and effectiveness of the two methods in approaching the final curves.)

As already noted, the intercorrelations among X_2, X_3, and X_4 were only moderate in this case. In a problem where the intercorrelations among the independent variables were quite high, the changes in the several regression curves as a result of the successive approximation process might be more marked than in the example just completed. In such a case the convergence toward the curves of best fit will be slower than where the intercorrelations are low, and a larger number of successive approximations will be required to determine the final curves.

If, after several approximations have been made, the new curves start swinging up and down over curves previously determined, the approximation has probably been carried far enough. Especially where the intercorrelations for two independent variables are very high, a rise in the slope of one curve will cause a fall in the slope of the other. In such a case the exact position of each of the two curves is indeterminate, and the zone within which the last two or three approximations vary will indicate something of the uncertainty as to the exact shape or location of each curve. As will be shown later (Chapter 17), the reliability of *any* net regression line or curve varies inversely with the extent to which the particular independent variable is correlated with the other independent variables. Where two variables are so closely correlated that the relation to the dependent variable may be ascribed to either independent variable or parceled out more or less arbitrarily between the two, their individual effect is indeterminate. Only by securing a large enough sample can the true influence of each be judged. When used with due regard to the logical significance of the curves obtained, any one of the several methods will tend to give results which are substantially the same—that is, which lie within the range of possible accuracy imposed by the facts of the particular sample.

DETERMINING STANDARD ERROR OF ESTIMATE AND THE INDEX OF MULTIPLE CORRELATION. The standard error of estimate may now be determined by first computing the value of $s_{z'''}$. This can be done most simply by scaling off, on Figure 16.7, the departures of the last adjusted values (the crosses) from the final trend curve. These departures are the z'''''s. Any errors which have been made in any of the successive graphic transfers will accumulate in these residuals. A more exact check can be made by reading off the estimated values for each observation from the final curves and adding them up to calculate the estimated X_1''' and z''', according to the same method used in Chapter 14. The z''' values as computed in this manner should agree closely with the z'''''s scaled from the final approximation chart. These calculations are shown in Table 16.2.

Column 10 of Table 16.2 gives the residuals as scaled off from the last approximation curve on Figure 16.7. Column 9 gives the residuals as computed in the usual way from the several curve readings. It is evident that the two columns agree very closely, the largest difference being only 0.4. This is an indication of the degree of accuracy maintained in the successive graphic transfers. In this case graph paper 8 by 10 inches was used in preparing the charts for Figures 16.2 to 16.7, and each of the transfers was double-checked. If higher accuracy in the mechanical process is desired, a still larger scale could be employed.

Taking the residuals in column 9 as the most accurate, we may now calculate their standard deviation (around their own mean). It works out at 2.88. This compares with a standard deviation for X_1 of 7.19.

Before computing $\bar{S}_{1.f(2,3,4)}$, we need the values for n and m. A simple parabola or hyperbola with two constants would probably represent $f_2'''(X_2)$ and $f_3'''(X_3)$. However, $f_4''(X_4)$ with its two inflections would probably require at least three constants. In addition, there is an a

Table 16.2

CALCULATION OF ESTIMATED X_1 FROM FINAL REGRESSION CURVES

Year, X_4 (1)	X_2 (2)	X_3 (3)	$f_2'''(X_2)$ (4)	$f_3'''(X_3)$ (5)	$f_4''(X_4)$ (6)	$\Sigma(f_2+f_3+f_4)$ $= X_1'''$ (7)	X_1 (8)	z''' (8–7) (9)	$z'''*$ (10)
1920	88.3	77.5	57.1	4.9	9.7	71.7	72.3	0.6	0.9
1921	47.5	60.2	67.8	−1.8	8.1	74.1	78.5	4.4	4.4
1922	71.3	58.5	60.5	−2.1	6.5	64.9	57.9	−7.0	−7.0
1923	88.3	67.0	57.1	−0.3	4.9	61.7	63.0	1.3	1.5
1924	69.0	70.8	61.0	1.0	3.4	65.4	63.7	−1.7	−1.8
1925	78.4	70.3	59.1	0.8	1.9	61.8	62.9	1.1	0.8
1926	88.0	70.8	57.2	1.0	0.3	58.5	60.3	1.8	2.1
1927	78.9	71.3	59.0	1.2	−1.6	58.6	59.6	1.0	1.3
1928	83.4	71.8	58.1	1.4	−3.7	55.8	55.2	−0.6	−0.5
1929	89.2	72.5	57.0	1.8	−5.4	53.4	51.5	−1.9	−1.7
1930	65.6	73.2	61.9	2.2	−6.3	57.8	58.6	0.8	1.0
1931	38.0	70.8	72.2	1.0	−6.9	66.3	65.6	−0.7	−0.7
1932	18.3	61.0	84.6	−1.7	−7.3	75.6	81.4	5.8	5.9
1933	28.7	59.0	77.3	−2.0	−7.5	67.8	65.0	−2.8	−2.8
1934	31.2	70.0	75.8	0.7	−7.4	69.1	64.6	−4.5	−4.1
1935	38.8	73.0	71.7	2.0	−7.0	66.7	65.4	−1.3	−1.1
1936	59.3	74.0	63.5	2.6	−6.4	59.7	61.1	1.4	1.3
1937	71.2	86.0	60.5	11.0	−5.4	66.1	65.6	−0.5	−0.8

* These are the values of z''' scaled off from Figure 16.7.

constant, represented by the mean of the z''''s. Altogether, then, it would probably take eight constants to fit mathematical curves to the regression functions graphically determined. Accordingly, $n = 18$ and $m = 8$. With these values, we can now compute $\bar{S}$ by equation (15.6).

$$\bar{S}^2_{1.f(2,3,4)} = \frac{ns^2_{z'''}}{n-m} = \frac{18(2.88^2)}{18-8} = 14.9299$$

$$\bar{S}_{1.f(2,3,4)} = 3.86$$

Similarly, by equation (15.7)

$$I^2_{1.234} = 1 - \frac{S^2_{1.f(2,3,4)}}{s_1^2} = 1 - \frac{(2.88)^2}{(7.19)^2}$$

$$= 0.8396$$

$$I_{1.234} = 0.916$$

The index of multiple correlation, 0.92, is moderately high. The adjusted standard error of estimate, 3.86, indicates that if it were possible to measure this same relationship from a very large sample drawn from the same universe, the errors in estimating steel costs for the observations in that large sample would *probably* have a standard deviation on the order of 3.86, rather than 2.88, per ton.[10]

ESTIMATING COST FOR NEW OBSERVATIONS. We can now use the data for 1938 and 1939, which we have disregarded to this point, to work out estimates for those years from the regression curves, by the same process that is shown in Table 16.2. The values are:

Year	X_2	X_3	$f_2''(X_2)$	$f_3''(X_3)$	$f_4''(X_4)$	X_1''	X_1	z''
1938	36.2	90.0	73.0	14.5	−4.3	83.2	80.5	−2.7
1939	60.7	89.7	63.1	14.2	−3.0	74.3	76.0	1.7

Just as in the similar example in Chapter 14, it is necessary to extrapolate two of the regression curves beyond the base data in making this estimate for subsequent years. In spite of the additional possibility of error which this introduces, both of the new estimates show residuals no larger than $\bar{S}_{1.f(2,3,4)}$. This indicates that the changes in steel costs during these next two years were in general related to the same factors as during earlier years and to about the same degree. (The student can check this conclusion by adding these two new observations to the original data, and re-analyzing the resulting sample of 20 observations.) If the trend or other factors were extrapolated much further, or if a sudden change in the conditions surrounding the industry were to occur, much larger errors of estimation might be experienced (as did happen later due to great changes in general price levels.)

RESTATING SHORT-CUT RESULTS FOR PUBLICATION. The same methods described on pages 234 to 240 of Chapter 14 can be used with curves obtained by the short-cut process, to prepare them for publication. There is a shorter method, however, which takes advantage of the fact that the curves obtained by the short-cut method are already in terms of a net value of X_1, for one variable, plus adjustments to that value for the other variables. All that is necessary is to determine the average value of the final z's and use this average as the a constant. (In the illustrative example just given, this average was only 0.07, and consequently

---

[10] See Chapter 20 for the application of sampling concepts and error formulas to time series.

was ignored.) Then the final functions are determined as follows (for the final curves of the illustrative problem):

$$F_2(X_2) = a + f_2'''(X_2)$$
$$F_3(x_3) = f_3'''(x_3)$$
$$F_4(x_4) = f_4''(x_4)$$

It is evident that, except for the slight adjustment of adding a to the first curve, these curves are the same as the final curves shown on Figures 16.5, 16.6, and 16.7.

Identifying "Joint" Relations by the Short-Cut Process. In some problems the relation between the variables is such that the independent variable cannot be explained fully by a regression equation which *adds* the regression of X_1 on variable X_2, to that on X_3, etc. Instead, in such cases the relation is so complex that the net change in X_1 with given changes in X_2 will vary with the associated values of X_3 or other variables. This type of relationship, designated "joint correlation," is discussed subsequently (Chapter 21). Where such correlation is present, it will show up in the process of examining the subgroups of observations in the first steps of the short-cut process.

The following empirical data will serve to illustrate the occurrence of joint correlation:[11]

Observation Number	X_1	X_2	X_3	X_4
1	216	9	4	6
2	160	10	8	2
3	140	2	7	10
4	264	4	11	6
5	30	5	2	3
6	56	7	1	8
7	5	1	5	1
8	16	2	2	4
9	70	2	5	7
10	126	7	6	3
11	180	10	3	6
12	280	5	7	8
13	120	3	4	10
14	25	1	5	5
15	224	4	8	7
16	120	6	0	2

[11] From Wilfred Malenbaum and John D. Black, The use of the short-cut graphic method of multiple correlation, *Quarterly Journal of Economics*, Vol. LII, p. 97, November, 1937.

The number of cases here is so small that it is difficult to eliminate the effects of X_3 and X_4, to determine the first approximation to the $X_1 X_2$ relation. An approximate grouping can be made, however, by classifying the observations into three groups, as follows:

1. Those with X_3 and X_4 both *larger* than their respective means.
2. Those with X_3 and X_4 both *smaller* than their respective means.
3. Those with X_3 and X_4 one above and one below their respective means.

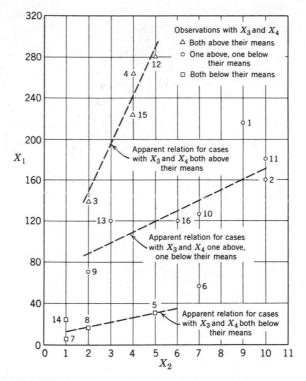

Fig. 16.8. Relation of X_1 to X_2 with observations classified on X_3 and X_4. When natural numbers are used, the net regression of X_1 on X_2 appears to shift with the accompanying values of X_3 and X_4.

This gives groupings with four observations (3, 4, 12, and 15) in the first group, four (5, 7, 8, and 14) in the second, and eight (1, 2, 6, 9, 10, 11, 13, and 16) in the third. Plotting each of these groups of observations, and drawing an approximate line through each, gives the results shown in Figure 16.8.

This figure differs from those we have examined previously (such as

Figure 16.3) in that the relations as shown by the several subgroups do not parallel one another at relatively constant distances, but instead diverge sharply. It appears, therefore, that the relation of X_1 to X_2 depends not only on the value of X_2 but also on the *associated values of X_3 and X_4*.

In this particular case the progressive nature of the relations shown on Figure 16.8 might lead us to suspect that the relation, instead of being an additive one, is a multiplying one. If that is the case, though it could not be represented adequately by an equation of the type:

$$X_1 = f_2(X_2) + f_3(X_3) + f_4(X_4)$$

it still might be represented by:

$$X_1 = [\phi_2(X_2)] [\phi_3(X_3)] [\phi_4(X_4)]$$

If that is the case, it can be determined by using the relation:

$$\log X_1 = f_2(\log X_2) + f_3(\log X_3) + f_4(\log X_4)$$

We can test whether this is likely to give a satisfactory fit by replotting Figure 16.8 on double logarithmic paper, or by plotting it on ordinary paper, substituting the logarithms of X_1 and X_2 for the natural values. Let us do the latter.

When that is done, the relations appear as shown in Figure 16.9. The three lines, fitted roughly to the three sets of observations, now appear more nearly parallel. In particular, the line of the upper group, which in Figure 16.8 made almost a 60-degree angle with the line for the lower group, is almost perfectly parallel to it in Figure 16.9. Apparently in this example the problem can be handled satisfactorily by the usual short-cut procedures, merely by transforming the variables from natural numbers to logarithms.

Where this transformation, or other simple transformations, do not serve to make the successive subgroups show approximately parallel relations, the methods of Chapter 21 must be employed instead.

Application of the Short-Cut Method to Large Samples. The short-cut method might be applied to samples too large for plotting the individual observations separately, by using a modification of the process of subgrouping and averaging illustrated in Chapter 23. The averages from Table 23.1, plotted in Figures 23.1 and 23.2, indicate quite well the final slope of the net regression lines. That is because the influence of the other independent variable was largely held constant by the process of subclassifying. In the same way the lines of averages from subgroups would tend to indicate the regression curves in problems where curves were needed. With a sufficient number of observations, the first approximation to each of the net regression curves might be obtained from charts

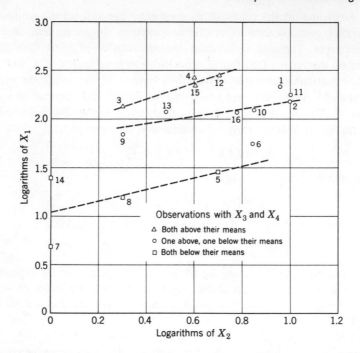

Fig. 16.9. When the logarithms of the data shown in Figure 16.8 are used, the net regression of X_1 on X_2 is found to be about the same, regardless of the accompanying values of X_3 and X_4.

of subaverages similar to Figures 23.1 and 23.2 on pages 390–391. These several first approximation curves could then be made the basis for working out estimated values of X_1 and residuals. The process of successive approximations could then be continued exactly as illustrated in Chapter 14. Since the first approximation curves would approach fairly near to the true net regressions, the number of approximations required to obtain the same closeness of fit would usually be less than by the earlier method.

Combination of Short-Cut Procedures and Mathematical Procedures. Both the short-cut method of this chapter and the longer successive-approximation method of Chapter 14 depend on graphic methods in arriving at the curves of best fit. Where especially high accuracy is desired, the final slope of the several curves can be checked by least squares, according to the methods set forth in Chapter 17 on pages 291 to 293.

Some investigators prefer to use the short-cut method to determine the approximate shapes of each of the several net regression curves, and then to determine the final net regressions by fitting algebraic curves capable of representing those several shapes. The technique for fitting these mathematical curves to several variables is also set forth in Chapter 14. If there is a logical basis to support the curves employed, there is value to this procedure. If the equations are simply selected empirically, however, the mathematical curves have no more meaning than the graphic ones, for the reasons already discussed fully in Chapter 6, except for greater ease and certainty in determining confidence intervals for each regression. It is true that any one fitting the same set of mathematical curves to the same data by the same method will get exactly the same result, to the fifth decimal place in the values of the constants, if desired. Curves obtained by different investigators by either graphic process, on the contrary, may vary slightly from one to another. But the identical constants obtained by the least-squares fit have only a fictitious accuracy, as compared with their standard errors, or with the zone of uncertainty within which the function can be determined from the given set of observations. Multiple regression curves are dependable only with respect to this confidence zone, rather than to the exact line (as explained in Chapter 17).

Summary

Under certain conditions first approximations to multiple regression lines or curves may be obtained directly from the original observations by a graphic process based on the comparison of individual observations, considering several variables simultaneously. This process eliminates the necessity of computing linear regressions by arithmetical means. Further, it substitutes graphic measurements for arithmetic calculations in correcting these curves to their final shape by successive approximations. It requires the researcher to examine his data more thoroughly and so to exercise thought and care in working out the relations and in interpreting their significance. Carefully used, it materially reduces the time required in determining multiple regression curves.

REFERENCES

Bean, L. H., A simplified method of graphic curvilinear correlation, *Jour. Amer. Stat. Assoc.*, Vol. XXIV, pp. 386–397, December, 1929.

Waite, Warren C., Some characteristics of the graphic method of correlation, *Jour. Amer. Stat. Assoc.*, Vol. XXVII, pp. 68–70, March, 1932.

Ezekiel, Mordecai, Further remarks on the graphic method of correlation, *Jour. Amer. Stat. Assoc.*, Vol. XXVII, pp. 183–185, June, 1932.

Malenbaum, W., and J. D. Black, The use of the short-cut graphic method of multiple correlation, *Quart. Jour. Econ.*, Vol. LII, pp. 66–112, November, 1937.

Bean, L. H., and Mordecai Ezekiel, The use of the short-cut graphic method of multiple correlation, Comment, and Further comment, *Quart. Jour. Econ.*, Vol. LV, pp. 318–346, February, 1940.

Wellman, H. R., Application and uses of the graphic method of multiple correlation, *Jour. Farm Econ.*, Vol. XXIII, pp. 311–316, February, 1941.

Waite, Warren C., Place of, and limitations to, the method, *Jour. Farm Econ.*, Vol. XXIII, pp. 317–322, February, 1941.

Working, E. J., and Geoffrey Shepherd, Notes on the place of the graphic method of correlation analysis, *Jour. Farm Econ.*, Vol. XXIII, pp. 322–323, February, 1941.

Foote, Richard J., and J. Russell Ives, The relationship of the method of graphic correlation to least squares, U. S. Dept. Agr., Bur. Agr. Econ., *Stat. and Agr.* No. 1, April, 1941.

Foote, Richard J., The mathematical basis for the Bean Method of Graphic Multiple Correlation, *Jour. Amer. Stat.Assoc.*, pp. 778–788, December, 1953.

Note 16.1 These discussions, and an address by Meyer A. Girshick summarized in the February, 1941, *Journal of Farm Economics*, have provided definite proof of the meaning of the graphic method. In linear multiple correlation the graphic method gives results which tend to approach the lines secured by a least-squares solution, even if the first approximations are purely arbitrary guesses, while the speed of convergence depends on the intercorrelation among the independent variables. The higher their intercorrelation, the slower tends to be the speed of the convergence.

Note 16.2 If the standard error of estimate, adjusted for the degrees of freedom, $\bar{S}_{1.23\ldots k}$, is calculated as each new set of approximation curves is completed, it will show whether the gain in closeness of fit is sufficient to offset any additional flexibility introduced in the curves. The validity of this test, however, depends upon the user's skill in estimating what value of m to employ.

SECTION V

Significance of Correlation and Regression Results

CHAPTER 17

The sampling significance of correlation and regression measures

Early in this book it was pointed out that when any statistical measure, such as an average, is determined from a sample selected from a universe under study, the true value of that measure in the universe might be different from the value shown by the sample. Methods were discussed which enable one to estimate how far the average from such a sample may vary from the true average, for a stated proportion of such samples. Such estimates enable one to judge how much confidence may be placed in an average calculated from a given sample.

Different Types of Sampling Models. The applicability of sampling concepts to correlation coefficients differs widely according to the nature of the universe from which the sample is selected and the manner in which sample values of the independent variables are obtained. The interpretation of regression coefficients is much less dependent upon the shape of the underlying universe (if any) or the manner in which sample values of the independent variables are chosen; however, estimates of the *reliability* of regression coefficients are influenced by these factors. For convenience of exposition we shall distinguish between two principal situations or "models"—the *correlation model* and the *regression model*.[1]

[1] Compare M. G. Kendall, Regression, structure and functional relationship, Part I, *Biometrika*, Vol. 38, pp. 11–25, June, 1951.

Kendall distinguishes between (1) situations in which values of the independent

The *correlation model* requires strictly random samples from normal bivariate or multivariate universes. This means that the joint frequency distribution of the two or more variables in the sample will be representative of the corresponding distribution in the universe; that the distribution of each variable will tend to follow the normal frequency curve; and further, that the standard deviations of Y values for all values or class intervals of X, will be the same within normal sampling fluctuations.[2] (If, as in the auto-stopping example of Chapters 3, 4, and 6, the s_y increases, both in the universe and in the sample, with increasing values of X, and the differences in s_y in different arrays can be corrected by using some transformation of Y such as, in this case, log Y, that is sufficient to satisfy the condition—but only if the transformed values are used in calculating the correlation coefficients.) For samples which meet the stated conditions, standard errors may be calculated for both correlation and regression *statistics* derived from the sample which will provide a sound basis for probability statements as to their nearness to the corresponding *parameters* for the universe.

The term *regression model* will be used here to imply that values of the independent variable or variables are selected in advance by the investigator, as is typical in controlled experiments, with no requirement that

---

variables are decided before the sample is drawn—the *determined* case; and (2) situations in which the values of the independent variables are randomly drawn from an underlying multivariate distribution—the *undetermined* case. Kendall's determined case corresponds to the regression model of this chapter. Kendall's undetermined case is somewhat broader than our correlation model, for the underlying multivariate distributions assumed by Kendall do not necessarily follow the normal curve.

Thus, in Kendall's undetermined case, both regression and correlation coefficients in a sample would be regarded as estimates of the corresponding parameters of an underlying universe which exists in nature or at least independently of any intentional influence on the part of the investigator who has drawn the sample. The standard errors of the regression coefficients have the same interpretation in this case as in both Kendall's determined case and in the undetermined normal multivariate distribution of our correlation model. However, as noted by Kendall in another place, very little is known about the sampling distribution of the correlation coefficient except in the case of normal universes. (See M. G. Kendall, *The Advanced Theory of Statistics*, Vol. 1, p. 346, Charles Griffin, London, 1943.) Hence, the interpretation of the correlation coefficient is uncertain unless both conditions of our correlation model are met, namely that the sample values of the independent variables be drawn randomly and that the underlying universe be normal.

See also George W. Snedecor, *Statistical Methods Applied to Experiments in Agriculture and Biology*, 5th ed., pp. 413–14, Iowa State College Press, Ames, Iowa 1956, for a distinction between models which is substantially identical with that used in the present chapter.

[2] Similarly, sample values of X will tend to be normally distributed with approximately equal standard deviations for all values or class intervals of Y.

the distributions of the independent variables in the sample will be representative of those in the universe—if indeed a "natural" underlying universe exists at all. In this model measures of correlation have only a very limited meaning, and estimates of the *reliability* of regression coefficients apply only to a distribution of samples each drawn or constructed under exactly the same conditions as the given sample, including the same range and relative frequency of values of the independent variables.

Reliability of Regression Coefficients

Simple Regression. We have already noted that estimates of regression coefficients are less affected by differences in model than are estimates of correlation coefficients. For both models, the reliability of the observed simple regression coefficient varies inversely with the standard deviation of the individual observations around the regression line, and directly with the square root of the number of cases in the sample. The standard deviation of regression coefficients from one sample to another can be estimated by the equation[3]

$$s_{b_{yx}} = \frac{\bar{S}_{y \cdot x}}{s_x \sqrt{n}} \tag{17.1}$$

The application of this equation may be illustrated by data drawn from a sampling experiment using a universe of known characteristics, and with normal distributions of the variables. The first sample, of 50 cases, gave the value $b_{yx} = 0.175$, with $\bar{S}_{y \cdot x} = 2.46$, and $s_x = 2.44$. Computing the value of $s_{b_{y \cdot x}}$ by equation (17.1)

$$s_{b_{yx}} = \frac{2.46}{2.44\sqrt{50}} = \frac{2.46}{17.25} = 0.143$$

The variable $t = (b - \beta)/s_b$ follows a "Student's" or t-distribution which, as noted in Chapter 2, approaches a normal distribution as the number of degrees of freedom in the sample estimate becomes large. In this expression β is the true (but unknown) regression coefficient in the universe, and b and s_b are values computed from a particular sample. The values of t from repeated samples are distributed symmetrically about zero even if the sample size is very small. Conventionally, probability statements are based on the normal distribution when the number

---

[3] In some textbooks, b_{yx} would be used to represent the regression coefficient as determined from the sample and β_{yx} would be used to represent the true parameter in the universe. In this notation, the value for β_{yx} as shown in Table 17.1 is 0.152. In consulting textbooks using this notation, we should not confuse this use of the β with the special definition given for it in Chapters 9 and 12, in equations (9.1) and (12.9).

of degrees of freedom in the sample estimate exceeds 30. If the number is equal to or only slightly more than 30, statements based on the normal distribution are quite accurate for t-values between zero and ± 2 but tend to exaggerate the level of significance actually attained for extreme values of t (less than -3 or more than $+3$). Given sample estimates with 30 or more degrees of freedom, in 2 samples out of 3, on the average, the observed regression coefficient will lie within one standard error of the true parameter. If in this case we say that the universe regression coefficient lies within the interval 0.175 ± 0.143, or between 0.032 and 0.318, we are making a statement of a type which, if made for a succession of such samples, will be wrong 1 time out of 3, on the average. Parallel statements for intervals of $\pm 2s_{b_{yx}}$ or $\pm 3s_{b_{yx}}$ are subject to the same interpretation as that given earlier, and to the same adjustments for small samples given in Table 2.3. However, $n - (m - 1)$ must be used in entering data in that table, instead of n, in order to allow for the additional degrees of freedom used up in determining the regression coefficients.

We can now illustrate the meaning of the estimated standard error for our one sample by comparing it with the other regression coefficients obtained in the sampling experiment mentioned. In that experiment, 5 samples were drawn at random from the universe for each of 3 sizes of samples—30, 50, and 100 observations. The values obtained are shown in Table 17.1.

Table 17.1

VALUES OF b_{yx} SHOWN BY SUCCESSIVE SAMPLES DRAWN FROM THE SAME UNIVERSE, WITH DIFFERENT SIZES OF SAMPLES

	30 Observations	50 Observations	100 Observations
	0.292	0.175	0.113
	0.012	−0.297	0.120
	−0.136	0.144	0.303
	−0.022	0.130	0.197
	0.449	0.167	0.132
Universe value	0.152	0.152	0.152

In this case, we see that the value for the universe, 0.152, lay within our confidence interval for $P = 0.67$. Also, of the 5 samples with $n = 50$, 4 had values of b_{yx} within the interval from our sample. If our sample had happened to be the second one, however, its b_{yx}, -0.297, would probably have differed from the universe value by more than twice its own standard error. The values in Table 17.1 also illustrate how the variation in successive sample values declines as the size of the sample is made larger. We could estimate the expected decline, of course, by recalculating equation (17.1) using 30 and 100 for n, and then comparing the estimated variation for samples of 30, 50, and 100 with the values shown in the table. (This is left as an exercise for the student.)

It will be noted from equation (17.1) that $s_{b_{yx}}$ varies inversely with s_x, the standard deviation of values of the independent variable in the sample. In the *correlation model*, s_x is an estimate of the true parameter σ_x in the underlying bivariate normal universe. (If strictly random samples were drawn from a "natural" universe which followed other than a normal distribution, s_x would still be an estimate of the parameter σ_x in that universe.) The *regression model* implies that values of X in successive samples are controlled or selected on the same principles as they were for the original sample. The estimated sampling variance of b is therefore an unbiased estimate of the variance of b's from such a set of successive samples. The first three pages of Chapter 18 give some arithmetic examples illustrating the principle that such a selection of X values does not change the value of b. Accordingly, if ordinary 95 per cent confidence intervals are used, the true regression coefficient in the universe is covered in 19 out of 20 samples, just as in the case of similar confidence intervals for universe means. However, as is also demonstrated in Chapter 18, purposeful selection of extreme values of X can reduce the standard error of b_{yx} very substantially relative to the value that would be obtained if the X values were selected so as to follow a normal frequency curve. Thus, for a given sample size, $s_{b_{yx}}$ may show much wider variations from one artificial universe to another than from sample to sample drawn from the same universe.

Net (Multiple) Regression. The standard error of a net regression coefficient may be estimated by the equation[4]

$$s_{b_{12.34\ldots m}} = \sqrt{\frac{\bar{S}^2_{1.234\ldots m}}{n s_2^2 (1 - R^2_{2.34\ldots m})}} \tag{17.2}$$

As with simple regression coefficients, the reliability of net regression coefficients is affected by the number of cases in the sample and the standard error of estimate. In addition, it is affected by how closely the given independent variable [X_2 as equation (17.2) is stated] can be estimated from the other independent variables (as X_3 and X_4). The more highly

[4] The standard errors of net regression coefficients can be determined at the same time as the regression coefficients, and as part of the same set of computations, by various modifications of this formula. See Appendix 2, pages 499 to 502, and 507 to 516.

the independent variables are interrelated among themselves, the less reliably can the net regression of X_1 upon any one of them be determined.

Again, we can illustrate the use of the equation by results from a sampling study. In this case 3 independent variables were used, and successive samples were drawn, 16 of 30 observations, 10 of 50, and 5 of 100. The first sample drawn of 30 observations gave values as follows: $b_{12.34} = 0.583$; $b_{13.24} = 0.366$; $b_{14.23} = 0.949$; $\bar{S}_{1.234} = 2.81$; $s_2 = 2.53$; and $R_{2.34} = 0.708$. Substituting the appropriate values in equation (17.2),

Table 17.2

DISTRIBUTION OF VALUES FOR NET REGRESSION COEFFICIENTS FOR REPEATED SAMPLES DRAWN FROM THE SAME UNIVERSE

Range of Values	30 Observations	50 Observations	100 Observations	True Value
Values for $b_{12.34}$:				
−0.79 to −0.60	1			
−0.59 to −0.40	0			
−0.39 to −0.20	1			
−0.19 to −0.	0	2		
0. to 0.19	2	1	1	
0.20 to 0.39	6	4	4	+0.320
0.40 to 0.59	4	3		
0.60 to 0.79	2			
Values for $b_{13.24}$:				
−0.19 to 0.	2			
0. to 0.19	3			
0.20 to 0.39	5	6	2	+0.377
0.40 to 0.59	2	2	2	
0.60 to 0.79	1	1	1	
0.80 to 0.99	2	1		
1.00 to 1.10	1			
Values for $b_{14.23}$:				
0. to 0.19				
0.20 to 0.39		1		
0.40 to 0.59	1	1		
0.60 to 0.79	8	2	2	
0.80 to 0.99	3	4	3	+0.824
1.00 to 1.19	4	1		
1.20 to 1.39		1		
1.40 to 1.59				

the value of $s_{b_{12.34}}$ is found to be 0.287. The observed regression may therefore be written as 0.583 ± 0.287. If this sample value departs from the true value no more than can be expected in 2 cases out of 3, the true value will lie within the confidence interval 0.296 to 0.870; or, if it departs no more than can be expected in 19 cases out of 20, the true value will lie in the interval from 0.009 to 1.157.

Table 17.2 shows the distribution of the values obtained from the various sets of samples, as compared to the true parameters. The true value for $b_{12.34}$ was 0.320, or within the first interval just given. It may be noted that 11 of the 16 samples of size 30 gave values for this coefficient within 0.287 of the true value, and all but one fell within 0.574 of it. Again this illustrates how the variability of statistics which tend to be distributed according to either normal or t-distributions may be estimated by appropriate error formulas, hence how the degree of confidence to be placed in conclusions from a given sample may be judged.

The qualifications that use of this error formula may impose upon regression results may be illustrated by a problem in which the theory of sampling was reasonably applicable, namely, the relation between the average amount of feed a herd of cows receives and the resulting milk production per cow. Table 17.3 shows these results for two different studies, regarding each set of observations as a sample that was random with respect at least to the dependent variable.

This table illustrates two points: first, that the net regressions are not very accurate even though the multiple correlation is 0.80 to 0.86; and second, that the reliability of the net regression differs from variable to variable, being much greater for some variables than for others. It is obvious that some of the net regression coefficients are not at all statistically significant, whereas others indicate the probable relationship within a fairly narrow range.

Thus for the percentage of lime, with the standard error as large as the regression coefficient, there is 1 chance in 3 that the sample net regression coefficient differs from the universe value by more than the sample regression itself, and 1 chance in 6 that the sample regression is of opposite sign from that in the universe. With the total digestible nutrients, on the other hand, with the standard error only 18 per cent of the observed value, there is but little chance that the observed value differs from the universe regression by more than 36 per cent, and very little chance that it differs as much as 50 per cent.

If the regression equation is to be used solely as a basis for making new estimates of the value of the dependent factor to be expected for given values of the independent factors, then the accuracy of the several net regression coefficients does not make such a great difference. Any

Table 17.3

STANDARD ERRORS OF PARTIAL REGRESSION COEFFICIENTS, IN PER CENT OF
THE VALUE OF THE COEFFICIENT*

Item	Wisconsin Study	Minnesota Study
Number of observations	95	77
Number of variables	10	8
Multiple correlation, adjusted for number of variables	0.805 ± 0.039	0.862 ± 0.034

Standard Error of Regression Coefficients†

Independent variable:		
Total digestible nutrients	17.8%	17.0%
Nutritive ratio	18.4%	14.1%
Per cent of protein "good"	42.0%	
Per cent of lime	99.9%	
Per cent summer feeding	25.9%	
Per cent silage	31.9%	20.3%
Fat test of milk	15.7%	5.5%
Per cent fall freshening	27.4%	17.5%
Value per cow	39.7%	
Age of cows		26.5%
Per cent grain in ratio		29.9%

* Mordecai Ezekiel, The application of the theory of error to multiple and curvilinear correlation, *Journal of the American Statistical Association*, Vol. XXIV, No. 165A, March, 1929, Supplement, p. 103.

† The coefficients are for the net regression of milk production on the factors stated. The original article gave the per cent figures in terms of the "probable error" (*P.E.* = 0.6745 of the standard error).

deficiency in one may be compensated for by an excess in another. (This does not hold true, however, if estimates are made for extreme values of variables whose regressions are subject to large errors. See Chapter 19 on this point.) But if the major interest is not in the total estimate, but in the changes in the dependent factor with changes in each particular independent factor, then the reliability of each particular regression coefficient becomes of real importance. In the illustration cited, for example, it would not do to know merely that the milk production per cow varied both with protein content and with lime, if it was desired to know how much to allow for protein and how much for lime in compounding a ration. Instead, the standard errors indicate that the influence of protein (as represented in the "nutritive" ratio) has been fairly accurately measured, whereas the influence of lime has not been accurately measured at all. Not much confidence therefore can be placed in the conclusions as to this latter factor.

In any regression study where the results are based upon a sample of observations drawn at random from a known universe, and where any importance is to be attached to the values found for the several regression coefficients, it is essential that the standard errors of each of those coefficients be determined and considered. As is illustrated in the examples just discussed, a sample may have a high multiple correlation and yet yield regression coefficients for some variables which are almost entirely the result of chance fluctuation, and therefore are not statistically significant. This may occur even with moderately large samples, such as the sample of 95 cases in the first example just considered. Computation, presentation, and discussion of the standard errors of the regression coefficients are therefore vital parts of any such multiple correlation study.

Regression Line. Not only may the observed *slope* of the regression line vary from the true slope, but the elevation of the line, as observed from a sample, may vary from the true elevation. Equation (17.1) has already indicated a way of determining the standard error of the regression coefficient, and so of estimating the probable range within which the true slope lies. The height of the regression line is most accurately determined for the mean estimated value $M_{y'}$ of the dependent factor, corresponding to the observed mean value of X, the independent factor. If we define the mean as

$$M_{y'} = a_{yx} + b_{yx}M_x$$

we may find its standard error by the formula

$$s_{M_{y'}} = \frac{S_{y \cdot x}}{\sqrt{n}} \tag{17.3}$$

The standard error of the whole regression line may now be determined from equations (17.1) and (17.3). We may illustrate by data from the cotton-yield problem used as an example in Chapter 8. With 14 observations, the values were $b_{yx} = 16.70$, $a_{yx} = -2.261$, $M_x = 1.97$, $\bar{S}_{yx} = 8.28$, $s_x = 0.73$, $M_y = M_{y'} = 30.64$, $s_y = 14.43$.

$$M_{y'} = -2.261 + (16.70)(1.97) = 30.64$$

$$s_{M_{y'}} = \frac{8.28}{\sqrt{14}} = 2.21$$

$$s_{b_{yx}} = \frac{8.28}{0.73\sqrt{14}} = 3.03$$

Since the estimated value Y' equals $M_{y'} + b(x)$, the standard error of the estimate for any value of x will include the standard errors of $M_{y'}$ and of $b(x)$. Standard errors are standard deviations; hence they should be summed by adding their squares. The standard error of Y', for any particular value of x, is therefore given by the equation[5]

$$s_{y'} = \sqrt{s_{M_{y'}}^2 + (s_{b_{yx}}x)^2} \qquad (17.4)$$

By using this relation, the calculation of the standard error of Y', for selected values of X, is shown in Table 17.4.

Table 17.4

Selected Values of X	Departures from Mean, x	Calculation of $s_{y'}$				
		$s_{b_{yx}}x$ $= 3.03x$	$(s_{b_{yx}}x)^2$	$s_{M_{y'}}^2$ $= 2.21^2$	$s_{y'}^2 =$ $(s_b x)^2$ $+ s_{My'}^2$	$s_{y'}$
0.97	−1.00	−3.030	9.1809	4.8841	14.0650	3.75
1.47	−0.50	−1.515	2.2952	4.8841	7.1793	2.68
1.97	0	0	0	4.8841	4.8841	2.21
2.47	0.50	1.515	2.2952	4.8841	7.1793	2.68
2.97	1.00	3.030	9.1809	4.8841	14.0650	3.75
3.47	1.50	4.545	20.6570	4.8841	25.5411	5.05
3.97	2.00	6.060	36.7236	4.8841	41.6077	6.45

There are 14 cases; subtracting the one extra constant involved in correlation determinations gives 13 as the number of observations with which to judge from Table 2.3 the significance of these standard errors. Taking values midway between those for 10 and for 16 cases, we find that a statement that the true values β_{yx} and $M_{y'}$ do not differ from the observed values by more than the calculated standard errors will be wrong for 34 out of each 100 such statements, on the average. Similarly, the statement that they do not differ by more than twice the calculated standard errors will be wrong for 7 out of 100 such statements, on the average. The chances are therefore 93 out of 100 that the departure of this regression line from the true line will not be larger than the confidence intervals just calculated. Plotting $2s_{y'}$ above and below the corresponding values

[5] Holbrook Working and Harold Hotelling, Applications of the theory of error to the interpretation of trends, *Journal of the American Statistical Association Papers and Proceedings*, Vol. XXIV, pp. 73–85, March supplement, 1929.

of Y', given by the regression line, shows this interval. These limits are plotted in Figure 17.1, together with the original observations and the regression line. The limits within which the universe regression "probably" lies could be shown in a similar manner for any other desired level of probability. As is evident in the figure, the probable true position of the line becomes very uncertain as the limits of the data are approached.

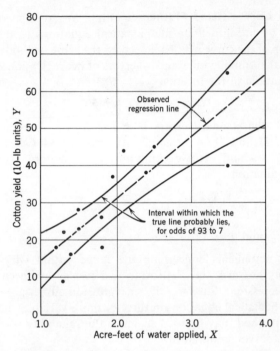

Fig. 17.1. Linear regression of cotton yield on irrigation water applied, and range within which the true relation probably lies.

In most studies of statistical relationships between variables the regression line is the most important result of the study. The confidence that can be placed in the line determined from a random sample is no greater than is indicated by the probable error of its slope, or the standard error zone of its position. Accordingly, the final statement of the regression equation should always indicate clearly the standard errors of the net regression coefficients and the number of observations on which the conclusions are based. This will serve to caution the reader of the extent to which the values may vary from the true value simply due to chance fluctuations of sampling, and so caution him not to attach more importance to them than their significance justifies. It is not customary in most fields to tabulate or chart values of the standard error zone of the regression line in published reports. However, the investigator should at least calculate this error zone for a few representative sets of values of the independent variables for his own guidance and reflection, and it would be appropriate to pass on to his readers at least a general statement as to the reliability of the regression line in a few broad ranges of values of the independent variables.

Regression Curves Fitted Mathematically. Where regression curves are obtained by fitting definite mathematical equations to the data, the standard error of the curve may be judged by the same methods previously presented for determining the standard errors of net regression coefficients. Thus, if a parabola of the formula

$$X_1 = a + bX_2 + b'X_2^2$$

is determined, the standard errors of b and b' may be determined by equations (17.2) and (17.4), treating X_2 and X_2^2 as two independent variables. This would involve computing the standard error zone of $f(X_1)$ by the equation

$$s_{1.2,2^2}^2 = \frac{\bar{S}_{1.2,2^2}^2}{n} + (s_b x_2)^2 + 2s_b s_{b'} x_2 u + (s_{b'} u)^2 \qquad (17.5)$$

where $U = X_2^2$, and $u = X_2^2 - M_{X_2^2}$

Probability statements concerning the range within which the true curve probably lies may then be formulated just as has been illustrated for a linear regression. Similarly, if net regression curves are determined by fitting mathematical equations in three or more variables, an extension of this same method may be used to judge the reliability of each of the net regression curves so obtained.[6]

Regression Curves Determined Graphically. For regression curves, either simple or multiple, determined by graphic processes, no such exact mathematical estimation of the standard error intervals is possible. Experimental studies summarized in earlier editions of this book have given some indication of the range of sampling errors in such curves, and how that may be estimated; but these approximations to the standard-error zone are not as reliable as the equations shown thus far for algebraically determined curves, and more work is needed on this problem.[7] The experimental work did show that the reliability of graphic curves varies inversely with the standard error of estimate for the whole sample, and apparently tended to vary inversely with the intercorrelation among the

---

[6] Henry Schultz, The standard error of a forecast from a curve, *Journal of the American Statistical Association*, Vol. XXV, pp. 139–185, June, 1930.

[7] See the second edition of this book, pages 327–39.

independent variables, just as is the case with net regression coefficients; but that in addition the reliability of the results varied with the thickness of observations along the regression curve for each independent variable, being more reliable where observations were reasonably frequent, and less reliable where the observations thinned out. Now that electronic calculators make elaborate computations much less burdensome, in cases where graphic curves have been fitted for lack of any theoretical reason to expect a given type of equation, the reliability of the regressions could be determined by the following process: select several sets of equations that could be reasonably expected to approximate the graphic curves, and fit them simultaneously by the method explained on pages 205 to 210; after determining the set that came nearest to reproducing the graphic curves with the smallest number of constants, determine the standard-error zone of each curve by the methods indicated in the preceding section. This effort would be worth making, however, only if it was desired to make inferences from the sample as to the universe; or if it was important to determine which net regression curves had been determined with so little significance that they should be excluded from the solution.

An alternative estimate of the reliability of graphic regression curves may be made by determining the regression equation (15.10):

$$X_1 = a_{1.2'3'4'} + b_{12'.3'4'}[f_2'''(X_2)] + b_{13'.2'4'}[f_3'''(X_3)] + b_{14'.2'3'}[f_4'''(X_4)]$$

To compute the new constants required in this equation, the functional readings corresponding to the independent variables are correlated with the original values of the dependent variable. Thus, in Table 14.15, the values read from the final curves, shown in the fourth, fifth, and sixth columns, would be substituted for the original independent variables in running the multiple correlation with X_1. If X_2', X_3', etc., are used to represent these transformed values, the data to be correlated for the first four observations would be:

X_2'	X_3'	X_4'	X_1	X_2'	X_3'	X_4'	X_1
7.4	11.7	12.3	24.5	8.4	12.2	12.2	27.9
7.9	13.0	11.8	33.7	8.8	9.9	12.2	27.5

If the net regression coefficients come out 1.0, that indicates that no change need be made in the curves. If any b comes out other than unity, however, the values read from the corresponding curve should be adjusted as indicated by the regression results. The adjustment may be worked out as follows:

In the same way that $f_2'''(X_2)$ was used to indicate the values read from the final set of approximation curves, let $f_2'''(x_2)$ represent the deviations of those readings for each variable from the average of all the readings for the particular variable. That is, for each observation

$$f_2'''(x_2) = f_2'''(X_2) - M_{f'''(X_2)}$$

The regression equation (15.10) may then be restated

$$x_1 = b_{12'.3'4'}[f_2'''(x_2)] + b_{13'.2'4'}[f_3'''(x_3)] + b_{14'.2'3'}[f_4'''(x_4)]$$

and the corrected functions will be as follows:

$$f_2(x_2) = b_{12'.3'4'}[f_2'''(x_2)]$$

$$f_3(x_3) = b_{13'.2'4'}[f_3'''(x_3)]$$

$$f_4(x_4) = b_{14'.2'3'}[f_4'''(x_4)]$$

The correction merely expands or contracts the curve, making all the high estimated values higher and all the low values lower, or vice versa.

When equation (15.10) is computed for the values in Table 16.2, the following multiple regression equation is obtained:

$$X'''' = -6.1099 + 1.0926[f_2'''(X_2)] + 1.1257[f_3'''(X_3)] + 1.1094[f_4'''(X_4)]$$
$$(0.1509) (0.3601) (0.1982)$$

As expected, all the b's come out close to 1.00, so only small change in the shape or slope of the curves would result from applying these corrections. The adjusted standard error of estimate for this new set of regressions, and the index of multiple correlation as shown by the $\bar{R}$ for the multiple regression equation, are practically unchanged from those calculated in Chapter 16, indicating that the slight shift of the curves has made only slight improvement in the over-all fit to the data, and that there is no great significance in the final changes in their shape. The fact that all three regression coefficients are three or more times their standard errors indicates that all three regression curves are significant.

Equation 15.10 thus provides a partial answer to the question of estimating the reliability of curvilinear net regressions obtained by graphic methods. It is only a partial answer, as more precise mathematical approximations to each net regression curve would use up two or more degrees of freedom and would require us to estimate two or more constants which determine the shape of each curve, such as $f_2''(X_2)$. The standard errors for equation (15.10) do not tell us whether or not the best or most logical shapes of $f_2'''(X_2)$, etc., have been attained; they simply show that the particular approximations to these shapes represented by $f_2'''(X_2)$, etc., do have a statistically significant association with X_1. (Other alternative ways of adjusting the final graphic regression curves by supplementary calculations, presented on pages 400 to 404 of the second edition, are omitted from this edition for lack of space.)

Reliability of Correlation Coefficients

Correlation Coefficients. For large random samples drawn from normal bivariate universes in which the true correlation ρ is not far from zero, the distribution of the observed correlations in successive samples will tend to be nearly normal, and the standard error of the coefficient may be approximately estimated by the formula

$$s_{r_{yx}} = \frac{1 - r_{yx}^2}{\sqrt{n - 1}} \tag{17.6}$$

If the sample is small, however, no simple equation will be adequate. Further, if the true value of the correlation is much smaller or larger than zero, the distribution of sample values will be badly skewed, as will be evident intuitively if we consider the probable distribution of values in a range from -1 to 1 with the mode (say) at 0.75. The reliability of the observed correlation coefficient must therefore be estimated by other methods than equation (17.6) in most cases.

Certain of Fisher's methods for determining the reliability of observed correlations may be put into simpler form for general use, as shown in Figure 17.2. This figure is based upon the idea that, although we cannot be sure of the true correlation existing in the universe on the basis of the correlation shown in a given sample, we can estimate a minimum value for the true correlation, with a given chance of being wrong. Figure 17.2 has been calculated, from Fisher's results, to show such probable minimum correlations in the universe, with the probability that the statements based on the figure will be wrong for 1 sample out of 20, on the average.[8] The results have been plotted for different sizes of samples and observed correlations. Thus, if a random sample of 20 gives an observed correlation of 0.70, the figure shows at a glance that we can say that the true correlation is greater than 0.44, with the expectation that such statements will be wrong only in 1 sample out of 20, on the average. Similarly, for an observed correlation of 0.55 with a sample of 35 cases, reading from the line for observed correlation at 0.55, and interpolating between $n = 30$

[8] For the source of Figs. 17.2 through 17.5, see Appendix 3, note 4.

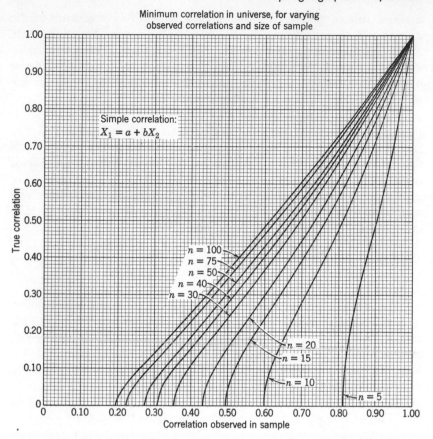

Fig. 17.2. Under conditions of random sampling, 1 sample out of 20, on the average, will show a correlation coefficient with a ± value as high as that "observed in sample," when drawn from a universe with the stated true correlation.

and $n = 40$, gives 0.32, which means that we can say that the true correlation is greater than 0.32, with the same degree of confidence. The figure can be used in a similar manner for any other size of sample up to 100, and any observed correlation.

Figure 17.2 deserves close study, for it tells a great deal about the sampling reliability, or, rather, unreliability, of correlation coefficients. The bottom line, for example, shows that, when samples are drawn from a universe where the true correlation is zero, 1 sample out of 20 will show a correlation as high as ±0.60, on the average, with samples of 10 cases; as high as ±0.49, with samples of 15 cases; and as high as ±0.35, even with samples of 30 cases. Similarly, if the samples are drawn from a universe where the true correlation is 0.50, 1 sample out of 20, on the average, will show a correlation as high as 0.81, with samples of 10; as high as 0.73, with samples of 20; and as high as 0.69, with samples of 30. Many other similar comparisons can be made readily. For example, if the true correlation is 0.80 and samples of 10 cases are used, 5 per cent of the samples will show correlations as high as 0.93. These facts do not take into account the tendency of many students to examine a number of possible independent variables and to select for more detailed study those which show the highest correlation with the independent factor. If that is done, the possible minimum correlation in the universe, corresponding to the correlation observed in the sample so selected, will be much lower than would be estimated from Figure 17.2.

Multiple Correlation Coefficients. The reliability of coefficients of multiple correlation varies not only with the correlation and the size of sample, but also with the number of independent variables. Fisher has developed an exact method for judging the significance of observed coefficients of multiple correlation.[9] Figures 17.3, 17.4, and 17.5 provide a simple method of applying his conclusions for multiple correlation coefficients, in the same way that Figure 17.2 provides for simple correlation coefficients. For problems involving 3, 5, and 7 independent factors, respectively, these figures show the approximate minimum true correlation that probably exists in the universe with any size of sample up to 100, and for any observed correlation, with the probability that the statements based on the figure will be right for 19 samples out of 20, on the average. Thus if, with 30 observations, a correlation of $R_{1.23456} = 0.80$ should be obtained, we can say that the true correlation (from Figure 17.4) is at least 0.58. Similarly, if for 50 observations a correlation of $R_{1.234} = 0.62$ were obtained, Figure 17.3 gives 0.42 as the probable minimum correlation in the universe. These conclusions, of course, are subject to the cautions given on pages 279 to 281 with respect to coefficients of simple correlation. Problems with 2, 4, or 6 independent variables may be considered by interpolating between the corresponding values given for 1, 3, 5, or 7 independent variables.

Considering the problem mentioned above, where a sample of 30 observations showed $R_{1.234} = 0.538$, Figure 17.3 gives a value of 0.16 as the probable minimum correlation. From the single sample we could then say that the true correlation is at least 0.16 in the universe from which the sample was drawn, with 1 chance in 20 of being wrong.

Figures 17.3, 17.4, and 17.5 show the possibilities of getting high correlations from a random sample, even when there is little or no correlation in the universe from which that sample was drawn. Thus, for 3

[9] R. A. Fisher, The general sampling distribution of the multiple correlation coefficient, *Proceedings of the Royal Society*, A, Vol. 121, pp. 654–673, 1928.

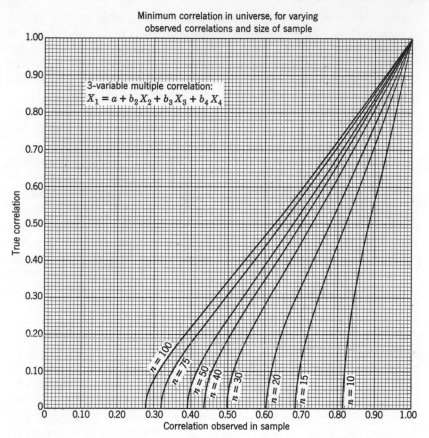

Minimum correlation in universe, for varying observed correlations and size of sample

3-variable multiple correlation:
$$X_1 = a + b_2 X_2 + b_3 X_3 + b_4 X_4$$

True correlation

Correlation observed in sample

$n = 100$ $n = 75$ $n = 50$ $n = 40$ $n = 30$ $n = 20$ $n = 15$ $n = 10$

Fig. 17.3. Under conditions of random sampling, 1 sample out of 20, on the average, will show a multiple correlation as high as that "observed in sample," when drawn from a universe with the stated true multiple correlation, in the case of multiple correlation with 3 independent variables.

independent variables, Figure 17.3 shows that, if samples of 15 observations are used, in 1 sample out of 20, $R_{1.234}$ will be as large as 0.69, even if the correlation in the universe is zero, and as large as 0.78, even if the true correlation in the universe is only 0.40. Similarly, if there are 7 independent variables, Figure 17.5 shows that, if samples of 20 cases are used, in 1 sample out of 20, on the average, $R_{1.2345678}$ will be as high as 0.79 with zero correlation in the universe, 0.85 with 0.50 in the universe, and 0.91 with 0.70 in the universe. Even with samples as large as 100 cases, $R_{1.2345678}$ in 5 per cent of the samples will be as high as 0.37 for samples drawn from a universe with zero correlation, and as high as 0.57 for samples drawn from a universe with 0.40 as the true correlation.

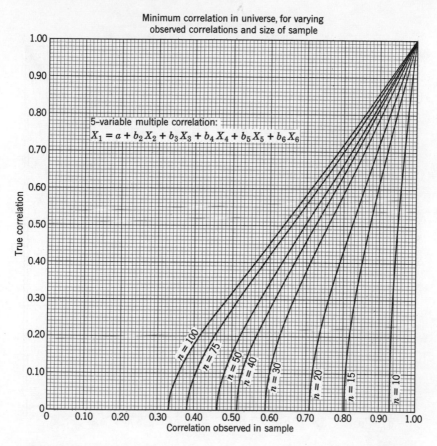

Fig. 17.4. Under conditions of random sampling, 1 sample out of 20, on the average, will show a multiple correlation as high as that "observed in sample," when drawn from a universe with the stated true multiple correlation, in the case of multiple correlation with 5 independent variables.

Figure 17.4 gives similar probabilities for 5 independent variables. Many other combinations of size of sample, true correlation in the universe, and observed correlation for 5 per cent of the samples are given in these figures.

If the several independent variables in the multiple correlation study had been selected by considering a large number of possible independent variables, and by retaining only those which showed the highest gross or net correlation with X_1, there is a much larger possibility of the correlation in the sample exceeding the true correlation in the universe by a wide margin. In fact, it is almost certain to be erroneously high. If error

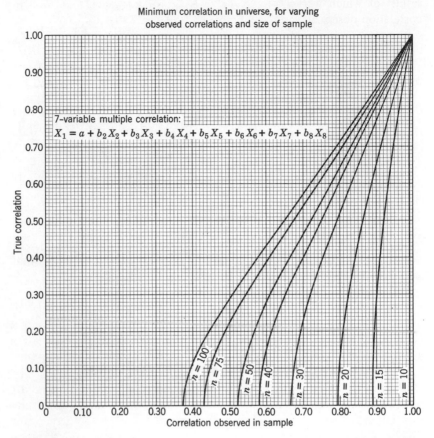

Fig. 17.5. Under conditions of random sampling, 1 sample out of 20, on the average, will show a multiple correlation as high as that "observed in sample," when drawn from a universe with the stated true multiple correlation, in the case of multiple correlation with 7 independent variables.

calculations are to be used to judge the sampling significance of the correlations or regressions observed, the variables must be selected purely on logical or deductive grounds (as discussed at length in Chapter 26) rather than on any such basis of empirical selection of those which show the apparent closest relation. It should always be remembered that, if the choice is purely empirical, the next following time period (in the case of economic, meteorological, or other time series) might readily reverse the order of apparent importance of the several variables.

EXAMPLES OF THE SAMPLING FLUCTUATION IN MULTIPLE CORRELATION COEFFICIENTS. In the sampling experiment referred to earlier, some results of which were given in Table 17.2, the coefficients of multiple correlation observed were distributed as shown in Table 17.5.

Table 17.5

DISTRIBUTION OF VALUES FOR MULTIPLE CORRELATION COEFFICIENTS FOR REPEATED SAMPLES DRAWN FROM THE SAME UNIVERSE (TRUE VALUE 0.563)

Range of Values	30 Observations	50 Observations	100 Observations
0.300–0.399	4	1	
0.400–0.499	3	5	
0.500–0.599	4	1	4
0.700–0.799	4	3	1
0.700–0.799	1		

This table illustrates how widely sample values of R may vary in small samples. In the first sample of 30 observations, for example, $R_{1.234} = 0.538$. But from Table 17.5, the sample selected might have happened to be one of the samples that gave values as low as near 0.300, or as high as over 0.700. From Figure 17.3, we estimate that such a correlation might be observed, in 1 sample out of 20, from a universe with a true correlation as low as 0.15. Table 17.5 illustrates the need for this caution in interpreting such coefficients from small samples.

Indexes of Correlation. No precise mathematical estimates are possible for the standard errors of simple and multiple indexes of correlation based on graphic curves. Where they are based on mathematically fitted equations, their reliability may be estimated by the same interpretation from Fisher as is given in Figures 17.2 to 17.5, using $(m - 1)$ in place of the number of independent variables. (For complex curves with many constants, this will involve going back to his original article to calculate the probable minimum value in the universe, for $P = 0.95$.) For graphic curves, the same interpretation may be made, on the assumption that, except for the degrees of freedom involved in fitting the curves, the correlation index is simply the correlation coefficient between X and the X' values calculated from the curve or curves, as at least a first approximation to its reliability. In view of all the qualifications as to the meaning of correlation coefficients or indexes, it is not usually considered advisable to make much effort to interpret their sampling significance.

All of these calculations of the sampling significance of r and R apply rigorously only to samples drawn from a natural universe of multivariate normal distribution (the correlation model).

Adjustments for Numbers of Observations and Constants

During the 1920's and 1930's—and occasionally more recently—research workers were sometimes overimpressed with the importance of results indicated by the values of r and R they obtained on the basis of studies with only small numbers of observations. As noted earlier, a straight line fitted to two observations will pass precisely through both. In this case $r = 1$ and $S = 0$, even though a larger number of observations from the same universe might show little or no correlation. With small numbers of observations the unadjusted values of r^2 or R^2 tend to over-state the proportion of variation in Y that is associated with X in the universe from which the sample was drawn. This is true even if the values of X are fixed by an experimenter; in this case the universe is one having the full array of values of Y for each of these fixed values of X. Only a single value of Y for each value of X appears in each sample.

Since the squares of the unadjusted values of the standard error of estimate and the standard deviation from a small sample also tend to be biased and to underestimate the true variance in the universe from which the sample was drawn, adjustments for these coefficients were introduced earlier in the text, in equations (2.1) and (7.3). These adjustments are

$$\bar{s}^2 = s^2\left(\frac{n}{n-1}\right) \qquad \text{and} \qquad \bar{S}_{y \cdot x}^2 = s_z^2\left(\frac{n}{n-2}\right)$$

The first of these equations gives an unbiased estimate of the square of the true parameter σ in the universe; the second gives an unbiased estimate of the parameter σ_z^2—the variation not associated with, or accounted for, by X. It seems logical, therefore, under some circumstances to calculate adjusted values for r^2 or R^2 that are consistent with the two measures above.

Derivation of the Adjustments. The true correlation ρ_{yx} in the universe of the preceding paragraph may be defined by the following relationship:

$$\rho_{yx}^2 = \frac{\sigma_{y'}^2}{\sigma_y^2} = \frac{\sigma_y^2 - \sigma_z^2}{\sigma_y^2} \tag{17.7}$$

Given the unadjusted value of r_{yx} from a particular sample, we can obtain an estimate of ρ_{yx}, which we will designate $\bar{r}_{yx}$, by substituting in equation (17.7) the sample values of s_y and S_y, adjusted by equations (2.1) and (7.3) to eliminate the bias previously noted. The equation then becomes

$$\bar{r}_{yx}^2 = \frac{s_y^2\left(\dfrac{n}{n-1}\right) - S_{y \cdot x}^2\left(\dfrac{n}{n-2}\right)}{s_y^2\left(\dfrac{n}{n-1}\right)} = 1 - \left[\frac{S_{y \cdot x}^2}{s_y^2}\right]\left(\frac{n-1}{n-2}\right) \tag{17.8}$$

Similarly, substituting the values of $\bar{s}_1^2$ and $\bar{S}_1^2$ in the equations for i_{12}^2, $R_{1.23\ldots k}^2$, and $I_{1.23\ldots k}^2$ [equations (7.8), (12.6), and (15.7)], the adjusted values are:

$$
\left.\begin{array}{l} \bar{i}_{12}^2 \\[2em] \bar{R}_{1.23\ldots k}^2 \\[2em] \bar{I}_{1.23\ldots k}^2 \end{array}\right\} = 1 - \left[\frac{S_{1.23\ldots k}^2}{s_1^2}\right]\left(\frac{n-1}{n-m}\right) = \left\{\begin{array}{l} 1 - (1 - i_{12}^2)\left(\dfrac{n-1}{n-m}\right) \\[1.5em] 1 - (1 - R_{1.23\ldots k}^2)\left(\dfrac{n-1}{n-m}\right) \\[1.5em] 1 - (1 - I_{1.23\ldots k}^2)\left(\dfrac{n-1}{n-m}\right) \end{array}\right.
$$

$$(17.9)$$

Equation (17.9) thus provides a formula where subscript k is the number of variables and m is the total number of degrees of freedom used up in fitting the regression equation, by which squares of the coefficients of multiple correlation and indexes of simple or multiple correlation can be adjusted to avoid consistently overestimating the closeness of relationship in the universe.

In using equations (17.8) and (17.9), if the value for the adjusted coefficient comes out a minus quantity, then 0 should be used for the adjusted value.

These adjusted values, like correlation coefficients generally, are of limited usefulness except when the sample has been randomly drawn from a normal universe, and has not been selected in such a way as to make the distribution of any variable in the sample different from that in the universe. (Chapter 18, following, gives arithmetic examples of how seriously the correlation in the sample may be changed from that in the universe if such artificial selection of the sample is permitted.)

The meaning of the adjusted values, $\bar{R}_{1.23\ldots k}^2$ or $\bar{I}_{1.23\ldots k}^2$, may also be explained as providing an unbiased estimate of the per cent of exact variance most probably associated with particular sets of values of the independent variables in a universe where those latter values remain fixed, but values of the dependent variable are subject to random variation.

If it is desired to adjust the coefficient of partial correlation, equation 17.9 may be used for that also.

Comments on Use of the Adjustments. The smaller the number of observations, the larger the number of independent variables considered, and the more complex the curves employed, the greater will be the tendency for the observed (unadjusted) standard error of estimate to understate the true error of estimate in the universe, and for the observed correlation, simple or multiple, to overstate the true correlation in the universe.

Where freehand methods have been used to determine net regression curves, there is particular need to be careful in appraising the relative importance of the relationship as measured by the unadjusted I^2. In such cases, particularly if the sample is small and the number of independent variables is large, $\bar{I}^2$ will give a more realistic estimate of the relationship in the particular universe from which the observations were drawn, or of the importance of the relation observed in the sample.

In current practice, the adjusted coefficients $\bar{r}$ and $\bar{i}$ are seldom used. When the values of X are set by an experimenter, this adjustment is implicit in the *analysis of variance* table which is generally used in appraising experimental results (see Chapter 23). The limited usefulness of the correlation coefficient in controlled experiments has already been noted. The reliability of the regression coefficient and the regression line can be tested without reference to r in both experimental and non-experimental situations (see Chapter 23 and earlier sections of this chapter).

When a sample has been drawn at random from a bivariate or multivariate normal distribution (or close approximation thereto) the investigator may be interested in a direct test of the reliability of r or R as an estimate of probable correlation in the universe. Figures 17.2 through 17.5, based upon the unadjusted values of the coefficients, provide such a test, as explained earlier.

Illustrations of the Adjustments as Shown by Sampling Experiments.
STANDARD ERRORS OF ESTIMATE. The tendency for the unadjusted values of correlation coefficients or indexes to be biased upward in small samples, and for standard errors of estimate to be biased downward when they are based on small samples, was shown very clearly in an experimental study, already referred to, made by the senior author. Two universes of known correlations were employed, and a number of samples were drawn for each of the three sample sizes 30, 50, and 100. The curvilinear net regressions were then determined for each sample separately (by the successive approximation method) and the standard deviations were worked out for the residuals in each case. The central values of these standard deviations of the residuals, for the samples of each size, were:

Number of Observations	Observed Standard Deviation of z*	
	Universe 1	Universe 2
30	1.95	1.53
50	2.18	1.64
100	2.21	1.72
Entire universe	2.40	1.80

* These values are the median values observed.

It is quite evident from these results that the unadjusted sample values tended to give standard errors of estimate smaller than that for the universe as a whole. The smaller the sample employed, the greater the overestimate of the reliability of the estimates.

The shape of the universe net regression curves indicated that about 10 constants would be needed to approximate the entire regression equation with a mathematical function. Using 10 for m in the equation

$$\bar{S}^2_{1.23\ldots k} = s^2_z \left(\frac{n}{n-m} \right) ,$$

and carrying the same adjustment through for all the observed values shown, adjusted standard errors of estimate were obtained as follows:

Number of Observations,	Value Used for	Universe 1		Universe 2	
n	m	Observed s_z	Adjusted $\bar{S}_{1.f(234)}$	Observed s_z	Adjusted $\bar{S}_{1.f(234)}$
30	10	1.95	2.39	1.53	1.87
50	10	2.18	2.43	1.64	1.83
100	10	2.21	2.33	1.72	1.82
Entire universe	$\cdots$	2.40	$\cdots$	1.80	$\cdots$

In each case the adjusted values came much nearer to agreeing with the true value for the universe than did the unadjusted values.

MULTIPLE CORRELATION INDEXES. The modal values of the indexes of multiple correlation for the repeated samples from each universe were found to be as follows, in comparison with the true index of correlation for the entire universe:

Number of Observations in Sample	Observed Index of Multiple Correlation in Samples Drawn from Same Universe
30	0.77
50	0.71
100	0.68
Entire universe	0.62

In every case the modal observed correlation exceeded the true correlation in the universe, and the smaller the size of the sample, the larger the excess. Adjusting these values by equation (17.9), results are obtained as follows:

Number of Observations, n	Value Used for m	Observed I	Adjusted $\bar{I}$
30	10	0.77	0.64
50	10	0.71	0.63
100	10	0.68	0.65
Entire universe	$\cdots$	0.62	$\cdots$

Here again the adjusted modal values are in much closer agreement with the universe value than were the unadjusted values.

Testing for Curvilinearity. The adjustments for degrees of freedom provide one indication as to whether a curvilinear regression line or lines really fit the observed data better than does a straight line or lines, or whether a complex curve gives a better fit than a simple curve. Unless the *adjusted* standard error of estimate is lower for the curve (or curves) than for the straight lines, any reduction in the unadjusted S from the linear equation may be regarded as a fictitious improvement in accuracy. As we take additional variables into account, or use up more degrees of freedom by employing more constants in the curves, we obtain a certain amount of spurious increase in the apparent correlation, and a spurious decrease in the unadjusted standard error of estimate. Correcting for n and m removes this spurious effect, and thus is of direct value, even if no inferences are drawn as to universe values. More precise tests of the significance of the improvement in correlation obtained by fitting curvilinear regressions are given in Chapter 23, pages 403 to 411.

Summary

Coefficients of simple and net *regression*, when determined from a sample, may vary more or less widely from the true value for the "natural" universe from which the sample was drawn (or for the artificially constructed universe which we described as the regression model). In either case, confidence intervals can be calculated which serve to indicate how much reliance can be placed on each regression coefficient observed.[10]

Coefficients of simple or multiple *correlation* have clear-cut sampling significance only in the case of samples based on the *correlation model*. For such samples, means are available for estimating the probable minimum correlation in the universe for $P = 0.95$, from the fact that 1 sample out of 20 might show a correlation as high as that obtained in the sample,

[10] For the sampling characteristics of β, see Technical Note 8 in Appendix 3.

even if the universe had only the specified correlation. When universes and samples are artifically constructed or selected (the regression model), correlation coefficients or indexes have no significance as a basis for inferences as to the underlying "natural" universe if any, but apply only to the very restricted universe of the regression model.

There is a tendency to upward bias in calculating coefficients or indexes of correlation from small samples or in problems which involve many variables or complex curves. Approximate adjustments are presented to correct for this bias. These are especially important, with small samples or many constants, in comparing the significance of alternative solutions which use up varying degrees of freedom.

The standard error of estimate is also subject to downward bias unless adjusted for degrees of freedom, but adjustments for that have already been presented in earlier chapters.

REFERENCES

Fisher, R. A., The general sampling distribution of the multiple correlation coefficient, *Proc. Roy. Soc.*, A, Vol. 121, pp. 655–673, 1928.

Ezekiel, M., The application of the theory of error to multiple and curvilinear correlations, *Proc. Amer. Stat. Assoc.*, Vol. XXIV, March, 1929.

———, A first approximation to the sampling reliability of multiple correlation curves obtained by successive graphic approximations, *Annals of Math. Stat.*, Vol. 1, September, 1930.

Schultz, Henry, The standard error of a forecast from a curve, *Jour. Amer. Stat.. Assoc.* pp. 139–185, Vol. XXV, No. 170, June, 1930.

Kendall, M. G., *The Advanced Theory of Statistics*, Vol. I, Chapters 14 and 15, Vol. II, Chapters 22 and 28, Charles Griffin, London, 1943.

CHAPTER 18

Influence of selection of sample and accuracy of observations on correlation and regression results

Selection of Sample

Methods of determining linear and curvilinear regressions, together with appropriate measures of their significance and accuracy, have been set forth in previous chapters. These methods do not yield results representative of the universe from which the sample observations have been drawn, however, if that sample is not truly representative of the particular relation being determined in the universe from which the sample is drawn. There are various ways in which the sample may fail to represent the universe, and the resulting extent to which the correlation constants will be biased will vary both with the character of the unrepresentativeness and with the individual coefficients. Each type of abnormality must therefore be treated separately.

The samples may be selected from the universe in such a way as to exclude all the observations falling beyond a certain value of a given variable, thus ruling out values either at one or at both extremes, or perhaps ruling out middle values and selecting only extreme ones. This may be done for either the dependent variable or the independent variable or variables, or for both together. Such a selection of observations produces certain specific effects upon the correlation constants. Under some conditions it may be very desirable to select the observations in this way, if the resulting aberrations in the correlation constants are recognized and allowed for.

A second and somewhat more difficult type of problem to deal with arises when there are errors of measurement in obtaining the values of one or more of the variables—such errors as might arise, for example, in estimating the total production of corn in the United States within a given year, or in working out, from a farmer's memory, what was the income on his farm the previous year. Here again the effect on the correlation constants will depend upon whether the errors are random or biased and upon which variable or variables are affected by the errors. A separate discussion must therefoie be given each case.

The clearest way of indicating the effect of these various departures from truly representative sampling may be by first stating the general principles involved, and then illustrating the way those principles work out by concrete illustrations. Except where specially stated otherwise, the discussion will apply solely to linear relations. The effects for curvilinear relations are in general analogous to those to be discussed. The illustrations here are purely in arithmetic terms. More gencral illustrations in algebraic terms would yield parallel conclusions.

Selection of Sample with Respect to Values of Independent Variable. If the sample is selected with respect to values of the independent variable, that will not tend to affect the slope of the regression line but will affect the value of the coefficient of correlation. If the selection is such that extreme values are rejected but intermediate ones are left in, the correlation will be lowered below that prevailing in the universe; if intermediate values are rejected and only extreme ones are used, the correlation will be raised above that prevailing in the universe. If the standard deviations of arrays are equal, the standard error of estimate will tend to remain the same, regardless of the selection.

These principles may be illustrated by the set of hypothetical data shown in Table 18.1.

Table 18.1

CORRELATION TABLE, SHOWING HYPOTHETICAL FREQUENCIES
AT SPECIFIED VALUES

Values of Y	Values of X				
	0	1	2	3	4
0	$\cdots$	$\cdots$	1	1	1
1	$\cdots$	1	2	2	1
2	1	2	4	2	1
3	1	2	2	1	
4	1	1	1		

For the data shown, $r = -0.47$, $s_x = s_y = 1.134$, and $b_{yx} = -0.50$. If now the values had been selected with reference to X, so as to exclude values below 1 or above 3, the number of observations would have been reduced from 28 to 22. For this restricted set of observations, $\bar{r} = -0.26$, $s_x = 0.739$, $s_y = 1.09$, but $b_{yx} = -0.50$. Computing the standard error of estimate, we arrive at $\bar{S}_{y \cdot x} = 1.02$ for the first case and 1.07 for the second. It is quite apparent that the correlation has been lowered by the restriction in the selection of the values of X; but the regression of y on x has not been changed at all, and the standard deviation of the residuals has been only slightly changed.

If now the selection is such that only extreme values of X are taken, say below 1 and above 3, the number of observations is reduced to 6. Computing the results for those values, we have $s_x = 2.00$, $s_y = 1.29$, $\bar{r} = -0.71$, but $b_{yx} = -0.50$! Also $\bar{S}_{y \cdot x} = 1.00$.

Bringing the three sets of results together for comparison, we have the following tabulation.

With X used as *independent* variable:

	s_x	s_y	$\bar{r}_{xy}$	$\bar{S}_{x.y}$	b_{yx}
All cases	1.13	1.13	−0.47	1.02	−0.50
Extreme values of X excluded	0.74	1.09	−0.26	1.07	−0.50
Only extreme values of X used	2.00	1.29	−0.71	1.00	−0.50

In the illustration, the observations are distributed very symmetrically. Were they distributed more irregularly the values of b and S might vary more widely between the three sets, but still would tend to be similar.

These three examples thus illustrate the principles stated before: that selection with respect to the independent factor does not tend to change the regression or the standard deviation of the residuals but does affect the correlation, lowering the correlation if it has lowered the dispersion of the independent factor and raising the correlation if it has increased the dispersion of that factor.

Selection of Sample with Respect to Values of Dependent Factor. Selection with respect to values of the dependent factor is more serious, in that it affects all the constants. According to whether the effect is to raise or to lower the standard deviation of the dependent factor, such selection tends to raise or lower both the regression coefficient and the coefficient of correlation from the value for the universe and likewise to raise or lower, respectively, the standard error of estimate.

These principles may be illustrated from the three examples just used, by regarding X as the dependent factor and Y as the independent factor and noting the influence of the selection with regard to the dependent factor X upon the regression of X on Y, b_{xy}. For the first case, with all values left in, $b_{xy} = -0.50$ and $\bar{S}_{x \cdot y} = 1.02$. For the second case, however, with extreme values of X left out, b_{xy} drops to -0.23 and $\bar{S}_{x \cdot y}$ becomes 0.73. For the third case, with only extreme values of X included, b_{xy} increases to -1.20 and $\bar{S}_{x \cdot y}$ becomes 1.55. Bringing these three sets together yields the following comparison.

With X used as *dependent* variable:

	s_x	s_y	$\bar{r}_{xy}$	$\bar{S}_{x \cdot y}$	b_{xy}
All cases	1.13	1.13	−0.47	1.02	−0.50
Extreme values of X excluded	0.74	1.09	−0.26	0.73	−0.23
Only extreme values of X included	2.00	1.29	−0.71	1.55	−1.20

These results indicate the extent to which selection with regard to the dependent factor may completely destroy the usefulness of all the results.

Selection of Samples with Reference to Values of Both Variables. Selection of cases with reference to values of both independent and dependent variables has an even greater effect upon the conclusions than the two cases already discussed, because selection of extreme values tends to exaggerate the correlation and regression, and of central values to lower both to an even greater extent than where the selection is with respect to the dependent factor alone.

If, in the data of Table 18.1, only those cases are selected in which values of X below 2 are associated with values of Y above 2, and in which values of X above 2 are associated with values of Y below 2, the observations are reduced to 10 cases, as follows:

Values of X	Values of Y	Number of Cases	Values of X	Values of Y	Number of Cases
0	3	1	3	0	1
0	4	1	3	1	2
1	3	2	4	0	1
1	4	1	4	1	1

For these values, $s_x = 1.48$, $s_y = 1.48$, $\bar{r}_{xy} = -0.90$, $b_{yx} = -0.91$, and $\bar{S}_{y \cdot x} = 0.68$.

It is evident that such selection raises both the correlation and the regression above the true value for the universe. This is to be expected, for this selection is equivalent to picking out the pairs of values which *do* show correlation with each other. Restricting the selection to paired values of above 1 for both variables, and below 3 for both variables, likewise would be picking out cases so as to eliminate all correlation. Such selection obviously destroys the value of the results.

Conclusions with Reference to Selection of Data. If an investigator is interested only in the regression line and not in the degree of correlation, and if the regression is truly linear, selection of data with reference to the independent factor (or factors) will not tend to change the slope of the regression line (or lines). Under those conditions selection of extreme cases of the independent factor may yield a reliable indication of the regression with many fewer observations than if the cases were selected at random. This principle is frequently applied in experimental or laboratory work, but is equally applicable in other types of investigations.

If the regressions are curvilinear, however, special selection of either extreme or central items of the independent variables forestalls the determination of the nature of the function, since curvilinear regressions can be determined only for the ranges of the independent factor within which observations have been secured. For such regressions, therefore, the nature of the function may be more accurately determined if the independent items are selected so as to be spread fairly uniformly through the whole range of values, thus affording a sufficient number of observations for accurate determination of the nature of the relation throughout the whole range. Selection purely at random frequently provides more observations than are needed for certain portions of the curve, and provides so thin a scattering of observations at other portions as to make its true position and shape quite indeterminate, as has been illustrated previously. Even if curvilinearity is only suspected, such a uniform distribution of values for the independent variable provides an improved basis for determining whether or not the regression is truly linear, as compared with an equal number of observations selected at random. At the same time, if the dependent factor is normally distributed, with its standard deviation similar for different arrays, selection with reference to the independent factor does not tend to change the standard error of estimate. (The regression model, as discussed in Chapter 17.)

If the primary interest, however, is not in the nature of the relations and in determining how closely values of the dependent factor may be estimated (regressions and standard error), but instead is in determining what proportion of the original variation in the dependent factor can be accounted for on the basis of the relations determined (correlation and determination), then anything other than random selection with reference to any factor will give estimates of the closeness of the correlation which either over- or underestimate the true correlation in the universe from which the sample is drawn. For most accurate results in such problems, the distribution of the dependent factor in the sample should be an accurate representation of the distribution in the universe from which the observations were drawn, and the only selection which would be justified would be aimed at securing such a sample. (See the correlation model, as discussed in Chapter 17.)

Since the correlation coefficient or index, and the parallel measures of determination, are of significance only with respect to the standard deviation of the observed values of the dependent factor, it follows that when the dependent factor has such an abnormal distribution that its standard deviation is of little value as a descriptive statistic, the measures of correlation also tend to be of little value. For any series which actually yielded such an extreme distribution as the dichotomous values used in the third case of those just illustrated, measures of correlation would have little significance except their formal mathematical definition. Yet the regressions and standard errors of estimate would tend to retain all their usual value and significance, so long as no selection had been made with reference to values of the dependent variable. In such a case, attempting to select the values of the dependent factor so as to make the series more nearly normal might seriously bias the regression results.

Accuracy of Observations

The data with which the statistician has to deal are frequently subject to errors of observation. If corn yields are being studied in relation to fertilizer applications, for example, farmers may be able to estimate the yield per acre on a given tract only to within 5 or 10 bushels of the true yield. If livestock prices are being studied, the market reporter may not be able to get his daily average nearer than within 10 or 25 cents per 100 pounds of the true average of all the sales for the day. Or if educational ratings are being studied, the instructor may not be able to grade the test papers nearer than to within 5 or 10 per cent of the grade each really deserves. All these illustrations are akin to the difficulties of the surveyor, who finds he cannot measure his angles more accurately than within a certain number of seconds; or of the astronomer, who finds his repeated observations disagree from each other by fractions of a second. But the errors of measurement are ordinarily tremendously greater in biological, economic, or social investigations than in physical observations; and for that reason statisticians must be particularly careful to use their data in such a way as to minimize the influence upon their conclusions of the errors which may be present.

Errors of observation may be such that they are not correlated with the value being observed, hence tend to fall equally above and below the true values throughout the range of the variable; or else they may be such that they are correlated with the variable, tending usually to make the observed value fall above the true value in the upper part of the range and below the true value in the lower part, or *vice versa*.

In correlation problems, there are two sets of true values involved, those for the dependent and independent variables; and there may also be two sets of errors, one tending to cause the observed values for the dependent variable to differ from the true values, and the other affecting the independent variable. The extent to which such errors, if present, modify or impair the results of correlation analysis depends both upon the type of the errors and the variables which they affect.

If the errors affect only values of the dependent factor, and if they are not correlated with the true values, their presence tends to lower the correlation and to increase the standard error of estimate, but does not tend to change the slope of the regression line from the true slope for the universe. If, however, uncorrelated errors are in the independent factor, that not only tends to lower the correlation and increase the standard error of estimate, but also tends to decrease the regression below the true value. Both of these cases may be illustrated from the same set of data used before.

Errors in the Dependent Variable. The data used in Table 18.1 may be modified by assuming some random error influences Y, making one-third of the values 1 unit higher, one-third 1 unit lower, and leaving one-third unchanged. With these changes, the data appear as in Table 18.2.

Table 18.2

CORRELATION TABLE, SHOWING HYPOTHETICAL FREQUENCIES AT SPECIFIED VALUES, WITH RANDOM ERRORS IN Y

Values of Y	Values of X				
	0	1	2	3	4
-1	$\cdots$	$\cdots$	1	$\cdots$	1
0	$\cdots$	1	$\cdots$	2	$\cdots$
1	1	$\cdots$	3	$\cdots$	1
2	$\cdots$	2	2	3	$\cdots$
3	1	2	3	1	1
4	$\cdots$	$\cdots$	1	$\cdots$	$\cdots$
5	1	1	$\cdots$	$\cdots$	$\cdots$

For these data, $\bar{r}_{yx} = -0.33$, $b_{yx} = -0.50$, and $\bar{S}_{y.x} = 1.46$. The introduction of the random error into Y has lowered the correlation from that of -0.47 for the original values and increased the standard error of estimate; but it has had no significant effect upon the regression of Y on X, the new value -0.50 being identical with the value of -0.50 for the original data in Table 18.1.

Errors in the Independent Variable. If, however, X is regarded as the dependent factor and Y as the independent, the regression coefficient for the new values, $b_{xy} = -0.28$, is found to be much reduced from that of -0.50 for the original values. Introducing even random errors into the observations of the independent factor markedly reduces the observed regression below the true value.

The errors considered to this point have all been random errors. If, instead, the errors are correlated with either of the factors, their presence would obscure the true relationship and bias any correlation constants which might be computed, tending to make them either too high or too low, depending on the inter-relations between the errors and the variables.

Errors in Both Variables. If random errors are associated with both variables simultaneously, their effects are a blending of those just illustrated, tending to reduce both the closeness of correlation and the regression below the true values. For example, if random errors of the same magnitude are introduced into X as well as Y of Table 18.1, the values appear as in Table 18.3.

Table 18.3

CORRELATION TABLE, SHOWING HYPOTHETICAL FREQUENCIES AT SPECIFIED VALUES, WITH RANDOM ERRORS IN BOTH X AND Y

Values of Y	Values of X						
	−1	0	1	2	3	4	5
−1	...	...	...	...	2		
0	...	...	1	1	1		
1	1	...	1	1	1	1	
2	...	1	2	2	1	1	
3	...	1	2	2	1	1	1
4	...	...	...	...	1		
5	...	1	1				

With these changes, the correlation is reduced to practically 0, the standard error is increased to 1.524, and the regression of Y on X is changed to -0.179. The comparison of these constants with those for the original data in Table 18.1 illustrates the extent to which the presence of random errors in the observed values of the variables may reduce the accuracy and effectiveness of correlation analysis.

DEALING WITH ERRORS IN BOTH VARIABLES. The methods of computing the regression line considered to this point are methods which take one variable as given, or independent, and the other variable as based upon it, or dependent. If it is known that all the errors of observation are random and are in one variable, and none are in the other, the effect of those errors may best be eliminated by considering the one with no errors as independent and the other as dependent. As has just been demonstrated, the regression line then obtained will be practically identical with that which would be obtained if no random errors at all were present.

In some cases it may be known that both variables are equally subject to random error with the same variance, yet it may be desired to obtain a regression line which most accurately expresses the relation between the two. That can be done by a special method, which fits the line on the condition that the sum of the squares of the departures of each observation *perpendicular* to the fitted line shall be made a minimum (in contrast to the usual condition that the sum of the squares of the *vertical* departures from the fitted line shall be made a minimum, with the dependent variable plotted as the ordinate). This special method involves an entirely different procedure for fitting the line, and is not given here.[1] Chapter 24 presents another and more modern method of dealing with a similar type of problem.

Errors of Observation in Multiple Correlations. The points which have been illustrated here for simple correlation are equally true for multiple correlation, both with respect to the influence of selection of sample and of the effect of errors of observation. The influence of errors of observation in multiple correlation problems may be illustrated by a case based on actual economic data.

Over the 17 years from 1907 through 1923, the monthly price of lambs shows a very high correlation with the price of wool and the price of dressed lamb. When X_2 is used for the price of wool, in cents per pound, X_3 for prices of dressed lamb, in cents per pound, and X_1 for prices of live lambs, in cents per pound, multiple correlation gives, for the 204 observations, $R_{1.23} = 0.991$ and $x_1 = 0.144x_2 + 0.354x_3$.

To test what effect random errors would have had on this correlation, two dice were thrown 204 times, giving random values from 2 to 12.

[1] Abraham Wald, Fitting of straight lines if both variables are subject to error, *Annals of Mathematical Statistics*, Vol. XI, No. 3, pp. 284–299, September, 1940. See also Albert Madansky, The fitting of straight lines when both variables are subject to error, *Jour. Amer. Stat. Assoc.*, Vol. 54, No. 285, pp. 173–206, March, 1959.

These values were then added to the successive values of the dependent, and a similar set of 204 values to the successive observations of one independent factor, to see what effect that would have on the results. In Table 18.4 the notation $X + e$ is used to designate the variables to whose values these "random errors" had been added.

Table 18.4

EFFECT OF INTRODUCING RANDOM ERRORS ON CORRELATION RESULTS

Independent Variables	Dependent Variable	Multiple Correlation	Regression Equation
X_2 and X_3	X_1	0.991	$0.144x_2 + 0.354x_3$
X_2 and X_3	$X_1 + e$	0.821	$0.112x_2 + 0.424x_3$
X_2 and $X_3 + e$	X_1	0.953	$0.163x_2 + 0.277x_3$
X_2 and $X_3 + e$	$X_1 + e$	0.804	$0.152x_2 + 0.306x_3$

These results illustrate the principles just set forth. The introduction of random errors into the dependent variable (X_1) reduces the correlation, and changes somewhat the size of the two regression coefficients. The errors in this case may not have been completely randomly distributed and uncorrelated with X_1, X_2, and X_3, even though determined by throws of dice.

But the second modification, where the error is introduced into the independent variable X_3 instead, is much more striking. The correlation is not reduced so much as in the first case, and the regression of X_1 on X_2 is changed only slightly from the original value—and increased as it happens. The net regression of X_1 on $X_3 + e$, however, is only three-fourths as large as was the net regression of X_1 on X_3, in spite of the fact that the error introduced was only enough to raise the standard deviation of X_3 from 6.14 to 6.64.

The final case, with errors introduced into both X_1 and X_3, shows the lowest correlation of any, as would be expected, even though the net regressions are not greatly changed.

Just how great an effect random errors may have upon the results depends upon the magnitude of the errors, the original variation in the variables, and the closeness of the inter-correlation. Equations can be derived to show how great a reduction in correlation errors of a given magnitude will produce, but they are of little practical use in economic work, since it is usually difficult enough to determine whether there are errors of observation or not, and much more so to determine what magnitude they have.[2] Reports or estimates of prices, or commodity production, or supply are nearly always subject to more or less error. The same is true of many other economic data. It may be slightly reassuring to know that observational errors even as large as those just considered still modify the regression results as little as these have been seen to do.

If there is known to be a large but random error in observing some variable, that variable may still be used as the dependent variable in a correlation study without making the regressions or estimating equation very far wrong, if determined with a large number of cases; but, on the other hand, any use of that variable as an independent variable will be certain to yield results which understate the actual relations.

Biased errors tend to make the results more or less in error, regardless of the variables to which they apply. If the errors tend either to magnify or to minimize the differences which actually exist, they will have a parallel effect on the regression coefficients if they apply to the dependent variables and an inverse effect if they apply to an independent variable. There are so many different types of bias, however, that no more definite statement of the effects can be laid down.

Random errors have the same type of effect in curvilinear correlation that they do in linear correlation, since if they are truly random they will tend to be balanced out along all the portions of the regression curve alike if in the dependent variable, or they will tend to confuse the relations along the curve if in the independent variable; and so they reduce the differences observed. About the only real difference between linearity and curvilinearity with regard to errors is that random errors in the dependent variable could be "balanced out" in the case of a straight-line regression with a somewhat smaller number of observations than would be necessary to secure valid results for a curvilinear regression.

Where, with random errors in the dependent factor, there are not enough cases available to "balance them out," the effect of the errors is to

---

[2] In the problem given, the significant values determining the effect of the errors are:

$$s_1 = 3.96 \qquad s_{1+e} = 4.74$$
$$s_3 = 6.14 \qquad s_{3+e} = 6.64$$

If the errors are in the dependent variable alone, the relations between the true and the apparent correlations are indicated by the equation:

$$R^2_{(1+e).23\ldots n} = \frac{(R^2_{1.23\ldots n})s_1^2}{s_1^2 + s_e^2} = R^2_{1.23\ldots n} \frac{1}{1 + (s_e^2/s_1^2)}$$

This gives what the new correlation would be if the errors were truly random, so that the new regression equation came out identical with the old. In the problem given, this gives an expected value for R of 0.827 as compared to the 0.821 actually obtained.

throw a varying amount of error into the conclusions, the exact amount of the error depending on how closely the errors approach being canceled out. The illustrative case, where with over 200 observations the regressions were still changed somewhat, probably indicates what may be obtained by a combination of slight departures from true "randomness" in the errors with a sample not quite large enough to eliminate entirely all the resulting instability. This may be nearer to what would usually happen in practice than the theoretically complete elimination of the errors in the dependent variable shown in the symmetrical arithmetic examples.

Summary

Modification of the observations from the true conditions, either by selection of the sample or by the presence of errors of observation, tends to alter the value of the coefficient of correlation. If the regression line or curve is of primary interest, however, its accuracy of determination may be increased by suitable selection of observations with respect to independent factors. The standard error of estimate likewise is little affected by selection with respect to the independent variables. Similarly, random errors of observation may not influence the regressions, if the factor they affect can be treated as the dependent factor and if enough observations are available to balance out the errors. These points hold true for multiple correlation problems as well as for two-variable problems.

CHAPTER 19

Estimating the reliability
of an individual forecast

Chapter 17 has indicated the kind of variability from sample to sample that may be expected in determining statistical constants, such as regression and correlation coefficients, and in fitting regression lines and curves. It has provided means of estimating, from the values obtained from a single sample, various indications of how far and how frequently the statistics from successive samples of the same size are likely to vary from the true values of the parameters in the universe from which the samples are drawn.

Reliability of an Individual Forecast

The practical statistician frequently has to deal with a quite different problem. Having taken a given sample, and having determined from that sample how the selected dependent variable is related to one or more independent variables, he then has the problem of drawing new observations of the same independent variable(s) from the same universe, and of estimating from those new values the most probable value of the dependent variable for the new cases. The standard error applicable to such an estimate is called the standard error of *forecast*. In ordinary usage, "forecast" suggests something in the future, like tomorrow's weather, or next year's corn yield. However, statisticians also use this term in connection with new observations from universes in which time is not a source of uncertainty.

For example, in a sample of children drawn at random from the school population of a given city, certain relations may be determined between their age and height and their weight. From these relations, how closely can we expect to estimate the weight of a new child, selected at random from the same population, once we know its age and height? In problems such as this, we are concerned with the possible difference between the estimated value X_1' and the actual value X_1, for new observations drawn from the same universe as the sample. We have calculated standard errors for the regression coefficient and line, the standard error in estimating X or X_1 in the sample under study, and adjustments of the standard error of estimate for "degrees of freedom" to obtain unbiased estimates of the variation about the true regression line in the parent universe. The present problem, however, involves the accuracy of estimates made from the line or curve *obtained from the sample*, in the light of the possible sampling *errors of that line*, as compared to the true line, plus the possible range of errors of the estimates around the true line. What we need, therefore, is a means of combining the standard error of the regression line with the standard error of estimate $S_{1.23...m}$.

Simple Regression. For a simple two-variable regression, the square of the standard error of a single estimate is given by the equation[1]

$$s_{Y'-Y}^2 = s_{M_{y'}}^2 + (s_{b_{yx}}x)^2 + \bar{S}_{y.x}^2 \tag{19.1}$$

Applying this equation to the illustration used previously, on page 288, we can tabulate the calculation of various values as in Table 19.1. Column (3), values of $s_{y'}^2$, is taken from Table 17.4, next to last column, since $s_{y'}^2 = s_{M_{y'}}^2 + (s_{b_{yx}}x)^2$.

Table 19.1

Selected Values of X (1)	Departures from Mean, x (2)	Calculation of $s_{y-y'}$			
		$s_{y'}^2$ (3)	$\bar{S}_{y.x}^2$ (4)	$s_{y-y'}^2 =$ (3) + (4) (5)	$s_{y-y'}$
0.97	−1.00	14.0650	68.62	82.6850	9.09
1.47	−0.50	7.1793	68.62	75.7993	8.71
1.97	0	4.8841	68.62	73.5041	8.57
2.47	0.50	7.1793	68.62	75.7993	8.71
2.97	1.00	14.0650	68.62	82.6850	9.04
3.47	1.50	25.5411	68.62	94.1610	9.70
3.97	2.00	41.6077	68.62	110.1977	10.50

The last column gives the standard errors of forecast for values of Y' estimated from new values of X drawn from the same universe. It is

---

[1] The derivation of this equation is given in Note 5, Appendix 3.

apparent from these values that standard errors for individual forecasts near the mean of X are but little larger than S_{yx}. Thus the standard error for the forecast of 22.3 for Y' when $X = 1.47$ is only $s_{y-y'} = 8.71$, as compared with $S_{yx} = 8.28$. The further the observed value of X departs from the mean, the larger the uncertainty of the individual forecast. Thus when $X = 3.97$, $s_{y'-y} = 10.50$. We can state this uncertainty of the estimate more simply by expressing the relation as follows:

$$\text{When} \quad X = 1.47, \qquad \overline{\overline{Y}} = 22.3 \pm 8.71$$
$$\text{When} \quad X = 3.97, \qquad \overline{\overline{Y}} = 64.0 \pm 10.50$$

Here we have introduced a new symbol, $\overline{\overline{Y}}$, to designate the probable range within which the true value will lie, for 2 estimates out of 3 on the average.

These standard errors of individual forecasts are interpreted in the same way as any other standard error, as indicating (for various selected multiples of the standard error) the proportion of a succession of such forecasts which can be expected to show departures from the true values of stated sizes for any specified degree of confidence. Thus, in the problem illustrated on pages 139 and 289, when yields are estimated for new plots with 3.97 feet of water applied, 2 out of 3 new observations, on the average, should show yields falling within 10.5 ten-pound units of the estimated yield. Table 2.3 should be used in calculating confidence intervals from this standard error, in the same way as before.

Multiple Regression. The equation for the standard error of an individual forecast made from a multiple regression equation is similar to that given for simple correlation, with the addition of expressions for the additional variables, as follows:

$$s_{x'_{1.234}-x_1}^2 = S_{1.234}^2 \left[1 + \frac{1}{n} + c_{22}x_2^2 + c_{33}x_3^2 \right.$$
$$\left. + c_{44}x_4^2 + 2c_{23}x_2x_3 + 2c_{24}x_2x_4 + 2c_{34}x_3x_4 \right] \quad (19.2)$$

In this equation x_2, x_3, and x_4 are the values of the independent variables from which the forecast is made, stated as departures from the respective means M_2, M_3, and M_4, as calculated in the original sample from which the regression equation was calculated. The c values for equation (19.2) are obtained by the simultaneous solution of the following equations:

$$\left. \begin{array}{l} (\Sigma x_2^2)c_{22} + (\Sigma x_2x_3)c_{23} + (\Sigma x_2x_4)c_{24} = 1 \\ (\Sigma x_2x_3)c_{22} + (\Sigma x_3^2)c_{23} + (\Sigma x_3x_4)c_{24} = 0 \\ (\Sigma x_2x_4)c_{22} + (\Sigma x_3x_4)c_{23} + (\Sigma x_4^2)c_{24} = 0 \end{array} \right\} \quad (19.3)$$

$$\left. \begin{aligned} (\Sigma x_2^2)c_{32} \ + (\Sigma x_2 x_3)c_{33} + (\Sigma x_2 x_4)c_{34} = 0 \\ (\Sigma x_2 x_3)c_{32} + (\Sigma x_3^2)c_{33} \ + (\Sigma x_3 x_4)c_{34} = 1 \\ (\Sigma x_2 x_4)c_{32} + (\Sigma x_3 x_4)c_{33} + (\Sigma x_4^2)c_{34} \ = 0 \end{aligned} \right\} \qquad (19.4)$$

$$\left. \begin{aligned} (\Sigma x_2^2)c_{42} \ + (\Sigma x_2 x_3)c_{43} + (\Sigma x_2 x_4)c_{44} = 0 \\ (\Sigma x_2 x_3)c_{42} + (\Sigma x_3^2)c_{43} \ + (\Sigma x_3 x_4)c_{44} = 0 \\ (\Sigma x_2 x_4)c_{42} + (\Sigma x_3 x_4)c_{43} + (\Sigma x_4^2)c_{44} \ = 1 \end{aligned} \right\} \qquad (19.5)$$

Solving these equations, $c_{23} = c_{32}$, $c_{24} = c_{42}$, and $c_{34} = c_{43}$. The coefficients of the equations to the left of the equality signs are identical in all three sets of equations and are also identical with those of the normal equations (11.9), used in determining the net regression coefficients. This makes it possible to compute the values for the c's at the same time as for the b's, with only slight additional work. (See Appendix 2, pages 499 to 502, and 522 to 524).

The c's for a large number of independent variables are obtained by an expansion of equations (19.3) to (19.5), setting up as many sets of simultaneous solutions as there are independent variables and placing the 1 on the right-hand side of the equations opposite the variable whose (Σx^2) occurs with the c_{mm}'s, just as for the second set of equations (19.4) above, 1 occurs to the right of the equation where $(\Sigma x_3^2)c_{33}$ occurs as one of the items on the left of the equality sign.

The standard error of the individual forecast will differ for each combination of values of the various independent variables. If these all fall about their means, $s_{x-x'}$ will be only slightly larger than $\bar{S}_{1.234}$. If one or more fall far from it, the standard error of the forecast will be correspondingly large.

For n variables, the general formula for the square of the standard error of the individual estimate is given symbolically by

$$s_{x'_{1.23\ldots m}-x_1}^2 = \bar{S}_{1.23\ldots m}^2 \left[1 + \frac{1}{n} + (c_2 x_2 + c_3 x_3 + \ldots + c_m x_m)^2 \right] \quad (19.6)$$

In expanding equation (19.6) for any number of variables, it *must be interpreted by the special condition* that $c_2 c_2 = c_{22}$, $c_2 c_m = c_{2m}$, etc.

The standard errors of individual estimates made from multiple regression equations, according to equations (19.2) or (19.6), can be interpreted in the same way as those from simple regression equations.

Curvilinear Regression. Where a simple or multiple curvilinear relation is determined by fitting mathematical regression equations, the standard error of individual estimates can be computed by an extension of equation (19.2). Thus if a cubic parabola has been fitted using

$$Y = a + b_2 X + b_3 X^2 + b_4 X^3$$

we can compute this equation most readily by writing it in the form

$$Y = a + b_2 X + b_3 U + b_4 V$$

where $U = X^2$ and $V = X^3$

The standard error of an individual estimate is then given by the equation

$$s^2_{y'_{xuv}-y} = \bar{S}^2_{y.xuv}\left[1 + \frac{1}{n} + c_{xx}x^2 + c_{uu}u^2 + c_{vv}v^2 \right.$$
$$\left. + 2c_{xu}xu + 2c_{xv}xv + 2c_{uv}uv\right]$$

Similar expansions are available for mathematical regression equations for two or more variables.[2]

The Applicability of a Regression Equation to an Extrapolation beyond the Observed Range

We have already seen examples, in Chapters 14 and 16, of how estimates might sometimes need to be made for new observations which lie beyond the range included in the original sample. We have also seen the possibility of exceptionally large errors of estimate when the formulas or curves are extrapolated in this way beyond the observed range. A rough rule-of-thumb has been given that estimates beyond the observed range should never be made, or, if they must be made, should be regarded as exceptionally hazardous. This present section will explore further the meaning of the statement "beyond the range of observation."

Where only two variables are concerned, there is no question as to the range covered in the original observations. Thus if we consider the data plotted in Figure 8.2, it is apparent at once that the independent variable X covers the range from 1.2 to 3.5. Any new values of X smaller or larger than those values would be beyond the observed range.

Where two or more independent variables are concerned, the situation is more complex. Thus the data of the example plotted in Figures 10.4 and 10.5 show that the acres range from 60 to 240, and the cows range from 0 to 18. Suppose a new observation were drawn from the same universe, with 225 acres and 17 cows. Would that observation be within the original range? At first it might seem that it would, since the number of acres falls within the original acreage range, and the number of cows within the original range for cows.

[2] Henry Schultz, The standard error of a forecast from a curve, Journal of the American Statistical Association, pp. 139–185, Vol. XXV, June, 1930.

Multiple regression, however, is concerned not merely with the relation of the dependent variable to each independent variable separately, but with the composite relation to all the independent variables together. Is the *combination* of 17 cows and 225 acres either exactly or approximately within the joint distribution of the original observations? The joint values for X_2 and X_3, which were represented in the original observations, are shown plotted on Figure 10.6. From this it is evident that the new combination lies well outside the observed joint distribution.

The original sample had some farms of between 200 and 250 acres, but none of them had more than 6 cows. It also had some farms of 15 or more cows, but none of them had more than 120 acres. The single original case that came anywhere near the new observation was a farm with 14 cows and 180 acres. Since the new observation lies well outside the *joint distribution* or combination of values represented in the original sample, any estimate made for it from a regression equation based on that sample is subject to a hazard beyond that given by the error formulas discussed earlier in this chapter. Those formulas give accurate values of the probable error of individual estimates only within the range represented by the original sample. Extrapolation of the regression equation or curves beyond that range, or combination of values, represents an extension into unknown fields, where sudden changes in the nature of the relations might conceivably occur. *A priori* knowledge of the relations, based on technical facts and theories, or on other evidence, may justify extrapolations of the curves. Estimates of error for such extrapolations are only as reliable as the assumptions on which the extrapolations are based.

Where there are three or more independent variables, it is still more difficult to determine whether a given new combination of values lies outside the joint distribution of the three or more variables in the original sample. In many cases this can be determined by careful checking of the new observation against dot charts of the correlations among the independent variables. Thus, suppose a new observation were drawn with 2 cows, 100 acres, and 4 men. Would this be within the range of the original observations?

Careful inspection of the observations tabulated in Table 11.3, on page 177, reveals that, although the combination of 2 cows and 100 acres is well within the observed joint distribution for those two variables, no such combination occurred with 4 men, or even with 3 men. The nearest values are one observation (No. 7) of 3 men with 6 cows and 170 acres and one other (No. 12) of 3 men with 15 cows and 120 acres. The new observation, of 4 men with 2 cows and 100 acres, would apparently involve much more human labor, to care for that many cows and acres, than was represented in the original observations, and therefore lies far outside the joint distribution represented in the sample for the three variables. It is quite possible that that much labor would represent a wasteful use, so that the additional men would be as likely to reduce the farm income as to increase it. An estimate of income for this new farm, based on the relations shown in the sample for quite different farms, might therefore be very sadly in error.

The rough process of comparing the new observation with the values of the independent variable for the original observations as illustrated above, may serve reasonably well for determining whether the new observation is or is not represented in the original sample. Mathematical methods are available for estimating the probability that an observation drawn at random from the universe represented by the original sample will deviate from the "central tendency" of its joint distribution farther than any specified combination of values of the independent variables (note article by Waugh and Been). Carrying through such calculations ordinarily would involve an amount of labor out of proportion to the value of the information obtained. For very exact work, or for estimates of very great importance, however, it might be worth working them out. This would be true especially where the new observation happened to fall at about the edge of the distribution zone of the previous observations, so that it was uncertain whether or not it would be safe to estimate the dependent variable from the relations previously observed.

When forecasts are being made from regression equations for time series, even more complicated issues are involved. These are considered in Chapter 20.

REFERENCES

Waugh, Frederick V., and Richard O. Been, Some observations about the validity of multiple regressions, *Stat. Jour. College of the City of New York*, Vol. 1, No. 1, pp. 6–14, January, 1939.

Schultz, Henry, The standard error of a forecast from a curve, *Jour. Amer. Stat. Assoc.*, pp. 139–185, No. 170, Vol. XXV, June, 1930.

Armore, Sidney J., and Edgar L. Burtis, Factors affecting consumption of fats and oils other than butter, in the United States, U. S. Dept. of Agr., *Agr. Econ. Res.* Vol. II, No. 1, pp. 7–9, January, 1950.

CHAPTER 20

The use of error formulas
with time series

This chapter is concerned with universes which are extended over time and in which both the original observations and subsequent new ones are measured at successive intervals of time. The corn yield example of Chapter 14 and the steel cost example of Chapter 16 both used time-series data.

Many problems important in economics and other social sciences involve measurements in time. But time-series problems also arise in the biological and physical sciences. In recent years statisticians have recognized the similarities between time-series problems in the various disciplines and have made considerable progress toward their solution.

Differences Between Time-Series and Other Types of Data. The attitudes of research workers toward regression analysis of time series have varied between widely separated extremes. In the early and middle 1920's many researchers were completely unaware of problems connected with the sampling significance of time series. Then, under the (partly misinterpreted) influence of articles such as Yule's on "nonsense-correlations," it became fashionable to maintain that error formulas simply did not apply to time series.[1] There was some implication that reputable statisticians should leave time series alone. But in some fields a large amount of data already existed in the form of time series; and the variables so recorded were frequently important in the theories of the

[1] Yule himself did not adopt a wholly negative attitude, but spelled out *certain situations* in which correlations between time series would be highly misleading. These did not preclude the existence of other situations in which correlations between time series would have the usual sampling significance.

G. U. Yule, Why do we sometimes get nonsense-correlations between time-series?—A study in sampling and the nature of time series, *Journal of the Royal Statistical Society*, Vol. 89, No. 1, pp. 1–64, 1926.

respective fields. Further, controlled experiments involving these variables would be impossible, or expensive and time-consuming. During the 1930's, therefore, some research workers continued to apply regression methods to time series but with considerable trepidation; the previous editions of this book defended the practice, partly on intuitive grounds.

All the error formulas presented in Chapters 17 and 19 assume that the sample has been drawn from a clearly defined universe. If the observations are completely random drawings from a bivariate or multivariate normal universe (the "correlation model"), all of the simple, partial, and multiple correlation coefficients, the simple and partial regression coefficients, and the standard error of estimate, calculated from the observations may be regarded as estimates of corresponding parameters of that universe. The regression and correlation coefficients from successive random samples would be distributed according to the error formulas given in Chapter 17.

If the set of observations results from a controlled experiment, error formulas apply strictly only to a universe in which the distribution of the values of the independent variables remain fixed in successive samples (the "regression model"). Values of the dependent variables for each given combination of values of the independents are distributed randomly in successive samples about an "expected" or universe value—a point on the universe regression curve or surface. In Chapter 18 it was demonstrated that the correlation coefficient is strongly affected by purposeful selection of observations with respect to values of the independent variable. It can be shown readily that error formulas are also affected by such selection. For from equation (17.1)

$$s_b = \frac{S_{y \cdot x}}{s_x \sqrt{n}}$$

If we select extreme values of x (which are deviations about a sample mean), s_x increases and s_b therefore decreases. If we select a narrow range of values of x, s_x decreases and s_b therefore increases, i.e., the sample regression coefficient becomes a less reliable estimate of the universe value. (We assume here a universe in which $y = \beta x + z$, where z is randomly distributed with zero mean and variance σ_z^2; $\rho_{xz} = 0$.[2] The adjusted standard error of estimate in any sample should then be an unbiased estimate of σ_z regardless of the manner in which x values are selected; similarly, the regression coefficient from any sample should be an unbiased estimate of β.)

The concern of applied research workers about time-series analysis in the early 1930's was based upon departures from the "correlation model" only. This model assumes that each observation in a sample is selected

---

[2] In this chapter, β is used in the sense of the true universe value of b.

purely at random from all the items in the original universe, so that a below-average value is just as likely to be followed by a high one as by another low one. If the successive months or years in a time series are regarded as successive observations, this assumption obviously may not hold true. For example, each successive item of a linear trend line is perfectly correlated with each preceding item. Each price of a given commodity on succeeding days or months may show some relationship to prices in the preceding period.

If the correlation between each item of a series and the item of the same series next following it in time is calculated by the usual methods, the resulting correlation coefficient is termed the *coefficient of autocorrelation*. Many time series show autocorrelations that differ significantly from zero, meaning that the basic conditions of simple sampling have probably not been met. This situation casts doubt upon the stability of correlation coefficients between two such series from one period to another, and hence upon the applicability of standard errors of correlation coefficients.

In addition to this technical fact (which applies to some time series but not to others), there was a broad philosophical objection to the idea that any sequence of time-series observations was in fact a sample from a definite, unchanging universe. When any phenomenon is sampled at successive intervals of time the universe being studied can never be *precisely* the same. Successive astronomical observations might differ, even if in imperceptible degrees, because of the loss of matter through radiation of energy from the various stars. Surveying measurements in successive years might differ because of slight geological shifting of the earth's surface, or because of erosion or other changes in the soil surface. Normal crop yields change because of improvements in the genetic make-up of the seed so that what would be normal yields for certain weather at one time become subnormal yields at another. Human populations, too, change constantly from dietary, genetic, and other causes so that the average relationship between height and weight may change with time. Industrial techniques change, tending to modify relationships, as in the steel-cost example; so do the wants of consumers and the types and qualities of goods.

Some of the possible changes in universes mentioned above will be trivial for most practical purposes. Among these are changes in astronomical observations and surveying measurements, where the real changes in the universe over a limited span of years might be small relative to the precision with which our instruments are able to measure them, and completely negligible in their effects upon regression coefficients between different variables in the universes. In general, universes subject to human intervention change more rapidly than those completely dependent upon natural processes. Even in universes involving human behavior or purposeful intervention, our knowledge of the subject matter involved may enable us to set reasonable limits upon the probable extent of changes in the universe over a given span of years.

To the extent that changes in the universe follow a steady rate of progression, they can sometimes be allowed for by trend factors (as shown in Chapter 14), by seasonal factors, simultaneously determined, or by progressive shifts in the regressions themselves. In such cases, forecasts of future changes depend upon a continuation of the same rate or degree of change. Where human intervention is possible, however, one can never be *completely* sure that a new event may not make a sudden change or break in the trend.

The sampling significance of regression analyses based on time series was clarified by Koopmans, Wold, and others in the late 1930's.[3] Clarification was achieved by shifting emphasis from the correlation to the regression model. Suppose that two variables, such as the supply and price of beef, are logically related by the equation $P = \alpha + \beta S + z$, where z reflects the random influences of other economic factors. Then if for any reason the supply of beef follows a cyclical pattern over time, the price of beef will trace out a similar cycle with the relative amplitude βS. Neither the supply nor the price observations will be random over time. But we can regard the successive values of S in the same light in which we regard the values of any independent variable in a controlled experiment. For example, we could have selected our samples of wheat in Chapter 6 in a specified "time" order, starting with those having high percentages of vitreous kernels and working down to those having low percentages of vitreous kernels. The percentages of such kernels in successive samples would show very high autocorrelation over "time," but this would not shake our confidence in the resulting estimates of the relationship between protein content and vitreous kernels.

The introduction of time into the wheat-protein example seems artificial and irrelevant, whereas the time sequence of observations on beef supplies seems natural and inescapable. However, if the parameters α and β remain constant over the entire period, the ordering of S over time is also irrelevant from a statistical viewpoint. The only requirement is that the *residuals* z be random with respect to time. If all of our observations are drawn from a universe defined by the equation $P = \alpha + \beta S + z$, time, as Wold puts it, "plays the secondary role of a passive medium."[4]

[3] Tjalling C. Koopmans, *Linear Regression Analysis of Economic Time Series*, Haarlem, De erven, F. Bohn, n.v., 150 pp., 1937.

[4] Herman Wold, *A Study in the Analysis of Stationary Time Series*, p. 1, Almqvist and Wiksells, Uppsala, 1938.

This would be true for equations in three or more variables if the partial regression lines or curves in the universe remained constant over the specified period.

The above argument does not, of course, mean that a novice in any field can go ahead and discover new truths by haphazardly regressing time series on one another. Instead, the researcher must consider carefully whether the series in question are logically related, and whether these logical relationships apply also to the trends and seasonals (if any); also, he must apply tests of autocorrelation to the residuals from the regression equation fitted.

The approach to sampling errors in terms of the regression model should dispel any mystical dread of time series as "somehow different" and substitute the more manageable questions: (1) Did the universe from which these observations were drawn remain sufficiently stable over the period to which they pertain; and (2) is the universe now sufficiently like that of the previous period that regression relationships based on that period still apply? The latter question is equally relevant to data from sample surveys or from controlled experiments.

For example, suppose we had drawn, in 1890, a random sample of adult males in the United States and determined their average height, average weight, and the regression of weight upon height. Inferences from the sample relationships to those in the universe would apply strictly only to the population in 1890. However, applications based on the 1890 sample would doubtless be made to the actual population existing in 1891 and later years—perhaps by clothing manufacturers who wanted to estimate appropriate proportions of the different sizes of ready-made garments. Assuming that a fair amount of time and expense were involved in repeating the 1890 sample, the decision as to how soon a new sample was needed would be a matter for professional judgment—the judgment of human biologists as to the probable extent of the effects of immigration of different groups since 1890, the effects of changes in diet, and perhaps the effects of changing attitudes toward obesity. The expense of rechecking the relationships for 1895 or 1900 would presumably be weighed against the possible scientific and/or economic benefits to be gained from any changes shown by the new analysis.

Relationships estimated from controlled experiments may also lose their relevance with the passage of time. Let us suppose that the relationship between cotton yields and irrigation water applied (Chapter 8) had been determined by an elaborate controlled experiment which sufficed to determine the regression relationship as of 1927 with a high degree of accuracy. Let us say that this experiment was applied to a particular strain of cotton and that no fertilizers were used. If the typical farming practices in 1927 were to use that particular strain of cotton and no fertilizer, the results of this experiment would be immediately useful to farmers in 1928. But as the varieties of seed in common use changed and levels of fertilizer use increased, agronomists in the area would have to decide on a judgment basis whether the 1927 experimental results were still applicable, and the direction and extent to which they should be modified in determining optimum irrigation rates. Sooner or later other controlled experiments would be needed to determine the net relation between cotton yields and applications of irrigation water with new cotton varieties, more fertilizer, and other changes in cultural practices. Perhaps joint functions would be needed, taking water and fertilizer applications into account simultaneously with the new varieties.

Hence, in any area subject directly or indirectly to human influence, no method of selecting observations will guarantee that the relations estimated in one time period will apply to the corresponding universes at some later date. If controlled experiments must be repeated from time to time as the relevant universe changes, it is reasonable to assume that time-series analyses also should be repeated occasionally using additional observations or, when sufficient time has elapsed, completely new observations. If the actual corn yield in 1957 is 20 bushels higher than that estimated from the 1890–1927 relationship, the discrepancy should not be charged to the fact that the relationship had been estimated from time series, but rather to the fact that the whole set of observations is out of date.

There are various ways in which time-series analyses can be tested and brought up to date. As new observations accumulate, we can use statistical tests to determine whether the residuals are "out of line" with the earlier relationship. For example, in extrapolating the corn-yield equation from 1928 through 1939 (Table 20.1) the sequence of five consecutive positive residuals from 1930 through 1934 would have furnished some grounds for suspecting that the estimates were throwing "significantly" below the new level of actual yields. If we use the analogy of tossing a coin, regarding a positive residual as heads and a negative residual as tails and remembering that the residuals should be randomly distributed over time if the regression equation is to remain applicable, then the probability of getting five successive positive residuals might be estimated[5] at $(\frac{1}{2})^5$, which equals $\frac{1}{32}$, or 0.03125. By 1934, then, we might have said that there was less than 1 chance in 20 that the former relationship was still completely applicable. The 1935 residual, again positive,

---

[5] The statements in this paragraph are illustrative rather than rigorous. For an introduction to more precise methods of estimating the probability of "runs" or sequences of different lengths see A. M. Mood, The distribution theory of runs, *Annals of Mathematical Statistics*, Vol. XI, No. 4, especially pp. 367–368, December, 1940.

would have reduced the odds to about 1 in 50 or less. But the damage
done by adhering to estimates from the 1890–1927 relationship would
still have been slight, as only one of these six positive deviations was
larger than the standard error of estimate and this one did not exceed two
standard errors of estimate. The standard error of forecast, which is the
theoretically appropriate basis for comparison, would give a still larger
error zone and lead to a still more tolerant appraisal of the importance of
the observed deviations.

Table 20.1

CORN YIELDS ESTIMATED FROM 1890–1927 REGRESSION RELATIONSHIPS
FOR THE YEARS 1928 TO 1939

| Year | Corn Yields | | |
	Estimated,* X_1'	Actual, X_1	Residual, $X_1 - X_1'$
1928	33.1	33.4	0.3
1929	33.7	31.5	−2.2
1930	24.3	25.8	1.5
1931	31.7	32.7	1.0
1932	32.7	35.4	2.7
1933	24.0	29.4	5.4
1934	17.4	18.9	1.5
1935	30.6	31.7	1.1
1936	10.6	18.5	7.9
1937	30.6	36.4	5.8
1938	31.4	36.5	5.1
1939	31.9	41.3	9.4

* Based on regression curves shown in Figure 14.10.

The large positive residual in 1936, 7.9 bushels, would have completed
the demonstration that some change had occurred in the earlier relationships. First, the probability of obtaining seven positive deviations in
succession would be less than 1 per cent if they were randomly distributed
in time. Second, the mean of the seven positive residuals differed from
the expected value (zero) by 2.8 standard errors of the mean. According
to Table 2.3 the probability of either a positive or a negative deviation of
this magnitude would be less than 0.05, and the probability of a positive
deviation of this magnitude would be less than 0.025. Finally, a test
based on variance analysis (see Chapter 23) indicates that the probability

of obtaining a single residual as large as that of 1936 from the 1890–1927 universe of residuals is less than 0.01. At this point, the evidence would be sufficient to justify reworking the entire analysis from 1890 through 1936, and perhaps checking it by means of a separate analysis for (say) 1910–1936. The principal adjustment as of 1936 would probably be an increase in the level of the time trend for recent years and an upward slope beginning in the early or middle 1930's. This would contrast with the slight downward slope of the time trend in Figure 14.10 during the 1920–1927 period.

Of course, the reliability of the new analysis would be limited by the standard errors of the constants of the regression curves just as was the old one. The standard error of estimate, 2.8 bushels, indicates that at best we cannot hope to forecast yields on the basis of rainfall, temperature, and trend within less than 2.8 bushels in 2 years out of 3 and that occasionally we must expect our estimates to be off by 5 bushels or more even with no change in the basic relationships.[6]

Presumably, a research worker interested in the corn-yield problem would also be familiar with the results of controlled experiments on new varieties, effects of fertilizer, and any other scientifically based information that might suggest whether, why, and in what direction the net regression curves and the residual trend in yield *might* be changing. This new information could be converted into new hypotheses as to the shapes of the net regression curves and time trend, which could then be tested against the data for recent years. In the latter part of Chapter 14 we discussed such information as an explanation of the trend in corn yields from 1928 through 1956.

Thus, the analysis of time series as a research tool is not distinctly different from the use of controlled experiments or random samples from a universe existing at some particular point in time. Even controlled experiments can be abused and misinterpreted by inept workers, and some aspects of the relevant universes may not be subject to experimental control by even the best scientists. The design of an experiment must take into account the probable sources of disturbances which might interfere with measuring the relationships involved. Otherwise, the estimated relationship between Y and X may be biased by the fact that another variable, Z, just happened to be associated with the values of X in the experiment.

For example, in an experiment to determine the effects of different levels of fertilizer application upon corn yields, an inept experimenter might design the study in such a way that the larger quantities of fertilizer happened to be applied to plots with higher than average basic soil

---

[6] As pointed out in Chapter 19, the error zone applicable to an individual forecast is somewhat larger than the standard error of estimate.

fertility or which had carried over larger residual quantities of chemical fertilizers from experiments made in the previous year. As an illustration of the problems encountered by top scientists, the designers of the elaborate fertilizer experiments mentioned in Chapter 23 made the following comment:

The predictions apply to particular soils in a particular year; production surfaces obtained under other rainfall and soil conditions can be expected to differ from those obtained in the experiments reported .... Traditional experimental procedures (wherein a few rates of one or more nutrients are applied) also refer to the rainfall, climatic, insect and crop conditions of the particular year.[7]

Some economic analyses have proved inadequate because technical problems, such as the effects of intercorrelation and the meaning of autocorrelation in the residuals, were not recognized. Economic statisticians could be excused for making such mistakes in the 1920's and 1930's; in the 1960's they can reasonably be expected to take these problems into account.

The Nature of "Randomness" in the Residuals from Time-Series Regressions. Before discussing practical tests of autocorrelation, we should perhaps say a few words more about the reasons for expecting and requiring random residuals in connection with a "successful" time-series regression analysis.

In part these reasons are the same as for regression analysis in general, which have been discussed implicitly and explicitly in a number of connections from Chapter 4 on. We are interested in estimating the "true" or "universe" relationship between a dependent variable (Y) and one or more independent variables. The basic assumption in least-squares regression analysis is that a systematic relationship of the following type exists in the universe: For every possible combination of values of the independent variables, there exists an *expected* value $E(Y|X_{1i}, X_{2i}, \ldots, X_{mi})$ which may be regarded as the arithmetic mean of all possible values of Y *given* that particular set (in our notation the ith set) of values of the independent variables. The individual values of Y for the given combination of X's will show only random deviations about the expected value. Most applications of least-squares regression analysis assume that the variance of the individual Y's (in the universe) about the relevant expected value is the same for all combinations of values of the X's. While the assumptions of uniform variance can be relaxed in certain specific ways[8]

[7] Earl O. Heady, John T. Pesek, and William G. Brown, Crop response surfaces and economic optima in fertilizer use, p. 325, *Iowa Agricultural Experiment Station Research Bulletin* 424, March 1955.

[8] See Hald, Anders, *Statistical Theory with Engineering Applications*, pp. 526–27, 551–52, and 627, John Wiley and Sons, New York, 1952.

the assumption of randomness of residuals about each segment of the regression surface cannot be modified.

If we fit a straight line to a set of observations in which the underlying relationship is really parabolic, we find that the line of averages of Y for different ranges of X departs markedly from the straight line and it does so in a systematic manner. Statistical tests as to whether a relationship is curvilinear are based upon the departure of such group averages from the straight line (see Chapter 23, page 405) relative to the departure that might be expected as a result only of the random forces responsible for variations of individual observations about each group average.

In the corn-yield example of Chapter 14, "time" was introduced as an explicit variable (X_2) in the regression function. This would be expected to go a long way toward randomizing the residuals with respect to time. If "time" had been left out of the analysis, and if the partial regression curves of yields on rainfall and temperature had come out as shown in Figure 14.10, the residuals might very well have shown significant auto-correlation, for they would still have contained the systematic "time" effects represented in the bottom section of that figure.

In many analyses involving time-series observations, it seems plausible that "time" *may* give rise to a special type of difficulty. In the regression model, two or more successive residuals may be influenced by common factors rather than by completely independent (random) ones. If the basic variables are essentially continuous over time, then in going from a high level (say) of temperature to a lower one we pass through all the intermediate values as well. The closer together in time we take our successive temperature readings the greater will be the autocorrelation between them (i.e., the correlation of each item in the series with its succeeding item). We have noted that such autocorrelation in the basic series does not invalidate error formulas. But if high autocorrelations do exist in the basic series it seems plausible that, for sufficiently short time lags, there could also be significant autocorrelation in the residuals.

Adjusting for Autocorrelation. Some of the earlier negativism concerning time series still surrounds the question, "What do you do if there *is* significant autocorrelation in the residuals?" From a good many texts and technical articles the main or only impression obtained is a negative one—"in this case the usual error formulas do not apply." Constructive answers—at least approximate ones—can be found in the literature as far back as 1935.[9]

The main point is that autocorrelated series give us less information per observation than do completely random ones. In connection with the

[9] See M. S. Bartlett, Some aspects of the time-correlation problem in regard to tests of significance, *Journal of the Royal Statistical Society*, Vol. XCVIII, pp. 536–543, 1935.

correlation model, Bartlett developed an approximate correction formula to deal with this fact. For example, we find that the temperature series of Chapter 14, during the period 1928–1956, is significantly autocorrelated at the 5-per cent probability level and autocorrelation in the rainfall series is just short of significance at the same level. The two coefficients are 0.368 and 0.231 respectively. According to Bartlett, this means that the standard error of the correlation coefficient between the two series is increased in approximately the ratio

$$\sqrt{\frac{1 + (.368)(.231)}{1 - (.368)(.231)}} = \sqrt{\frac{1.085}{0.915}} = 1.089$$

as compared with the usual formula. This could be interpreted also as meaning that the information contained in our 29 annual observations is equivalent to only $29/(1.089)^2 = 24$ strictly independent or random observations.

Wold (1953) takes a similarly constructive approach with respect to the meaning of autocorrelation in the residuals from time-series regression equations.[10] Assuming that the universe is essentially stable over a given time period, and that the autocorrelation coefficient between successive residuals is ρ_1, that between residuals lagged two time units is ρ_2, and so on, the standard errors of the partial regression coefficients are increased relative to the usual formula in the ratio $\sqrt{1 + 2\rho_1 + 2\rho_2 + \ldots}$. If we assume that $\rho_2 = \rho_1^2$, $\rho_3 = \rho_1^3$, and so on, the terms beyond ρ_1 will normally be quite small.

If we treat the temperature series (in which $\rho_1 = 0.368$) as if it were a series of residuals, ρ_2 (on our assumptions) would equal 0.135 and ρ_3 would equal 0.050. If we consider only the term in ρ_1, the standard errors are increased in the ratio 1.318; if we include ρ_2 as well they are increased in the ratio 1.416. The 29 autocorrelated "residuals" would then give us about the same level of accuracy in estimating regression coefficients as would $29/(1.416)^2 = 14$ strictly random residuals.

We turn now to some commonly used tests for the presence of significant autocorrelation in residuals and to a method of dealing with autocorrelation that is sometimes useful in economic time series, and might also apply to some time series encountered in other disciplines.

Testing for Autocorrelation in the Residuals. In the early 1940's, two tests for autocorrelation were developed. These involve, respectively, calculation of (1) a coefficient of autocorrelation, and (2) the ratio of the "mean-square successive difference" to the variance. The second of these, called von Neumann's ratio, is based upon somewhat less restrictive

[10] Wold, Herman, *Demand Analysis*, p. 211, Equation 7, John Wiley and Sons, New York, 1953.

assumptions and since 1950 has been generally used in the analysis of economic time series. It is applicable, of course, to time series arising in other fields as well.[11]

Calculation of the *coefficient of autocorrelation* may be illustrated using residuals from the steel-cost analysis (Table 16.3). The residual for 1921 is entered beside the residual for 1920, and so on, as shown in Table 20.2. The residual for 1920 is entered beside that of 1937, which is not

Table 20.2

TESTING FOR AUTOCORRELATION IN RESIDUALS FROM A REGRESSION LINE OR SURFACE: (1) USING THE COEFFICIENT OF AUTOCORRELATION*

Year	Residuals, z_t	Lagged Residuals, z_{t+1}	z_t^2	$(z_t)(z_{t+1})$
1920	0.9	4.4	0.81	3.96
1921	4.4	−7.0	19.36	−30.80
1922	−7.0	1.5	49.00	−10.50
1923	1.5	−1.8	2.25	−2.70
1924	−1.8	0.8	3.24	−1.44
1925	0.8	2.1	.64	1.68
1926	2.1	1.3	4.41	2.73
1927	1.3	−0.5	1.69	−0.65
1928	−0.5	−1.7	.25	0.85
1929	−1.7	1.0	2.89	−1.70
1930	1.0	−0.7	1.00	−0.70
1931	−0.7	5.9	.49	−4.13
1932	5.9	−2.8	34.81	−16.52
1933	−2.8	−4.1	7.84	11.48
1934	−4.1	−1.1	16.81	4.51
1935	−1.1	1.3	1.21	−1.43
1936	1.3	−0.8	1.69	−1.04
1937	−0.8	0.9	0.64	−0.72

* The residuals are those from the steel-cost example, column (10) of Table 16.3

illogical if we assume that *all* the residuals represent random drawings from the same universe. If the residuals are derived from mathematically

[11] In fact, von Neumann's ratio was originally developed in connection with ballistics experiments in which non-random changes in wind velocity and other variables interfered with attempts to measure the random error component in successive shots fired from a weapon—i.e., the intrinsic level of accuracy of the weapon itself. The generality of the method was immediately recognized by von Neumann and others.

fitted regressions, the means of z_t and z_{t+1} will be zero, and the coefficient of autocorrelation is calculated as follows:[12]

$$r_a = \frac{\Sigma z_t z_{t+1}}{\Sigma z_t^2} = \frac{-47.12}{149.03} = -0.3162 \qquad (20.1)$$

Residuals from a graphic analysis may not sum exactly to zero; in such cases the sums of squares and crossproducts are adjusted to a deviation-from-mean basis by the usual formulas (8.1). In the present example, $Mz_t = -0.0722$ (note that the means of z_t and z_{t+1} are identical), and the corrected value of r_a is

$$r_a = \frac{-47.214}{148.936} = -0.3170$$

This value is compared with tabulated values of the 5-per cent and 1-per cent probability levels of a theoretical distribution of autocorrelation coefficients derived by random sampling from a universe having a true autocorrelation of zero (Table 20.3).[13] For a sample of 18 observations, we estimate by interpolation that values of r_a smaller in absolute value than -0.425 could occur by chance as often as 5 times in 100, so the observed value, $r_a = -0.317$, does not indicate a significant degree of autocorrelation. We assume, therefore, that the usual error formulas are applicable to regression coefficients and forecasts derived from the steel analysis *if* the underlying relationships among the variables continue as they were in 1920–1937.[14]

A similar test applied to residuals from the linear-regression analysis for corn yields (Table 14.1) gave an r_a value of -0.104; the 5 per cent probability level for samples of 38 observations is -0.287, so we conclude that the usual error formulas apply to regression coefficients, and forecasts derive from the analysis of corn yields *if* the underlying relationships continue as they were during 1890–1927.

Another Test Is Given by von Neumann's Ratio. The calculation of von Neumann's ratio for the steel example is shown in the Table 20.4.

---

[12] Since, in this case, the same items appear in series z_t and z_{t+1}, $s_{z_t} = s_{z_{t+1}}$, and equation (20.1) is identical with equation (8.3) for the simple correlation coefficient between z_t and z_{t+1}.

[13] Note that usage since 1950 applies the term *autocorrelation* to the case treated by Anderson; *serial correlation* is usually defined as correlation between current values of one series and lagged values of another series.

[14] Note that the formulas cited from Bartlett and Wold involved the *universe* values of autocorrelation coefficients. In the present paragraph our test of the sample coefficient leads us to accept the hypothesis that autocorrelation in the universe is negligible or zero; if so, Wold's adjustment factor reduces to 1.

Table 20.3

5 AND 1 PER CENT SIGNIFICANCE POINTS FOR THE COEFFICIENT OF
AUTOCORRELATION (CIRCULAR DEFINITION)*

Sample Size,† N	Positive Tail		Negative Tail	
	5 Per Cent Level	1 Per Cent Level	5 Per Cent Level	1 Per Cent Level
5	0.253	0.297	−0.753	−0.798
6	0.345	0.447	−0.708	−0.863
7	0.370	0.510	−0.674	−0.799
8	0.371	0.531	−0.625	−0.764
9	0.366	0.533	−0.593	−0.737
10	0.360	0.525	−0.564	−0.705
11	0.353	0.515	−0.539	−0.679
12	0.348	0.505	−0.516	−0.655
13	0.341	0.495	−0.497	−0.634
14	0.335	0.485	−0.479	−0.615
15	0.328	0.475	−0.462	−0.597
20	0.299	0.432	−0.399	−0.524
25	0.276	0.398	−0.356	−0.473
30	0.257	0.370	−0.324	−0.433
35	0.242	0.347	−0.300	−0.401
40	0.229	0.329	−0.279	−0.376
45	0.218	0.313	−0.262	−0.356
50	0.208	0.301	−0.248	−0.339
55	0.199	0.289	−0.236	−0.324
60	0.191	0.278	−0.225	−0.310
65	0.184	0.268	−0.216	−0.298
70	0.178	0.259	−0.207	−0.287
75	0.174	0.250	−0.201	−0.276
80	0.170	0.246	−0.195	−0.271
85	0.165	0.239	−0.189	−0.263
90	0.161	0.233	−0.184	−0.255
95	0.157	0.227	−0.179	−0.248
100	0.154	0.221	−0.174	−0.242

* Adapted, with the kind permission of the editor, from R. L. Anderson, Distribution of the serial correlation Coefficient, *Annals of Mathematical Statistics*, Vol. 13, No. 1, pp. 1–13, 1942, with corrections, and recalculation of values for $N = 80$ to 100, suggested by Anderson.

Autocorrelation is presumed to be present in the population if the computed value of the coefficient of autocorrelation *exceeds* the value at the preselected significance level for the particular sample size and at the appropriate tail of the distribution. Use the positive tail for positive values of r_a and the negative tail for negative values of r_a.

Table 20.4

TESTING FOR AUTOCORRELATION IN RESIDUALS FROM A REGRESSION
LINE OR SURFACE: (2) USING THE RATIO OF THE MEAN-SQUARE
SUCCESSIVE DIFFERENCE TO THE VARIANCE*

Year	Residuals (1)	Successive Differences (2)	z_t^2 (3)	$(z_{t+1} - z_t)^2$ (4)
1920	0.9	· · ·	0.81	· · ·
1921	4.4	3.5	19.36	12.25
1922	−7.0	−11.4	49.00	129.96
1923	1.5	8.5	2.25	72.25
1924	−1.8	−3.3	3.24	10.89
1925	0.8	2.6	.64	6.76
1926	2.1	1.3	4.41	1.69
1927	1.3	−0.8	1.69	.64
1928	−0.5	−1.8	.25	3.24
1929	−1.7	−1.2	2.89	1.44
1930	1.0	2.7	1.00	7.29
1931	−0.7	−1.7	.49	2.89
1932	5.9	6.6	34.81	43.56
1933	−2.8	−8.7	7.84	75.69
1934	−4.1	−1.3	16.81	1.69
1935	−1.1	3.0	1.21	9.00
1936	1.3	2.4	1.69	5.76
1937	−0.8	−2.1	0.64	4.41

* Residuals same as in Table 20.2.

It is sometimes called "the ratio of the mean-square successive difference to the variance," which describes the manner in which it is calculated.

In the use of von Neumann's ratio the mean-square successive difference is obtained by summing the $n - 1$ values of column (4) and dividing by $n - 1$; no adjustment is made for the fact that $\dfrac{\Sigma(z_{t+1} - z)}{n - 1}$ may differ from zero. The adjustment when $\Sigma z / n \neq 0$ (which can occur only in

† For values of N above 100, use the following formulas to determine the significance points:

For the 5 per cent significance level $\quad -\dfrac{1}{N - 1} \pm \dfrac{1.645}{\sqrt{N + 1}}$

For the 1 per cent significance level $\quad -\dfrac{1}{N - 1} \pm \dfrac{2.326}{\sqrt{N + 1}}$

graphic analyses) is usually trivial and will be disregarded here. We have, in this example,

$$\frac{\delta^2}{s_z^2} = \frac{\dfrac{\Sigma(z_{t+1} - z_t)^2}{n-1}}{\dfrac{\Sigma z_t^2}{n}} \tag{20.2}$$

$$= \frac{\dfrac{389.41}{17}}{\dfrac{149.03}{18}} = \frac{22.91}{8.29} = 2.764$$

From the table of von Neumann's ratio (Table 20.5), we find that values larger than 2.895 could occur as frequently as 5 per cent of the time in samples of 18 observations from a non-autocorrelated universe; we therefore reject the hypothesis that the residuals from the steel regression are significantly autocorrelated.

The corresponding test for residuals from the linear-regression analysis of corn yields (Table 14.1) is as follows:

$$\frac{\delta^2}{s_z^2} = \frac{\dfrac{1058.02}{37}}{\dfrac{479.35}{38}}$$

$$= \frac{28.596}{12.615} = 2.267$$

Entering the table with a sample size of 38, we find that values of 2.589 or more could occur 5 per cent of the time in random drawings from a non-autocorrelated universe. Thus, we conclude that there is no significant autocorrelation in residuals from the corn analysis, and that the usual standard-error formulas apply to its regression coefficients and to individual forecasts for observations drawn from the same universe as those of 1890–1927.

Effects of Using First Differences Instead of Original Values. Regression analysis, mathematical or graphic, can be applied just as readily to series of first differences as to any other sets of numerical values. In a time series of n original values, X_t, there will be $n-1$ first differences, $X_{t+1} - X_t$. If two or more economic time series are intercorrelated as the result of trends which may not reflect logical or causal relations between them, the use of first differences will typically reduce intercorrelation and increase the probability that the regression coefficients obtained

Table 20.5

5 AND 1 PER CENT SIGNIFICANCE POINTS FOR THE RATIO OF THE
MEAN-SQUARE SUCCESSIVE-DIFFERENCE TO THE VARIANCE*

N	Values of K		Values of K'		N	Values of K		Values of K'	
	$P=0.01$	$P=0.05$	$P=0.05$	$P=0.01$		$P=0.01$	$P=0.05$	$P=0.05$	$P=0.01$
4	0.8341	1.0406	4.2927	4.4992	33	1.2667	1.4885	2.6365	2.8583
5	0.6724	1.0255	3.9745	4.3276	34	1.2761	1.4951	2.6262	2.8451
6	0.6738	1.0682	3.7318	4.1262	35	1.2852	1.5014	2.6163	2.8324
7	0.7163	1.0919	3.5748	3.9504	36	1.2940	1.5075	2.6068	2.8202
8	0.7575	1.1228	3.4486	3.8139	37	1.3025	1.5135	2.5977	2.8085
9	0.7974	1.1524	3.3476	3.7025	38	1.3108	1.5193	2.5889	2.7973
10	0.8353	1.1803	3.2642	3.6091	39	1.3188	1.5249	2.5804	2.7865
11	0.8706	1.2062	3.1938	3.5294	40	1.3266	1.5304	2.5722	2.7760
12	0.9033	1.2301	3.1335	3.4603	41	1.3342	1.5357	2.5643	2.7658
13	0.9336	1.2521	3.0812	3.3996	42	1.3415	1.5408	2.5567	2.7560
14	0.9618	1.2725	3.0352	3.3458	43	1.3486	1.5458	2.5494	2.7466
15	0.9880	1.2914	2.9943	3.2977	44	1.3554	1.5506	2.5424	2.7376
16	1.0124	1.3090	2.9577	3.2543	45	1.3620	1.5552	2.5357	2.7289
17	1.0352	1.3253	2.9247	3.2148	46	1.3684	1.5596	2.5293	2.7205
18	1.0566	1.3405	2.8948	3.1787	47	1.3745	1.5638	2.5232	2.7125
19	1.0766	1.3547	2.8675	3.1456	48	1.3802	1.5678	2.5173	2.7049
20	1.0954	1.3680	2.8425	3.1151	49	1.3856	1.5716	2.5117	2.6977
21	1.1131	1.3805	2.8195	3.0869	50	1.3907	1.5752	2.5064	2.6908
22	1.1298	1.3923	2.7982	3.0607	51	1.3957	1.5787	2.5013	2.6842
23	1.1456	1.4035	2.7784	3.0362	52	1.4007	1.5822	2.4963	2.6777
24	1.1606	1.4141	2.7599	3.0133	53	1.4057	1.5856	2.4914	2.6712
25	1.1748	1.4241	2.7426	2.9919	54	1.4107	1.5890	2.4866	2.6648
26	1.1883	1.4336	2.7264	2.9718	55	1.4156	1.5923	2.4819	2.6585
27	1.2012	1.4426	2.7112	2.9528	56	1.4203	1.5955	2.4773	2.6524
28	1.2135	1.4512	2.6969	2.9348	57	1.4249	1.5987	2.4728	2.6465
29	1.2252	1.4594	2.6834	2.9177	58	1.4294	1.6019	2.4684	2.6407
30	1.2363	1.4672	2.6707	2.9016	59	1.4339	1.6051	2.4640	2.6350
31	1.2469	1.4746	2.6587	2.8864	60	1.4384	1.6082	2.4596	2.6294
32	1.2570	1.4817	2.6473	2.8720					

* Adapted, with the kind permission of the editor, from B. I. Hart, Significance levels for the ratio of the mean square successive difference to the variance, *Annals of Mathematical Statistics*, Vol. 13, No. 4, p. 446, 1942.

At the given level of significance and the appropriate sample size (N), a computed K is indicative of positive autocorrelation if it falls below the critical value of K, and is indicative of negative autocorrelation if it exceeds the corresponding critical value of K'; if it falls between the two critical values, no evidence of autocorrelation is present.

will represent meaningful relationships.[15] If there is positive autocorrelation in the residuals from an analysis based on original values, residuals from the corresponding first-difference analysis typically show lower and/or non-significant autocorrelation.

Slutsky and others have observed that the *cumulative sums* of a series of random numbers will trace out cyclical, or positively autocorrelated, patterns. For example, suppose we have a random sequence composed of the values 1 and -1. The cumulative sum of these values will fluctuate around zero. But if it reaches a value of 1 there is a 50 per cent chance that the next observation will raise it to 2. If the sum attains the value of 2, the next observation will change it to 3 or 1, both of which contribute to positive autocorrelation. If the cumulative sum were -2 at the end of n observations, the next observation would change it to -3 or -1, again contributing to positive autocorrelation.

By analogy, the residuals $z_{t+1}, z_{t+2}, z_{t+3}$ may be thought of as a cumulative sum of first differences $(z_{t+1} - z_t, \ z_{t+2} - z_{t+1}$, etc.) added to the first residual z_t. If the actual residuals are autocorrelated, their first differences may still be random. This is not inevitably true, but it is frequently found to be true in practice.

Some examples reported by Cochrane and Orcutt, based upon the work of Richard Stone, are shown in Table 20.6.

Referring to Table 20.5, we find that the 1-per cent and 5-per cent points for testing positive autocorrelation in samples of 19 observations are 1.0766 and 1.3547—i.e., values *lower* than these are indicative of positive autocorrelation. Of the 13 analyses based on original values, 10 showed significant autocorrelation at the 5-per cent level; 3 of these were significant at the 1-per cent level. It is noteworthy that the inclusion of a linear time trend in the analyses based on original data did not eliminate autocorrelation; 6 of the 8 analyses that included a trend factor showed significant autocorrelation.

The results for the 11 first-difference analyses are strikingly different. Not one of them shows significant autocorrelation; one veers toward *negative* autocorrelation, but not significantly so.

Although the above results based on original data may not be typical of those which might be obtained in other time-series studies, the effect of first-difference transformations when the residuals are positively autocorrelated will quite generally be in the direction shown.

---

[15] The common-sense meaning of a regression between first differences is as follows: If (say) high original values of price are associated with low original values of consumption and vice versa, it follows logically that a *change* from a high price to a low price will be associated with a *change* from a low-consumption to a high-consumption figure. In strictly random samples linear regression coefficients estimated from first differences will closely approximate those obtained from the original values.

Table 20.6

VALUES OF VON NEUMANN'S RATIO FOR A NUMBER OF DEMAND STUDIES FOR THE UNITED KINGDOM, 1920–1938

Commodity	Number of Parameters	von Neumann's Ratio	
		Original Data	First Differences
Beer	3	1.28	1.86
	4	1.13	2.01
	4 + time	1.23	
Spirits	3 + time	1.26	2.63
Telegrams	3	1.24	1.61
	4 + time	1.10	1.65
Imported wine	4	1.49	1.84
Communication services	3 + time	0.71	2.05
	4 + time	0.70	2.11
Lard	3 + time	0.90	2.06
Margarine	4	1.26	1.80
	4 + time	2.02	
	5 + time	2.31	2.31
Mean value of δ^2/s_z^2		1.28	1.99

Source: D. Cochrane, and G. H. Orcutt, Application of least-squares regression to relationships containing autocorrelated error terms, *Journal of the American Statistical Association*, Vol. 44, No. 245, pp. 32–61, March, 1949.

Testing for Changes in Regression Relationships over Time. Some general considerations and tests based upon the randomness and size of residuals have already been outlined. Where a regression surface has been fitted mathematically, suspected changes in the true relationship may be tested by adding a "discontinuity variable" which takes the value 0 for all years prior to a given date and the value 1 for all subsequent years. This variable will have a mean and a standard deviation both between 0 and 1; the usual tests of significance can be applied to the net regression of the dependent variable upon the "discontinuity variable."

A significant net regression coefficient implies a significant change in the relationship from one period to the other. As the same net regression coefficients between the dependent variable and the "real" independent variables are applied to the estimates for both periods, a significant discontinuity coefficient implies that the residuals before and after the given date probably represent drawings from different universes.

For example, we might be uncertain as to whether the observed values of certain economic variables after World War II may be regarded as coming from the same statistical universe as those for 1922–1941. A significant discontinuity coefficient would lead us to answer this question in the negative. We might then proceed to fit separate regression equations to each of the two periods and compare the results. In some cases it may be only the constant term, *a*, that is significantly different in the two periods; if so, data for the two periods may be combined for the purpose of estimating the *b* values, but with the discontinuity variable added to reflect the shift in the value of the constant term.

Further Comments on "Practical" Forecasting in Time Series. In Chapter 19 we described the error formulas applicable to forecasts of individual values of a dependent variable for new observations not included in the sample on which the regression equation was based. The error zone for such forecasts may prove to be fairly wide, particularly for unusual combinations of values of the independent variables, and in no event can it be smaller than the standard error of estimate. In random sampling from a stable universe, 5 observations in 100, on the average, will depart from the regression estimate by more than two standard errors of forecast.

The same probability statements apply to forecasts from regression equations based on time series *if the relevant universe remains stable over time*. In deciding this, particularly for cases in which the observed time series are short, the statistician must draw on the theory of his field and his experience with other time-series analyses of similar variables, as well as upon the evidence in the particular set of observations at hand. If large numbers of observations are available, as with certain types of meteorological data, the stability of the universe can be investigated by fitting the same regression function to the data for each of several non-overlapping periods.

Some economists have found diagrams such as Figure 20.1 helpful in considering the ways in which a universe extended over time may be changing, and in appraising which of its parameters may have been most seriously altered. This particular diagram refers to a universe in which the observations are annual totals or averages for the entire United States. If we were interested in a universe of weekly observations of retail beef prices and quantities purchased by consumers in a particular town, our diagram would include a somewhat different set of factors and some of the arrows representing directions of influence might be reversed.

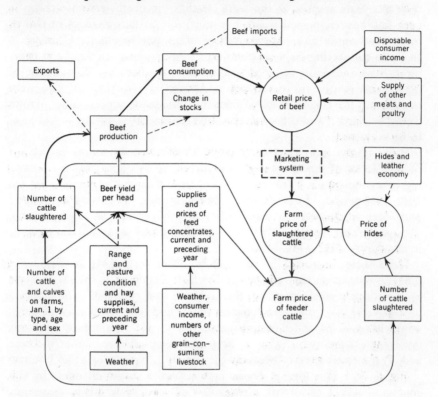

Fig. 20.1. The demand and supply structure for beef. Arrows show direction of influence. Heavy arrows indicate major paths of influence, which account for the bulk of the variation in current prices. Light solid arrows indicate definite but less important paths; dashed arrows indicate paths of negligible, doubtful, or occasional importance. (Karl Fox, Analysis of demand for farm products, United States Dept. of Agriculture, *Technical Bulletin* 1081, p. 34, Sept., 1953.)

Suppose that we had determined the simple regression of "farm price of slaughter cattle" upon the "number of cattle and calves on farms, January 1" for some period such as 1922–1941, and that the relationship was close enough to have considerable forecasting value during those years. Does it appear likely that this same regression line will still apply in the 1950's?

At any given time, each arrow in Figure 20.1 stands for a regression coefficient (or line or curve) connecting two variables. If we follow only the heavy arrows, the diagram implies that there are five major links in the chain connecting the two variables in which we are interested. Stability of the regression between "farm price" and "January 1 numbers" over the years implies at the least (1) that the coefficients attaching to each link have remained nearly constant; or (2) that changes in two or more coefficients have approximately offset one another. Changes in the links connecting disposable consumer income and the supply of other meats with the retail price of beef could also affect the fluctuations of farm price relative to variations in January 1 numbers. If one were considering the lagged effect of price on subsequent supply, other arrows would be needed and other relationships besides those shown would have to be examined.

Some of the individual links could be checked by simple inspection of time series of the two variables involved, or by comparing a freehand regression based on a few recent years with a similar regression for the 1922–1941 period. Thus, the link between "number of cattle slaughtered" and "beef production" involves only a question of fact as to whether the average carcass weight of cattle slaughtered today is about the same as it was during 1922–1941.

If we were interested in the stability from 1922–1941 to date of a universe consisting only of the one link just described, it should be quite easy to decide whether or not this universe had changed. The stability of a universe involving "beef consumption" and the "retail price of beef" might have to be tested simultaneously with the stability of universes involving (1) the retail price of beef and disposable consumer income; and (2) the retail price of beef and the supply of other meats and poultry.

Figure 20.1 is a logical device rather than a statistical one. In this context it serves mainly as a check list of ways in which a time-series universe could change. It is almost certain that *some* elements in a universe as complex as the whole of Figure 20.1 will change over a twenty-year period. But changes in a few elements may leave regression relationships in many of the "subuniverses" of Figure 20.1 substantially unaltered.

Summary

In this chapter we have discussed the applicability of standard-error formulas to regression analyses based on time series; we have shown how these can still be applied in many cases, and presented some statistical tests for determining whether or not a particular set of time-series observations meets the requirements for such application. We also pointed out that knowledge of the theory of the subject matter in question and experience with other time-series analyses of similar variables are important in deciding whether a particular set of time-series observations can be regarded as a sample from a stable universe. Our position is that the special difficulties and uncertainties associated with time-series analysis have frequently been exaggerated; that the strictly statistical problems of time series, particularly those of autocorrelation, can be dealt with scientifically; and that the results of sample surveys or controlled experiments may also get out of date with the passage of time.

REFERENCES

Yule, G. U., Why do we sometimes get nonsense-correlations between time-series?—A study in sampling and the nature of time series, *Jour. Roy. Stat. Soc.*, Vol. 89, No. 1, pp. 1–64, 1926.

Koopmans, Tjalling C., *Linear Regression Analysis of Economic Time Series*, Haarlem, De erven, F. Bohn, n.v., 150 pp., 1937.

Bartlett, M. S., Some aspects of the time correlation problem in regard to tests of significance, *Jour. Roy. Stat. Soc.*, Vol. XCVIII, pp. 536–543, 1935.

Wold, Herman, in association with Lars Jureen, *Demand Analysis. A Study in Econometrics.* John Wiley and Sons, New York, 1953. Especially Chapter 2, pp. 43–45, and Chapter 13, pp. 209–213.

Anderson, R. L., The problem of autocorrelation in regression analysis, *Jour. Amer. Stat. Assoc.*, Vol. 49, No. 225, p. 113, March, 1954.

Durbin, J., and G. S. Watson, Testing for Serial Correlation in Least Squares Regression, *Biometrika*, Vol. 38, nos. 1–2, 1951, pp. 159–177.

Miscellaneous
Special Regression Methods

CHAPTER 21

Measuring the relation between
one variable and two or more
others operating jointly

In working out the change in one variable with changes in other variables up to this point we have assumed that the relation of the dependent factor to each independent factor did not change, no matter what combination of other independent factors was present. In the case of the yield of corn, for example, as worked out in Chapter 14, we assumed that the effect of a given change in rainfall upon the yield was the same, no matter what was the temperature for the season. The significance of this assumption may be shown by combining the estimate for rainfall with the estimate for temperature, and plotting the combined influence of the two variables. In Table 14.19 we already have this combined influence worked out, so all we have to do is to plot it. Figure 21.1 shows the resulting figure. In this figure inches of rainfall are read along the right-hand edge of the bottom of the cube, degrees of temperature along the left-hand edge, and the yield along the vertical edge. The yield for any combination of temperature and rainfall is shown by the distance the upper surface of the solid figure is above the point of intersection of the corresponding values in the base plane.[1]

[1] The way this figure is made may be thought of as follows: Suppose we drew a series of charts of the estimated differences in yield with differences in rainfall, with one chart for an average temperature of 70°, one for 72°, one for 74°, etc. Then if we cut these charts off at the yield line, and arrange them one back of the other, at even distances, we have a figure looking much like Figure 21.1. The lines sloping

Inspecting Figure 21.1, we can now see what is meant by saying that the changes in yield are assumed to be the same for each change in rainfall, no matter what the temperature. As shown in the figure, the maximum yield with a temperature of 70 degrees is obtained at about 12 inches of rain—and that is also the rainfall which produces a maximum yield with a temperature of 72, 74, or 78 degrees. Each curve has the same shape, and the only difference is their elevation above the base. On looking

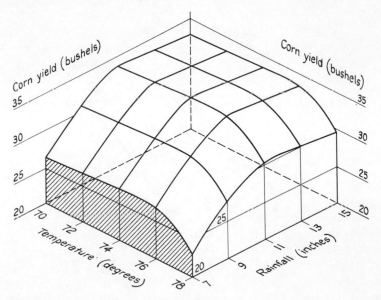

Fig. 21.1. Probable yield of corn for various specific combinations of rainfall and temperature, from multiple curvilinear regression.

at it the other way, we find that the same is true of temperature. With 9 inches of rainfall the maximum yield is obtained at about 75 degrees, and the maximum is also at 75 degrees with other levels of rainfall. This relation necessarily follows the assumptions made in measuring it. Figure 21.1 merely shows the estimate we get by the use of equation (14.1):

$$X_1 = a + f_2(X_2) + f_3(X_3)$$

In working out these estimates we simply add together the estimated value for X_2 and the estimated value for X_3. It does not make any difference what the value of X_3 is, the changes in X_1 assumed to accompany

across the surface from left to right represent what would be the tops of this series of charts. (In this figure the estimates are charted for all combinations of the two variables, even for some not represented in the sample and not shown in Table 14.19.)

particular changes in X_2 are the same—and that is what the figure shows.

Only a little reflection is needed to indicate that Figure 21.1 may not tell the whole truth of the relation of yield to rainfall and temperature. It is quite possible that the crop can use more rain in a hot season than in a cool one, so that the rainfall which will produce the maximum crop may be higher in a season of high average temperature than in a season of low temperature. If that is really the case, equation (14.1) is unable to

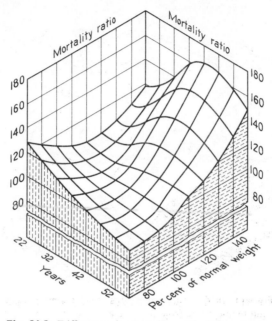

Fig. 21.2. Differences in mortality with differences in weight, for men of various ages (each in percentage of average mortality for that age). Illustration taken from an article by Andrew Court.

express the relationship, for, that equation assumes that the change in yield with rainfall is the same, no matter what the temperature.

An extreme illustration of a changing relationship is shown in Figure 21.2. This figure, which is based on actuarial investigations,[2] shows the differences in mortality among men from the usual rate, for differences in weight at different ages. Taking the 22-year line, for example, we see that men who are much over normal weight have a much higher mortality than normal for that age. Then as the weight is less the mortality is less, until at normal weight there is only normal mortality. But as the

<hr>

[2] *Medico-Actuarial Investigations*, Vol. II, p. 24, 1913.

weight drops still more, the mortality increases again, until below 80 per cent of the normal weight the mortality is more than 20 per cent in excess of normal.

The relation is different for 52-year-old men, however. For them the mortality is also higher for those above normal weight and decreases as normal weight is reached. But as the weight falls below normal the mortality continues to decrease, until for men who are only 70 per cent of

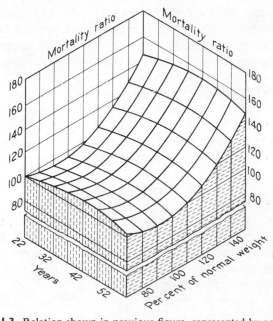

Fig. 21.3. Relation shown in previous figure, represented by equation
$$X_1 = f_2(X_2) + f_3(X_3).$$

normal weight, the mortality is more than 15 per cent *below* the normal for that age. For ages intermediate between these two, the change is also intermediate—as is shown in the chart, 27 years is similar to 22, but not so marked, and the line for 47 years is similar to that for 52. At 42 years, there is apparently little difference in mortality anywhere between 70 per cent of normal weight and 100 per cent, but there is an even more marked rise in mortality with overweight than at higher or lower ages.

The previous methods of analysis would be quite incapable of dealing adequately with the relationship shown in Figure 21.2. Were equation (14.1) used to represent this relation, the higher mortality with lower weights for young men would tend to balance out the lower mortality for the older men at the same weight. In fact, the erroneous conclusion might be reached that the age does not affect the relation between weight and mortality. Figure 21.3 shows the results of an attempt to represent this relation by the methods previously discussed. It is quite obvious that this representation leaves out many of the important facts.

Use of "Joint Functions" to Show Combined Effects. What is needed in both the corn-yield problem and the mortality problem is some way of determining what the yield in the one case, or the mortality in the other, is most likely to be for any given *combination* of the two independent variables. That is quite different from asking for the separate effect of each one. Obviously, a small change in one independent factor will be expected to be accompanied by only a small change in the dependent, so that all the estimated yields (or mortalities) will be expected to lie along a continuous surface; but the surface will be free to warp or change its shape in different portions like the surface shown in Figure 21.2, instead of being held rigidly to the same shape in each dimension, like the surfaces in Figures 21.1 or 21.3. Mathematically, such a changing relation between one variable and two or more others is known as a *joint functional relation*, and may be indicated by the equation:

$$X_1 = f(X_2, X_3) \tag{21.1}$$

This is read simply that "X_1 is a joint function of X_2 and X_3." That means only that, for any combination of values of X_2 and X_3, there will be some particular value of X_1. Equation (21.1) is therefore capable of representing either a relation such as that shown in Figure 21.1, or the more complex relation shown in Figure 21.2.

The problem of determining the extent to which corn yield varies with the joint effect of temperature and rainfall may be said to be one of determining the functional relation of yield to the two other factors, according to the relation shown in equation (21.1).

Determining a Joint Function for Two Independent Variables

Where only two independent variables are concerned, the joint functional relation may be determined quite simply, if a large enough number of observations is available.

The process may be illustrated by data from a different problem, shown in Table 21.1. The observations are from a field study of haystack dimensions in the Great Plains area, made in the late 1920's. The very rapid introduction of pickup hay-baling machines since 1945 has largely eliminated the former practice of selling hay loose in the stack. (The number of pickup hay balers in the United States increased elevenfold

Table 21.1

DATA TAKEN FROM NEBRASKA ROUND STACKS MEASURED IN 1927 AND 1928*

Volume (cubic feet)	Circum-ference (feet)	"Over" (feet)	X_2†	X_3†	X_1†	X_1'	z
2853.00	69.0	37.00	0.139	0.168	0.455	0.485	−0.030
2702.00	65.0	36.50	0.113	0.162	0.432	0.445	−0.013
3099.00	73.0	38.50	0.163	0.185	0.491	0.545	−0.054
1306.00	62.5	26.50	0.096	0.023	0.116	‡0.125	−0.009
2294.00	70.0	35.00	0.145	0.144	0.361	0.440	−0.079
2725.00	68.0	36.50	0.133	0.162	0.435	0.465	−0.030
3309.00	71.0	39.25	0.151	0.194	0.520	0.550	−0.030
2790.00	64.0	36.75	0.106	0.165	0.446	0.440	0.006
2756.00	62.0	38.50	0.092	0.185	0.440	0.455	−0.015
5237.92	80.0	43.00	0.203	0.233	0.719	0.700	0.019
3149.82	67.0	37.60	0.126	0.175	0.498	0.480	0.018
5498.46	79.0	44.60	0.198	0.249	0.740	0.745	−0.005
3397.83	66.0	38.00	0.120	0.180	0.531	0.480	0.051
3007.56	62.0	36.80	0.092	0.166	0.478	0.445	0.033
4574.29	79.0	41.10	0.198	0.214	0.660	0.650	0.010
6228.59	73.0	48.00	0.163	0.281	0.794	‡0.815	−0.021
2318.64	63.0	30.20	0.099	0.080	0.365	0.270	0.095
3176.71	68.0	37.75	0.133	0.177	0.502	0.490	0.012
2352.31	70.0	32.50	0.145	0.112	0.371	0.375	−0.004
2174.44	69.0	31.62	0.139	0.100	0.337	0.340	−0.003
2694.72	73.0	34.50	0.163	0.138	0.431	0.440	−0.009
3333.53	70.0	37.25	0.145	0.171	0.523	0.495	0.028
4328.92	78.5	40.00	0.195	0.202	0.636	0.610	0.026
2115.04	67.0	31.25	0.126	0.095	0.325	0.320	0.005
2489.08	66.5	33.75	0.123	0.128	0.396	0.390	0.006
2296.65	64.5	32.38	0.110	0.110	0.361	0.340	0.021
3117.21	65.5	37.58	0.116	0.175	0.494	0.470	0.024
4088.36	74.0	40.33	0.169	0.206	0.612	0.600	0.012
4180.88	72.0	40.50	0.157	0.207	0.621	0.590	0.031
2318.19	63.0	33.00	0.099	0.119	0.365	0.345	0.020
1946.90	58.0	31.00	0.063	0.091	0.289	0.265	0.024
2479.89	61.0	36.50	0.086	0.162	0.394	0.405	−0.011
3174.80	73.0	37.00	0.163	0.168	0.502	0.505	−0.003
2151.54	64.0	33.00	0.106	0.119	0.333	0.355	−0.022
3475.68	73.0	39.50	0.163	0.197	0.541	0.575	−0.034
4393.08	71.0	42.00	0.151	0.223	0.643	0.620	0.023
2819.50	69.0	35.00	0.139	0.144	0.450	0.435	0.015
3703.49	70.0	38.50	0.145	0.185	0.569	0.525	0.044

* Acknowledgment is due W. H. Hosterman, of the Bureau of Agricultural Economics, U. S. Department of Agriculture, for the use of these data.

† $X_2 = \log_{10}$ (circumference) − 1.700, stated to three decimal places.

$X_3 = \log_{10}$ ("over") − 1.4, stated to three decimal places.

$X_1 = \log_{10}$ (volume) − 3.0, stated to three decimal places.

‡ Estimated by extrapolation of the surface shown on Fig. 21.4.

Table 21.1 (*Continued*)

Volume (cubic feet)	Circumference (feet)	"Over" (feet)	X_2†	X_3†	X_1†	X_1	z
2742.81	72.5	34.50	0.160	0.138	0.438	0.440	−0.002
3002.40	66.0	35.50	0.120	0.150	0.477	0.430	0.047
1854.19	69.0	30.50	0.139	0.084	0.268	0.295	−0.027
1982.07	62.0	31.00	0.092	0.091	0.297	0.295	0.002
2470.86	65.0	33.50	0.113	0.125	0.393	0.375	0.018
1203.15	60.1	26.25	0.079	0.019	0.080	‡0.120	−0.040
2843.84	71.0	36.00	0.151	0.156	0.454	0.465	−0.011
2636.25	66.0	36.00	0.120	0.156	0.421	0.440	−0.019
1998.39	65.0	32.00	0.113	0.105	0.301	0.335	−0.034
2005.03	64.0	32.00	0.106	0.105	0.302	0.330	−0.028
2568.76	66.0	35.00	0.120	0.144	0.410	0.420	−0.010
2161.18	65.0	32.50	0.113	0.112	0.335	0.345	−0.010
2112.20	67.0	32.00	0.126	0.105	0.325	0.345	−0.020
3009.33	65.0	38.00	0.113	0.180	0.478	0.475	0.003
1992.24	63.0	31.00	0.099	0.091	0.299	0.295	0.004
2746.98	70.0	34.00	0.145	0.131	0.439	0.415	0.024
2238.27	64.0	35.00	0.106	0.144	0.350	0.400	−0.050
1747.47	67.0	30.00	0.126	0.077	0.242	0.275	−0.033
2863.91	67.0	36.00	0.126	0.156	0.457	0.450	0.007
3593.47	72.0	39.00	0.157	0.191	0.555	0.550	0.005
2435.48	62.0	35.00	0.092	0.144	0.387	0.385	0.002
2430.18	63.0	34.00	0.099	0.131	0.386	0.370	0.016
2590.07	67.0	35.00	0.126	0.144	0.413	0.425	−0.012
3577.68	70.0	41.00	0.145	0.213	0.554	0.585	−0.031
3299.24	73.0	40.00	0.163	0.202	0.518	0.585	−0.067
1986.14	64.0	32.50	0.106	0.112	0.298	0.310	−0.012
3109.04	68.0	38.00	0.133	0.180	0.493	0.500	−0.007
2821.56	71.0	37.00	0.151	0.168	0.450	0.495	−0.045
2932.24	67.0	38.00	0.126	0.180	0.467	0.490	−0.023
3304.63	69.0	38.00	0.139	0.180	0.519	0.505	0.014
2565.46	72.0	35.00	0.157	0.144	0.409	0.445	−0.036
4509.93	74.0	41.33	0.169	0.216	0.654	0.625	0.029
4804.01	81.0	42.00	0.208	0.223	0.682	0.680	0.002
4241.80	75.0	40.75	0.175	0.210	0.627	0.620	0.007
4516.10	69.2	43.25	0.140	0.236	0.655	0.630	0.025
5011.62	77.5	43.10	0.189	0.234	0.700	0.695	0.005
2110.73	65.0	31.50	0.113	0.098	0.324	0.320	0.004
2775.70	76.0	34.60	0.181	0.139	0.443	0.450	−0.007
3927.90	72.0	39.00	0.157	0.191	0.594	0.550	0.044
4212.77	80.0	41.50	0.203	0.218	0.624	0.665	−0.041
3562.64	78.5	38.50	0.195	0.185	0.552	0.575	−0.023

See footnotes on first page of table.

Table 21.1 (Continued)

Volume (cubic feet)	Circum-ference (feet)	"Over" (feet)	X_2†	X_3†	X_1†	X_1'	z
2853.96	75.0	35.50	0.175	0.150	0.455	0.475	−0.020
3294.38	69.0	38.00	0.139	0.180	0.518	0.505	0.013
1689.54	63.0	30.50	0.099	0.084	0.228	0.280	−0.052
2228.84	62.0	33.00	0.092	0.119	0.348	0.340	0.008
2362.61	64.0	34.00	0.106	0.131	0.373	0.355	0.018
3088.28	68.0	38.50	0.133	0.185	0.490	0.510	−0.020
3820.79	70.0	40.00	0.145	0.202	0.582	0.560	0.022
3126.64	63.0	36.90	0.099	0.167	0.495	0.435	0.060
3624.75	71.0	38.45	0.151	0.185	0.559	0.530	0.029
3023.97	73.0	36.50	0.163	0.162	0.480	0.490	−0.010
6045.42	79.0	47.00	0.198	0.272	0.781	‡0.805	−0.024
3100.11	64.0	37.00	0.106	0.168	0.491	0.445	0.046
3378.07	70.0	38.00	0.145	0.180	0.529	0.510	0.019
3040.29	77.0	35.00	0.186	0.144	0.483	0.465	0.018
2252.16	65.0	32.50	0.113	0.112	0.353	0.345	0.008
3552.61	76.0	37.00	0.181	0.168	0.551	0.520	0.031
2635.90	66.0	34.50	0.120	0.138	0.421	0.405	0.016
3201.41	71.0	35.50	0.151	0.150	0.505	0.455	0.050
2590.21	69.0	35.00	0.139	0.144	0.413	0.435	−0.022
3743.55	76.0	38.25	0.181	0.183	0.573	0.560	0.013
3858.03	73.0	39.50	0.163	0.197	0.586	0.575	0.011
3829.44	74.0	39.75	0.169	0.199	0.583	0.585	−0.002
2556.44	66.0	33.00	0.120	0.119	0.408	0.365	0.043
3119.07	69.0	36.00	0.139	0.156	0.494	0.460	0.034
2122.38	65.5	32.00	0.116	0.105	0.327	0.335	−0.008
2921.92	69.0	36.00	0.139	0.156	0.466	0.460	0.006
2936.35	72.5	34.50	0.160	0.138	0.468	0.435	0.033
2427.66	76.0	33.00	0.181	0.119	0.385	0.405	−0.020
2069.38	65.0	31.50	0.113	0.098	0.316	0.320	−0.004
1899.54	72.0	30.00	0.157	0.077	0.279	0.285	−0.006
4289.28	78.5	40.50	0.195	0.207	0.632	0.630	0.002
2407.39	67.5	32.50	0.129	0.112	0.381	0.360	0.021
3097.99	66.0	35.50	0.120	0.150	0.491	0.430	0.061
3893.67	75.5	39.25	0.178	0.194	0.590	0.585	0.005
2238.66	68.0	31.75	0.133	0.102	0.350	0.340	0.010
2314.79	64.0	33.10	0.106	0.120	0.364	0.355	0.009
2667.07	66.0	34.70	0.120	0.140	0.426	0.410	0.016
2582.07	68.0	33.50	0.133	0.125	0.412	0.395	0.017
3426.50	75.0	37.00	0.175	0.168	0.535	0.520	0.015
2307.34	60.0	33.40	0.078	0.124	0.363	0.335	0.028
3960.41	76.0	39.30	0.181	0.194	0.598	0.585	0.013

See footnotes on first page of table.

between 1945 and 1956.) However, the data still provide an excellent illustration of the general problem of joint functional relations.

At the time of the study, farmers in this area ordinarily sold their hay unbaled and in the stack. It was therefore necessary to estimate the quantity of hay in each stack. Two measurements, which could be made readily with only a rope, were usually employed—the perimeter around the base of the stack and the "over," or the distance from the ground on one side of the stack over the center to the ground on the other.

The observations shown in Table 21.1 are all for round stacks. These stacks vary in height and shape to some extent, however, so their volume cannot be computed from the basal circumference by any simple mathematical rule. The volumes shown in the table were computed from careful surveying measurements of all the dimensions of each stack—much more exact measurements than a farmer would be able to make in practice. The problem is to establish the average volume for specified circumferences and "overs," so the farmers might use these two measurements, and also to determine how reliable are these estimates.

The volume will tend to be some function of the basal area times the height. The basal area is a function of the square of the basal circumference; the "over" is a function of both the basal diameter and the height—but attempts to separate the two have been unsuccessful. It is obvious that because of the multiplying nature of the relations,

$$\text{volume} = f(\text{circumference})(\text{over}).$$

Such a relationship may be approached by use of the relation

$$\log_{\text{volume}} = f(\log_{\text{circumference}}) + f(\log_{\text{over}})$$

Attempts to determine the relationship by this equation, however, have not been fully successful. The shape of the stacks apparently shifts with changes in size. The problem is evidently one where the relation may best be expressed by a joint function such as

$$\text{volume} = f(\text{circumference, over})$$

Such a relation could be determined directly from the data by the methods which will presently be described. It is evident that the correlation surface would have a marked upward slope as the two dimensions increased together, even if the usual volume formulas applied. The work may be somewhat simplified by stating each variable as a logarithm and then determining the joint relation according to the equation

$$\log_{\text{volume}} = f(\log_{\text{circumference}}, \log_{\text{over}})$$

The logarithms (to base 10) are entered in Table 21.1, designated as X_2, X_3, and X_1. (1.7 has been subtracted from the logarithm for circumference, 1.4 from the logarithm for "over," and 3.0 from the logarithm for volume.)

The joint function may be determined either by fitting some appropriate algebraic equation, or by graphic processes. Only in rare cases will there be a good logical basis for judging the form of the joint function to be expected. In most cases, therefore, even the algebraic equation must be selected with some reference to its ability to represent the type of joint function shown in the data, as empirically determined by some form of graphic examination. The methods of determining the joint function will therefore first be illustrated for the graphic method, and appropriate mathematical equations to represent various forms of joint surfaces will be considered later.

Subgrouping and Averaging the Observations. The first step is to classify the observations according to X_2, and subclassify according to X_3, and determine the averages of X_1, X_2, and X_3 for each group. It is not worth while to make too many groups. Four groups each way would give 16 subgroups, and 5 each way would give 25. If the cases were uniformly distributed through 25 subgroups, that would make less than 5 cases to a group, which is rather thin for a satisfactory average. The cases will not necessarily be distributed uniformly, so it may be best to try the fivefold classification. The results are shown in Table 21.2.

Table 21.2

NUMBER OF HAYSTACK OBSERVATIONS, CLASSIFIED ACCORDING TO X_2 AND X_3
(LOGARITHMS OF CIRCUMFERENCE AND "OVER")

X_3 Values	X_2 Values				
	Under 0.090	0.090– 0.119	0.120– 0.149	0.150– 0.179	0.180 and over
Under 0.100	2	7	3	1	
0.100–0.139	1	14	10	3	2
0.140–0.179	1	8	17	8	2
0.180–0.219	$\cdots$	2	10	14	7
0.220 and over	$\cdots$	$\cdots$	1	2	5

There is a marked correlation between X_2 and X_3, so a few groups have 10 or more reports, whereas 15 out of the 25 have under 5. Preliminary examination of the data indicates that a unit change in X_3 is generally accompanied by a larger change in X_1 than is a unit change in X_2. Accordingly we may decide to halve the groups in the central portion of the range of X_3, making the class intervals with respect to that variable under 0.100, 0.100–0.119, 0.120–0.139, 0.140–0.159, 0.160–0.179, 0.180–0.199, 0.200–0.219, and 0.220 and over. With 5 classes for X_2, this will give a 40-group classification—but with many of the "cells" vacant. Averaging X_2, X_3, and X_1 for each of the resulting groups gives means as shown in Table 21.3.

Table 21.3

HAYSTACK DATA: AVERAGE X_2, X_3, AND X_1, FOR OBSERVATIONS CLASSIFIED BY X_2 AND X_3

X_3 Values	Number of Cases	Mean X_2	Mean X_3	Mean X_1
		X_2 under 0.090		
Under 0.100	2	0.071	0.055	0.185
0.100–0.119				
0.120–0.139	1	0.078	0.124	0.363
0.140–0.159				
0.160–0.179	1	0.086	0.162	0.394
		X_2 0.090–0.119		
Under 0.100	7	0.102	0.081	0.278
0.100–0.119	10	0.107	0.112	0.332
0.120–0.139	4	0.106	0.127	0.379
0.140–0.159	2	0.099	0.144	0.369
0.160–0.179	6	0.105	0.167	0.473
0.180–0.199	2	0.103	0.183	0.459
		X_2 0.120–0.149		
Under 0.100	3	0.130	0.085	0.278
0.100–0.119	6	0.132	0.108	0.362
0.120–0.139	4	0.130	0.131	0.417
0.140–0.159	12	0.129	0.149	0.440
0.160–0.179	5	0.135	0.171	0.483
0.180–0.199	8	0.135	0.181	0.515
0.200–0.219	2	0.145	0.208	0.568
0.220 and over	1	0.140	0.236	0.655

Table 21.3 (Continued)

X_3 Values	Number of Cases	Mean X_2	Mean X_3	Mean X_1
		X_2 0.150–0.179		
Under 0.100	1	0.157	0.077	0.279
0.100–0.119				
0.120–0.139	3	0.161	0.138	0.446
0.140–0.159	4	0.159	0.150	0.456
0.160–0.179	4	0.163	0.167	0.492
0.180–0.199	9	0.161	0.193	0.558
0.200–0.219	5	0.167	0.208	0.606
0.220 and over	2	0.157	0.252	0.719
		X_2 0.180 and over		
0.100–0.119	1	0.181	0.119	0.385
0.120–0.139	1	0.181	0.139	0.443
0.140–0.159	1	0.186	0.144	0.483
0.160–0.179	1	0.181	0.168	0.551
0.180–0.199	3	0.186	0.187	0.574
0.200–0.219	4	0.198	0.210	0.638
0.220 and over	5	0.199	0.242	0.724

Fitting a Joint Function Graphically. The most rapid method of getting an approximate shape of the surface is by drawing contour lines just as surveyors draw contours in preparing a topographic map. A chart is prepared as shown in Figure 21.4, with values of X_2 as the abscissa and with X_3 as the ordinate. (In making this chart, a sheet of cross-section paper ruled 10 lines to the inch, and 16 by 24 inches large, was used, in order to enter and read the values with satisfactory accuracy.) A dot is then entered corresponding to average values of X_2 and X_3 for each group in Table 21.3. The average value of X_1 is then written in by the dot for each group, and enclosed in parentheses when the group has less than 3 cases. Dotted lines are drawn in, roughly separating the dots into those having X_1 values below 300, below 350, etc., as nearly as possible, and leaving intermediate values at corresponding distances between the lines where possible. (Values for the lines are indicated at the end of each line.) Solid lines are then drawn in, corresponding as closely as possible to the dotted lines but spaced as regularly as possible across the chart, and with similar shapes, so as to give a smooth continuous surface without "bumps," while conforming to the general shape of the topographic surface

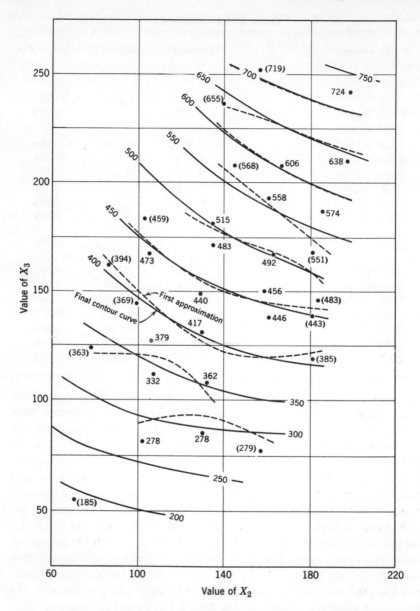

Fig. 21.4. Mean values of X_1 for group mean values of X_2 and X_3, and original and smoothed contours fitted to group averages.

indicated by the dotted lines. (This process of smoothing can be checked by reading off values of X_1 corresponding to successive given values of X_2 with X_3 constant, or of X_3 with X_2 constant, smoothing these values graphically, and then drawing the corresponding smoothed contours in on the chart.) Even with only freehand smoothing by eye, however, as used in drawing the solid lines shown in the figure, reasonably good results can be obtained.

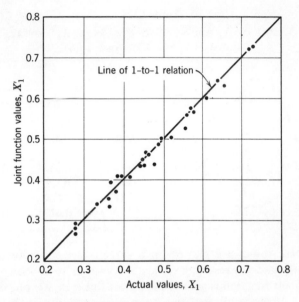

Fig. 21.5. Estimated values of X_1' read from Fig. 21.4 related to actual mean values of X_1.

Values of X_1 read from the freehand solid lines entered in Figure 21.4 correlate with the original values of X_1 for the group averages as shown in Figure 21.5. Apparently there is a close and one-to-one correspondence between the original values and the smoothed surface, with a very high correlation.

An alternative method of determining the surface graphically is by plotting the M_1 values against the M_2 values for each set of groups according to given values with respect to X_3, fitting lines or curves freehand to each of these sets of data (representing successive intercepts or slices across the function), smoothing estimated values read from these fitted lines for given values of X_3 across the other axis, and continuing this process until no further changes in the smoothed function seem needed. This process, illustrated in detail in previous editions of this book (second edition, pages 382 to 387) has little advantage over the contour method and is far slower, and therefore has been omitted here.

The shape of the joint function described by the solid contour lines on Figure 21.4 may be seen by reading off values for given combinations of X_2 and X_3, and showing these values in tabular or graphic form. Table 21.4 gives them, read from the chart by interpolation between the contours, scaling off the distances perpendicular to the nearest contours.

Table 21.4

VALUES OF X_1 FOR SPECIFIED VALUES OF X_2 AND X_3,
ESTIMATED FROM CONTOUR LINES

X_2	X_3				
	0.80	0.100	0.150	0.200	0.250
0.100	0.270	0.312	0.405	0.487	$\cdots$
0.120	0.280	0.328	0.430	0.518	$\cdots$
0.140	0.286	0.340	0.448	0.548	0.677
0.160	0.290	0.349	0.463	0.578	0.713
0.180	$\cdots$	$\cdots$	0.477	0.600	0.732
0.200	$\cdots$	$\cdots$	$\cdots$	0.617	0.747

When these values are plotted on a three-dimensional diagram, with X_1 as the ordinate, the resulting surface is found to be as shown in Figure 21.6. The joint function is seen to be almost linear in any one dimension, but warped upwards, so that the slope corresponding to $b_{12.3}$ is much steeper at high values of X_3 than it is at low values of X_3. This warped surface could not be fully represented by a plane of any type.

In determining the surface shown in Figures 21.4 and 21.6, the data were subclassified into a 5×8 grouping. Out of the 40 possible subgroups, 1 or more observations occurred in 31. Discarding the one observation with $X_3 = 0.055$ and $X_2 = 0.071$, both far below the range of the other observations (though with a value of X_1 not inconsistent with them), there were 30 group averages on which to base the graphic analysis to see if there was any consistent indication of the presence of a joint function. As shown in the two figures, this gave a definite indication of the presence of a joint function which could not be represented by adding two net functions for X_2 and X_3.

In this case, with 119 usable observations, and with high multiple correlation so that the individual values of X_1 fell close to the joint surface (as will be seen shortly), it was possible to use a large number of subgroups in getting many averages to which to fit the surface. Where the number of observations is smaller, or the multiple correlation lower, a smaller number of groups would have to be used. Even a 2 × 2 classification with 4 groups, or a 3 × 3 with 9, would serve to give some indication of whether or not a joint function was present. In such cases, if the group averages indicate a warped surface, not only should group averages be plotted on the topographic chart on which the contours are to be drawn (Figure 21.4), but the original observations should also be plotted, separately

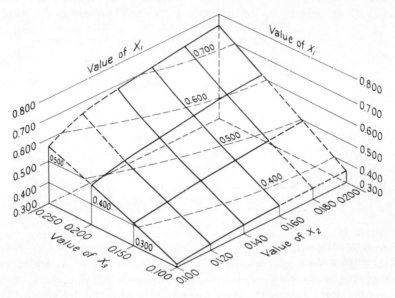

Fig. 21.6. Estimated volume of X_1 for specific combinations of X_2 and X_3, from smoothed contours.

designated, and the contour lines fitted with some reference to the individual observations as well as the group averages. This is especially necessary if the surface of the function appears to be sharply curved rather than linear, as the average of points along a convex or concave curve will of course not lie on the curve.

Where the number of observations is quite small, but the correlation is high, the contours may be fitted directly to the individual observations without making averages at all. An illustration of such a problem is shown on page 373.

Fitting a Joint Function Algebraically. In the same way that definite equations can be determined by least squares to represent curvilinear net regressions, certain types of joint functional surfaces can be represented by algebraic equations of types which can be fitted by least squares. After the kind of surface present has been determined by fitting contours graphically and plotting the surface so found, it is then much easier to choose an appropriate equation. In some cases, the equation can be deduced logically and used as a model of the relation expected. In the haystack-volume problem, Figure 21.6 shows that the joint function is of a type where the regression of X_1 on X_3 is substantially linear for any given value of X_2, but the slope of the regression $b_{13.2}$ changes as the values of X_2 change. If it is assumed that the slope changes at a constant rate with changes in X_2, this condition may be expressed in the relation

$$X_1 = a + b(c + dX_2)X_3$$

Multiplied out, this becomes

$$X_1 = a + bcX_3 + bdX_2X_3$$

which may be stated

$$X_1 = a + eX_3 + g(X_2X_3) \tag{21.2}$$

The values of a, e, and g may then be determined by the usual methods of linear multiple correlation, with X_3 and the values of the product (X_2X_3) used as the independent factors.

If it is assumed that X_1 varies with X_2, other than through its influence on $b_{13.2}$, an additional term may be added to the equation, making it

$$X_1 = a + eX_3 + g(X_2X_3) + hX_2 \tag{21.3}$$

Determining the values of the four constants of equation (21.3) from the haystack data, and working out estimated values of X_1 for specific combinations of values of X_2 and X_3, we would arrive at substantially the same joint functional surface as was determined by the contour method.

Other algebraic equations which have shown good ability to represent joint functions are as follows:

$$X_1 = a + b_2X_2 + b_3X_3 + b_4(X_2 + X_3)$$

(This equation is useful with a plane rotated so as to show higher values diagonally across the joint surface.)

For production functions, where two fertilizer constituents are each applied in varying quantities, these equations have proved useful:

$$X_1 = a + b_2X_2 + b_3X_3 + b_4(X_2^2) + b_5(X_3^2) + b_6(X_2X_3)$$

$$X_1 = a + b_2X_2 + b_3X_3 + b_4\sqrt{X_2} + b_5\sqrt{X_3} + b_6\sqrt{X_2X_3}$$

In fitting these equations, the values are first determined for the complete set of terms as shown, and then when certain of the coefficients show values insignificant as compared to their own standard errors, the solutions are recomputed omitting these terms.

The last two and other equations were extensively tested in an experimental study of the results of applying varying quantities of phosphate and nitrogen to corn, red clover, and alfalfa.[3] The equations used were deduced from the hypothesis that in applying fertilizer in successive units of the same size, the law of diminishing returns would hold, so that (1) each additional unit of each fertilizer would add less to the yield than the preceding unit, up to some maximum point; and (2) that the combined result of two fertilizer ingredients together would be greater than the sum of the effect of each one separately. Both equations, one based on parabolas and one on square roots, satisfy these two logical conditions, and provided two alternative equations to represent the model. (A third form not shown here, did not quite meet these two conditions.)

The authors found that the square-root form of equation gave the best fit to the experimental data for all three crops. For red clover and alfalfa, certain terms gave results not statistically significant, and the final equations therefore eliminated these terms.

Stating total yield (Y) in bushels or tons per acre, phosphate (P) in pounds of P_2O_5 per acre, and potassium (K) and nitrogen (N) in pounds per acre, the three selected equations were:

Corn
$$Y = -5.682 - 0.316N - 0.417P + 6.3512\sqrt{N}$$
$$+ 8.5155\sqrt{P} + 0.3410\sqrt{NP}$$

Red clover
$$Y = 2.468 - 0.003947P + 0.02834\sqrt{K} + 0.127892\sqrt{P}$$
$$- 0.000979\sqrt{KP}$$

Alfalfa
$$Y = 1.837 - 0.0014K - 0.0050P + 0.061731\sqrt{K}$$
$$+ 0.173513\sqrt{P} - 0.001440\sqrt{KP}$$

The yield-response surfaces described by the equations for corn and for alfalfa are shown in Figures 21.7 and 21.8.

With the relations determined in an algebraic equation, it is relatively simple to estimate new values of the dependent variable for various values of the independent variables, and to determine confidence intervals for each constant. Various other mathematical operations can also be performed. For example, the authors calculated the "replacement rates" of one nutrient by another, at various points on the joint surface, and, by applying assumed costs and prices of the several variables, calculated

---

[3] Earl O. Heady, John T. Pesek, and William G. Brown, Crop response surfaces and economic optima in fertilizer use, *Iowa State College Agricultural Experiment Station Research Bulletin* 424, March, 1955.

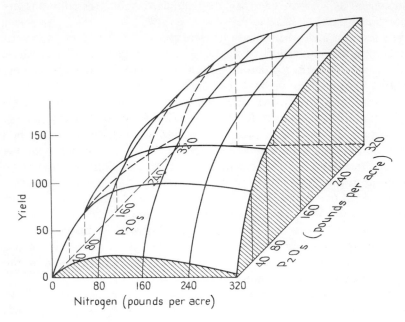

Fig. 21.7. Response of corn yields to varying quantities of nitrogen and phosphate, as shown by fitted joint function (from bulletin by Heady *et al.*, *op. cit.*)

probable least-cost combinations and economic optima in the use of fertilizers.[4]

Determining the Standard Error of Estimate and Index of Multiple Correlation. With the joint function fitted by an algebraic equation, the standard error of estimate and the index of multiple correlation with respect to the joint function can be calculated by the usual equations, (15.6) and (15.7). Similarly, confidence intervals of any given probability around the fitted joint surface can be calculated by an extension of the methods of Chapter 17, and the reliability of individual forecasts by the methods of Chapter 19. These confidence calculations are, of course, applicable only where the observations represent a sample randomly drawn from a defined universe,—the same one for which estimates are to be made. Where, as in the case of results based on experimental data such as those shown in Figures 21.7 and 21.8, the data are selected (experimentally or otherwise) with respect to the values of the independent variables, the correlation index has sampling significance only in the very special "universe" in which those same values of the independent variables

[4] Earl O. Heady, *et al.*, *op. cit.*, pp. 321–25.

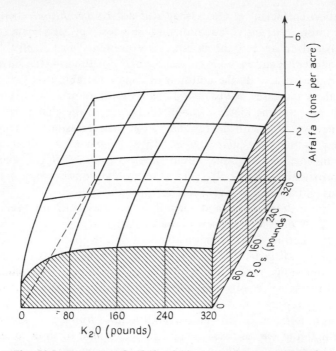

Fig. 21.8. Response of alfalfa yields to varying quantities of potassium and phosphate, as shown by fitted joint function (from bulletin by Heady *et al., op. cit.*)

remain fixed for all possible samples. The standard error of estimate may still give a reasonably accurate forecast of errors to be expected in new observations drawn from the more general universe, in which the frequencies of various values of the independent variables may differ from those in the original sample.[5] (See Chapter 17.) If the closeness of fit for joint surfaces fitted to the same data by several different equations is to be determined, either the multiple correlation index or the standard error of estimate will serve to indicate which ones will give relatively satisfactory results. For such cases it is important to include the adjustments for degrees of freedom [equations (12.2), (15.6), and (17.9)] if the several equations tested have varying numbers of terms and if n is not very large—say, 100 cases or less.

It should be noted that the equation which gives the highest adjusted

[5] Least-squares regression analysis typically assumes that the variance of residuals is uniform throughout the universe represented by the regression function. On this assumption, σ_z^2 is independent of the particular values chosen for the independent variables.

correlation coefficient or index may still not be *significantly* better than one or more alternative equations. Some tests of significance in this sense, based on analysis of variance, are presented in Chapter 23. The adjustments of $\bar{R}$ and $\bar{I}$ merely correct for bias in estimating the corresponding universe values. If the addition of a new variable using one degree of freedom increases R^2 by more than the amount $(1 - R^2)/(n - m)$, $\bar{R}^2$ will rise. But the increase could still be due to the chance correlation of the new variable with the residuals from the regression involving other variables in this *particular sample*. In a sample of 20 observations a fourth independent variable would have to increase $R^2_{1.2345}$ over $R^2_{1.234}$ by approximately $3(1 - R^2_{1.234})/(n - m)$ to be considered a significant improvement at the 5 per cent level of probability.

This more demanding (and scientific) test of significance serves as a further curb on the tendency to particularize regression functions too closely to what may be random features of one sample. While the danger is sometimes alleged to be greatest when freehand curves are used, the same temptations arise when two or more mathematical equations are fitted to the same set of observations. But if an apparent gain has been achieved by testing a number of alternative additional terms in the equation, or a number of alternative additional variables in the multiple correlation, and one of those tested has been found to show an apparent gain with odds, say, of 19 out of 20 by the tests *applied to that one variable alone*, then the true odds of it being significant are far less. For example, if ten alternative additional terms have each been tested in turn, and one of them has been found to show a gain with this much significance, the true odds for its being significant are not 19 out of 20, but only 10 out of 20—since any one of the ten tested might, by chance, have shown the same level of significance.

When the joint function has been determined graphically, as in the haystack problem, the same general method is followed as with graphic multiple regression curves—working out estimated values, X_1', from the graphic function, calculating the z's and their standard error, and then calculating the two statistics from this. Performing this process for the haystack problem, estimated values for each observation in the sample are read off from the contour lines in Figure 21.4, interpolating linearly for distances between the adjacent contours. The resulting values of X_1' are shown to the nearest 5 units in the third decimal place in the next-to-the-last column of Table 21.1. (More exact values could be read by scaling the distances in millimeters from the original large chart, and this was done in reading the values for Table 21.4; but it was felt this would give an impression of spuriously high accuracy for the individual estimates—though it might have reduced the size of s_z slightly.) The value of

$z(= X_1 - X_1')$ is then calculated, as shown in the final column of the table, and so is the value of s_z. The standard deviation of the original values of X_1 is 0.1265, whereas s_z is 0.0281. In this case the regression surface accounts for a large part of the variation in volume. The standard error in estimating X_1 from the joint function may therefore be stated

$$S_{1.f(2,3)} = s_z = 0.0281$$

Similarly, the index of multiple correlation for X_1 as a joint function of X_2 and X_3 may be computed by the usual equation[6]

$$I_{1.f(2,3)}^2 = 1 - \frac{s_z^2}{s_1^2} = 1 - \frac{(0.0281)^2}{(0.1265)^2}$$

$$= 1 - 0.047342 = 0.952658$$

$$I_{1.f(2,3)} = 0.976$$

The joint function therefore explains 95 per cent of the observed variance in volume.

The sampling significance of the values of $I_{1.f(2,3)}$, whether determined by graphic or algebraic fitting, is discussed in Chapter 17.

It is evident that the volume of a round haystack may be very closely estimated from the rough farm measurements of circumference and "over." The standard error of estimate, 0.0281, indicates that the logarithm of volume can be estimated to ± 0.0281 of the true logarithm for two-thirds of the observations, and to ± 0.0562 for 95 per cent of them. Taking the antilogs of these values, 0.0281 to $\bar{1}.9719$ and 0.0562 to $\bar{1}.9438$, we find that the confidence intervals are 93.7 and 106.6 per cent for $P = 0.67$, and 87.9 and 113.7 per cent for $P = 0.95$.

Stating the Conclusions Shown by the Joint Function. After the joint relations have been determined by either method, the regression surface may be stated in simpler terms by preparing tables showing the expected values of X_1 for stated combinations of X_2 and X_3. In this problem, that involves determining the logarithms of X_2 and X_3 for the selected values, reading off from the charts the corresponding estimated value for the logarithm of X_1, and finding its antilogarithm. Carrying out this process, we obtain the values shown in Table 21.5.

A similar table for the corn yields corresponding to Figure 21.7 is shown in Table 21.6.

---

[6] In view of the large number of observations, the adjustments for n and m are ignored here.

Table 21.5

AVERAGE VOLUME OF ROUND HAYSTACKS FOR DIFFERENT COMBINATIONS OF CIRCUMFERENCE AND "OVER"

Circumference	"Over" (feet)			
	30	34	38	42
	Cubic feet	*Cubic feet*	*Cubic feet*	*Cubic feet*
60 feet	1,760	2.240		
65 feet	1,860	2,430	2,990	
70 feet	1,910	2,660	3,270	4,070
75 feet	· · ·	2,690	3,510	4,470
80 feet	· · ·	· · ·	· · ·	4,730

Table 21.6

PREDICTED TOTAL YIELDS FOR SPECIFIED COMBINATIONS OF FERTILIZER APPLIED TO CORN*

P_2O_5	Nitrogen (*Pounds per acre*)				
	0	80	160	240	320
Pounds per acre			*Bushels per acre*		
0	−5.7†	25.8	24.0	16.8	6.8
80	37.1	95.9	105.4	106.9	104.1
160	35.3	105.4	119.6	124.6	124.9
240	26.1	104.8	122.6	130.4	133.0
320	13.1	99.2	120.0	130.1	134.7

* From Heady *et al.*, *op. cit.*, p. 305.

† As published; charted as zero.

Determining a Joint Function for Three or More Independent Variables

There are several different ways by which joint functions involving three or more independent variables may be fitted. In some cases it may be desirable to allow for the joint influence of two variables while simultaneously eliminating or holding constant the net effect of one or more additional independent variables. Thus in the corn-yield problem of Chapter 14 it might be desirable to determine the joint relation of yield to rainfall and temperature, while simultaneously allowing for the upward tendency in yields during the period studied. This might be done by determining the relation according to the general equation

$$X_1 = f_{2,3}(X_2, X_3) + f_4(X_4) \tag{21.4}$$

This relation may be worked out either algebraically or graphically. Algebraically, it would simply mean adding one or more appropriate terms in X_4 [$b_4 X_4$; or $b_4 X_4 + b_4'(X_4^2)$; or $b_4 X_4 + b_4' X_4^3$; etc.] to the equations such as those previously considered. In view of the large number of terms in the regression equation, it would be especially important to eliminate those terms whose coefficients did not show significant values, one by one, and to solve the equations and recalculate the values of the remaining constants with those terms omitted.

If worked out graphically, it may be done by combining the graphic contour method for two independent variables with the method of successive approximations for multiple curvilinear regressions as discussed in Chapters 14 and 16. The steps would be (1) to determine the usual net regression curves for each of the independent variables according to the simpler equation,

$$X_1 = a + f_2(X_2) + f_3(X_3) + f_4(X_4)$$

and then (2) to subclassify the residuals according to the values of X_2 and X_3 and enter them on a chart like Figure 21.4, for the group averages, or for individual observations if the number was not large enough to give satisfactory group averages. If the chart showed indications of the presence of a joint relation, contours would then be fitted to them. The previous residuals would then be adjusted to take account of this joint relation. If the variance had been significantly reduced, the residuals would then be averaged with respect to the remaining independent factor X_4, to see if the net curve for that factor would need to be changed now that the joint relation to the two other factors had been allowed for. This process of successive approximation would be continued until the final shape of the curve for X_4, and the joint surface for X_2, X_3 had been best determined. The final function $f_{2,3}(X_2, X_3)$ would be calculated by adding the values for the earlier curves $f_2(X_2)$ and $f_3(X_3)$ to the values from the new joint function *for the residuals*, for selected values of X_2, X_3.

Contour Fitting for Three Independent Variables with a Small Sample. An example of three independent variables fitted with a joint function to two of them, and with only 16 observations, will illustrate this process in part. This is based on data relating to the yield of potatoes and rainfall in Maine, in July and August. The preliminary multiple regression analysis had given a value for trend in yields (X_4) as shown in the fifth column of Table 21.7.

Table 21.7

WEATHER CONDITIONS AND YIELD OF POTATOES IN MAINE

Year	Rainfall to August 1 (July doubled)	Rainfall August 1 to September 15	Yield	Adjustment for Trend*	Yield Adjusted for Trend	Estimated Yield	Residual
	X_2	X_3	X_1		X_1	$f(X_2, X_3)$	z
	Inches	Inches	Bushels	Bushels	Bushels	Bushels	Bushels
1913	13.17	3.66	220	+26	246	248	−2
1914	11.33	4.08	260	+27	287	260	27
1915	15.96	4.12	179	+31	210	229	−19
1916	15.46	3.77	204	+33	237	236	1
1917	17.77	5.53	125	+31	156	155	1
1918	18.09	3.87	200	+22	222	220	2
1919	12.25	5.41	230	+17	247	248	−1
1920	13.29	7.62	177	+15	192	196	−4
1921	7.82	6.11	298	+13	311	323	−12
1922	16.40	5.12	187	+12	199	197	2
1923	10.61	3.51	258	+9	267	278	−11
1924	9.10	6.13	315	+7	322	308	14
1925	11.30	5.38	250	+5	255	262	−7
1926	9.60	5.60	290	+3	293	297	−4
1927	13.98	6.02	232	+1	233	226	7
1928	15.45	6.45	220	−1	219		

* Simultaneously determined while allowing for trend. See F. V. Waugh, Methods of forecasting New England potato yields, U. S. Department of Agriculture, Bureau of Agricultural Economics, Mimeographed report, February, 1929.

The data are plotted in Figure 21.9, with the yield adjusted for trend used as the dependent factor. Drawing in contours so as to separate years of similar yields, we find that a very peculiar type of surface is indicated—one that changes elevation very rapidly between the combination of high early rainfall and low late rainfall, and high early rainfall and high late rainfall. When these results are used to forecast the yield in 1928 (which year, it will be noted, was not plotted or used in determining the contours) a yield of about 180 bushels is indicated. This is only in fair agreement with the final yield of 219 bushels, determined several months after the climatic data were available to give the forecast stated.

Reading off the estimated values for each year shown, the estimated adjusted yields as shown in the next-to-the-last column of Table 21.7 are obtained. The standard deviation of the residuals, shown in the next column, is 10.6 bushels, whereas the s of the yield adjusted for trend is 63.0. If five constants are assumed to be necessary to represent the surface mathematically, the standard error of estimate would be 13.0 bushels and the index of correlation for the surface indicated by the contours would be 0.98. If it is assumed that the trend line was fairly accurately projected, the standard error of estimate indicates that an error as great as that in 1928 would be likely to occur only very rarely. The fact of high correlation and of low standard error for the period covered could be judged from the closeness with which the contours fit the individual observations on the contour chart, Figure 21.9, in the same way that closeness of observations to the regression line indicates high correlation in the case of simple correlation.

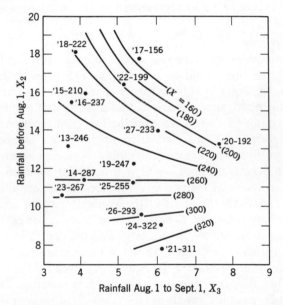

Fig. 21.9. Yield of potatoes for years of specified rainfall before August 1 and after August 1, and contours fitted directly to the data.

In this case, the new residuals after adjusting for the joint function (last column of Table 21.7) do not indicate that any further shift in the net trend line is needed.

Determining Joint Functions in *k* Variables. Given a large enough sample, and sufficiently high correlation, it should be possible to determine joint functions in three or more independent variables. Such a relation would be expressed by the equation

$$X_1 = f_{2,3,4,\ldots k}(X_2, X_3, X_4, \ldots X_k) \tag{21.5}$$

Such a complex relationship would be far too elaborate to represent on any single diagram, though a solid diagram could be drawn for three independent variables by "transparent" drawing (or the use of several colors) which superimposed several different surfaces, for different values of X_4, on a solid diagram such as Figure 21.6.

An approximation to a full complex joint function could be made by using several subordinate joint functions in one regression equation. Thus if corn yields were to be explained by five independent variables, $X_2 = $ rain in July, $X_3 = $ temperature in July, $X_4 = $ rain in August, $X_5 = $ temperature in August, and $X_6 = $ time, the yield might be well explained by a set of relations represented by the equation

$$X_1 = f_{2,3}(X_2, X_3) + f_{4,5}(X_4, X_5) + f_6(X_6) \tag{21.6}$$

Such a relationship, involving two sets of joint functions and one single net regression function, could be fitted by an extension of the methods already discussed, either by solving for an appropriate algebraic equation, or fitting by successive approximations by the method of graphic approximations and contour fitting for each of the two joint functions.

The more complicated case indicated by the more general equation (21.5) would involve even greater difficulties. Where an appropriate type of equation has been deduced to fit a given joint function, it can be expanded to represent as many more variables as the conditions of the case require, and within the limits of the availability of the data to determine the necessary constants. The arithmetic labor becomes increasingly heavy, and unless the sample is very large and the correlation quite high, many of the constants will show values without statistical significance. These can then be dropped, and the equations solved for the terms which are statistically significant. For three independent variables, for example, the square-root equation used in the corn-fertilizer example would become

$$X_1 = a + b_2 X_2 + b_3 X_3 + b_4 X_4 + b_5 \sqrt{X_2} + b_6 \sqrt{X_3} + b_7 \sqrt{X_4}$$
$$+ b_8 \sqrt{X_2 X_3} + b_9 \sqrt{X_2 X_4} + b_{10} \sqrt{X_3 X_4} \tag{21.7}$$

Obviously this ten-constant equation could be fitted effectively only to a very large number of observations. Similar expansions can be made for the various other types of joint functions described. For more than three independent variables, they would become very cumbersome indeed.

To determine the general function (21.5) graphically for even three independent variables would require a very large number of observations, since a threefold classification would be needed. If only 4 groups were used for each variable, 64 subgroups would be possible. Not unless there were sufficient observations so that, say, 3 to 5 might fall in each subgroup, on the average, could such a relation be determined with any degree of accuracy, unless the correlation was very high indeed. If the joint correlation were perfect, one case to a subgroup would be sufficient to indicate the nature of the function. Somewhat similar requirements would apply if the surface were fitted algebraically to obtain significant values of the many constants.

With three independent variables, successive smoothing in three dimensions would be involved. The process might be simplified by dividing the observations into several groups according to one variable, determining the functional relation to the other two independent variables separately for each group, and then smoothing the results for the different groups together to determine the change in joint function with changes in the first variable.

Figure 21.10 illustrates some results of this sort, for a four-dimensional joint function. These results were obtained from an analysis of 190 observations of sales of individual lots of apples. The records were first separated into those for each of the 5 sizes of apples, and the joint functional relation of price to amount of insect injury and amount of scab determined separately for each size. These results were then smoothed between apples of different sizes, to make the "surface" of the imaginary four-dimensional solid diagram show a gradual continuous change over every dimension.[7]

The four-dimensional relationship

$$X_1 = f(X_2, X_3, X_4)$$

may be visualized by a composite diagram,[8] as illustrated in Figure 21.10. This shows that the presence of defects reduced the price of large apples much more than the price of small ones, and that either defect alone reduced the value of apples of any size materially, whereas both defects together reduced the price only slightly more than one alone.

[7] This illustration is from an analysis supplied by Frederick V. Waugh. For a more elaborate study of the same type, see John R. Raeburn, Joint correlation applied to the quality and price of McIntosh apples, Cornell University Agricultural Experiment Station *Memoir* 220, March, 1939.

[8] It is also possible to show the relations on one diagram by superimposing the four surfaces one on top of the other, and drawing all four as if they were transparent sheets. This can be done more elegantly by making a solid model, with the four surfaces represented by plastic sheets, each of a different color.

The only limitation to the number of variables which could be considered jointly is the number of observations available. Where it is possible to determine joint relation, that affords a very satisfactory statement of the relationship, since the real relation is not obscured by assumptions hidden in the regression equation used.

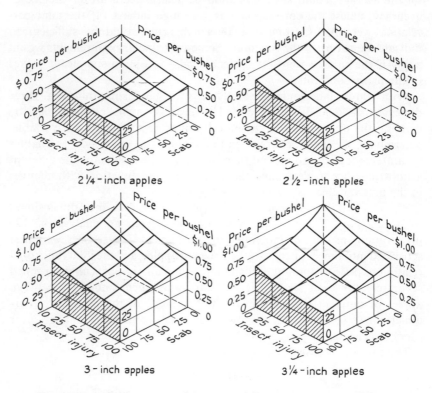

Fig. 21.10. Average price of apples of given sizes, for various combinations of amount of insect injury and amount of scab.

Just as the accuracy of net regression curves is influenced by the distribution of the observations along various portions of the curve and the standard error of estimate, so the reliability of a joint regression surface (such as that shown in Figure 21.10) would depend not only on the standard error of estimate, but also on the number of cases falling within each portion of the area. Where the joint regression surface is determined mathematically, its reliability can be estimated by an extension of the equations presented in Chapters 17 and 19. Methods of estimating the standard errors of a surface determined graphically have not yet been developed.

Summary

This chapter has developed means by which the relation between one variable and two others operating jointly may be determined, either where no other variables are concerned or where one or more additional independent variables are taken into account. Methods are also discussed for measuring the influence of three or more independent variables operating jointly; but the increased number of observations necessary for such determinations restricts the field of usefulness of this type of analysis.

REFERENCES

Ezekiel, Mordecai, The determination of curvilinear regression "surfaces" in the presence of other variables, *Jour. Amer. Stat. Assoc.*, Vol. XXI, pp. 310–320, September, 1926.

Waugh, Frederick V., The use of isotropic lines in determining regression surfaces, *Jour. Amer. Stat. Assoc.*, Vol. XXIV, p. 144, June, 1929.

Court, Andrew T., Measuring joint causation, *Jour. Amer. Stat. Assoc.*, Vol. XXV, No. 171, pp. 245–254, September, 1930.

Bruce, Donald, On possible modifications in the Ezekiel method of curvilinear multiple correlation, Typewritten manuscript, filed in the Library, Agr. Mktg. Serv., U. S. Dept. Agr.

Earl O. Heady, John T. Pesek, and William G. Brown, Crop response surfaces and economic optima in fertilizer use, *Iowa State College Agr. Expt. Sta. Res. Bul.* 424, March, 1955.

CHAPTER 22

Measuring the way a dependent variable changes with changes in a qualitative independent variable

It is sometimes necessary to determine the change in one variable associated with changes in a qualitative independent factor; i.e., one which varies in ways that cannot be measured quantitatively. Thus if one is studying the effect of various factors affecting the values of individual farms, one might wish to include the kind of road on which the farm was located. Yet different kinds of roads, such as arterial highway, all-weather, gravel, or dirt, cannot be stated in the same way that the measurements for continuously variable factors can.

Measuring Simple Correlation with a Qualitative Variable. Where a single qualitative factor is to be considered as an independent variable, and a quantitative factor as the dependent, the regression relation may be determined by grouping the observations according to the category of the qualitative factor, and calculating the average value of the dependent factor for each group.

The intensity of correlation between the quantitative dependent factor and a qualitative independent factor is measured by the *correlation ratio*, which corresponds to the correlation coefficient or index in that it measures the proportion of the variation in the dependent factor explained by its association with the independent factor. Using n_0 to represent the number of cases in each successive group according to the independent factor, and M_0 to represent the mean of the dependent variable Y in each such group, the correlation ratio η_{yx} is defined

$$\eta_{yx} = \sqrt{\frac{\Sigma[n_0(M_0^2)] - n(M_y)^2}{ns_y^2}} \tag{22.1}$$

The calculation of η_{yx} may be illustrated using some of the data given in Table 22.1. X_5 in this table shows the way eggs are packaged for sale, in 3 types; and X_1 is the price of eggs per dozen. The first step is to average the prices for each type of sale, with the following results:

Method of Sale, X_5	Number of Cases	Average Price for That Method, M_{1_0} *Cents per dozen*
A—Sold without carton	17	47.94
B—Sold in carton, but unbranded	33	52.91
C—Sold in carton with brand name	24	52.08

Apparently, eggs in cartons sold on the average for more than eggs not in cartons, but in this case adding a brand name to the carton did not appear to make much difference in price, and if anything, brought a lower price.

The calculations are then performed as follows:

X_5 Group	Number of Cases	$\Sigma_0 X_1$	M_{1_0}	$(\Sigma_0 X_1)(M_{1_0})$ $= (M_{1_0})^2 n_0$
A	17	815	47.941	39,071.92
B	33	1,746	52.909	92,379.11
C	24	1,250	52.083	65,103.75
Summations	74	3,811	51.500	196,554.78

To calculate η_{yx} we also need the value ns_1^2, which is calculated separately to be 3,228.50.

Equation (22.1) then becomes

$$\eta_{yx} = \sqrt{\frac{\Sigma n_0(M_0^2) - n(M_y)^2}{ns_y^2}} = \sqrt{\frac{196,554.78 - (51.500)(3,811)}{3,228.50}}$$

$$= \sqrt{\frac{288.28}{3,228.50}} = \sqrt{0.0893} \quad \text{so} \quad \eta_{yx} = 0.30$$

The correlation ratio, 0.30, indicates that there is some relation between the kind of package and the price, but a rather low one, with the relation explaining only 9 per cent at best of the variance in egg prices from store to store. (If adjustment were made for n and m by the usual formula, with $m =$ the number of groups used, the relation would be still lower.)

Measuring Multiple Correlation with One or More Qualitative Independent Variables, and Other Quantitative Variables. In the egg case, other variables such as quality, weight, and color of the eggs may also influence the price, and more or less obscure the true relation to the method of packaging. Such relationships may be explored by an extension of the method of multiple correlation. This involves first solving for the relations to the quantitative variables; then calculating the unexplained residuals, studying their relation to the qualitative variable or variables, and then readjusting the relations found with the other independent variables after allowing for this relation. While this could be done either by using algebraic equations or by successive graphic approximations, the method will be illustrated here, using the method of successive approximations for both qualitative and quantitative variables.

The data given in Table 22.1 are from a study of the relation of various quality factors to the retail price of eggs.[1] The factors shown in the table are X_2, an index of the interior quality of the eggs; X_3, the weight of each dozen in ounces; X_4, the number of white eggs in each dozen; X_5, the type of carton the eggs were sold in; and X_1, the price per dozen, in cents. Net curvilinear regressions have been determined for the three quantitative factors by the successive approximation method, and estimated prices have been worked out by the regression equation

$$X_1' = a' + f_2(X_2) + f_3(X_3) + f_4(X_4)$$

The residuals, z''', obtained by subtracting these estimated prices from the observed prices, X_1, are also shown in the table.

Determining the Net Influence of the New Variable. The first step in determining the net regression of X_1 on X_5 is to group the residuals from the previous curves, z''', according to the new factor X_5, and determine the average for each group. This gives results as follows:

Value of X_5	Average of z''
A—no carton	−2.5
B—carton	+0.9
C—carton and brand name	+0.6

These results show that, after making allowances for the size, color, and quality, eggs in unmarked cartons sold 3.4 cents above those sold in bulk, on the average, but those with branded cartons sold only 3.1 cents above eggs in bulk. (This contrasts with an average difference of practically 5 cents a dozen in the simple averages, before the effects of the

---

[1] Charles B. Howe, Some local market price characteristics which affect New Jersey egg producers; factors influencing the retail price of eggs. N. J. Agr. Expt. Sta. Bul. 150 1927.

Table 22.1

DATA FOR EGG PROBLEM, WITH A NON-QUANTITATIVE INDEPENDENT VARIABLE

Independent Variables				Dependent Variable,	z'''	$f(X_5)$	z'''
X_2	X_3	X_4	X_5*	X_1			
21	23	4	C	35	−7.3	+0.6	−7.9
35	24	12	C	45	−8.4	+0.6	−9.0
26	23	12	B	55	3.4	+0.9	2.5
27	24	12	B	55	3.3	+0.9	2.4
31	22	12	A	50	−1.8	−2.5	0.7
35	24	12	C	44	−9.4	+0.6	−10.0
28	23	12	C	60	8.2	+0.6	7.6
41	23	12	B	50	−4.8	+0.9	−5.7
28	26	2	C	45	−1.6	+0.6	−2.2
24	23	11	B	52	4.6	+0.9	3.7
28	20	12	C	45	−5.5	+0.6	−6.1
49	24	12	C	55	−3.6	+0.6	−4.2
30	24	12	C	55	2.4	+0.6	1.8
48	23	12	B	60	1.9	+0.9	1.0
19	22	9	C	45	1.8	+0.6	1.2
22	23	3	A	45	1.7	−2.5	4.2
33	25	12	C	60	6.6	+0.6	6.0
26	24	12	C	59	6.9	+0.6	6.3
35	23	12	B	55	2.1	+0.9	1.2
20	23	12	B	50	−0.9	+0.9	−1.8
25	25	12	B	55	2.6	+0.9	1.7
46	24	12	B	60	2.5	+0.9	1.6
30	26	1	B	45	−3.2	+0.9	−4.1
24	24	12	B	55	3.1	+0.9	2.2
48	23	12	B	60	1.9	+0.9	1.0
17	22	12	C	55	4.8	+0.6	4.2
18	22	12	A	45	−5.3	−2.5	−2.8
41	24	12	C	55	−0.3	+0.6	−0.9
30	25	12	C	67	14.0	+0.6	13.4
19	24	2	B	53	8.3	+0.9	7.4
47	24	0	B	55	0.9	+0.9	0.0
32	24	12	B	55	2.2	+0.9	1.3
26	24	12	B	49	−3.1	+0.9	−4.0
38	24	12	A	42	−12.2	−2.5	−9.7
29	23	12	B	42	−9.9	+0.9	−10.8
24	24	0	A	45	−2.9	−2.5	−0.4
37	25	12	A	40	−14.3	−2.5	−11.8

* A designates "sold without carton," B "sold in carton but unbranded," and C "sold in carton with brand name."

Table 22.1 (Continued)

Independent Variables				Dependent Variable, X_1	z'''	$f(X_5)$	z'''
X_2	X_3	X_4	X_5^*				
36	23	12	A	48	−5.1	−2.5	−2.6
10	23	0	B	47	1.2	+0.9	0.3
35	24	12	C	59	5.6	+0.6	5.0
22	22	12	B	52	1.2	+0.9	0.3
29	21	12	B	55	4.0	+0.9	3.1
16	23	0	B	40	−6.5	+0.9	−7.4
6	22	3	B	40	−1.0	+0.9	−1.9
31	23	12	B	55	2.8	+0.9	1.9
26	23	12	B	55	3.4	+0.9	2.5
36	21	12	B	60	7.8	+0.9	6.9
39	22	12	B	55	1.4	+0.9	0.5
42	23	12	B	60	4.8	+0.9	3.9
36	24	12	C	60	6.4	+0.6	5.8
47	22	12	B	60	2.8	+0.9	1.9
27	24	12	C	55	2.8	+0.6	2.2
31	22	12	A	50	−1.8	−2.5	0.7
26	22	11	A	40	−7.2	−2.5	−4.7
45	23	12	A	60	3.5	−2.5	6.0
18	25	12	C	45	−6.6	+0.6	−7.2
35	24	12	C	50	−3.4	+0.6	−4.0
21	23	12	C	55	4.0	+0.6	3.4
44	23	12	A	60	3.9	−2.5	6.4
48	24	12	A	55	−3.6	−2.5	−1.1
33	24	12	A	55	2.0	−2.5	4.5
47	24	12	C	55	−3.1	+0.6	−3.7
16	22	5	A	45	3.9	−2.5	6.4
32	25	0	B	50	0.8	+0.9	−0.1
45	25	12	B	55	−2.4	+0.9	−3.3
46	23	12	B	57	0.0	+0.9	−0.9
32	24	12	C	55	2.2	+0.6	1.6
16	23	1	C	41	−4.2	+0.6	−4.8
30	25	1	C	50	2.3	+0.6	1.7
24	22	0	A	42	−5.0	−2.5	−2.5
44	24	11	B	50	−2.6	+0.9	−3.5
25	22	12	B	49	−2.1	+0.9	−3.0
16	23	0	A	45	−1.5	−2.5	1.0
31	24	8	A	48	3.2	−2.5	5.7

* A designates "sold without carton," B "sold in carton but unbranded," and C "sold in carton with brand name."

other variables were allowed for.) These results cannot be accepted as the final effect of package on price without first raising the question whether the curves previously determined to show the influence of the other factors might be changed somewhat were the type of package taken into account. Whether this will be true or not depends upon whether there is any correlation between the new factor and the factors previously considered, or whether they are quite independent of each other. This can be determined by sorting the other factors according to the values of X_5, and determining their averages for each group. The results are:

Value of X_5	Averages of Other Independent Variables			Number of Cases
	X_2	X_3	X_4	
A —no carton	30.6	23.1	8.6	17
B—carton	31.6	23.2	9.6	33
C—carton and brand	29.9	23.8	10.2	24

There does seem to be some correlation between X_5 and the other variables. Apparently the eggs sold in unmarked cartons are, on the average, of the best quality and of medium size; the eggs sold in cartons under brand names are of larger size, but are not of such high quality, on the average; whereas those sold in bulk average medium in quality but low in size.[2] Accordingly, the curves previously determined for the change in price with differences in size and in quality may have included some portion of the effect really associated with cartons instead. Now that at least an approximate measure has been obtained of the influence of carton on price, the previous curves may be modified by taking this factor also into account.

Taking Account of the Non-Quantitative Variable in Estimating X_1 and z. In Table 22.1 in the column headed $f(X_5)$ the approximate influence of differences in carton on price are entered, the averages found in the tabulation on page 380 being used. Since these values would be added to the previous estimated values of X_1 to obtain the new estimates, they are subtracted from the previous residuals (z''') to obtain the revised residuals. The last column shows these new values, z''''. Before using these new values to see if any changes are necessary in the other regression curves we may first determine how much the standard error of estimate has been reduced by taking X_5 into account. This could be determined directly by computing the standard deviation of the new z'''' values; but a much shorter method is available, using the same principle employed

[2] The exact correlation between X_5 and X_2, X_3, and X_4, respectively, can be computed by use of the correlation ratio.

in calculating the correlation ratio. By the use of this method, the $s_{z'''}$ may be computed from the $s_{z''}$ by the formula

$$s_{z'''}^2 = \frac{ns_{z''}^2 - [\Sigma(n_0M_0^2) - n(M_{z'''})^2]}{n} \tag{22.2}$$

The necessary computations are:

X_5	$M_{z''}$ $(= M_0)$	Number of Cases $(= n_0)$	n_0M_0	$n_0(M_0)^2$
A	−2.5	17	−42.5	106.25
B	0.9	33	29.7	26.73
C	0.6	24	14.4	8.64
Sums			1.6	141.62

$$M_{z''} = \frac{1.6}{74} = 0.0216$$

So

$$s_z^2 = \frac{74(5.06)^2 - (141.62 - 0.04)}{74} = 23.69$$

$$s_{z'''} = 4.87$$

Computing the standard error for estimates based on X_5 and the other variables, we must recognize that the value of m has been increased by three by the introduction of the new factor; so, whereas m was assumed to equal 8 previously, it now equals 11. Adjusting the values of 5.06 for $s_{z''}$ and 4.87 for $s_{z'''}$ by equation (15.6), we find $\bar{S}_{1.f(2,3,4)} = 5.36$, and $\bar{S}_{1.f(2,3,4,5)} = 5.27$. Apparently the introduction of X_5 as a factor has had as yet but slight effect on the accuracy with which egg prices might be estimated.

Making Further Successive Approximation Corrections. It is still possible that the regressions for the other factors might be modified now that X_5 has been approximately allowed for. Consequently the values of z'''' are classified according to the values of X_2, X_3, and X_4, and the averages computed for each group. The averages in Table 22.2 suggest that the curve for $f_2(X_2)$ might be modified slightly, so as to rise more steeply in the portion up to $X_2 = 36$ and less steeply thereafter. Table 22.3 does not indicate any consistent relation between X_3 and z'''', so no further change in $f_3(X_3)$ is indicated. Table 22.4 indicates that the curve for $f_4(X_4)$ might also be altered slightly, so as to have a somewhat steeper slope.

If $f_2(X_2)$ and $f_4(X_4)$ were modified as suggested, a new estimated value

Table 22.2

AVERAGE VALUES OF z''' FOR CORRESPONDING X_2 VALUES

X_2 Values	Number of Cases	Average of X_2	Average of z'''
0–14	2	8.0	−0.9
15–19	9	17.2	−0.2
20–29	23	25.1	+0.1
30–39	24	33.5	+0.3
40–49	16	45.5	−0.1

Table 22.3

AVERAGE VALUES OF z''' FOR CORRESPONDING X_3 VALUES

X_3 Values	Number of Cases	Average of z'''
20	1	−6.1
21	2	5.0
22	13	0.1
23	23	0.2
24	25	−0.1
25	8	0
26	2	−3.2

Table 22.4

AVERAGE VALUES OF z''' FOR CORRESPONDING X_4 VALUES

X_4 Values	Number of Cases	Average of X_4	Average of z'''
0	7	0	−1.3
1– 2	5	1.4	−0.4
3– 5	4	3.8	+0.2
8–11	5	10.0	+0.5
12	53	12.0	+0.2

of X_1 might then be worked out, using these new curves and the previous curve for $f_3(X_3)$, and using the values for $f_5(X_5)$ already entered in Table 22.1. The new z's based on these new estimates might then be classified with respect to X_5, to determine if any change need be made in the values for $f_5(X_5)$ worked out on page 380. If any material change were found necessary in X_5, the residuals might be corrected accordingly, and then averaged with respect to X_2, X_3, and X_4, to see if any further changes would be needed in their values. This process of successive approximation should be continued until no further significant change was indicated in any of the curves, or until the $\bar{S}_{1.f(2,3,4,5)}$ showed no further reduction.

It does not seem worth while, in this problem, to carry out the additional steps just outlined. In a problem where the non-quantitative factor is an important one, however, and where it is significantly correlated with the other independent variables, the determination of the net function for that factor should be carried through a sufficient number of approximations to measure the final net effect of each factor as accurately as possible.

Taking the preliminary results shown in Table 22.2 as the final measure of the influence of type of container on price, it appears that eggs sold in an unmarked carton brought, on the average, 3.4 cents more per dozen than eggs of the same quality, size, and color sold in bulk, and 0.3 cent more than eggs sold in a carton with a brand name. (This last result might reflect the experience of consumers with branded eggs of poor quality as indicated in the tabulation on page 383, which might tend to make them sell at a discount even when they were of equal quality.) The improvement in closeness of fit may be measured by the slight reduction in the adjusted standard error of estimate, or by the increase in the index of multiple correlation. Computing the adjusted indexes of multiple correlation corresponding to the standard errors of estimate before and after the type of carton is allowed for, by equation (17.9), $\bar{I}_{1.234} = 0.59$; and $\bar{I}_{1.2345} = 0.62$. The corresponding indexes of determination, 35 and 38 per cent, indicate that taking into consideration the differences in the carton has increased the proportion of egg prices which can be explained by 3 per cent of the original variance, after due allowance is made for the additional constants introduced.

When the stricter criterion of "significant improvement" based on variance analysis is applied (see Chapter 21, pages 366 to 368, and Chapter 23, pages 403 to 411), however, it appears that this increase could have been due to the chance characteristics of this particular sample. The same percentage increase in explained variance based on 200 observations rather than 74 would have been significant in the sense that it would be expected to occur by chance in only 1 sample out of 28 of the larger size.

The first approximation to the regression on non-quantitative factors can be made directly from the residuals from the linear multiple regression equation, instead of waiting until after approximate regression curves are determined for the other factors. In case a non-quantitative factor is a very important one, it may be roughly included in the net linear regressions by designating successive groups by a numerical code which approximates the expected influence of the variable. Then if the true influence is of a different order from the expected influence, that fact will show up when the first approximation curves are worked out. (For the non-quantitative factor the averages of residuals must be interpreted as discrete points for each class, however, rather than as a continuous function.) Thus it might have been tentatively assumed that eggs in branded cartons would sell above eggs in unbranded cartons, and both would sell well above eggs in bulk. The bulk eggs could then have been designated by 1; the unbranded cartons by 3; and branded cartons by 4. The net linear regression would have been positive; but the residuals would have revealed that eggs in branded cartons really averaged lower in price (other factors equal) than in unbranded cartons.

This technique is also useful in time-series analysis to determine the net seasonal movement (as in a price series), while simultaneously allowing for the influence of other variables. For monthly data, this would involve 12 groups, one for each month. Smoothing the values through the year in a continuous curve would reduce somewhat the number of degrees of freedom used up.

Where there are two or more qualitative factors involved, the methods presented here may be extended to deal with all of them simultaneously by successive approximations.

Summary

Where an independent factor is not a continuous variable, but may be classified into two or more groups, the regression of a dependent factor on it may be determined with respect to each group, and measures of regression and simple correlation can be calculated. Where other independent factors are also involved, net relations with the qualitative factor or factors may also be determined, while holding other factors constant by the usual multiple correlation process. Standard errors and indexes of multiple correlation may be worked out to include the effects of non-quantitative independent factors as well as for continuously variable factors.

CHAPTER 23

Cross-classification and the analysis of variance

Introduction. In earlier editions, *cross-classification and averaging* was presented as a method of analysis that stopped short of formal multiple regression but embodied the basic idea of net regression lines or curves. In the present edition the concept of net regression has already been presented on a mathematical basis; the "drift lines" of the short-cut graphic method have also been presented as approximations to the final net regression curves. When large numbers of observations are available, the idea of studying the relationship between X_1 and X_2 *within* each of a number of subclasses of X_3 is intuitively obvious. Most data of the census type are presented in the form of cross-tabulations, so that frequencies and average levels of a "dependent" variable can be obtained for different combinations of values (usually classes or ranges) of two or three "independent" variables. Average incomes of workers cross-classified by age and years of schooling would be an example. These are all quantitative factors. In addition there may be non-quantitative subdivisions—the first three variables may be available for male and female workers separately and for each state or region.

Sometimes simple methods are overlooked, so the first section of this chapter will present an example of analysis by means of cross-classification and averaging. The remainder will present some basic principles of the analysis of variance and discuss its relationship to regression analysis. While analysis of variance has been developed primarily in connection with agricultural and biological experiments, its use has spread over the whole range of experimental sciences including, in recent years, some applications to the social sciences. For example, the formulas used in Chapter 22 to estimate the effects of a qualitative independent variable are based upon variance analysis concepts.

Analysis by Cross-Classification and Averaging

Analysis by averages where there are two independent variables involves classifying the records first by one variable, then breaking each of the resulting groups into several smaller groups according to the values of the second variable. If a third independent variable were to be considered, these groups would be broken up into still smaller groups, according to

Table 23.1

CROSS-CLASSIFICATION OF REPORTS ACCORDING TO SIZE OF FARM AND SIZE OF DAIRY HERD

Size of Farm	Size of Dairy Herd								
	Under 6 Cows			6 to 11 Cows			12 Cows and Over		
	Acres	Cows	Income	Acres	Cows	Income	Acres	Cows	Income
	Number	Number	Dollars	Number	Number	Dollars	Number	Number	Dollars
50–99 acres	...	...	...	80	6	610	60	18	960
	...	...	...	...	...	...	70	17	1,020
	...	...	...	...	...	...	90	12	800
	...	...	...	...	...	...	80	15	800
Total	...	...	...	...	...	...	300	62	3,580
Average	...	...	...	80	6	610	75	15.5	895
100–149 acres	120	1	590	100	9	900	110	12	880
	...	...	...	110	6	740	120	15	1,080
	...	...	...	...	...	...	110	16	1,130
Total	...	...	...	210	15	1,640	340	43	3,090
Average	120	1	590	105	7.5	820	113	14.3	1,030
150–199 acres	160	0	700	170	6	820	180	14	1,260
	...	...	...	160	7	860	160	12	980
Total	...	...	...	330	13	1,680	340	26	2,240
Average	160	0	700	165	6.5	840	170	13	1,120
200 acres and over	220	0	830	240	7	960			
	230	2	760						
	220	2	760						
Total	670	4	2,350						
Average	223	1.3	783	240	7	960			

the values of the third variable. Then the values of the dependent variable, as well as each of the independent variables, would be averaged for each subgroup. This process is known as subclassification or cross-classification.

Cross-Classification for Three Variables. In the problem presented in Chapter 10 there were two independent variables—number of cows and number of acres. The records would therefore need to be classified into groups both according to the number of cows and the number of acres

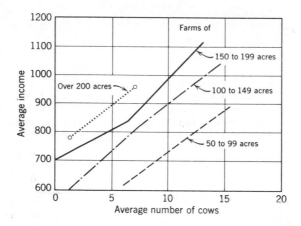

Fig. 23.1. Difference in average income with difference in number of cows, for farms grouped by size of farm.

on each farm. Since there is such a small number of records the groups should not be made too small. Let us take three groups for cows, and four groups for the size of farm, or twelve possible groups in all. The records are classified into twelve groups, and totals and averages computed for each, as in Table 23.1.

None of these groups has a sufficient number of farms represented to make the averages particularly significant; yet even so a certain regularity in the averages can be observed. In each column the average income increases as the size of farm increases, though there is but little difference in the average number of cows from group to group; similarly across each line of averages the income increases as the number of cows increases, though there is but little difference in the average size of farm from group to group. The average incomes from Table 23.1 are charted in Figures 23.1 and 23.2, first for differences in the number of cows with farms of similar sizes, and then for differences in the number of acres, with farms of similar numbers of cows.

Both figures show the tendency for income to increase with an increase in the independent variable, when the effect of the other variable is held fairly constant by the grouping process. In Figure 23.1 the lines show about the same general slope for each of the four groups, though there are some irregularities. Figure 23.2 similarly shows about the same general change in income with a given change in the size of the farm, no matter what is the number of cows; but here the irregularities from group to group are more striking.

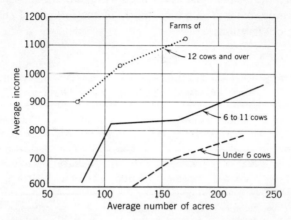

Fig. 23.2. Difference in average income with difference in number of acres, for farms grouped by numbers of cows.

Individual differences as large as those shown might be due simply to random differences in sampling and therefore have no real meaning. But when a series of small samples shows consistent trends, as these do, they are more meaningful than each taken separately.

Although the subgroup averages show the general effect of changes in one variable, such as cows, upon income, with the effect of the other variable, acres, removed, they cannot be considered to show the specific effect of specific differences. For example, much more evidence would be needed to prove that, between 75 and 100 acres, a change of 1 acre has much greater effect upon income on farms with 6 to 11 cows than on farms with 12 cows or more, even though the lines in Figure 23.2 appear to indicate this. All that is really proved is that on farms of both numbers of cows there is a tendency for income to increase with an increase in the number of acres.

The averages in Table 23.1 may be summarized for publication in a form similar to Table 23.2. The number of cases represented in each average is included to prevent the reader from placing an undue amount of confidence in an average based on a small number of observations. In addition, the reader should be given some indication of the approximate standard error of each average.

The very small number of cases included in each of the groups is strikingly brought out in Table 23.2. Even if there were five times as many farms to deal with—100 in all—if they were distributed in the same manner, the largest group would have only 20 cases, and all the rest would have 15 or less, which, under ordinary conditions, would be hardly enough for really significant averages. This subsorting technique is most useful with census-type material in which many of the individual cells contain 100 or more observations.

Table 23.2

DIFFERENCE IN AVERAGE INCOME FOR FARMS OF DIFFERENT SIZES AND WITH DIFFERENT SIZES OF DAIRY HERD

Size of Farm	Under 6 Cows in Herd		6 to 11 Cows in Herd		12 Cows or Over in Herd	
	Size of Group	Average Income	Size of Group	Average Income	Size of Group	Average Income
	Number of farms	*Dollars*	*Number of farms*	*Dollars*	*Number of farms*	*Dollars*
50 to 99 acres	...	...	1	610	4	895
100 to 149 acres	1	590	2	820	3	1,030
150 to 199 acres	1	700	2	840	2	1,120
200 to 249 acres	3	783	1	960	...	...

Average Differences between Matched Subgroups. After the observations have been grouped and averaged as shown in Table 23.2, average differences in the dependent variable (as here, dollars of income), with given differences in each independent variable, can be roughly determined while holding constant the other independent variable or variables. This involves determining the average differences between the averages for the dependent variable for matched groups. The computations are shown in Tables 23.3 and 23.4.

From these results it appears that increasing the number of cows from under 6 to between 6 and 11, without changing the size of farm, was accompanied by an average increase of $182. Increasing the cows further to over 12 cows was accompanied by a further increase of income of $258. Similarly, increasing the size of farm from under 99 acres to 100–149 acres, without changing the number of cows, was accompanied

Table 23.3

CHANGE IN AVERAGE INCOME BETWEEN GROUPS MATCHED FOR SIZE OF FARM

Size of Farm	A Under 6 Cows	B 6 to 11 Cows	C Increase $(B - A)$	D Over 12 Cows	E Increase $(D - B)$
Acres	*Dollars*	*Dollars*	*Dollars*	*Dollars*	*Dollars*
50– 99	· · ·	610	· · ·	895	285
100–149	590	820	230	1,030	210
150–199	700	840	140	1,120	280
200–249	783	960	177	· · ·	· · ·
Average change with cows	· · ·	· · ·	182	· · ·	258

Table 23.4

CHANGE IN AVERAGE INCOME BETWEEN GROUPS MATCHED FOR NUMBER OF COWS

Number of Cows	A 50–99 Acres	B 100–149 Acres	C Increase $(B - A)$	D 150–199 Acres	E Increase $(D - B)$	F 200–249 Acres	G Increase $(F - D)$
	Dollars	*Dollars*	*Dollars*	*Dollars*	*Dollars*	*Dollars*	*Dollars*
Under 6	· · ·	590	· · ·	700	110	783	83
6 to 11	610	820	210	840	20	960	120
12 or over	895	1,030	135	1,120	90	· · ·	· · ·
Average change with acres	· · ·	· · ·	173	· · ·	73	· · ·	102

by an increase of $173 in income. A further increase to 150–199 acres was accompanied by a further average increase of $73 in income, and to 200–249 acres, by $102 more income. (In this discussion "increase" in size or cows has been used to designate differences between results for farms of different sizes or with different numbers of cows.) These rough measurements of differences in the dependent variable with differences in one independent variable, while holding a second independent constant by subsorting, may be compared with results obtained by the more exact methods set forth in earlier chapters.[1]

---

[1] In computing Tables 23.3 and 23.4, no attention was paid to weighting the results according to the number of cases falling in each group, or to the sampling reliability of each average. For a discussion of the first of these points see page 395.

This same method may be applied to get the average difference between matched subgroups, where two or more other independent variables are held constant by the grouping.

Limitation of Cross-Classification for Many Variables. This small example illustrates one fundamental difficulty with the method of sub-classification and averaging—the large number of cases required for conclusive results. Though there are only two independent variables involved, and the records are classified into only three groups one way and four the other, apparently 100 cases or more would be required for really significant results. If it had been desired to subclassify the records according to two more additional variables—say number of men employed and number of hogs kept—that would have greatly increased the number of records necessary. If each of the groups already shown had been further subdivided into 3 groups for men and 4 groups for hogs, that would have increased the number of possible groups to 108. Where over 100 records would have been needed in the first case to give results at all reliable, probably 500 or more would be needed with this further classification.

The method of subclassification and averaging has further short-comings; it provides no measure of how *important* the relation shown is as a cause of variation in the factor being studied, or of how closely that factor may be estimated from the others on the basis of the relations shown. The method of subclassification and averaging thus does not determine the relationships where many variables are involved so satis-factorily as does multiple regression.

Fitting Regressions to Group Means. Sometimes, particularly with census data or with sample surveys based on a thousand or more cases, we may have access to the cell averages and the number of observations in each cell but not to the original observations or to estimates of standard deviations or variances within cells. If the dependent variable and one or more of the classifying variables are quantitative, an obvious way of summarizing the data is to fit simple or multiple regressions to the cell means. (We use actual cell means of the classifying variables if they are available, otherwise midpoints of class intervals.)

If the cell frequencies are fairly large, the cell means will show much less variability than would the individual observations. Correlation coefficients between two series of cell means will sometimes be very high. Also, we can calculate standard errors of regression coefficients, treating each mean as a single observation. But the sampling significance of these measures is by no means clear *unless* we have some idea as to the variance of the original observations.

The effects of grouping can be illustrated roughly as follows: Suppose we have a regression based on 900 original observations, and find that $S_{y \cdot x} = 100$ and $r^2 = 0.20$. The variance per observation about the line would be $S_{y \cdot x}^2 = 10,000$ and the explained variance 2,500. Now, suppose the observations are grouped into 9 class intervals of X with 100 observations in each interval. The "standard error of estimate" about a regression line fitted to the 9 group means will now be on the order of 10 rather than 100, while the standard deviations of X and of Y' will be about the same as before. Thus, the unexplained variance (per group mean) is about 100; the explained variance remains about 2,500, and r^2 between the group means is about 0.96. It is important to note that 0.96 is not a measure of the "success" of the analysis in the same sense as a similar value obtained in relating original observations. If one has any knowledge at all concerning the order of magnitude of the standard deviations of original observations in a particular study or even in other studies of the same variables (such as family income and consumption of particular foods), one can convert the standard error of a regression coefficient based on group means to a rough estimate of the standard error that might have been obtained from ungrouped data. One or two such calculations would provide a rough basis for appraising the reliability of other regression coefficients derived from the same survey or census.

Certain other refinements may be needed to obtain results comparable to those from ungrouped data. If the variability of the dependent variable within each cell is assumed constant but the numbers of observations in different cells vary a great deal, the situation will be improved if the squares and cross-products involving each cell mean are weighted by the number of observations in the cell. Further, if we knew that the variance of the dependent variable differed from cell to cell according to some function of the dependent variable, we might (1) attach less weight to the classes having greater than average variability; or (2) transform the original means into logarithms or other functions that should make the residual variance roughly the same about each section of the regression line.

The Analysis of Variance

In earlier chapters we have distinguished between the "correlation model" and the "regression model" with particular reference to the different interpretations of the correlation coefficient in the two cases. We have noted that regression coefficients and standard errors of estimate, however, have the same interpretation in both models.

The correlation model was originally applied to non-experimental data—random samples from a universe in which all variables were normally distributed. The regression model grew out of applications in which the values of the independent variables could be selected as part of an experiment. Particularly in connection with agricultural and biological experiments, there developed an extensive theory of experimental design. Part of this theory related to situations in which the independent factors were qualitatively rather than quantitatively different. For example, does one variety of wheat have a significantly higher yield than another? Is one insecticide significantly more effective than another? Regression analysis is not adapted to such cases. However, out of the problem of comparing the effects of qualitatively different "independent" factors grew a very powerful tool called "analysis of variance." Moreover, the theory underlying analysis of variance is sufficiently general to include regression analysis as a special case. In this section we shall consider only certain aspects of variance analysis that are closely related to regression problems.

Basic Principles of Variance Analysis. The sum of squares of a set of observations about their mean can be represented as the sum of two independent sums of squares—specifically, in simple linear regression analysis, a sum of squares of deviations of regression values from their mean and a sum of squares of deviations about the regression line. We have also noted that the first sum of squares has a single degree of freedom, and the second has $n - 2$ degrees of freedom, n being the total number of observations to which the regression line is fitted.

In discussing tests of significance, we introduced the t-ratio, or ratio of a coefficient to its standard error. For the simple regression coefficient, this ratio can be written as follows, (from equations (5.5) and (17.1)):

$$t_b = \frac{b}{s_b} = \frac{\Sigma xy}{\Sigma x^2} \div \frac{\bar{S}_{y \cdot x}}{s_x \sqrt{n}}$$

$$t_b = \frac{\Sigma xy}{\Sigma x^2} \div \frac{\sqrt{\dfrac{\Sigma z^2}{n-2}}}{\sqrt{\dfrac{\Sigma x^2 \cdot n}{n}}} = \frac{\Sigma xy}{\Sigma x^2} \cdot \frac{\sqrt{\Sigma x^2}}{\sqrt{\dfrac{\Sigma z^2}{n-2}}}$$

$$= \frac{\Sigma xy}{\bar{S}_{y \cdot x} \sqrt{\Sigma x^2}} \tag{23.1}$$

In this form, the t-ratio is used to estimate the probability that an observed value of b might have been obtained by chance in random sampling from a population in which the true regression coefficient was zero.

A closely related measure is basic to the analysis of variance. "A variance" is equal to a sum of squares divided by the appropriate number of degrees of freedom. In simple regression, the sum Σy^2 can be divided into two parts or components, $b^2\Sigma x^2$ and Σz^2—that is,

$$\Sigma y^2 = b^2\Sigma x^2 + \Sigma z^2$$

The first component has a single degree of freedom and the second has $n-2$ degrees of freedom, so the corresponding variances are $b^2\Sigma x^2$ and $\Sigma z^2/n-2$, or $S_{y\cdot x}^2$. It is intuitively plausible that a measure of the significance of the regression relationship between y and x can be based upon the ratio of these two variances,

$$F = \frac{b^2\Sigma x^2}{S_{y\cdot x}^2}$$

Substituting $b = \dfrac{\Sigma xy}{\Sigma x^2}$ in this expression we obtain

$$F = \frac{(\Sigma xy)^2}{(\Sigma x^2)^2} \cdot \frac{\Sigma x^2}{S_{y\cdot x}^2} = \frac{(\Sigma xy)^2}{S_{y\cdot x}^2(\Sigma x^2)} = t^2 \qquad (23.2)$$

It will be noted at once that F is equal to t^2, or $t = \sqrt{F}$.

The equality $t = \sqrt{F}$ holds only when the total sum of squares is divided into two, and only two, parts. This occurs in simple linear regression analysis and in testing the significance of the difference between two means, two standard deviations, and so on. But the F-ratio is also applicable to situations in which a total sum of squares is divided into three or more components. The question asked in each case is as follows: Could the variances under comparison have been obtained by random sampling from the same population? This question is answered, of course, on the basis of probability rather than certainty. Values of F covering a wide range of paired values of degrees of freedom have been tabulated by Snedecor and are reproduced here with the kind permission of Dr. Snedecor and the Iowa State College Press.

Applications of Variance Analysis: (1) *Difference between the Effects of Two "Treatments."* The following data are taken from the study used earlier by Heady, Pesek, and Brown:[2] Twenty-eight plots of ground are selected at random in a field and are planted to corn of a single variety. Eighteen of the plots receive no fertilizer, and 10 receive nitrogen fertilizer at the rate of 40 pounds N per acre. The yield of corn obtained on each plot is carefully measured. Our problem is to discover whether the application of fertilizer at the specified rate has a significant effect upon corn yields.

[2] Earl O. Heady, John T. Pesek, and William G. Brown, Crop response surfaces and economic optima in fertilizer use, *Iowa State College Agricultural Experiment Station Research Bulletin* 424, March 1955, p. 330.

Table 23.5

5 Per Cent and 1 Per Cent Points for the F-Distribution*
5 per cent in light face, 1 per cent in bold face

n_2 (degrees of freedom for denominator)	n_1 (degrees of freedom for numerator)											
	1	2	3	4	5	6	8	10	12	20	50	100
1	161	200	216	225	230	234	239	242	244	248	252	253
	4,052	**4,999**	**5,403**	**5,625**	**5,764**	**5,859**	**5,981**	**6,056**	**6,106**	**6,208**	**6,302**	**6,334**
2	18.51	19.00	19.16	19.25	19.30	19.33	19.37	19.39	19.41	19.44	19.47	19.49
	98.49	**99.01**	**99.17**	**99.25**	**99.30**	**99.33**	**99.36**	**99.40**	**99.42**	**99.45**	**99.48**	**99.49**
3	10.13	9.55	9.28	9.12	9.01	8.94	8.84	8.78	8.74	8.66	8.58	8.56
	34.12	**30.81**	**29.46**	**28.71**	**28.24**	**27.91**	**27.49**	**27.23**	**27.05**	**26.69**	**26.35**	**26.23**
4	7.71	6.94	6.59	6.39	6.26	6.16	6.04	5.96	5.91	5.80	5.70	5.66
	21.20	**18.00**	**16.69**	**15.98**	**15.52**	**15.21**	**14.80**	**14.54**	**14.37**	**14.02**	**13.69**	**13.57**
5	6.61	5.79	5.41	5.19	5.05	4.95	4.82	4.74	4.68	4.56	4.44	4.40
	16.26	**13.27**	**12.06**	**11.39**	**10.97**	**10.67**	**10.27**	**10.05**	**9.89**	**9.55**	**9.24**	**9.13**
6	5.99	5.14	4.76	4.53	4.39	4.28	4.15	4.06	4.00	3.87	3.75	3.71
	13.74	**10.92**	**9.78**	**9.15**	**8.75**	**8.47**	**8.10**	**7.87**	**7.72**	**7.39**	**7.09**	**6.99**
7	5.59	4.74	4.35	4.12	3.97	3.87	3.73	3.63	3.57	3.44	3.32	3.28
	12.25	**9.55**	**8.45**	**7.85**	**7.46**	**7.19**	**6.84**	**6.62**	**6.47**	**6.15**	**5.85**	**5.75**
8	5.32	4.46	4.07	3.84	3.69	3.58	3.44	3.34	3.28	3.15	3.03	2.98
	11.25	**8.65**	**7.59**	**7.01**	**6.63**	**6.37**	**6.03**	**5.82**	**5.67**	**5.36**	**5.05**	**4.96**
10	4.96	4.10	3.71	3.48	3.33	3.22	3.07	2.97	2.91	2.77	2.64	2.59
	10.04	**7.56**	**6.55**	**5.99**	**5.64**	**5.39**	**5.06**	**4.85**	**4.71**	**4.41**	**4.12**	**4.01**

Table 23.5 (Continued)

n_2 (degrees of freedom for denominator)	n_1 (degrees of freedom for numerator)											
	1	2	3	4	5	6	8	10	12	20	50	100
12	4.75	3.88	3.49	3.26	3.11	3.00	2.85	2.76	2.69	2.54	2.40	2.35
	9.33	**6.93**	**5.95**	**5.41**	**5.06**	**4.82**	**4.50**	**4.30**	**4.16**	**3.86**	**3.56**	**3.46**
14	4.60	3.74	3.34	3.11	2.96	2.85	2.70	2.60	2.53	2.39	2.24	2.19
	8.86	**6.51**	**5.56**	**5.03**	**4.69**	**4.46**	**4.14**	**3.94**	**3.80**	**3.51**	**3.21**	**3.11**
16	4.49	3.63	3.24	3.01	2.85	2.74	2.59	2.49	2.42	2.28	2.13	2.07
	8.53	**6.23**	**5.29**	**4.77**	**4.44**	**4.20**	**3.89**	**3.69**	**3.55**	**3.25**	**2.96**	**2.86**
18	4.41	3.55	3.16	2.93	2.77	2.66	2.51	2.41	2.34	2.19	2.04	1.98
	8.28	**6.01**	**5.09**	**4.58**	**4.25**	**4.01**	**3.71**	**3.51**	**3.37**	**3.07**	**2.78**	**2.68**
20	4.35	3.49	3.10	2.87	2.71	2.60	2.45	2.35	2.28	2.12	1.96	1.90
	8.10	**5.85**	**4.94**	**4.43**	**4.10**	**3.87**	**3.56**	**3.37**	**3.23**	**2.94**	**2.63**	**2.53**
30	4.17	3.32	2.92	2.69	2.53	2.42	2.27	2.16	2.09	1.93	1.76	1.69
	7.56	**5.39**	**4.51**	**4.02**	**3.70**	**3.47**	**3.17**	**2.98**	**2.84**	**2.55**	**2.24**	**2.13**
40	4.08	3.23	2.84	2.61	2.45	2.34	2.18	2.07	2.00	1.84	1.66	1.59
	7.31	**5.18**	**4.31**	**3.83**	**3.51**	**3.29**	**2.99**	**2.80**	**2.66**	**2.37**	**2.05**	**1.94**
50	4.03	3.18	2.79	2.56	2.40	2.29	2.13	2.02	1.95	1.78	1.60	1.52
	7.17	**5.06**	**4.20**	**3.72**	**3.41**	**3.18**	**2.88**	**2.70**	**2.56**	**2.26**	**1.94**	**1.82**
100	3.94	3.09	2.70	2.46	2.30	2.19	2.03	1.92	1.85	1.68	1.48	1.39
	6.90	**4.82**	**3.98**	**3.51**	**3.20**	**2.99**	**2.69**	**2.51**	**2.36**	**2.06**	**1.73**	**1.59**
1,000	3.85	3.00	2.61	2.38	2.22	2.10	1.95	1.84	1.76	1.58	1.36	1.26
	6.66	**4.62**	**3.80**	**3.34**	**3.04**	**2.82**	**2.53**	**2.34**	**2.20**	**1.89**	**1.54**	**1.38**

* Abridged, with permission of author and publisher, from George W. Snedecor, *Statistical Methods*, 5th. ed., Iowa State College Press, Ames, 1956.

The values resulting from the experiment are tabulated below, the symbol Y being used to represent an actual yield and y a deviation from the mean yield for a specified group.

Item	Treatment 1: No Fertilizer	Treatment 2: Nitrogen (40 pounds per acre)	Total, Treatments 1 and 2
Observations (n)	18	10	28
Mean yield (M)	26.65	60.66	38.7964
ΣY^2	14,794.71	43,982.22	58,776.93
nM^2	12,784.00	36,796.36	42,144.56
Σy^2	2,010.71	7,185.86	16,632.37
Degrees of freedom	17	9	27

The analysis of variance for these data is as follows:

Source of Variation	Degrees of Freedom	Sum of Squares	Mean Square
Total	27	16,632.37	$\cdots$
Difference between treatments	1	7,435.80	7,435.80
Variation within treatments	26	9,196.57	353.71

$$F = \frac{7,435.80}{353.71} = 21.02$$

The total sum of squares about the mean of the 28 observations is 16,632.37. The sums of squares about the individual means are 2,010.71 and 7,185.86, totalling 9,196.57. The difference of 7,435.80 (equals 16,632.37 − 9,196.57) is due to the departures of the two group means from the mean of the entire set of 28 observations. According to the logic of the experiment, this term is an estimate of the effect of the difference in treatment—nitrogen versus no nitrogen—accorded to the two groups of plots. The variation *within* each group is presumably due to a large number of factors of essentially random incidence independent of the fertilizer treatment.

As it happens, the *sum* of squares within groups exceeds that arising from the difference in treatments. But the mean square column shows that the effect of the difference in treatments *per degree of freedom* is 21.02 times as large as the corresponding variation within groups. Referring to Table 23.5, we find that for degrees of freedom $n_1 = 1$ and $n_2 = 20$, an F-ratio greater than 8.10 would occur by chance only 1 time in 100 experiments if the treatment actually had no effect upon corn yields. The probability of obtaining an F-ratio as great as 21.02 if the treatment were ineffective is negligibly small; we conclude therefore that nitrogen fertilizer applied at the rate of 40 pounds per acre has a highly significant effect upon corn yields.

The expected value of F would be 1.0 if the difference in treatments had no systematic effect upon yields. For in that event the variance of the group means about the general mean would reflect the same random forces as does the variance of each subset of observations about its group mean. The classification of observations on the basis of the amount of nitrogen applied would then be irrelevant; it would not contribute significantly to an explanation of the variance of all the observations about the general mean.

The analysis-of-variance table presented above suggests certain obvious analogies with regression analysis. The total sum of squares has been partitioned into "explained" and "unexplained" components. If the two treatments were two different varieties of corn we could say that 45 per cent of the observed variation in yields was attributable to the difference in varieties. In the current instance the treatments can be quantified (as zero and 40 pounds of nitrogen respectively) and we can actually fit a regression line to the 28 observations or draw a freehand line between the two points (26.65, 0) and (60.66, 40) with a slope of 0.85025 bushels per pound of nitrogen.

The student may gain a more complete understanding of the relation between variance analysis and regression analysis by actually fitting the regression line just mentioned. The basic data are given in Table 23.6.

The discontinuous values of X are quite typical in controlled experiments. The required values are as follows:

$$\Sigma Y = 1{,}086.3 \qquad M_y = 38.7964$$

$$\Sigma X = 400 \qquad M_x = 14.2857$$

$$\Sigma YX = 24{,}264 \qquad \Sigma X^2 = 16{,}000$$

From these we find that

$$b_{yx} = \frac{24{,}264 - 1{,}5518.56}{16{,}000 - 5{,}714.28} = 0.85025$$

the same value obtained by simply drawing a line through the points previously indicated. The constant a is given by

$$a = 38.7964 - 0.85025(14.2857)$$

$$= 26.65$$

This is the mean of the 18 observations for plots which received no fertilizer $(X = 0)$. When $X = 40$, $Y = 26.65 + 40(0.85025) = 60.66$,

Table 23.6

CORN YIELDS RELATED TO QUANTITY OF NITROGEN FERTILIZER APPLIED

Corn Yield, Y	Nitrogen Fertilizer Applied, X	Corn Yield, Y	Nitrogen Fertilizer Applied, X
Bushels per acre	Pounds per acre	Bushels per acre	Pounds per acre
24.5	0	32.4	0
6.2	0	27.4	0
26.7	0	5.3	0
29.6	0	17.9	0
22.1	0	23.9	40
30.6	0	11.8	40
44.2	0	60.2	40
21.9	0	82.5	40
12.0	0	96.2	40
34.0	0	80.7	40
37.7	0	81.1	40
34.2	0	51.0	40
38.0	0	79.5	40
35.0	0	39.7	40

Source: Heady *et al.*, *op. cit.*

the mean of the 10 observations for plots which received 40 pounds of nitrogen.

Referring back to the analysis of variance table, it is clear that the variation within treatments is identical with the sum of squared deviations about the regression line and the variation due to the difference between treatments is equal to that "explained" by the regression line. The sum of sqaares about the general mean has 27 degrees of freedom. As the regression line must pass through this general mean, only one additional degree of freedom is used up in determining the *slope* of the regression line, *b*. Hence the 26 degrees of freedom shown for "variation within treatments" check with the 26 degrees of freedom left in the residuals about a least-squares regression line fitted to 28 observations.

Applications of Variance Analysis: (2) Differences among the Effects of Three or More "Treatments." Suppose now that, in addition to the 28 plots already noted, we have applied 80 pounds of nitrogen per acre to 10 other randomly selected plots. We can again use variance analysis to answer a question that is more appropriate where the treatments are *qualitatively* different: Do variations in the amount of nitrogen applied have a significant effect upon corn yields?

The appropriate analysis of variance in this case is as follows:

Source of Variation	Degrees of Freedom	Sum of Squares	Mean Square
Total	37	46,476.79	· · ·
Differences among treatments	2	24,288.06	12,144.03
Variation within treatments	35	22,188.73	633.96

$$F = \frac{12,144.03}{633.96} = 19.16$$

Referring to Table 23.5, we note that for $n_1 = 2$ and $n_2 = 30$ degrees of freedom an F-ratio greater than 5.39 would be expected to occur by chance only 1 time in 100 experiments if in reality there were no relationship between corn yields and amounts of nitrogen applied.

The analogy with regression analysis can also be extended to this case. We could fit a second degree parabola precisely to the three points (26.65, 0), (60.66, 40), and (86.62, 80), 86.62 bushels being the mean yield obtained on the 10 plots which received 80 pounds of nitrogen. The "variation within treatments" would be identical with the sum of squared deviations about this parabola, and the difference between this amount and the sum of squared deviations about the general mean would be "accounted for" by the parabolic relation between group-mean yields and quantity of nitrogen applied. This analogy could obviously be extended to any number k of groups receiving quantitatively different "treatments" of a single factor *if* we are willing to conceive of a kth-order parabola passing through the k points of group means. (The significance of such a parabola would be no greater than that of a series of straight lines connecting each successive group mean to the one preceding it, as all the yield observations would be concentrated at the k discrete levels of fertilizer application.)

Applications of Variance Analysis: (3) Testing the Significance of Additional Terms in a Simple Regression Equation. Suppose that we have the data presented in the first and third columns of Table 23.7; we assume for present purposes that the X and Y values represent only 9 individual observations. We have fitted successively a straight line and a second-order parabola, and wish to determine whether the new term in X^2 has significantly increased the proportion of variance attributable to regression.

The least-squares regression line and parabola are respectively:

Straight line: $Y' = 56.37 + 0.19792X$

Parabola: $Y' = 34.54 + 0.66598X - 0.0014627X^2$

Table 23.7

RELATION BETWEEN AVERAGE YIELDS OF CORN OBTAINED AND QUANTITIES OF
NITROGEN FERTILIZER APPLIED

Nitrogen Applied, X	Number of Plots	Average Yield of Corn, Y	Sum of Squared Yields, $\Sigma Y_i^2 *$
Pounds per acre		*Bushels per acre*	
0	18	26.65	14,794.71
40	10	60.66	43,982.22
80	10	86.62	88,022.40
120	10	103.59	123,820.19
160	18	104.09	218,485.05
200	10	97.66	117,912.18
240	10	101.79	124,570.93
280	10	106.03	133,183.85
320	18	105.27	222,421.15

Source: Heady *et al.*, *op. cit.*
* Based on the individual observations.

The following values are also obtained:

$$\Sigma X = 1,440 \qquad M_x = 160$$

$$\Sigma Y = 792.36 \qquad M_y = 88.04$$

$$\Sigma X^2 = 326,400 \qquad M_x^2 = 36,272$$

$$\Sigma Y^2 = 75,681.3122 \qquad \Sigma YX = 145,777.60$$

$$\Sigma YX^2 = 33,669,728$$

These result in the following adjusted sums of squares and cross-products:

$$\Sigma y^2 = 5921.94$$

$$\Sigma yx = 19,000$$

$$\Sigma yx^2 = 4,929,248$$

The variance in yields explained by the straight-line regression may be calculated as $0.19792 \, \Sigma yx = 3,760.48$, and that explained by the parabolic regression curve may be computed from equation (12.2) as

$$0.66598 \, (\Sigma yx) - 0.0014627 \, (\Sigma yx^2) = 12,653.62 - 7,210.01 = 5,443.61.$$

The difference between either of these values and Σy^2 would be the sum of squared deviations around the corresponding regression function, while the difference between 5,433.61 and 3,760.48 is the additional variance explained by the parabolic regression.

The analysis of variance to determine the significance of the new term in X^2 in the parabolic equation is conducted as follows:

Source of Variation	Degrees of Freedom	Sum of Squares	Mean Square
Total	8	5,921.94	...
Linear regression	1	3,760.48	3,760.48
Additional variation accounted for by parabolic regression	1	1,683.13	1,683.13
Variation around parabolic regression	6	478.33	79.2

$$F = \frac{1,683.13}{79.72} = 21.11$$

Referring to Table 23.5 for $n_1 = 1$ and $n_2 = 6$ degrees of freedom, we find that the probability of a value of F greater than 13.74 occurring by chance would be less than 0.01 (1 per cent) if the regression of yield upon the additional term in the universe were really zero. It follows also that the parabola gives a significantly better representation of the relationship between Y and X than does the straight line. This process could be extended to determine whether a third-order parabola gives a significantly better fit than the second-order parabola already fitted.

Applications of Variance Analysis: (4) Testing for Curvilinearity of Regression. Let us now take account of the fact that each Y value in Table 23.7 is an average of 10 or 18 observations. Assume that we have fitted a straight line to the 9 mean values, weighting each of the 9 corresponding values of Y, Y^2, X, X^2, and XY by the number of observations (10 or 18) represented by the particular group mean. As a result of the weighting pattern, this line differs slightly from that of the preceding section, the new equation being

$$Y' = 52.11514 + 0.212208X$$

The variance due to regression is obtained by computing the squared deviations of the 9 possible values of Y' about the general mean and weighting each of them by the number of individual cases (10 or 18) located at that particular value of X.

The analysis of variance for the current example is as follows:

Source of Variation	Degrees of Freedom	Sum of Squares	Mean Square
Total	113	242,707	$\cdots$
Within groups	105	149,366	1,422.53
Due to linear regression	1	61,676	61,676
Deviation of means about linear regression	7	31,665	4,523.57

$$F = \frac{4,523.57}{1,422.53} = 3.18$$

Referring to Table 23.5, we note that for $n_1 = 6$ and $n_2 = 100$, a value of F greater than 2.99 would be obtained by chance only 1 time in 100 samples if the true regression in the universe were linear.

The logic of the above test may need some clarification. The variance within groups is the "pooled" variance of individual yields about their respective group means. This variance presumably results from random factors independent of the quantity of nitrogen applied. If the group means deviated from the regression line in consequence only of these same random factors, the expected variance of the means about the regression line would be equal to the variance of individual yields about the group means.[3] In this case, then, the expected sum of squares due to deviations of group means about the regression line would be $7(\bar{s}_y)^2$, or $7(1,422.53) = 9,957.71$. Their actual contribution was 31,665, or 3.18 times the expected value.

As such a value would be highly improbable under our present assumption, we conclude that other, non-random, factors are primarily responsible for the deviations of group means about the regression line. This is equivalent to saying that the expected mean values of Y for given values of X *in the universe* follow some (unspecified) curvilinear pattern. We might then proceed to fit an appropriate type of curve to the data.

Applications of Variance Analysis: (5) Significance of Two or More "Principles of Classification."   In certain types of experiments, we are

---

[3] Note again that a variance is a sum of squares divided by the appropriate number of degrees of freedom. While the group means would cluster more closely around the regression line than would the individual observations according to the relation $s_m^2 = \left(\dfrac{s_y}{n}\right)^2$, the squared deviation of each group mean from the regression line would be multiplied by n, the number of observations on which it was based, to arrive at the contribution of their deviations to the total sum of squares. On the average, then, each mean might be expected to contribute approximately $(s_y)^2$ to this sum of squares. But two degrees of freedom are taken up in fitting the regression line to the 9 group means, so the exact expected contribution of the group means would be $7(s_y)^2$.

interested in determining whether each of two or more factors is significantly associated with a given variable. In some cases the factors are qualitative rather than quantitative—for example, we might have 3 varieties of corn and 3 different insecticides, and apply each insecticide to each variety. A complete experiment might then involve 9 plots or some multiple of 9. We wish to analyze the 9 observed yields to determine whether either or both of the two "principles of classification"—differences in varieties and differences in insecticides—are significantly associated with differences in yields.

Some assumed data corresponding to such an experiment are shown in Table 23.8.

The total variation is the sum of squared deviations of the 9 individual yields about the general mean, $M_y = 85.078$. The variation due to differences in insecticides is measured by the squared deviations of the 3 group means in the right-hand column about the general mean. (Each of these squared deviations must be multiplied by 3, as each group mean represents 3 individual observations). The variation due to differences in varieties is measured by the squared deviations of the 3 group means in the bottom row about the general mean, multiplied by 3 as before. The computation of these values will be left to the student.

The analysis of variance for this example is as follows:

Source of Variation	Degrees of Freedom	Sum of Squares	Mean Square
Total	8	11,012.38	$\cdots$
Differences among insecticides	2	4,973.42	2,486.71
Differences among varieties	2	4,928.94	2,464.47
Residual, or "interaction"	4	1,110.02	277.50

For insecticides: $$F = \frac{2,486.71}{277.50} = 8.96$$

For varieties: $$F = \frac{2,464.47}{277.50} = 8.88$$

Referring to Table 23.5, we note that for $n_1 = 2$ and $n_2 = 4$ degrees of freedom, values of F larger than 6.94 would occur by chance only 5 times in 100 if either of the "principles of classification" were in reality not associated with corn yields. To reduce this probability to 1 in 100 ($P = 0.01$), F would have to be 18.00 or larger, so the results of our experiment are significant at the 5-per cent, but not at the 1-per cent, level for each of the two factors.

The above example is analogous to multiple regression with two independent variables. One could, of course, apply each of 3 different fertilizers to each of the insecticide-and-variety combinations of our example and apply each of 2 or more different methods of plowing to each of the 27 combinations of 3 factors—and so on. These experiments would be analogous to multiple regression with 3 or 4 independent variables.

Table 23.8

HYPOTHETICAL DATA ON CORN YIELDS RESULTING FROM COMBINATIONS OF 3 VARIETIES AND 3 INSECTICIDES

Insecticides	Varieties:			Totals	Means
	B_1	B_2	B_3		
A_1	40.60	55.88	59.07	155.55	51.849
A_2	58.63	119.83	123.75	302.21	100.736
A_3	56.72	128.15	123.07	307.94	102.646
Totals	155.95	303.86	305.89	765.70	85.078
Means	51.986	101.286	101.962	85.078	

One peculiarity of the analysis of variance just summarized is the use of the socalled "interaction term" in the denominator of the F-ratio. Presumably random factors not related to insecticides or varieties are responsible for a portion of the observed variation in yields. If the effects of insecticides and of varieties upon yield are strictly independent of one another, the interaction term gives us an estimate of the level of other (random) effects. But it may be that there are *joint* effects of insecticides and varieties in addition to the independent ones; if so the true random component would be smaller than the interaction term. Hence, use of the interaction term in the above fashion gives a conservative estimate of the F-ratio. If we replicated our experiment two or more times we could gain a better estimate of the random component from the variance of individual yields about the mean obtained for each of the 9 combinations of insecticides and varieties.

Applications of Variance Analysis: (6) Relation to Multiple Curvilinear Regression. The following experimental study by Heady, Pesek, and Brown (referred to earlier in Chapter 21) is classically suited to our present purpose. The complete set of 114 observations is included in Table 23.9, the structure of which clearly suggests a multiple regression of

Table 23.9

EXPERIMENTAL YIELDS OF CORN FOR VARYING LEVELS OF FERTILIZER
APPLICATION ON CALCAREOUS IDA SILT-LOAM SOIL IN WESTERN IOWA IN 1952
(yields are in bushels per acre)*

P_2O_5 (pounds)	Nitrogen (pounds)								
	0	40	80	120	160	200	240	280	320
0	24.5	23.9	28.7	25.1	17.3	7.3	16.2	26.8	25.1
	6.2	11.8	6.4	24.5	4.2	10.0	6.8	7.7	19.0
40	26.7	60.2			96.0	95.4			81.9
	29.6	82.5			107.0	95.4			76.4
80	22.1		99.5		115.9		112.4		129.0
	30.6		115.4		72.6		125.6		82.0
120	44.2			119.4	113.6			114.9	124.6
	21.9			97.3	102.1			129.2	83.0
160	12.0	96.2	102.2	133.3	129.7	105.7	130.5	123.6	135.6
	34.0	80.7	108.5	124.4	116.3	115.5	124.3	142.5	122.7
200	37.7	81.1			128.7	140.3			136.0
	34.2	51.0			109.3	142.2			118.2
240	38.0		97.2		127.6		121.1		130.9
	35.0		107.8		125.8		114.2		144.9
280	32.4			129.5	134.4			130.0	124.8
	27.4			125.2	127.6			141.9	114.1
320	5.3	79.5	116.9	135.7	122.9	138.7	127.3	131.8	127.9
	17.9	39.7	83.6	121.5	122.7	126.1	139.5	111.9	118.8

* Two figures are shown in each cell since the treatments were replicated (i.e., two plots received the same fertilizer quantities and ratios).

Source: Earl O. Heady, John T. Pesek, and William G. Brown, Crop response surfaces and economic optima in fertilizer use, *Iowa State College Agricultural Experiment Station Research Bulletin*, 424, March 1955, page 330.

yields upon two different plant nutrients, which we will refer to as nitrogen (N) and phosphate (P). As each combination of quantities of N and P was applied to two different plots, the variance of these two yields about their mean, with one degree of freedom for each of the 57 combinations, provides an estimate of the random error component in yields, independent of the effects (additive, joint, or both) of the fertilizer elements.

The meaning of partial or net regression is also implicit in Table 23.9. For each of 9 levels of phosphate application, a net regression line or curve can be fitted to the observations on corn yield and quantity of nitrogen applied. Similar net regression curves can be derived for the other plant nutrient. That this is a *regression* rather than a *correlation* model is evidenced by the selection of widely spaced values of the independent factors and a heavy concentration of observations toward the extreme ends of their ranges. A random sample from a normal trivariate universe would give a large number of observations near the point of means of the two independent factors and a very small number near the extremes.

The authors proceeded to fit the following multiple curvilinear regression joint function by least squares (as noted earlier in Chapter 21):

$$Y = -5.682 - \underset{(0.040)}{0.316N} - \underset{(0.040)}{0.417P}$$

$$+ \underset{(0.8676)}{6.3512\sqrt{N}} + \underset{(0.8680)}{8.5155\sqrt{P}} + \underset{(0.0385)}{0.3410\sqrt{NP}}$$

The figures in parentheses are standard errors of the regression coefficients. The *t*-ratios range from 7.32 to 10.44, all highly significant at the 1-per cent probability level.

The analysis of variance associated with the complete experiment is as follows:

Source of Variation	Degrees of Freedom	Sum of Squares	Mean Square
Total	113	242,707	$\cdots$
Treatments	56	233,811	4,175
Due to regression	5	222,828	44,566
Deviations from regression	51	10,983	215
Among plots treated alike	57	8,896	156

$$F = \frac{44,566}{156} = 286$$

The regression is, of course, highly significant. The variance of deviations of treatment means about the regression is small enough to be accounted for by the same forces responsible for variation among plots treated alike. Thus there is no point in searching for a better regression surface; while another equally good *might* be found it would be impossible to show that it gave a *significantly* better fit to the data.

By way of contrast, the authors had fitted a different curvilinear joint regression surface about which the variance of treatment means was 625; the variance among plots treated alike remained, of course, at 156. The corresponding *F*-ratio, 4.01, would occur slightly less often than 1 time in 100 experiments if the deviations of treatment means from the regression surface were due exclusively to the same random factors that caused yields to vary among plots treated alike. It followed logically that one or more other regression surfaces might be found that would better express the true (universe) relationship between corn yields and applications of nitrogen and phosphate.

Summary

The relation of one variable to several others may be approximately determined by detailed cross-classification of the observations according to combinations of class intervals of all the independent variables. Very large total numbers of observations are required to make the group averages accurate, however, for the number of groups increases rapidly with additional variables. Simple or multiple regression equations can be fitted to the individual observations, or to the group averages, preferably weighting each average by the number of observations on which it is based. Correlation coefficients based on group averages are often much higher than those based on the original observations, as the averages suppress much of the random variation present in the latter; correlation coefficients and standard errors of regression coefficients based on grouped data must be interpreted with care.

The methods of variance analysis can be used to break down the sum of squared deviations of a variable about its mean into a number of additive components. Ordinarily these methods are used in connection with experimental data, and the design of the experiment determines the number of relevant components and the logical significance of each. Variance analysis is particularly suited to situations in which the independent "variables" or "principles of classification" are qualitative, as (for example) different varieties of wheat or different insecticides. In fact, the method of dealing with a qualitative independent variable in multiple-regression analysis, presented in Chapter 22, is essentially an adaptation of variance-analysis concepts, as is the *correlation ratio*, discussed in the same chapter.

Historically, variance analysis has been applied chiefly to discontinuous or grouped data. When the principles of classification were quantitative, the analysis has generally stopped (explicitly or implicitly) with discontinuous "lines of averages" connecting the group means. *All* the variation of group averages of the dependent variable about its general mean is presumed to be explained by the independent variables. A continuous regression function fitted to the same data would not pass precisely through all the group averages (except in rare or trivial cases), so the proportion of total variation "explained" by the regression function would be somewhat smaller than that attributed in variance analysis to differences among group means.

The historical distinction between the two approaches is not required by the underlying statistical theory, and today statisticians dealing with data from controlled experiments often fit regression functions to the group means and appraise the goodness of fit of the functions using significance tests customary in variance analysis. These tests lead to the same estimates of probability levels as do the standard-error formulas applicable to regression constants. Both approaches may be regarded as based upon the same general theory of least-squares estimation. As in regression analysis, attempts to measure the separate contributions of the different independent factors to total variance often encounter the difficulty of *covariance*, which is essentially intercorrelation among the independent variables.

The methods of variance analysis may also be used to test the significance of the improvement in fit shown by a curved regression over a linear one, of additional constants in fitting a curve, or of an additional variable in a multiple regression and correlation problem.

REFERENCE

Williams, E. J., *Regression Analysis*, John Wiley and Sons, Inc., 1959.

CHAPTER 24

Fitting systems of two
or more simultaneous equations

Introduction. In 1943, Haavelmo introduced a drastically different method of statistical analysis for estimating relationships among economic time series.[1] Although this method was designed to handle problems in the field of economics, similar problems may well exist in other disciplines such as biology and physiology.

A complete description of the computations involved in handling fairly large sets of simultaneous equations is beyond the scope of this book. Friedman and Foote[2] have published a handbook which is recommended to those who are interested in applying this approach to any but the simplest cases. Here we will simply examine the logic and mathematics on which the method is based, as illustrated by a two-equation model.

In earlier chapters we discussed two possible interpretations of regression equations. The relation between protein content of wheat and the percentage of vitreous kernels was presented as purely descriptive or empirical —it did not express a "causal mechanism" by means of which an increase in the percentage of vitreous kernels would inevitably (except for random errors) lead to an increase in the protein content. In the auto-stopping example, however, we found strong logical grounds for expecting the speed of a car to affect the distance to stop, and for the relationship to follow a second-degree parabola. Although changes in automobile design

---

[1] Trygve Haavelmo, The statistical implications of a system of simultaneous equations, *Econometrica*, Vol. 11, pp. 1–12, January, 1943.
———, The probability approach in econometrics, *Econometrica*, Vol. 12, Supplement, 118 pp. July, 1944.
[2] Joan Friedman, and Richard J. Foote, Computational methods for handling systems of simultaneous equations. U. S. Department of Agriculture, *Agriculture Handbook* No. 94, 109 pp. illus., November, 1955.

and road surfaces might alter the numerical values of the coefficients, the general form of the relationship would remain parabolic. On even firmer grounds the trajectory of an artillery shell would follow a parabolic "law" except as modified by wind resistance or other additional influences.

These two interpretations have their counterparts in the analysis of economic time series. Thus, we might find that a set of variables such as the number of hogs on farms on January 1, the January level of steel production, and the January index of wholesale prices (all commodities) showed a high degree of correlation with the average retail price of pork during January–March. This would be a purely empirical relationship, though possibly a useful one. Economic theory does not provide us with any *simple* interpretations of the net regression coefficients in this case, although some rather long and roundabout chains involving correlations with a series of intermediate variables might by hypothesized.

The other interpretation might be illustrated by a regression analysis relating the per capita consumption of pork in a given year to the retail price of pork, the retail price of beef, and the per capita disposable income of consumers. The income series and the meat prices might be deflated by an index of retail prices of all consumer goods and services; an index of retail prices of all consumer goods and services *other than* pork and beef, also deflated by the index of all "consumer prices," might be included as an additional variable. Economic theory tells us that these variables in these forms logically belong in a consumer demand function for pork; it tells us which of the net regression coefficients should be positive and which negative, and places definite limitations upon the shapes of the net regression curves. A relationship of this type is sometimes called a "behavior equation"—i.e., consumers change their purchases of pork because they *experience* changes in their incomes and in the retail prices of pork, beef and other things. These variables have immediate logical connections with consumer decisions, whereas consumers have no direct experience of changes in steel production, hog numbers, and wholesale prices.

In the terminology of simultaneous equations analysis, our second regression represents an attempt to estimate "structural coefficients"; if our estimating procedure is appropriate the equation as a whole will be a "structural equation." Conceptually, it is quite different from the empirical estimating equation of our first example.

In brief, the object of the simultaneous-equations method is to determine "structural equations"; its proponents argue that, in most situations involving economic time series, "structural equations" cannot be estimated satisfactorily by the least-squares, single-equation methods presented in the first twenty-two chapters of this book.

Basic Concepts, Problems, and Definitions.   We are all familiar with the economic concept that the price of a commodity is determined by the intersection of a supply curve and a demand curve. But price can also be regarded in an "active" sense as bringing the quantity supplied and the quantity demanded into balance—in fact, equality—with each other. If we could perform some controlled experiments in which we fixed prices at various levels and measured separately the quantities supplied (sold) by producers and the quantities demanded (purchased) by consumers at each price, the regression of quantities *sold* upon price would give us the supply curve (with a positive slope, typically) and the regression of quantities *purchased* upon price would give us the demand curve (with a negative slope). Note that the quantity supplied by producers would not necessarily equal the quantity demanded by consumers under the conditions of this experiment; to actually hold the price at some predetermined level we must stand ready to buy surpluses from producers and to avert possible shortages by selling to consumers from stocks under our control.

In practice we may be unable to perform such experiments. Even if we could, we would still be loath to forego the information contained in time-series observations for earlier years, and these observations did not arise from a controlled experiment.

Let us assume that we have only one measure of quantity for each year—i.e., the quantity supplied by producers is identical with the quantity purchased by consumers—and call the price-quantity observations in two successive years (P_1, Q_1) and (P_2, Q_2). Suppose we note that P_2 has increased sharply over P_1. If the two points were on the same *supply* curve, Q_2 should be larger than Q_1, for producers would find it profitable to turn out a larger quantity at the higher price. If the two points were on the same *demand* curve, Q_2 should be smaller than Q_1, for consumers would buy a smaller quantity at the higher price.

To be perfectly concrete, suppose we find that P_2 is 20 cents higher than P_1 and that Q_2 is 10 units higher than Q_1, so the ratio

$$\left(\frac{P_2 - P_1}{Q_2 - Q_1}\right) = \frac{20}{10} = 2.00.$$

Evidently demand has increased so much that consumers will buy 10 more units in spite of the 20-cent increase in price; we can be sure that the demand curve has shifted upward. If the supply curve has not shifted at all, the ratio 2.00 is an unbiased estimate of the slope of the supply curve. But (working solely from the time series observations) we do not know whether it has shifted or not. The 10-unit increase in Q *could* have resulted from an upward shift of 10 cents in the level of a supply curve having a true slope of 1.00—or a 15-cent shift in a supply curve having a true slope of 0.50. And each of these alternatives would be consistent with various combinations of slopes and shifts of the (equally unknown) demand function. (We assume P to be measured along the vertical, and Q along the horizontal, axis of a diagram.)

In the universes to which least-squares regression equations apply, we can reduce the standard errors of our estimates of universe parameters indefinitely by increasing the number of observations in our sample. But (in simple linear regression) any pair of observations will give us an unbiased estimate of the regression slope. The present problem is different in that each pair of observations gives us (in general) a *biased* estimate of the universe parameters. We cannot improve our estimates of these parameters by simply increasing the number of *biased* estimates implicit in each pair of observations. It is different also in that we are trying to estimate *two* universe regressions from a single set of values of P and Q— one regression which typically has a positive slope and one which invariably has a negative slope. These are not the two "elementary regressions" of least-squares theory, for, from the arithmetic of regression analysis, the coefficients

$$b_{pq}\left(= \frac{\Sigma pq}{\Sigma q^2}\right) \quad \text{and} \quad b_{qp}\left(= \frac{\Sigma pq}{\Sigma p^2}\right)$$

must both have the same sign or both be zero. Our problem cannot be solved by choosing either P or Q as "the" dependent variable. The economics of the situation implies that P (as an annual average) and Q (as an annual total) have equal claim to this status if both producers and consumers are free to adjust the quantities they sell or buy during the course of the year.[3]

The first step in a simultaneous-equations analysis is to "specify the

---

[3] If we thought of producers as offering a certain definite quantity q_1 for sale in Week 1, consumers would pay a price P_1 corresponding to that quantity on their demand function. Then, suppose that producers determine the amount they will offer in Week 2 on the basis of the "lagged" supply function,

$$q_2 = a_2 + b_2 P_1$$

or, more generally,

$$q_t = a_2 + b_2 P_{t-1}$$

If the quantity q_2 so determined, differs from q_1, consumers will pay a price P_2, different from P_1, so that, from Week 1 to Week 2, the price will have moved along the consumer demand function according to the formula

$$(P_2 - P_1) = \frac{1}{b_1}(q_2 - q_1) \quad \text{or} \quad (q_2 - q_1) = b_1(P_2 - P_1)$$

If this weekly response mechanism operated throughout the year, then from Week 52 in Year 0 to Week 52 in Year 1 there would have been 52 supply responses, each model" by which the observations were generated. This is equivalent to defining the universe with respect to which a given set of observations has sampling significance. In the present case the model will be specified as follows:

Demand curve: $$Q = a_1 + b_1 P + u \tag{24.1}$$

Supply curve: $$Q = a_2 + b_2 P + v \tag{24.2}$$

In these equations, Q stands for quantity and P for price; u and v are "disturbances," random with respect to time and having an expected value (universe mean) of zero. Both equations contain the same two variables. With least-squares methods we can calculate the regression of Q on P and the regression of P on Q—two different equations—but there is absolutely no basis for identifying one of these as a supply curve and the other as a demand curve.

The disturbance u in the present model cannot be treated as a random error in either Q or P; rather, it causes random shifts in the *level* of the demand curve. Similarly, v causes random shifts in the level of the supply curve. As the curves shift about, their points of intersection constitute the values of P and Q in the successive time periods. (The values of P and Q are assumed to be measured without error.) If the demand curve remained fixed for several periods all the points of intersection would lie on the demand curve, which could then be estimated from the P and Q values by least squares. If the demand curve shifted while the supply curve remained fixed, the points of intersection would lie on the supply curve; in this case the least-squares relation between P and Q would give us the supply curve. In general, both curves are likely to shift and the least-squares regression of Q on P would represent neither a supply nor a demand curve but some uninterpretable mixture of the two.

If an infinite number of observations were drawn from the universe defined by equations (24.1) and (24.2), the expected (universe) value of the least-squares regression of Q on P would be:

$$B = \frac{b_1 \sigma_v^2 - (b_1 + b_2)\sigma_{uv} + b_2 \sigma_u^2}{\sigma_v^2 - 2\sigma_{uv} + \sigma_u^2} \tag{24.3}$$

---

relating q_t to P_{t-1} according to the coefficient b_2, and 52 price adjustments, each relating P_t to q_t according to the coefficient b_1.

This model suggests how the relationship between *annual* price and quantity observations may represent a combination of supply and demand adjustments such that neither supply nor demand effects can be estimated separately. (However, using weekly data, on the assumptions of this footnote, we could estimate the separate functions quite readily.)

where $\sigma_{uv} = \rho_{uv}\sigma_u\sigma_v$. This can be demonstrated by writing equations (24.1) and (24.2) with each variable stated in terms of deviations from its mean:

$$q = b_1p + u \tag{24.4}$$

$$q = b_2p + v \tag{24.5}$$

Solving these equations simultaneously for p and q in terms of the disturbances u and v we obtain

$$p = \frac{v - u}{b_1 - b_2} \tag{24.6}$$

$$q = \frac{b_1v - b_2u}{b_1 - b_2} \tag{24.7}$$

The least-squares regression of q on p is

$$B = \frac{\Sigma qp}{\Sigma p^2} = \frac{\Sigma[(b_1v - b_2u)(v - u)]}{\Sigma(v - u)^2} \tag{24.8}$$

If we carry out the multiplications and summations indicated and divide by the total number of observations, we obtain equation (24.3). If the demand curve has not shifted, $\sigma_u = 0$ and $B = b_1$; if the supply curve has not shifted, $\sigma_v = 0$ and $B = b_2$. But even if one of these special cases has actually existed we may not know it; in general, if equations (24.1) and (24.2) constitute the true and complete model, there is no way in which b_1 and b_2 can be estimated from the data. In current terminology, neither equation is "identifiable" and the system as a whole is "underidentified."

The situation can be resolved, however, if each equation happens to contain a different "predetermined variable." A predetermined variable is one whose values are known or causally determined *before* the current time period or by factors outside of the immediate supply-demand model. (The outside factors are also referred to as "exogenous variables.") Statistically, these new variables (predetermined, including exogenous) are treated as given numbers, analogous to the independent variables of least-squares regression analysis. In contrast, P and Q, called "endogenous variables," are analogous to the dependent variables of least-squares regression; their values are dependent upon those of the predetermined variables acting in conjunction with the random disturbances.

Suppose, then, that the true model is as follows:

Demand function: $Q = a_1 + b_1P + c_1Y + u$ $\qquad$ (24.9)

Supply function: $Q = a_2 + b_2P + c_2Z + v$ $\qquad$ (24.10) where u and v again are random disturbances, caused by a large number of more or less remote events in the economic system, none of which taken separately is of much consequence or has *measurable* effects on our particular demand-supply system. For concreteness, let us say that Y is consumer income and Z is an index of factors, known or logically determined prior to January 1, which affect the quantity of pork that will be supplied during the ensuing year. We shall treat Y as an exogenous variable, which influences the endogenous variables P and Q but is not influenced by them. This is a plausible approximation, as consumers spend only about 2 per cent of their incomes on pork and, conversely, not more than 2 per cent of consumer income is derived from activities closely associated with pork production. We might reasonably assume that some 98 per cent of the variation in Y is determined by factors other than the price and quantity of pork and that, *if* our results are biased by the fact that Y is not 100 per cent exogenous, the bias will not exceed 2 per cent.

The values of P and Q, then, are dependent on those taken by Y, Z, u, and v. We shall therefore solve equations (24.9) and (24.10) for P and Q as functions of the predetermined variables and the disturbances.[4] This solution leads to the following two equations:

$$P = A_1 + B_1 Y + C_1 Z + d_1 \qquad (24.11)$$

$$Q = A_2 + B_2 Y + C_2 Z + d_2 \qquad (24.12)$$

These equations are called the "reduced form" of the structural model, or "reduced-form equations." As indicated in Appendix 3, Note 6, the coefficients are definite algebraic combinations of the coefficients of the two structural equations; d_1 and d_2 are combinations of the disturbances (u and v) and the structural coefficients b_1 and b_2.

Each of the reduced-form equations contains a single dependent variable; the values of Y and Z are given numbers, like independent variables in the "regression model" of least squares; and d_1 and d_2 have the same statistical properties as the random residuals of least-squares regression, being uncorrelated (by assumption) with Y and Z. Therefore each of the reduced-form equations can be fitted separately by the familiar method of least squares. Then, as we know precisely what combination of the structural coefficients is included in each coefficient of the reduced-form equations, we can derive the structural coefficients by a few arithmetic operations on those of the reduced form.

Specifically, $b_1 = C_2/C_1$ and $b_2 = B_2/B_1$. Knowing b_1 and b_2, we can

---

[4] See Appendix 3, Note 6, for the details of this solution.

Table 24.1

BASIC DATA FOR "JUST-IDENTIFIED" MODEL OF SUPPLY AND DEMAND FOR PORK, UNITED STATES, 1922–1941

Year	Original Arithmetic Values				Logarithms				First Differences of Logarithms			
	Retail Price of Pork, P_t	Consumption of Pork, Q_t	Disposable Personal Income, Y_t	"Predetermined Elements in Pork Production," Z_t	P_t	Q_t	Y_t	Z_t	P_t	Q_t	Y_t	Z_t
	Cents per pound	Pounds per capita	Dollars per capita	Pounds per capita								
1922	26.8	65.7	541	74.0	1.428	1.818	2.733	1.869	...	...	...	...
1923	25.3	74.2	616	84.7	1.403	1.870	2.790	1.928	−0.025	0.052	0.057	0.059
1924	25.3	74.0	610	80.2	1.403	1.869	2.785	1.904	0.000	−0.001	−0.005	−0.024
1925	31.1	66.8	636	69.9	1.493	1.825	2.803	1.844	0.090	−0.044	0.018	−0.060
1926	33.3	64.1	651	66.8	1.522	1.807	2.814	1.825	0.029	−0.018	0.011	−0.019
1927	31.2	67.7	645	71.6	1.494	1.831	2.810	1.855	−0.028	0.024	−0.004	0.030
1928	29.5	70.9	653	73.6	1.470	1.851	2.815	1.867	−0.024	0.020	0.005	0.012
1929	30.3	69.6	682	71.2	1.481	1.843	2.834	1.852	0.011	−0.008	0.019	−0.015
1930	29.1	67.0	604	69.6	1.464	1.826	2.781	1.843	−0.017	−0.017	−0.053	−0.009
1931	23.7	68.4	515	68.0	1.375	1.835	2.712	1.833	−0.089	0.009	−0.069	−0.010
1932	15.6	70.7	390	74.8	1.193	1.849	2.591	1.874	−0.182	0.014	−0.121	0.041
1933	13.9	69.6*	364	73.6†	1.143	1.843	2.561	1.867	−0.050	−0.006	−0.030	−0.007
1934	18.8	63.1*	411	70.2†	1.274	1.800	2.614	1.846	0.131	−0.043	0.053	−0.021
1935	27.4	48.4	459	46.5	1.438	1.685	2.662	1.667	0.164	−0.115	0.048	−0.179
1936	26.9	55.1	517	57.6	1.430	1.741	2.713	1.760	−0.008	0.056	0.051	0.093
1937	27.7	55.8	551	58.7	1.442	1.747	2.741	1.769	0.012	0.006	0.028	0.009
1938	24.5	58.2	506	58.0	1.389	1.765	2.704	1.763	−0.053	0.018	−0.037	−0.006
1939	22.2	64.7	538	67.2	1.346	1.811	2.731	1.827	−0.043	0.046	0.027	0.064
1940	19.3	73.5	576	73.7	1.286	1.866	2.760	1.867	−0.060	0.055	0.029	0.040
1941	24.7	68.4	697	66.5	1.393	1.835	2.843	1.823	0.107	−0.031	0.083	−0.044

* Excludes quantities purchased and distributed under Government emergency programs. Per capita consumption including these quantities was 70.7 pounds in 1933 and 64.4 pounds in 1934.

† Adjusted by subtracting quantities purchased for Government emergency programs.

calculate $c_1 = B_1[-(b_1 - b_2)]$ and $c_2 = C_1(b_1 - b_2)$. The values a_1 and a_2 may be obtained by solving the following simultaneous equations:

$$-b_2a_2 + b_2a_1 = A_1(b_1 - b_2)(-b_2)$$
$$b_1a_2 - b_2a_1 = A_2(b_1 - b_2)$$

Adding the two equations, we can solve for a_2 as:

$$a_2 = \frac{A_1(b_1 - b_2)(-b_2) + A_2(b_1 - b_2)}{(b_1 - b_2)}$$

for at this stage all items in the right-hand term are known numbers. Then, dividing the first equation through by b_2, we obtain

$$a_1 = a_2 - A_1(b_1 - b_2)$$

It is important to note that this method of estimation flows directly from the specification of the model in equations (24.9) and (24.10). Without this specification it is quite unlikely that a researcher interested in demand curves and supply curves would have fitted either of the reduced-form equations. However, if he had been interested in *forecasting* P and Q for the coming year on the basis of information available as of January 1 he might very well have fitted these equations, using the variables Z and Y', Y' representing an advance forecast of Y. But he would not have called either of the equations a demand curve or a supply curve; they would simply have been prediction equations without structural significance.

Illustration of a "Just-Identified" Model. The model embodied in equations (24.9) and (24.10) is said to be "just identified" because it permits us to obtain a single unique estimate for each of the structural coefficients. This contrasts with the *underidentified* case previously mentioned in which some of the structural coefficients cannot be estimated at all, and with the "overidentified" case in which some of the coefficients can be estimated in two or more ways yielding different values.

As no new *statistical* calculations are involved in the just-identified case, we shall present only the final equations obtained in an attempt to derive simultaneous demand and supply equations for pork. The same symbols are used as in equations (24.9) and (24.10); the basic data were first differences of logarithms of annual observations for the period 1922–1941 (see Table 24.1).

The reduced-form equations are as follows:

$$P = -0.0101 + 1.0813\,Y - 0.8320Z \qquad R^2 = 0.893$$
$$(0.1339) \quad\;\; (0.1159)$$

$$Q = 0.0026 - 0.0018\,Y + 0.6839Z \qquad R^2 = 0.898$$
$$(0.0673) \quad\;\; (0.0582)$$

The numbers in parentheses are standard errors of the net regression coefficients. The coefficients of the structural equations are calculated as follows:

$$b_2 = \frac{-0.0018}{1.0813} = -0.0017$$

$$b_1 = \frac{0.6839}{-0.8320} = -0.8220$$

$$c_1 = 1.0813\{-[-0.8220 - (-0.0017)]\}$$
$$= 1.0813(0.8203) = 0.8870$$

$$c_2 = -0.8320(-0.8203) = 0.6825$$

To compute the constant terms a_1 and a_2, we solve the following two simultaneous equations:

$$0.0017a_2 - 0.0017a_1 = (-0.0108)(-0.8203)(.0017) = 0.0000$$

$$-0.8220a_2 - (-0.0017)a_1 = (0.0026)(-0.8203) = -0.0021$$

Adding the two equations, we obtain

$$-0.8203a_2 = -0.0021 + (0.0017)(0.0089) = -0.0021$$

$$a_2 = \frac{-0.0021}{-0.8203} = 0.0026$$

Then, substituting $a_2 = 0.0026$ into the first equation divided by 0.0017, we have $0.0026 - a_1 = 0.0089$, or $a_1 = -0.0063$. This completes the set of coefficients for the two structural equations. Consequently, the structural equations can be written as follows:

Demand function: $Q = -0.0063 - 0.8220P + 0.8870Y + u$

Supply function: $Q = 0.0026 - 0.0017P + 0.6825Z + v$

Because the data were in logarithmic form, the coefficients represent *approximately* the percentage changes in Q associated with changes of 1 per cent in each of the other variables. In particular, the coefficients of P are estimates of the elasticities of demand and (simultaneous or concurrent) supply for pork. Although standard errors of the structural coefficients are not presented here, such errors (appropriate for large samples) can be computed.

For each observation values of u and v are computed from the following forms of the last two equations:

$$u_t = Q_t + 0.0063 + 0.8220P_t - 0.8870Y_t$$

$$v_t = Q_t - 0.0026 + 0.0017P_t - 0.6825Z_t$$

The subscript t simply refers to the particular year for which the disturbances are being calculated. The disturbances for each year, and their successive differences, are shown in Table 24.2. The von Neumann ratios prove to be as follows:

Demand function:
$$K = \left(\frac{.008017}{.003553}\right)\left(\frac{19}{18}\right)$$

$$= 2.256403(1.055556) = 2.381760$$

Supply function:
$$K = \left(\frac{.007122}{.003181}\right)\left(\frac{19}{18}\right)$$

$$= 2.238919(1.055556) = 2.363304$$

Referring to Table 20.5 for $N = 19$ observations, we find no evidence of significant autocorrelation in either set of disturbances.

Table 24.2
DISTURBANCES COMPUTED FROM STRUCTURAL EQUATIONS

Year	Disturbances in Demand Function		Disturbances in Supply Function	
	u_t	$u_{t+1} - u_t$	v_t	$v_{t+1} - v_t$
1923	−0.013	. . .	0.009	. . .
1924	0.010	0.023	0.013	0.004
1925	0.020	0.010	−0.005	−0.018
1926	0.002	−0.018	−0.008	−0.003
1927	0.011	0.009	0.001	0.009
1928	0.002	−0.009	0.009	0.008
1929	−0.010	−0.012	0.000	−0.009
1930	0.022	0.032	−0.014	−0.014
1931	0.003	−0.019	0.013	0.027
1932	−0.022	−0.025	−0.017	−0.030
1933	−0.014	0.008	−0.004	0.013
1934	0.024	0.038	−0.031	−0.027
1935	−0.016	−0.040	0.005	0.036
1936	0.010	0.026	−0.010	−0.015
1937	−0.003	−0.013	−0.003	0.007
1938	0.014	0.017	0.019	0.022
1939	−0.007	−0.021	0.000	−0.019
1940	−0.014	−0.007	0.025	0.025
1941	−0.010	0.004	−0.003	−0.028

Comparison of Structural Equations with Their Least-Squares Counterparts. One can scarcely argue with the logic of the method of reduced forms if two or more endogenous variables are in fact simultaneously determined. But universal application of this approach to economic time series has been opposed on both pragmatic and theoretical grounds.

The pragmatic argument is based upon comparisons of published structural equations with least-squares relationships involving the same variables. In most cases the least-squares coefficients have been very close to those of the structural equations.

For example, the demand function for pork has been fitted by least squares in two different ways. When Q is treated as the dependent variable we obtain

Least squares: $Q = -0.0049 - 0.7205P + 0.7646Y \qquad R^2 = 0.903$
$$(0.0594) \quad (0.0967)$$

Structural: $Q = -0.0063 - 0.8220P + 0.8870Y + u$

In each case the structural coefficient is within less than two standard errors of the least-squares coefficient; the differences are not statistically significant.

When P is treated as the independent variable, we obtain

Least squares: $P = -0.0070 - 1.2518Q + 1.0754Y \qquad R^2 = 0.956$
$$(0.1032) \quad (0.0861)$$

For easier comparison, we divide all terms in the structural demand equation by the coefficient of P (using the positive sign) and transpose P and Q to opposite sides of the equality sign, obtaining

Structural: $P = -0.0077 - 1.2165Q + 1.0791Y + \left(\dfrac{u}{0.8220}\right)$

The coefficients of the two equations are almost identical—they differ by small fractions of one standard error. Evidently the least-squares equation with price dependent gives an excellent approximation to the structural demand function. And the logic of the simultaneous-equations approach supports the choice of price as the dependent variable in our least squares demand function *if* there is no simultaneous response of supply to price.[5]

[5] See Karl A. Fox, Structural analysis and the measurement of demand for farm products, *Review of Economics and Statistics*, Vol. XXXVII, No. 1, pp. 57–66, February, 1954.

The corresponding comparison for the supply function (with Q dependent in the least-squares equation) is as follows:

Least squares: $Q = 0.0022 - 0.0788P + 0.6090Z$ $R^2 = 0.910$
                                    (0.0522) (0.0734)

Structural: $Q = 0.0026 - 0.0017P + 0.6825Z + v$

The coefficient of P in the least-squares supply function is non-significant according to the usual criteria and has a negative sign, whereas normally we would expect a true supply function to show a positive response of quantity to price.

If we discard the price variable from the least-squares supply function as non-significant we obtain simply

Least squares: _ $Q = 0.0025 + 0.6841Z$ $r^2 = 0.898$
                                    (0.0857)

This is almost identical with the structural equation, recognizing that the coefficient of P in the latter is negligibly small and would prove to be statistically non-significant if its standard error were calculated.

When a model of the United States economy during 1929–1952, involving 15 simultaneous equations with a total of 51 structural coefficients, was compared with its least-squares counterpart, the following results were obtained:[6]

Ratio of Difference between Coefficients to Standard Error of the Structural Coefficient	Between Constant Terms	Between Net Regression Coefficients	Total
0 —0.49	7	14	21
0.50—0.99	5	6	11
1.00—1.49	2	8	10
1.50—1.99	1	2	3
2.00—2.99	0	4	4
3.00 and over	0	2	2
Total	15	36	51

Thirty-two of the least-squares coefficients were within one standard error, and 45 within two standard errors, of the corresponding structural

[6] From Karl A. Fox, Econometric models of the United States, *Journal of Political Economy*, Vol. LXIV, No. 2, p. 131, April, 1956. See also Table 3, pp. 135–136.

The original structural model will be found in L. R. Klein and A. S. Goldberger, *An Econometric Model of the United States, 1929–1952*, North-Holland Publishing Company, Amsterdam, 1955.

coefficients. (The standard errors of the *differences* between coefficients obtained from two random samples of the same size drawn from the same universe would average about 1.414 times as large as the standard error of either coefficient taken separately; the factor 1.414 is equal to the square root of 2). If we apply this analogy it appears that no more than 6, and possibly only 2, of the differences would be significant at the 5-per cent level.

The similarities between the simultaneous-equations model and its least-squares counterpart extend to the standard errors of the coefficients (most of which were within 10 per cent of one another) and the extent of autocorrelation in corresponding equations. Seven of the structural equations showed significant positive autocorrelation; the least-squares counterparts of six of these seven equations also showed significant positive autocorrelation. An earlier model by Klein included 37 coefficients; 26 of these were within one standard error and 32 within two standard errors of their least-squares counterparts.[7]

In the simple two-equation model for pork the similarity of the least-squares demand function to the corresponding structural equation has both theoretical and statistical explanations. In the first place, 90 per cent or more of the variation in pork *production* was attributable either to variables which were predetermined as of January 1 or to exogenous factors, such as the effects of weather and disease upon the number of pigs saved per litter. Under 1922–1941 conditions, most hogs were marketed at the age of 8 or 9 months; the gestation period for pigs is about 4 months. With this built-in lag of 12 or 13 months between sow breeding and hog slaughter (equals pork production), hog producers had very little latitude to change the *current* year's production of pork in response to the *current* year's price of pork or hogs. Hence, there was little reason to expect a significant net regression coefficient between the current price of pork and the current quantity of pork in the supply equation. As a matter of fact, this structural coefficient was non-significant, being based on a non-significant coefficient in the reduced form of the model. Moreover, as already noted, the structural coefficient was negative, whereas one would ordinarily expect an increase in price to induce an increase in supply.

If we discard P from the supply equation as non-significant, we are left with the least-squares equation

$$Q = 0.0025 + 0.6841Z$$
$$(0.0857)$$

[7] Both the structural equations and the least-squares equations for this model are presented in Lawrence Klein, *Economic Fluctuations in the United States*, 1921–1941, pp. 108–114, New York, John Wiley and Sons, 1950.

as a basis for estimating current pork consumption from a composite of predetermined factors affecting the production of pork. This new estimating equation also helps to explain why the least-squares demand function is so much like its structural counterpart. For if 90 per cent of the variation in Q is associated with a predetermined variable its statistical properties will be very similar to those of a predetermined variable. If any bias is involved in treating Q as a predetermined variable in the demand equation it would seem intuitively that the bias would not exceed 10 per cent.

On this argument, our least-squares demand equation fitted with price as the dependent variable is roughly consistent with the theory underlying the simultaneous-equations method. In fact, a structural equation which includes only one endogenous variable, the others being predetermined or exogenous, is called a "uniequational complete model." In this model a least-squares regression with the endogenous variable dependent gives us the best possible estimates of the structural coefficients. A good many least-squares demand functions for perishable foods can be rationalized on this basis.

The least-squares method is also applicable to the situation in which quantity sold (and consumed) in the current period is a function of price during the preceding period. The model here is

Demand curve: $\qquad\qquad P_t = a_1 + b_1 Q_t + u$ $\qquad\qquad\qquad$ (24.13)

Supply curve: $\qquad\qquad Q_t = a_2 + b_2 P_{t-1} + v$ $\qquad\qquad\qquad$ (24.14)

If neither u nor v is significantly autocorrelated, the least-squares regression of P_t on Q_t is an appropriate estimate of the demand curve, and the least-squares regression of Q_t on P_{t-1} is an appropriate estimate of the supply curve.

Wold[8] and others have argued that most economic problems to which simultaneous-equations methods have been applied could actually be expressed in terms of equations such as (24.13) and (24.14) above. In Wold's terminology, each equation would express a "unilateral causal dependence" and could be fitted separately by least-squares with the logically dependent variable in the dependent position. The complete set of equations would constitute a "recursive system" or "recursive

[8] See Herman Wold and Lars Jureen, *Demand Analysis*, New York, John Wiley and Sons, 1953.

See also Mordecai Ezekiel, Statistical analysis and the laws of price, *Quarterly Journal of Economics*, Vol. XLII, pp. 199–225, February, 1928. This article demonstrated, by economic and graphic analysis, the same point which Wold and Jureen demonstrated mathematically a quarter of a century later.

For a demonstration of a "recursive model," see Ezekiel, The cobweb theorem, *Quarterly Journal of Economics*, Vol. LII, pp. 255–280, February, 1938.

model." To achieve this unilateral causal dependence, the time unit of the model would be chosen to correspond to "the intervals of production planning." For many agricultural products the logical interval is a year; in general, the planning interval might vary from industry to industry. Wold's scheme appears reasonable when applied to individual commodities; its practicality has not been tested in more extensive models such as the Klein-Goldberger model of the United States economy mentioned on page 425.

Simultaneous-Equations Methods for "Overidentified" Models. The basic problem of *overidentification* can be illustrated in a two-equation demand and supply model for beef. The structural equations are assumed to be

Demand: $\qquad Q = a_1 + b_1 P + c_1 Y + d_1 W + u$ $\qquad$ (24.15)

Supply: $\qquad Q = a_2 + b_2 P + c_2 Z + v$ $\qquad\qquad$ (24.16)

where Z is an estimate of beef production based on wholly predetermined variables and W is consumption of other meats, assumed predetermined for present purposes. The reduced form of this model is

$$P = \left(\frac{a_2 - a_1}{b_1 - b_2}\right) - \left(\frac{c_1}{b_1 - b_2}\right) Y + \left(\frac{c_2}{b_1 - b_2}\right) Z$$
$$- \left(\frac{d_1}{b_1 - b_2}\right) W + \left(\frac{v - u}{b_1 - b_2}\right) \qquad (24.17)$$

$$Q = \left(\frac{a_2 b_1 - a_1 b_2}{b_1 - b_2}\right) - \left(\frac{c_1 b_2}{b_1 - b_2}\right) Y + \left(\frac{c_2 b_1}{b_1 - b_2}\right) Z$$
$$- \left(\frac{d_1 b_2}{b_1 - b_2}\right) W + \left(\frac{b_1 v - b_2 u}{b_1 - b_2}\right) \qquad (24.18)$$

We can estimate b_1 as a ratio of the coefficients of Z in the two equations. But b_2 can be estimated in two ways, one as a ratio of the coefficients of Y and the other as a ratio of the coefficients of W. If the two estimates of b_2 differ, we can also obtain different estimates for a_1, a_2, c_1, c_2, and d_1.

The reduced-form equations fitted to logarithms of the variables during 1922–1941 by least squares are as follows, neglecting the constant terms:

$$P = A_1 + 0.8185\,Y - 0.8521Z - 0.4346W + f_1(u, v) \qquad R^2 = 0.87$$
$$(0.1061) \quad\;\; (0.1877) \quad\;\; (0.1631)$$

$$Q = A_2 + 0.0509\,Y + 0.8801Z - 0.0899W + f_2(u, v) \qquad R^2 = 0.87$$
$$(0.0545) \quad\;\; (0.0964) \quad\;\; (0.0838)$$

Then $b_1 = 0.8801/-0.8521 = -1.0329$; this is our structural estimate of the elasticity of demand for beef. The two alternative values of b_2 are

$$b_{2(y)} = \frac{0.0509}{0.8185} = 0.0622$$

and

$$b_{2(w)} = \frac{-0.0899}{-0.4346} = 0.2068$$

Methods are available for resolving this apparent conflict and obtaining unique "best estimates" for each of the coefficients of an overidentified model. However the computations are quite extensive and will not be demonstrated here.

Some pragmatic comments are also warranted in the present case. First, neither of the reduced form coefficients in the numerators of the ratios from which b_2 is calculated differ significantly from zero; one is slightly smaller than its standard error and the other is only slightly larger. Starting from such unpromising materials it seems likely that any compromise estimate of b_2, which must be analogous to a weighted average of $b_{2(y)}$, and $b_{2(w)}$, will also be non-significant. If we discard Y and W from the second reduced-form equation we obtain a least-squares estimate of Q as a function of the composite predetermined variable Z:

$$Q = A_2' + 0.8878Z + f_2'(u, v) \qquad r^2 = 0.85$$
$$(0.0900)$$

If we estimate c_1 and d_1 on the assumption that $b_2 = 0$, we obtain

$$c_1 = -(0.8185)(-1.0329) = 0.8454$$

and $\qquad d_1 = -(0.4346)(-1.0329) = -0.4489$

Our structural demand function would then be

$$Q = a_1' - 1.0329P + 0.8454Y - 0.4489W + u$$

Transposing P and Q and dividing through by the coefficient of P, the structural equation becomes

$$P = a_1'' - 0.9682Q + 0.8185Y - 0.4346W + u'$$

The same variables fitted by least-squares with P in the dependent position give the following result:

$$P = a_1''' - 1.0645Q + 0.8815Y - 0.5247W \qquad R^2 = 0.95$$
$$(0.1179) \quad (0.0620) \quad (0.0887)$$

The respective coefficients differ by about one standard error in each case, or about 0.7 standard error of their differences *if* the regressions had been fitted to two random samples from the same universe. The similarities between the two equations can be rationalized on the same basis as in the case of pork.[9]

A full demonstration of computational methods for the overidentified case is beyond the scope of this book. Friedman and Foote (*op. cit.*, footnote 2) present complete computations for a lumber demand and supply model that is formally identical with our model for beef. Twenty pages (pages 30 to 49) are required to present the computations and describe the successive steps involved; much more space would be required to explain the logic of each step. The mathematical derivation of these methods is also much more complex than that of the normal equations for least-squares regression. Some of the more accessible presentations are listed in the chapter references.

However, a recapitulation of the central ideas of the simultaneous-equations approach may be helpful:

1. Variables are classified into two categories, (1) endogenous and (2) exogenous and/or predetermined.

2. One or more of the relationships to be estimated contain at least two endogenous variables. As these variables are jointly *dependent*, it is illogical to treat some of them as *independent* variables in least-squares regression equations.

Thus, if we denote endogenous variables by Y's and exogenous and/or predetermined variables by Z's, the single-equation and the simultaneous-equations models can be distinguished as follows:

Single: $\qquad Y_1 = f_1(Z_1, Z_2, \ldots, Z_k)$

Simultaneous: $\qquad (Y_1, Y_2, \ldots, Y_j) = f_j(Z_1, Z_2, \ldots, Z_k)$

In the simultaneous model we are dealing with the "regression" of a *set* of dependent variables upon a set of independent variables.

[9] The structural equations for the beef model have been estimated by Joan Friedman, using the simultaneous-equations method theoretically appropriate in the overidentified case. The equations, with standard errors of the respective coefficients, are as follows:

Demand: $\qquad P = a_1'' - 0.97Q + 0.89Y - 0.47W$
$\qquad\qquad\qquad\quad (0.24) \quad (0.13) \quad (0.18)$

Supply: $\qquad Q = a_2'' + 0.09P + 0.94Z$
$\qquad\qquad\qquad\quad (0.06) \quad (0.11)$

As expected, the coefficient of P in the supply function was non-significant. All other coefficients of the least-squares counterpart equations differed from corresponding coefficients of the structural equations by much less than one standard error of the latter coefficients.

3. The number of structural equations in a model must equal the number of endogenous variables. If there is only one endogenous variable, the model can be estimated by a single least-squares equation. Also, if any individual equation in a simultaneous system contains only one endogenous variable, this equation can be estimated by least squares.

4. The classification of variables must be based upon a logical analysis of their nature with respect to the economic model in question. Variables should not be reclassified for computational convenience or to meet the mathematical requirements for "identification." Conceivably, analysis of a model from the standpoint of economic theory might indicate that it is "underidentified" in the real world and that there is no way to estimate it from the data. To proceed with any kind of statistical estimation in this case would be mischievous and misleading.

5. It is assumed that all variables are measured without error. The exogenous and/or predetermined variables are not correlated with the random disturbances; the effects of these disturbances can appear, therefore, only in the movements of the endogenous variables.

Note, however, that the disturbances are attached to *equations* rather than to individual variables. This is sometimes emphasized by writing structural equations in the form

$$Y_1 + b_2 Y_2 + \ldots + b_j Y_j + c_1 Z_1 + \ldots + c_k Z_k + a = u$$

Any one of the Y's could be written in the extreme left position and all terms divided through by its coefficient, including the value of u for each observation, as was shown on page 424 when P was transferred to the traditional "dependent" position in place of Q. This operation does not alter the time pattern of u (except by a scale factor), and is in sharp contrast with the results of fitting least-squares equations with different variables in the dependent position. The latter almost always changes the time pattern of the residuals, which are regarded as attaching only to *the* dependent variable.

6. Although most of the empirical studies so far published have yielded structural coefficients not very different from those of single least-squares equations containing the same variables, there may be situations in which the results will be quite different and in which the structural coefficients will be more nearly correct. For many farm products it is not possible to increase the current year's production substantially in response to the current year's price. Annual observations for some industrial commodities should provide better tests of the effectiveness of simultaneous-equations methods.

Some Cautions About the Use of Simultaneous-Equations Methods. In earlier chapters we have discussed the many problems that arise in interpreting individual regression equations. Each regression is an application of statistical method to a particular subject matter problem. The choice of variables and, in many cases, types of functions to be fitted flow from one's knowledge of the subject matter. The precision with which each variable is measured may have important implications for the analysis. If a researcher works intensively with certain types of data for a number of years he may develop sound judgments concerning levels of measurement error, probable degrees of intercorrelation, and probable levels of unexplained variance in new regression analyses of such data. A person of less experience and insight applying the same basic technique to the same data may seriously mislead himself and others as to the meaning of his results.

Many applications of the simultaneous-equations approach up to this time have been marked by preoccupation with economic and statistical theory and marred by lack of cumulative and intimate knowledge of the data to which these were applied. If a "wrong" variable is included in one equation it may seriously distort the coefficients of other equations; a "right" variable subject to large measurement errors may have similar consequences. It seems clear that the earlier proponents of this approach overestimated the degree of simultaneity with which certain variables were determined and were therefore unduly pessimistic concerning the prevalence and magnitude of "least-squares bias" in applied research.

During the next decade or two, some researchers may build up the combined knowledge of techniques *and* data that is needed for adequate evaluation of the simultaneous-equations approach in various applied fields. The wider availability of electronic computers will make applications of the method feasible without consuming the researcher's time and energy in extensive computations. But difficulties in interpreting the results will continue for many years. And if Wold and others prove correct in their advocacy of recursive models, least-squares methods will remain dominant in dealing with non-experimental as well as experimental observations.

Summary

In this chapter we have presented the basic logic of the simultaneous-equations method of estimating economic relationships. This method is logically appropriate for estimating "structural" or causal relationships where the values of two or more variables are *jointly dependent* within each time unit for which observations are available.

The single-equation least-squares method to which most of this book is devoted is a valid application of the simultaneous-equations theory whenever a "structural" equation happens to contain only one logically dependent ("endogenous") variable. In some other cases single-equation methods can be made to provide close approximations to the desired "structural" coefficients.

Though designed primarily for the analysis of time-series data in economics, there may well be problems in the biological sciences to which the simultaneous-equations method is applicable. Many more empirical studies will be needed to establish clearly the practical value of this method in various areas of research.

REFERENCES

Friedman, Joan, and Richard J. Foote, Computational methods for handling systems of simultaneous equations, U. S. Dept. Agr. *Agriculture Handbook* 94. 109 pp., illus., November, 1955.

Marschak, Jacob, *Statistical Inference in Dynamic Economic Models*, by Cowles Commission Research Staff Members and Guests, edited by Tjalling C. Koopmans, with introduction by Jacob Marschak, pp. 1–50, Monograph 10, John Wiley and Sons, New York, 1950.

Hood, William C., and Tjalling C. Koopmans (eds.), *Studies in Econometric Method*, Cowles Commission Monograph 14, John Wiley and Sons, New York, 1953.

Koopmans, Tjalling C., Statistical Estimation of Simultaneous Economic Relations, *Jour. Amer. Stat. Assoc.*, Vol. 40, pp. 448–466, December, 1945.

Hildreth, Clifford, and F. G. Jarrett, *A Statistical Study of Livestock Production and Marketing*, John Wiley and Sons, New York, 1955.

Foote, Richard J., A comparison of single and simultaneous equation techniques, *Jour. of Farm Econ.*, Vol. 37, pp. 975–990, December, 1955.

Uses and Philosophy of Correlation and Regression Analysis

CHAPTER 25

Types of problems to which correlation and regression analysis have been applied

This chapter examines various types of research problems to which regression and correlation analysis have been applied, types of logical analysis made, and some of the pitfalls and difficulties encountered, and the kinds of conclusions reached.

When such a review was first presented in 1930, 77 individual studies were listed. In 1941, even a brief sampling of additional studies brought the number up to 95. In 1959, such studies have become so numerous that the titles alone would fill a book this large. In several fields (crop forecasting, supply and demand analysis, consumption studies, marketing studies, hydrology, forest mensuration, for example) whole books have been written about research methods in each field.

Brief History. The methods of correlation and regression analysis were first developed by students of heredity, notably Karl Pearson. The professional journal in this field, *Biometrika*, contains the original papers establishing the method, and many studies using it in the field of heredity. These include such studies as the relation of the stature of children to that of their parents. The very term "regression" itself comes from this initial use. When it was found that very tall or very short parents tended to have children who were on the average less tall or short, this was described as a tendency to "regress toward the mean," and the line describing this was called "the regression line."

The early studies were mainly limited to simple correlation, and put most emphasis upon the closeness of relation. Little use was made of the method as a practical research tool before World War I, although the theory had long been worked out, and outlined in standard statistical texts, especially in Yule's early editions (2).[1] Toward the third decade quantitative studies on the workings of the economic system were stimulated by Wesley C. Mitchell's pioneer work on Business Cycles (3), and a few students began to use the tool for this purpose, notably Henry L. Moore (4) and Henry A. Wallace (5). This was further encouraged in agricultural economics by the followers of Henry C. Taylor and John D. Black. Following the pioneer realistic price study on potatoes by Holbrook Working (6) in 1922, and studies under Howard Tolley on farm-management data, use of the method greatly expanded, with increased emphasis on regression.

This was facilitated by the introduction, early in the 1920's, of speedier and better controlled methods of computing the constants, as described earlier in this book, especially in Chapter 11. The use of the method spread rapidly in a number of fields, farm management, farm prices, commodity markets, crop estimating, forestry mensuration, general supply and demand analysis, and finally (in the late 1930's) to the inter-relations of the economic system as a whole (macroeconomic aspects), as analyzed by Hansen, Keynes, Tinbergen, and others.

A whole group of young research workers cooperated in the further development of correlation methods during the 1920's, notably Bradford B. Smith, Andrew Court, Louis H. Bean, and Frederick V. Waugh. As the use of these and related methods of more exact hypothesis and analysis spread in economic fields, a name was coined for the quantitative measurement of economic phenomena, and the international *Econometric Society* came into existence, with its own journal, *Econometrica*. A new language was soon developed in this field. (A few of these terms, such as "model," "exogenous" and "endogenous," and "identification," have been explained in this text.) In other fields, notably in psychology and education, a vigorous application and proliferation of the basic methods took place, though rather as a parallel and independent development. In subsequent years, the regression method has been applied widely in natural and physical sciences, social sciences, and in commercial and industrial uses.

The kinds of problems to which the method has been applied will now

[1] Numbers in parentheses refer to references at the end of this chapter.

be briefly reviewed, with somewhat fuller sampling of the earlier simpler studies.

Applications in Agricultural Technology

Weather Conditions and Crop Yields. The influence of varying weather conditions on crop yields is an obvious cause-and-effect relationship, with but little possibility of circular reasoning. Many efforts were made to identify and measure the relationships, even before regression and correlation were well understood. Some early studies, based on experimental yields, when analyzed later by simple correlation, showed $d = 0.86$ with a single weather variable (7). Analyses with more weather factors are generally needed to explain variations in crop yields, as illustrated by the corn-yield problem in Chapter 14, from Misner (8), and the potato-yield problem of Chapter 22, from Waugh and associates (9). Other studies include early ones by Moore on corn (10), cotton yields (11), and spring-wheat yields (12). Selection of variables was generally based upon farmers' experience, and upon that of workers in research stations.

An exhaustive review by Sanderson (13) of such work in many different regions and countries emphasized that many early studies used unduly flexible regression curves, and selected independent variables by retaining only those out of the many examined which showed the highest correlation with the dependent factor. This involved the danger that with such samples variables might be retained whose high correlation was due to chance fluctuations rather than true relationship. This resulted in regression equations which often did not perform well when used for subsequent forecasting. More recent carefully controlled studies have given better results, as for example on wheat yields in Canada, which gave high correlation ($\bar{I}$ of 0.81 to 0.94), and performed well when tested in forecasting subsequent yields (14). Other recent weather-yield studies include one in Peru (15) on potato yields.

Fisher, studying wheat yields at Rothamstead, pointed out that it really made little difference to the growth of a crop whether a given rain occurred on April 30 or May 1. Yet, if weekly periods were considered for all different factors, the number of constants in the regression equation might be excessive. He therefore applied a differential regression method determining the rate of change in yield with rate of change in rainfall throughout the growing period. With only a few constants, the resulting smooth curve showed that the maximum effect of rainfall on yield was in autumn and in spring. With rainfall distribution the only weather variable considered, correlations ranged from 0.32 to 0.63 (16). Sanderson developed a simplified method of making the calculations (17) and modifications to introduce joint functional relations where needed. The method has the disadvantage, however, that as yet it cannot distinguish between the effects of too little and too much rain *at each given season*; i.e., it cannot allow for curvilinear effects.

Physical Relations between Input and Output. Another type of problem particularly important in agricultural research is to determine the production function; i.e., the physical relation between the quantity of input in production and the resulting output.

In one pioneer study the gain in weight of beef steers was related to the quantities of each of several different feeds supplied, the length of feeding period, and to initial weight (18). Curvilinear regressions were found, and they showed a marked tendency towards diminishing returns.

Parallel analysis has been made of milk production as related to feed inputs and other variables. In most of these studies (19–22) the feeds used in milk produced have been considered on a herd-average basis. In one study with data for individual cows, the results agreed quite well with those based on herd averages (23).

Similar regression studies of the influence of physical input upon output have been made for potatoes (24, 25), cotton (26), and other crops.

In one study of cotton, mixed fertilizer, nitrate of soda, calcium arsenate applied, and the fertility of the land (as indicated by the yield of other crops, notably corn) were considered as separate variables. The results illustrate some of the logical problems which arise in regression analyses. In the year studied there was a heavy weevil damage on untreated fields, and the applications of arsenate increased the yield very materially. But these results show nothing of the effect of poison on yield when weevil infestation is slight. It would be necessary to repeat the study over several years with varying weevil damage on untreated fields and relate the differences in the effectiveness of poison to climatic and other factors, before it would be possible to judge whether or not it would pay to use poison in any particular year—and the prices both of poison and of cotton would enter into the final consideration.

Modern examples of such analyses (173) applied to fertilizer experimental results fitted by a joint function were shown in Chapter 21, with functions fitted algebraically, and only statistically significant terms included (Figures 21.7 and 21.8).

Other recent applications to agricultural production problems reflect increased specialization in research. Dot charts and simple correlations and regressions are widely used in interpreting experimental results, partial and multiple correlations and net regressions less frequently. In crop production, such analyses have been applied to such diverse problems as the relation of artificially produced wheat hybrids to their parents (27), of physical characteristics of different oat varieties to their yields (28), and the effect of weather conditions on the staple length of cotton (29); in livestock production, to studies of livestock physiology, such as reactions of animals to varying temperatures (30), prenatal development rates in lambs (31), factors affecting lamb weights, both hereditary and environmental (32) (with four independent variables and two joint functions), the effects of age and nutrition on hormones in cows (33), the physiology of swine digestion (34); and in livestock feeding, to determine the optimum levels of supplying different forms of sulfur in feeding lambs (35), and factors affecting lamb weights at weaning (36).

Relation of Physical Characteristics of Samples to Chemical Characteristics. Regression analysis has been widely used in determining how chemical properties could be estimated from observable physical properties. The example of protein content of wheat estimated from vitreous kernels (Chapter 6), was taken from comprehensive studies (37) in which weight, vitreous kernels, and the region of the country where grown were all found significant. The volume of bread is related to the gluten content of wheat and flour (38). The digestible composition of meat can be judged from the proportion of visible fat (39); the tenderness of the cooked meat from beef fat (40) or other characteristics of muscle fibers (41); or digestible starch from the per cent of crude fiber (42). Regression analysis may thus be used to generalize from tests under carefully controlled conditions, and to develop more rapid methods of estimating under everyday application, as in grading grain commercially or estimating nutritive value in the home.

Physical Appearance and Productivity. Earlier, crops and livestock were often selected for breeding on the basis of their outward size, shape, etc. Correlation studies have investigated the assumptions on which these practices were based. Studies of dairy cows by Gowen (43) indicated that the physical conformation of dairy cows generally had little relation to productive ability; and studies of corn showed that size and shape of corn kernels, ears, and plants had little relation to actual yielding ability (44, 45). These studies indicated that many of the time-honored points stressed in agricultural show competitions and in breeding selection had no utilitarian significance, and led to a new stress on performance records, rather than on physical appearance. Other work on animal conformation showed that scores by qualified judges had only a limited relation to the actual measurements, also only limited agreement between judges in different regions (46). Even the progeny testing of bulls under experimental conditions has been found inadequate to predict results under actual farm conditions, and "the repeatability of the special station results in the field are very low, even in the high herds." (47)

Application to Physical Inter-Relations in Other Fields

There are many other scientific and practical fields where regression and correlation analysis help both in the testing of hypotheses and in deriving regression equations by which difficult-to-measure or unknown variables can be estimated closely from those which can be readily measured or others already known.

Estimating Missing Water-Flow Records. With economic development, the question frequently arises as to whether the flow of a given river is sufficient to establish an irrigation system or hydroelectric development, or how large a dam and storage basin is needed to prevent floods. The best dam site is often at a point where few if any rainfall or water-flow records have been kept. As soon as the possibility begins to be examined, a stream-gauging station will usually be established, so that records for a few years—say 5 to 10 years—become available. Hydrologists have turned to regression analysis to extend these too-short records. Using weather or water-flow data at other locations, and relating them to the available data at the point desired, a regression equation is obtained to extend the record with a known degree of accuracy. One simple example was given in Chapter 5. Two general methods of approach are used (48); to correlate the reported flow at the desired point with that at other points on the same river, higher and lower if possible; or to correlate it with the flow on other nearby streams or rivers. Differences of elevation and configuration, as well as any local climatic features, also must be considered. (For example, rainfall may be heavy on one side of a divide, but light on the other, so that being on the same side would be more important than actual distance.) Multiple regression studies indicate which stations give the best estimate of stream flow at a desired point. A second approach is to estimate the probable basin runoff from the available records on rainfall, temperature, snow cover, topography, etc. These are combined to estimate runoff by elaborate engineering calculations, which themselves involve correlation studies at various points. By proper design of the analyses, estimates based solely on calculated runoff from weather records over the period when local stream-flow records were not kept could be compared with corresponding estimated stream flows based solely on flows at other points during those years, and an independent check could thus be obtained on the confidence with which those estimates can be used. Such a check would reduce the danger of the estimates being less reliable than their standard errors indicated, for reasons stated on pages 297 and 436.

The problem of estimating water flow has an extensive literature of its own among hydrologists, engineers, and other workers in this field (48), including combining records of stations for differing but overlapping periods of time (49), and combining measures of rainfall, runoff, snow cover, etc. with stream-flow records (50).

Applications in Industrial Engineering. Regression applications in engineering concern relations which could not be measured readily in the laboratory, such as the effect of local water characteristics on the amount of deposits inside water or steam pipes. They could also be applied to relating the weathering of different paints to the varying weather conditions to which they were exposed. Other applications in engineering have been suggested in recent literature (169). Studies are reported in the Netherlands of the relation of temperature to gas consumption (170), and in Brazil of an equation which related ultrasonic velocity in organic liquids to their refractive index and molecular structure (171). From the published work, applications in industry seem to be much less frequent than in other fields. That may be because here, as with industrial price studies, the work that is done is usually kept for private use within the business or corporation.

Estimating Volumes of Irregular Objects. The example in Chapter 22 is an illustration of deriving practical approximations to measurements that can be made exactly only with great difficulty and expense. Estimating the volume of usable lumber in a standing tree is another important application. Height of a tree and diameter of circumference near the base can easily be measured, but the exact content of usable lumber can be exactly determined only after the tree has been felled, trimmed, and the trunk then measured with great care. This problem is of great practical importance in practical logging, forest surveys, and forest experimental plots, which depend on accurate estimates of size and growth *without cutting* the trees. Multiple regressions have had significant success in deriving formulas and alignment charts for various species that are easy to use in practice and that have a high degree of reliability. Work in this field contributed to the early development of multiple-regression methods (51), and subsequent work has produced a sizeable special literature (52–54), with special methods of its own (181).

Applications in Astronomy. Astronomy, an exact science, calculated the movements of the various heavenly bodies by fitting equations to the observed positions, obtaining almost precise fits. The use of fitted equations was routine in this field long before they began to be used in less exact sciences through regression methods. In modern times, however, with spectroscopic and other new information, galaxies number in the thousands and millions; with the Einstein hypotheses and atomic-energy discoveries, statistical methods began to be applied to such problems as general laws of brightness, shift in color, apparent speed and distance, population of galaxies in space, and the whole idea of the ever-expanding universe. Studies in this field now use regression methods extensively. but the literature is too vast to give many references (55–57). The latest application is in computing the orbit of a "sputnik" from data reported by many different observers.

Applications in Agricultural Economics

Most of the problems considered to this point deal with relationships in existing universes or ones likely to remain substantially unchanged over the relevant time. But many problems in economics and other social sciences deal with universes that are subject to human influence and change at least gradually over time. Non-experimental data that exist only in the form of unique observations for each successive interval of time are particularly hard to appraise in terms of random sampling from a clearly defined universe.

The conceptual advances made in recent years by Koopmans, Wold, and others concerned with time series have been noted in Chapter 20. The sample-universe relationship can be established reasonably well in economic data, time series or other, *if* one is careful to define (and sample from) a universe that does not change materially over a specified period. For example, although *trends* in corn yields have no sampling significance with respect to later trends, the universe of net relations between rainfall, temperature and *yield deviations from trend* may very well remain constant, or nearly so, for many years. Similarly, in a regression equation explaining changes in the price of a farm commodity, one coefficient might remain nearly constant for many years; another might change gradually over time for fairly clear reasons; and a third might change quite radically over a five-year period again for clear or at least plausible reasons. The reasons may be demonstrable only if a dense network of current (and accurate) economic data exists for the commodity.

It is not surprising that some of the earlier regression analyses of problems in agricultural economics, particularly those based on time series, suffered from ambiguities in defining the universe with respect to which inferences from a given set of observations were justified. On the other hand some of the analyses made in this field during the 1920's would meet all present-day standards of workmanship and statistical reliability.

Farm Values as Related to Farm Characteristics and Other Factors. Estimates of the sales value of farms are needed for levying taxes, securing loans, or setting an offering price. Objective methods for making such estimates are therefore very important. Such estimates involve (1) the values of individual farms at any given time as affected by their character-istics and their rights to share in public subsidies or payments; and (2) changes in the general level of farm prices, with shifts in farm prices and taxes, and public controls, subsidies, etc. There is a specific universe, for the first, for each given period of time, and a large universe of farms which can be sampled. The second, with changing prices, costs, and public laws and regulations, involves time-series analysis. Since there are millions of farms in a large country such as the United States, operating under different physical, social, and even economic conditions, great geographic variations are available for study, and analyses can be made simultaneously in time and space.

The simpler problem was attacked first in a pioneer study by Haas (58). Actual sales prices of a large sample of farms in a given region were obtained, and also relevant facts such as distance from town, value of buildings, proportion of crop land, fertility of the soil, and type of road on which the farm fronted. Trends in land values were first eliminated, and the adjusted acre prices were related to the other factors mentioned by linear multiple correlation. Acre values estimated from the independent factors had a standard error of $19 per acre, with $R = 0.81$. As assessed valuations showed a much larger error, it was suggested that the impartial regression equation might be substituted for the less reliable human judg-ment in assessing individual farms for taxation purposes. A similar analysis, made much later with the same objectives (185), gave parallel results.

In another study a joint function was found for farm dwellings, with an expensive dwelling adding more to the value of large farms than of small ones. The net relation of value to type of road was determined by the method of Chapter 22. Preliminary work on this study, with farm values stated on a per-acre basis, gave a linear correlation of $R = 0.98$. This was due to the presence of a few very small farms, which showed farms and building values per acre both running into thousands of dollars. With these farms excluded, the linear multiple correlation dropped to $R = 0.64$, indicating the extent of the previous spurious correlation. Using curvi-linear relations and joint functions, the final analysis showed $I = 0.77$. With 368 observations available, more complex methods could be employed than would be feasible in most cases (59).

The second question was dealt with in a study by Chambers over the long period of rising land values prior to 1920. Farm-land values then reflected not only changes in farm incomes, but also discounted the expected continued rise in incomes. Further analysis covering the 1929 depression, World War II and the subsequent inflation period, and the changing public interventions, would have to examine far more complex issues.

Relation of Farm Organization to Farm Income. The combination of enterprises which will produce the best returns to farmers in a given locality has been much studied using data from farm surveys, detailed farm accounts, and other sources. Initially, the data were usually examined by simple sorting, often based on the dependent variable, with differences in farm income ascribed to concomitant differences in the average values of the other variables. Application of multiple correlation to such data gave different and more specific conclusions. Studies based on surveys in Pennsylvania (59), Iowa (60), and Virginia (61) considered such variables as farm size, crop acreage, size of important livestock enterprise (number of cows or hogs), efficiency of crop production and livestock production, and capital investment. About half the variation in earnings from farm to farm was usually explained. The dominant influence was usually the size and efficiency of the major enterprise, such as the corn and hog enterprise on Iowa hog farms; tobacco acreage, yield and quality, on Virginia tobacco farms; and the number of cows and efficiency of milk production on Pennsylvania dairy farms. Many similar studies were made in other areas.

The value of such statistical studies is, however, distinctly limited. The results hold true only for the year of the data, due to fluctuations in yields and prices. Each individual farm is a different entity, and the combination of enterprises which produces the best results *on the average* will not necessarily be the best for any one individual farm. If it were possible to observe one farm under a hundred different types of organization, but with the same price and yield conditions, and record the resulting profit secured under each organization, it would then be possible to determine what combination would yield maximum returns for that farm under the stated conditions. This can, however, be estimated by the farm-budget method of analysis (56, 62), which computes the probable income under alternative combinations of enterprises on a specified farm, and with varying intensities of operation and with selected combinations of prices and costs. Such an analysis can provide a guide to the most profitable farm organization for any desired combination of conditions (57). This method has therefore largely replaced over-all regression analysis based on large numbers of farms. The latter served to identify the major factors involved (172) and still assists in the input-output analysis (18) needed in farm-budget estimates. Since 1950 a more exact method of analysis, linear programming, has been applied, most extensively by Earl Heady (178, 182), in choosing input-output combinations that will maximize net income.

Efficiency of Organization of Marketing Units. Regression analysis has been applied in studying the efficiency with which individual market facilities—elevators, flour mills, milk-receiving stations, etc.—are organized and operated. Data from a sample of enterprises are analyzed with respect to the relation of direct and overhead costs per unit, to size, capacity utilized, organization of physical facilities, etc. Such studies provide managers and cooperative or business concerns with indications of points to watch for the most efficient results (63, 64).

Relation of Commodity Prices to Economic Conditions

Most price studies involve time-series analysis, with only a single observation available for any given period of time. However, there is usually enough continuity in the way that individuals react in the aggregate that fairly stable results can be secured from such series. Where the change in reaction is continuous and progressive, that trend itself can be made one variable in the analysis.

Price studies may be separated broadly into (1) those which follow the usual single-equation regression technique of taking one factor as dependent upon several other variables considered as given or independent, and then determining how the first may be estimated from the others; and (2) those that take two or more factors as simultaneously interdependent, and investigate the most probable nature of that interdependence by a system of equations simultaneously determined. Several other factors differing from equation to equation may, however, be held constant or taken into account at the same time (see Chapter 24).

When single-equation studies are used, prices may be examined with any one of several factors regarded as dependent, notably (1) central market (wholesale) prices; (2) consumption; and (3) production.

(1) Factors Affecting Central Market Prices. The earliest studies were those investigating the effect of production or total supply upon price, at some representative market. These studies take as independent variables factors such as production or supply for the season, and current general business activity, which are either predetermined or independent of the market price, and then study their apparent net influence upon the price of the given product. This avoids the circular reasoning that may be involved if prices of one product are related to independent variables, such as prices of competing products, which may in turn be influenced by the price which is to be explained.

ANNUAL PRICES. The simplest price studies are those which relate the market price for an agricultural commodity to the supply for a marketing year. Early work by Moore (65) indicated the general relation of price to supply for corn, hay, oats, potatoes, and cotton, with changing conditions eliminated by first differences. Subsequent work on potatoes (66–68), oats (69), and cotton (70, 71) included price levels, trends in demand, carryover, and prices of competing products as independent factors influencing price along with supply. With relatively short periods on which to base the analyses, exceedingly high correlations were secured in many cases—frequently above $\bar{R} = 0.95$. Advance forecasts, however, met with variable success, working well sometimes, and other times missing by wide margins.

More extensive recent work in the same field has covered all important commodities in many countries, and has included testing equations from the interwar period when "transplanted" to the postwar period. Fox based an extensive set of price analyses for United States farm and food products on data for 1922–1941, and found that most of them continued to forecast reasonably well during the postwar period. In many commodities the effect of governmental intervention had to be taken into account (72, 72a, 72b).

MONTHLY PRICES. With monthly prices considered and a larger number of individual observations, more elaborate studies are possible. There are, of course, serious questions as to how completely the successive monthly prices of a staple commodity are really independent of each other, i.e., how far they are affected by autocorrelations. Since many conditions are constantly changing, there are usually some elements of independence. An early cotton-price study by Smith (73) attempted to measure separately the influence of actual and of prospective supply on price, and the shifting of the two regressions through the crop year, using joint functions. A study of hog prices (74) developed both an empirical forecaster of prices, and an economic interpretation of the influence of market receipts, storage stocks, competing products, and business conditions on prices. The forecasting equation, obtained by selecting the variables with the highest correlation and the best lags out of many tested, and then fitting very flexible freehand curves to these variables, was subject to the weaknesses discussed on pages 290 to 298 of Chapter 17. Subsequent analyses have materially modified many of the conclusions. Monthly price forecasts of hog prices based on this study were less accurate than were forecasts which took more elements into consideration, and held more strictly to the original hypothesis (75). Parallel studies of monthly swine prices in Germany by Hanau (76) yielded reasonable results and gave forecasts which worked well in practice. A study of monthly prices of dressed lamb (77) gave a correlation of 0.98 for the seventeen years studied. The regression equation served to estimate monthly prices for a long period of years afterwards, from new current data for the independent variables, with practically no increase in the standard error of estimate. The independent variables were per capita supplies, price level, prices of competing meats, and business activity, and trend and seasonal factors simultaneously determined. Monthly price studies have also been made for perishable and canned fruits (187–189).

WEEKLY OR DAILY PRICES. For very perishable products, price studies need to deal with the average prices for each week or even for every day. Early studies of this type are those of watermelons by Hedden and Cherniack (78) and of peaches by Kantor (79) for New York City. Both studies took into account variation in demand during the week. Temperature had a marked influence on watermelon prices.

Similar studies for shorter or longer time intervals have since been published for other perishable products, including studies on canteloupes by Rauchenstine (80), on Bartlett pears by Hoos and Shear (81), on Louisiana strawberries by Mehren and Erdman (83), and on canned and citrus fruits by Hoos (187–189).

(2) The Effect of Price upon Consumption. This was examined in many early studies, especially of milk prices upon consumption (86). Later intensive studies have investigated demand curves in particular markets both by time-series analyses, and more generally by studies in space between regions or countries. Where ultimate consumption is being examined, retail price should logically be used to study consumption responses (72, 84, 85). In the United States differences in real income in time, or per capita in space, have been included as independent variables, and logarithmic transformations have been used to yield separate measures of the elasticity of consumption with respect to prices and real income (87). Extensive modern work on this field has been done by Stone in England (88), and by Fox and others in the U. S. Department of Agriculture. Wold and Jureen's rigorous study of demand provided a specially detailed examination of the hypotheses involved and of appropriate methods, in the light of these hypotheses (89). Independent measures of the income elasticity of demand derived from family budget studies agree moderately well with those derived from time-series and intraregional or intercountry studies, but in some cases the range has been wide (90). Work has also been done in Canada on factors affecting consumption of beef and pork (91), and meat generally.

Early studies for milk (92) showed that the other factors had a larger influence upon consumption than did prices. With cotton, price alone, and an upward trend in demand, almost completely determined world consumption (93), but United States consumption was also influenced by industrial activity. Later work has taken into account competitive synthetic fibers, as these become increasingly important. For potatoes, demand is very inelastic, and in years of low prices producers feed much larger quantities to livestock or allow them to go to waste. When the price falls very low, much of the supply is left in the ground undug (93). This "reservation demand" by producers involves a concurrent adjustment of supply to price, which might need analysis by the methods of Chapter 24 (72, pp. 10–11).

Studies of both cocoa and sugar consumption (94, 87) have been made in different countries with respect to the effect of both prices and average incomes on consumption during the same period of time. Similar studies

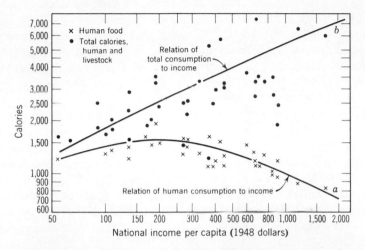

Fig. 25.1. Per capita consumption of calories per day in various countries in relation to national income per capita. Crosses indicate consumption for human food; solid dots, consumption for food, live-stock feed, and other purposes. (Food and Agriculture Organization of the United Nations, *State of Food and Agriculture*, p. 89, 1957.)

have been made on coffee consumption (95). The relation of expenditure on different foods to total family income has been studied for different income groups, at the same place and time, at various places, and in different countries (96) (Figures 25.1 and 25.2). Dot charts have been drawn to show different countries at several successive times. In many of these studies, the consistency between comparisons in time, in space, and between income groups, has usually, but not always, been remarkably good (90). Parallel work on consumption has been done in many countries and on many commodities (97, 98, 192, 194).

(3) *Relation of Changes in Production to Prices and Other Factors.*
In studying changes in output, prices in one period are related to acreage or production in some subsequent period or periods, with a lag depending on the length of the production process and on the time it takes producers to respond. An early study related cotton acreage to price for the previous

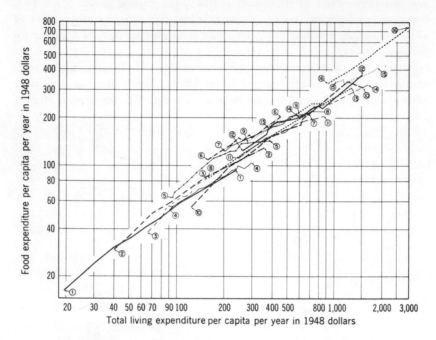

Fig. 25.2. Annual food expenditures per capita in various countries, for various levels of income per capita in each country.

1. India, 1954. Farindabad Township.
2. Ceylon, 1952–1953. Total population.
3. Ghana, 1955. Kumasi.
4. Japan, 1954. Towns of over 50,000 inhabitants.
5. Portugal, 1950–1951. Porto.
6. Portugal, 1948–1949. Lisbon.
7. Austria, 1952–1955. Towns of over 10,000 inhabitants.
8. Ireland, 1951–1952. Total towns and villages, farmers usually excluded.
9. Finland, 1950–1951. Urban married couples.
10. Panama, 1952–1953. Panama City.
11. Switzerland, 1936–1937. Workers and employees.
12. Sweden, 1952. Total population.
13. Sweden, 1948. Urban families with children.
14. Canada, 1948. Total non-agricultural population.
15. Canada, 1953. Five large towns.
16. United States, 1950. Large cities.

seasons (99). Subsequent experience showed, however, that continued high prices for two seasons had greater influence than for a single season (100). With hogs it took eighteen months for price changes to be reflected substantially in market receipts (75). Corn prices were as important as hog prices in affecting hog production. Studied separately for different type-of-farming areas, there were marked differences in responses in different areas to the corn/hog price ratio, depending on the position of the hog enterprise in the farming system (101). Milk production responded to changes in the milk-price/feed-cost ratio (102), with a short-time lag, which might reflect the intensity of feeding, and a long-time one, possibly due to changes in numbers of cows (103). Egg production reflects changes in egg prices, feed costs, and improving technology (184). The acreage of potatoes reflects prices for two years preceding, with important regional differences (100). In England, wheat acreage reflected purchasing power of the preceding crop (104). For minor crops, prices of the major crop of the region must also be considered. Thus sweet-potato acreage is influenced by cotton prices, and flax acreage by wheat prices. In some cases, it has been found that yields or per-acre returns for preceding years must be considered, as well as prices alone. The general price level of competitive products or of all commodities has also usually been considered. Because of the lag between price and production response in agriculture, the single-equation analysis has usually proved quite satisfactory for measuring the relationship.

It has often been found useful to state the subsequent acreage or production as an absolute or relative first difference. Often a very high price will not call forth any larger increase in production in the following year than will a moderately high price, perhaps owing to the inability of producers to expand operations more than a certain extent in any one year. With acreage controls and other new elements in the situation, these early results cannot be readily checked against the postwar data, except in a few cases, but the effects of the interventions may, however, be estimated (175, 177). Hog cycles have continued in the United States despite corn controls and price supports. Nerlove (179, 197) recently emphasized that farmers adjust to expected prices rather than past prices, and he tried to infer from historical data on prices and acreage change what these expected prices may have been. His exploratory studies for corn, wheat, and cotton imply that prices expected are only about half as variable as actual prices preceding planting, and that elasticities of supply with respect to expected prices are about twice as large as those obtained by relating the current year's acreage to prices of the preceding one or two years.

Comprehensive Studies of the Interactions of Supply, Demand, Income, and Price. A number of modern studies have devoted attention to describing the whole complex of forces and chain of events affecting all aspects of the pricing process, and their interplay with each other, particularly the "demand and price structure" series of the U. S. Department of Agriculture, initiated by Fox and Cavin and supervised by Foote. Fox (72) published diagrams similar to Figure 20.1 for major livestock products and various crops. Each diagram contains ten to twenty-five variables with arrows indicating directions of influence and expected relative importance of the different channels of influence (net regression coefficients). · The diagrams typically implied that some elements could be estimated by single-equation methods, but that others should logically be estimated by the simultaneous-equations methods of Chapter 24 (see also 193). Such studies include one on wheat (105) based on six structural equations which provide a complete hypothesis as to the direction and nature of the influence of all the variables in the system. Many were fitted by both the least-square (single-equation) and the simultaneous-equations approaches, with very little difference in the results from the two methods. Other parallel studies have been made on the feed-livestock economy (106), corn and total feed concentrates (107), dairy products (191), food fats and oils (108), and coffee (183). A comparable German review used both single equations and simultaneous equations, and includes an interesting summary of the nature of all the cross-elasticities among commodity groups (109). Sugar was similarly studied in Austria (199).

Factors Related to Price Margins. Other studies consider prices at different points in space or at different steps in the marketing process. Regression analysis has been used to measure the relative influence of changes in freight rates, location of supplies, and price level on the margin between potato prices in Minneapolis and New York (111). Since production affects *wholesale* prices most directly, *farm prices* affect production, and *retail prices* affect consumption, a complete explanation of the chain of price-supply-consumption events must also explain and analyze the links between the several market stages, farm-wholesale-retail. This phase of the problem seems in general to have been less intensively studied by means of regression analysis.

Relation of Characteristics of Individual Lots of a Commodity to Prices. The price studies discussed treat changes in prices from time to time, for lots of the commodity of uniform or of average quality and usually at one selected stage of the marketing process. Except where different observations are made in space, only one unique observation can be drawn from each successive time period. In determining why different lots of the same commodity, sold within a given period and at the same stage of the marketing process, should sell for different prices, sampling theory is more directly applicable. There is a true universe—all the sales of the specified kind taking place within the specified period—and as large a sample as is desired can be secured, up to the limits of the universe. Studies of farm prices are one example of this type, and the relation of the price of apples to size, insect injury, and scab, is another (Figure 21.10).

One study related the prices of different lots of asparagus to the length of green color, stalks per bunch, and uniformity (112). The conclusions presented effectively in pictorial style (see Fig. 25.3), had a marked influence on producers' practices and led to further experiments on how to produce asparagus with the desirable qualities (113). Similar studies, one of which was used as an example in Chapter 22, related to quality factors affecting egg prices per box in New York and Philadelphia (114) and in Wilmington (115).

Premiums for high-quality lots may vary from time to time with differences in the relative supply of different qualities. The conclusions apply in the universe from which the observations were drawn under the existing supply of the different sizes and qualities. They might not apply in another universe with different circumstances. Other studies, therefore, examine how premiums or discounts vary with differences in supplies of each quality. In crop years when high-protein wheat is very scarce, wheat with high protein content commands a marked premium, and that factor is much more important than weight; when high-protein wheat is more plentiful but much of the wheat is underweight, the weight factor becomes more important, and the protein premium becomes of much less significance (116). These studies used over 1,000 individual cars per year for several years, for monthly data, and fitted two joint functions for month and protein content, and for month and weight per bushel.

Similar studies on cotton related variations in supplies of each staple length to premiums. Premiums for peaches changed with the supply of each competing variety, but premiums for large sizes persisted despite increased supplies, though they were somewhat reduced (79).

These last two groups of studies illustrate concretely how universes changing in time may be subjected to statistical analysis, and how conclusions of value for new sets of circumstances may be reached. When conditions change in time with differences in supply of different sizes, qualities, or varieties, or with recurring changes in demand from day to day through the week, or from month to month through the year, factors related to the changes may be determined and allowed for. Just how far the conclusions will hold subsequently depends upon how adequately the real causes of the changes from time to time have been determined, and how much unaccountable dynamic or evolutionary change there has been and may be.

The application of statistics to demand and price analysis has resulted in several comprehensive books in the field. The pioneer study by Henry Schultz (117) has now been followed by texts by Shepherd (118), Fox (72b), and Wold and Jureen (89), and, on a less technical level, by Thomsen and Foote (119) and Waite and Trelogan (120).

Increased understanding of the real forces affecting supply, demand, and price has made it possible to prepare statements on the outlook for

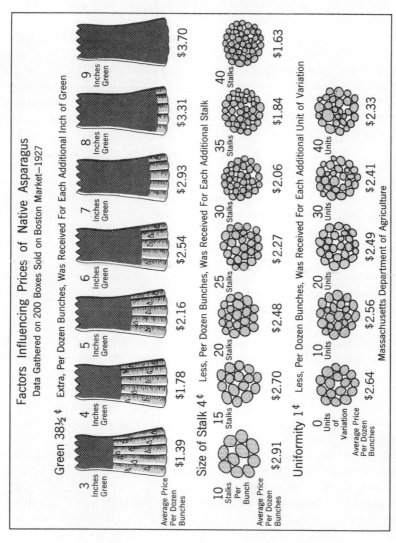

Fig. 25.3. A pictorial presentation of conclusions reached by a multiple-correlation study. (From Frederick V. Waugh.)

individual commodity situations, including forecasts of prospective change for as long as one to three years ahead. Forecasts of this sort have been published annually or more frequently by the U. S. Department of Agriculture for more than 35 years, with an accuracy on *direction* of movement of over 75%, and more limited forecasts for international commodity markets, published annually by the Food and Agriculture Organization of the United Nations for nearly a decade, have shown a comparable accuracy (168). The application of quantitative methods, including correlation and regression analysis, to price-making forces has thus begun to give economic analysis some of the ability to predict future events which is the ultimate test of real science.

Applications in General Economic and Industrial Studies

Correlation methods have been applied to economic problems in the general economic field in ways generally parallel to those in agriculture. With relatively less work done by public agencies, and more by private businesses and corporations for their own use, the amount of published studies in the non-agricultural areas is as a whole smaller than in agriculture, but the amount is still very large. Only a few illustrative examples can be given here.

Production Functions for Industries. The relation of volume of output to average cost per unit is an important consideration both in economic theory and in industrial organization. One example drawn from this field was used in Chapter 16. The cost function for such products as steel, hosiery, and furniture, has been studied for certain large concerns (121–123), using multiple regressions in some cases to examine total cost or per unit cost, as affected by percentage of capacity operated, wage rates, price levels, or the upward trend in labor efficiency. Another (unpublished) simple correlation analysis by the War Production Board of World War II showed that when a new model of tank, ship, or airplane first went into production, the labor per unit was very high, but as more units were produced in the same plant, efficiency increased and labor hours dropped, with a straight-line relation when both variables were expressed in logarithms. Studies of industrial production functions were summarized by Hans Staehle (201). Paul Douglas pioneered in studies of labor productivity (130).

Price-Making Forces for Industrial Commodities. The same methods used earlier with farm-product prices were later applied to the analysis of industrial demand. Steel (124), automobiles (125, 176), houses (126), and ships (127) have been the subjects of some of these studies. In fields where free competition does not prevail, but the dominance of a few large firms produces monopolistic competition, supply and demand relations may operate quite differently from the way they operate under freer atomistic competition (128). The situation is further complicated by the development of powerful labor unions in many industries which may at times raise their wages with little regard to increases in productivity. Great care is then necessary in setting up any statistical analyses in ways that represent the actual market situation. Regression has also been used in judging to what extent increases in productivity in various industries were reflected in corresponding wage increases or price decreases (129).

Determination of Utility Rates. A series of studies for public and private streetcar, bus, and other utility concerns related the density of traffic to fares, frequency of service, levels of employment and income, and motor vehicles per capita. These analyses served to forecast the probable effect on the volume of traffic and earnings from proposed changes in rates or service, with good accuracy. Sometimes they have shown that a proposed rate increase would so reduce traffic as to cause a loss, and a reduction in fare would increase net income. Since these studies are made as a commercial service, they have not been published. Parallel studies have been made of the effect of rates and other factors on the demand for taxicabs in Amsterdam (131), and on the relation of the density of automobiles in different countries to income per capita and taxation per car (132), by Tinbergen and his associates.

Factors Affecting International Trade. Studies of factors affecting the volume and direction of international trade have measured elasticity of import demand with respect to price in various countries over various periods of time, and in some cases, elasticity with respect to income. A recent study in this field (190) came to the conclusion that "the price mechanism works powerfully and pervasively in international trade," postwar as well as prewar.

Size Standards for Children's Clothes. A study of appropriate size standards for children's clothes examined body dimensions for thousands of children, together with their age, sex, and race. Height and girth at hips were found most important in judging size as a whole, with age having no net effect. A new system of clothes sizes, based on the joint distribution of these two measurements was recommended to clothing manufacturers (133).

Sales Quotas for Local Districts. Corporations planning sales campaigns, advertising budgets, or location of branch offices, need to estimate the sales potential of individual counties or other local units. Facts about the county, such as population, income, value of farm production, etc., are related to past sales by multiple regressions, and used to estimate future prospects. Naturally the weighting equations are very different if farm machinery, automobiles, or motion pictures are studied. Meat sales and quotas for different branch houses have been studied by such methods, for example (134).

Other Industrial Applications. Some other industrial applications are the use of multiple regression to forecast pig-iron production (135), and to determine how freight rates vary with terminal charges, length of haul, expense of operation, etc. (136).

Multiple regressions are extensively used in analyzing stock prices as related to business conditions. Business concerns sell their products a considerable period before the resulting profits or losses are reflected in their profits reports and dividend payments. Multiple-regression equations based on factors such as production or sales, prices, wages, or general business activity, with appropriate lags, often yield reasonably reliable forecasts of subsequent changes in stock prices, at least relative to the general market level. Multiple-regression equations for individual stocks are advertised for sale on the financial pages of the leading newspapers. Wider economic changes or unsuspected political developments may upset forecasts based on such regression equations, but if the study has been made over a period of time long enough to include such events, those risks, too, should be reflected in the observed standard error of estimate.

Application of Correlation to "Macroeconomic" Relations. An interesting application from the point of view of economic theory is measuring over-all or "macro" aspects of the behavior of the economic system, in line with the theories of Currie, Keynes, Hansen, and others. Hansen used a simple regression to show the Keynsian savings function for the American economy (137). Ezekiel made a general regression study of the savings, investment, and consumption for the United States, which yielded a quantitative statement of the under-employment equilibrium during the 1930's (138). This study handled an equilibrium problem by a series of single-equation solutions with lagged variables showing how past events influenced subsequent developments (i.e., business profits and subsequent investment). Such a set of equations is now called a "recursive system" by econometricians. There have been many other studies on similar lines, using more adequate data on national income, gross national product, and flow of funds in the economy (200). One notable study (139) demonstrated the so-called "Duesenberry effect," of the consumption function shifting in a series of steps and tending to hold to the previous high level when anything caused a down-turn in consumer income. Much parallel quantitative work has been done in England, Holland, and other European countries (140, 141, 198).

Policies based in part on such statistical studies may have contributed to the fact that during the decade following World War II there was at no time a sharp industrial depression of the sort that happened twice after World War I, and that such mild recessions as did occur were soon corrected. Regression analysis has also been applied to explain technological change and economic development (195, 196). New problems have arisen—those of creeping inflation and of rising price levels around the world—despite farm surpluses and steadily rising output of goods and services in nearly all countries. Such problems, too, are being subjected to careful theoretical and quantitative analysis, which may help the nations concerned to find effective solutions to them, too.

Correlation and Regression Methods in Political Science and Politics

Statistical analysis has been extensively applied to political behavior. The Gallup Poll, the Roper Poll, and others have become almost household words. Along with these, regression analysis has been used by Bean and others to establish many political relationships, such as those between votes by states and by the nation, and to develop the predicting reliability of opinions or votes in particular areas (142, 143). Regression methods have also been used in detailed studies of political structure and behavior in particular cities or localities (144).

Correlation Methods in Psychology and Education

In educational and psychological investigations correlation and regression methods are applied to the study of such problems as the relations of grades in different subjects, scores on different mental tests (146), or the relation of scores on mental tests to success in the schoolroom (147) or in later life (148).

Studies have also been made of the relation of mental and physical characteristics to success in different occupations (149), and the relation between civil-service salaries and a battery of job characteristics (180), which provided a basis for rating other jobs by the regression equation.

In psychological problems correlation analysis has been used primarily to measure closeness of relationship. One study reached the conclusion that even in groups of the same economic and social status, there is a small negative correlation between number of children per family and intelligence (150). In another study, a given test was repeated with varying time to complete it, and it was concluded that the test determined power alone, rather than speed (151). Here basing the conclusion on r (0.76) instead of d (0.58) led to overstressing the significance of the observed correlation. Other applications of correlation or partial correlation in psychological research (152–154) illustrate the usual tendency to depend on correlation coefficients rather than regressions as the means of expressing relationship.

Correlation methods have also been applied to borderline problems of psychology, sociology, and political science. A study of factors influencing the attitudes of mothers toward sex education indicated that previous environment was unimportant, but that there was a significant correlation between their opinion and the amount of sex education received by their children (155). A study of the opinions of college students showed that high prejudice, high misinformation, low grades, and conservatism were associated; and likewise low prejudice, good grades, low misinformation, and radicalism were associated. The correlations were low, however (156).

Some workers in psychology and education felt that the standard methods of correlation were not applicable to their data, and this has led to the development of various alternative methods, such as the Spearman "foot-rule correlation" for ranked data (157), and Kendall's rank correlation measures (158). They believed that discontinuous relationships and non-measurable phenomena are so frequent that in certain cases psychological laws may be valid only if stated in ordinal terms. Thus it appears that there is a hierarchy of needs (thirst, hunger, etc.) which are manifest under certain conditions, and which can be definitely stated in a ranked or priority listing, but not in satisfactory quantitative measurements. The rank-ordering of events, and the calculation of rank correlations (and sometimes partial and multiple correlations) based on such events, are often adequate for the decisions workers in these fields need to make (159). More recent methods also include an inverted test, by which the psychological state of a subject is based on rankings he has made (160).

Other workers in psychology and education, including Spearman (161) and Thurstone (162), have investigated the problem of the factors involved in intelligence. The factor approach has also been widely applied in studies of other traits, such as perception, personality, and learning. Although these investigations have led into involved calculations and highly refined mathematics, their actual significance is still in doubt.

One interesting recent application of correlation methods is that of determining the significance of the Rorschach test, where psychiatric difficulties of patients are judged on their interpretation of a series of irregular ink blots. Studies were made of agreement among a number of trained psychologists in judging the same patients by the same test results and other standard information. Six skilled psychologists rated 36 patients with a variety of psychological and psychiatric difficulties on 10 traits, and correlations were calculated for the ratings of each judge with every other judge on each trait. The 150 coefficients ranged from −0.116 to +0.716, with a median about 0.30. The judges thus disagreed widely in their ratings of patients by this test (163).

Tests of Correlation and Regression Results

It has been possible to verify or revise some earlier studies by applying them to later data, or by comparing them with analyses for longer periods. Studies of the response of milk production to prices received, for example, gave results quite different from those of earlier studies, and led to the conclusion that factors important while the industry was expanding in a given region, did not have the same significance after maturity was reached (164), i.e., the economic response was irreversible. More intensive analysis of the organization of typical farms provided conclusions as to the long-run response of production to price (165). In a different case, the response of milk production to variations in feed input on farms was later tested by a series of feeding experiments. The analysis of these results (166, 174), showed a net relation of milk output to feed input which agreed rather well with the earlier net regression (167), based on cow-test association records.

Enough has been presented to illustrate the wide range of problems in which the use of regression analysis sheds new light on actual relationships. These illustrations indicate the necessity for careful logical analysis, and the need both for good theoretical knowledge of the field in which the problem lies and for thorough technological knowledge of the elements involved in the particular problem.

Only a few of the significant statistical studies in any one field have been included. In many cases an important study has not been referred to because the point was already covered, or an unimportant study has been mentioned because of its pertinence to a particular topic. This discussion should therefore not be regarded as a critical evaluation of the work in any of the fields touched upon. Instead, the comments are intended solely to develop the variety, complexity, and significance of the problems to which regression and correlation analysis may be applied, and the care and thought which are even more necessary than the statistical computations if the results are to be of lasting value.

REFERENCES

1. Pearson, Karl, The law of ancestral heredity, *Biometrika*, Vol. II, pp. 211–236, 1903,
    ———, and Alice Lee, On the laws of inheritance in man, I. Inheritance of physical characters, *Biometrika*, Vol. II, pp. 357–462, 1903.
2. Yule, G. Udny, An Introduction to the Theory of Statistics, 6th edition, Chapter XII, Charles Griffin, London, pp. 229–253, 1922.

3. Moore, Henry L., *Economic Cycles, Their Law and Cause*, pp. 63–134, Macmillan, 1914.

——, *Forecasting the Yield and Price of Cotton*, Macmillan, 1917.

4. Wallace, Henry A., *Agricultural Prices*, Wallace Publishing Co., pp. 224, Des Moines, Iowa, 1920.

5. Mitchell, Wesley C., *Business Cycles*, Univ. Calif. Press, 1913.

6. Working, Holbrook, Factors determining the price of potatoes in St. Paul and Minneapolis, *Univ. Minn. Agr. Expt. Sta. Tech. Bul.* 10, 1922.

7. Sanderson, Fred H., *Methods of Crop Forecasting*, Harvard Univ. Press, 1954. (Includes, in pp. 234–35, summary and further analysis of data from P. I. Brounoff, Russian Meteorological Office, 1908–1912.)

8. Misner, E. G., Studies of the relation of weather to the production and price of farm products, I. Corn, Cornell Univ., Dept. of Agr. Econ., mimeographed publication, March, 1928.

9. Waugh, Frederick V., Chester D. Stevens, and Gustave Burmeister, Methods of forecasting New England potato yields, U. S. Dept. Agr., Bur. Agr. Econ. mimeographed report, February, 1929.

10. Moore, Henry L., *Economic Cycles, Their Law and Cause*, pp. 35–44.

11. Smith, Bradford B., Relation between weather conditions and yield of cotton in Louisiana, *Jour. Agr. Res.*, Vol. XXX, No. 11, pp. 1083–1086, June 1, 1925.

12. Patton, Palmer, Relationship of weather to crops in the plains region of Montana, *Mont. Agr. Expt. Sta. Bul.* 206, 1927.

13. Sanderson, *op. cit.*, pp. 57–78, 109–118, 181–238.

14. Wilson, C. F., and A. D. Holmes, Influence of precipitation and temperature on wheat yields in the prairie provinces, 1921–1940, Dom. Bur. of Stat., Canada, *Quart. Bul. of Agr. Stat.*, July–September, 1941.

15. Calzada, Jose, Estudio del clima del valle del Rimac, su influencia sobre el cultivo de papa y prónosticos, Min. of Agr. del Peru, Direccion General de Agricultura, *Divulgaciones e Informaciones No.* 6, 1955.

16. Fisher, R. A., The influence of rainfall upon the yield of wheat at Rothamsted, *Phil. Trans. B.*, vol. CCXIII, pp. 89–142, 1924.

17. Sanderson, *op. cit.*, pp. 239–254.

18. Tolley, H. R., J. D. Black, and M. J. B. Ezekiel, Input as related to output in farm organization and cost-of-production studies, *U. S. Dept. Agr. Bul.* 1277, pp. 7–12, 1924.

19. Misner, E. G., Relation of the composition of rations on some New York dairy farms to the economics of milk production, *Cornell Univ. Agr. Expt. Sta. Memoir* 64, 1923.

20. Vernon, J. J., and others, Factors affecting returns from the dairy enterprise in the Shenandoah Valley, *Va. Agr. Expt. Sta. Bul.* 257, pp. 32–42, 1927.

21. Ezekiel, M. J. B., P. E. McNall, and F. B. Morrison, Practices responsible for variations in physical requirements and economic costs of milk production on Wisconsin dairy farms, *Wis. Agr. Expt. Sta. Res. Bul.* 79, 1927.

22. Pond, George, and Mordecai Ezekiel, A study of some factors affecting the physical and economic costs of butterfat production in Pine County, Minn., *Univ. Minn. Agr. Expt. Sta. Bul.* 270, 1930.

23. Johnson, Sherman E., J. O. Tretsen, Mordecai Ezekiel, and O. V. Wells, Organization, feeding methods, and other practices affecting returns on irrigated dairy farms in Western Montana, *Univ. Mont. Agr. Expt. Sta. Bul.* 264, 1932.

24. Hardenburg, E. V., A study, by the crop survey method, of factors influencing the yield of potatoes, *Cornell Univ. Agr. Expt. Sta. Memoir* 57, 1922.

25. Reference 18 above, pp. 16–18.

26. Westbrook, E. C., and others, An economic study of farm organization in Sumter County, *Ga. State College of Agr. Bul.* 324, pp. 82–87, December, 1927.

27. Bell, G. D. H., Mary Lupton, and Ralph Riley, Investigations in the Triticinae, III. The morphology and field behaviour of the A_2 generation of interspecific and intergeneric amphlidiploids, *Jour. Agr. Science*, Vol. 46, part 2, pp. 199–231, August, 1955.

28. Grafius, J. E., The relationship of stand to panicles per plant and per unit area in oats, *Agron. Jour.*, Vol. 48, pp. 460–62, October, 1956.

29. Hanson, R. G., E. C. Ewing, and E. C. Ewing, Jr., Effect of environmental factors on fiber properties and yields of Deltapine cottons, *Agron. Jour.*, Vol. 48, pp. 573–581, December, 1956.

30. Casady, R. B., J. E. Legates, and R. M. Myers, Correlations between ambient temperatures varying from 60°–95° F. and certain physiological responses in young dairy bulls, *Jour. Agr. Science*, Vol. 15, pp. 141–152, February, 1956.

31. Joubert, D. M., A study of the pre-natal growth and development in the sheep, *Jour. Agr. Science*, Vol. 47, pp. 382–428, August, 1956.

32. Barnicoat, C. R., and others, Milk secretion studies with New Zealand Romney lambs, *Jour. Agr. Science*, Vol. 48, pp. 9–34, October, 1956.

33. Armstrong, David T., and William Hansel, The effect of age and plane of nutrition in growth hormone and thyrotropic hormone control of pituitary glands of Holstein heifers, *Jour. Animal Science*, Vol. 15, pp. 640–649, August, 1956.

34. Castle, Elizabeth J., and M. E. Castle, The rate of passage of food through the alimentary tract of pigs, *Jour. Agr. Science*, Vol. 47, pp. 196–204, April, 1956.

35. Albert, W. W., and others, The sulphur requirement of growing-fattening lambs in terms of methionine, sodium sulphate, and elemental sulphur, *Jour. Agr. Science*, Vol. 15, pp. 559–569, May, 1956,

36. de Baca, R. C., and others, Factors affecting weaning weights of cross-bred spring lambs, *Jour. Amer. Animal Sciences*, Vol. 16, pp. 667–678, August, 1956.

37. Shollenberger, J. H., and Corinne F. Kyle, Correlation of kernel texture, test weight per bushel, and protein content of hard red spring wheat, *Jour. Agr. Res.* Vol. 35, No. 12, pp. 1137–1150, Dec. 15, 1927.

38. Coleman, D. A., H. B. Dixon, and H. C. Fellows, Comparison of some physical and chemical tests for determining the quality of gluten in wheat and flour, *Jour. Agr. Res.*, Vol. 34, No. 3, pp. 241–246, Feb. 1, 1927.

39. Chatfield, Charlotte, Proximate composition of beef, *U. S. Dept. Agr. Circular* 389, 1926.

40. Cover, Sylvia, O. D. Butler, and T. C. Cartwright, The relationship of fatness in yearling steers to juiciness and tenderness of broiled and braised steak, *Jour. Animal Science*, Vol. 15, pp. 464–472, May, 1956.

41. Wang, Hsi, and others, Extensibility of single beef muscle fibers, *Jour. Animal Science*, Vol. 15, pp. 97–108, February, 1956.

42. Smith, Allan N., and Brynmor Thomas, The nutritive value of *Calluna vulgaris*, IV. Digestibility at three, seven and fourteen years after burning, *Jour. Agr. Res.*, Vol. 47, pp. 468–475, August, 1956.

43. Gowen, John W., Studies on conformation in relation to milk producing capacity in cattle, *Jour. Dairy Science*, Vol. III, No. 1, January, 1920; Vol. IV. No. 5, September, 1921.

43a. ———, Conformation and milk yield in the light of the personal equation of the dairy cattle judge, *Maine Agr. Expt. Sta. Bul.* 314, 1923.

44. Wolfe, T. K., A biometrical analysis of characters of maize and of their inheritance, *Va. Agr. Expt. Sta. Tech. Bul.* 26, 1924.

45. Richey, Frederick D., A statistical study of the relation between seed-ear characters and productiveness in corn, *U. S. Dept. Agr. Bul.* 1321, 1925.

46. Brown, C. J., and others, Relationship between conformation scores and live animal measurements of beef cattle, *Jour. Agr. Science*, Vol. 15, No. 3, pp. 911–921, August 1956.

47. Robertson, Alan, and I. L. Mason, The progeny testing of dairy bulls: a comparison of special station and field results, *Jour. Agr. Science*, Vol. 47, pp. 371–381, August, 1956.

48. Extending Stream-Flow Records, a handbook for hydrologists, U. S. Dept. of Interior, Geological Survey, Water Resources Branch, mimeographed report, September, 1947.

49. Harbeck, G. E., Combining two independent estimates, Dept. Interior, *Water Resources Bul.*, pp. 6–8, Feb. 10, 1947.

50. Folse, J. A., A new method of estimating stream flow, Carnegie Institute of Washington, *Pub.* 400, 1929.

50a. Kohler, M. A., and R. K. Linsley, Predicting the runoff from storm rainfall, U. S. Dept. Comm., *Weather Bur. Res. Paper* 34, September, 1951.

50b. Langbein, W. B., and C. H. Hardison, Extending stream flow data, *Proc. Amer. Soc. Civil Engineers*, Vol. 81, paper 826, November, 1955.

50c. Langbein, W. B., Stream gaging networks, *Pub.* 38, *de l'Assoc. d' Hydrologie, Assemblée générale de Roma*, tome III, pp. 293–303.

51. Bruce, Donald, On possible modifications in the Ezekiel method of curvilinear multiple correlation, Typewritten ms., filed in the Library, Agr. Mktg. Serv. U. S. Dept. Agr., 19 pp., about 1927.

52. Bruce, D., and L. H. Reineke, Correlation alignment charts in forest research, *U. S. Dept. Agr.*, *Tech. Bul.* 210, February, 1931.

53. Bruce, Donald, and Francis Schumacher, *Forest Mensuration*, 3rd edition, pp. 217–74, 305–25. McGraw-Hill, New York, 1950.

54. Ray, D. F., The Clements growth prediction charts for residual stands of mixed conifers in California, U.S.D.A. Forest Service, *Calif. Forest and Range Expt. Sta. Tech. Paper*, 9, May, 1955.

55. Pettit, Edison, Ultra-solar radiation, *Proc. Nat. Acad. Sciences*, Vol. 13, p. 380, 1927.

56. Hutson, J. B., Farm budgeting, *U.S.D.A. Farmers Bul.* 1564, July, 1928; revised, March, 1933.

57. Heady, Earl O., Glenn L. Johnson, and Lowell S. Hardin, *Resource Productivity, Returns to Scale, and Farm Size*, Iowa State College Press, 1956.

58. Haas, G. C., Sale price as a basis for farm land appraisal, *Univ. Minn. Agr. Expt. Sta. Tech. Bul.* 9, 1922.

59. Ezekiel, Mordecai, Factors affecting farmers' earnings in Southeastern Pennsylvania, *U. S. Dept. Agr. Bul.* 1400, pp. 39–60, 1926.

60. Taylor, C. C., A statistical analysis of farm management data, *Jour. Farm Econ.*, Vol. V, pp. 153–162, June, 1923.

61. Vernon, J. J., and M. J. B. Ezekiel, Causes of profit or loss on Virginia tobacco farms, *Va. Agr. Expt. Sta. Bul.* 241, 1925.

62. Pond, George, and Jesse W. Tapp, A study of farm organization in southwestern Minnesota, *Minn. Bul.* 205, 1923.

Pond, George, A study of dairy farm organization in southeastern Minnesota, *Minn. Agr. Expt. Sta. Tech. Bul.* 44, 108 pp., 1926.

63. Black, John D., and Edward S. Guthrie, Economic aspects of creamery organization, *Univ. Minn. Agr. Expt. Sta. Tech. Bul.* 26, 1924.

64. Schoenfeld, William A., Some economic aspects of the marketing of milk and cream in New England, *U. S. Dept. Agr. Circ.* 16, pp. 24–29, 1927.

65. Moore, Henry L., *loc. cit.*

66. Working, Holbrook, ref. 6.

67. ———, Factors affecting the price of Minnesota potatoes, *Minn. Agr. Expt. Sta. Tech. Bul.* 29, 1925.

68. Waugh, Frederick V., Forecasting prices of New Jersey white potatoes and sweet potatoes, *N. J. State Dept. of Agr. Circ.* 78, 1924.

69. Killough, Hugh B., What makes the price of oats, *U. S. Dept. Agr. Bul.* 1351, 1925.

70. Smith, Bradford B., The adjustment of agricultural production to demand, *Jour. Farm. Econ.*, Vol. VIII, No. 2, pp. 163–165, April, 1926.

71. Bean, Louis H., Some interrelationships between the supply, price, and consumption of cotton, U. S. Dept. Agr., Bur. Agr. Econ., mimeographed report, April, 1928.

72. Fox, Karl A., The analysis of demand for farm products, *U. S. Dept. Agr. Tech. Bul.* 1081, September, 1953.

72a. ———, Factors affecting the accuracy of price forecasts, *Jour. Farm Econ.*, Vol. XXXV, pp. 323–340, August, 1953.

72b. ———, *Econometric Analysis for Public Policy*, Iowa State College Press, 1958.

73. Smith, Bradford B., Factors affecting the price of cotton, *U. S. Dept. Agr. Tech. Bul.* 50, 1928.

74. Haas, G. C., and Mordecai Ezekiel, Factors affecting the price of hogs, *U. S. Dept. Agr. Bul.* 1440, 1926.

75. Ezekiel, Mordecai, Two methods of forecasting hog prices, *Jour. Amer. Stat. Assoc.*, Vol. XXII, pp. 22–30, March, 1927.

76. Hanau, Arthur, Die Prognose der Schweinepreise, *Vierteljahrshefte zur Kunjunkturforschung*, Sonderheft 7, Institut für Konjunkturforschung, Berlin, February, 1928.

77. Ezekiel, Mordecai, Factors related to lamb prices, *Jour. Pol. Econ.*, Vol. XXXV, No. 2, April, 1927.

78. Hedden, W. P., and Nathan Cherniack, Measuring the melon market, Prelim. report (mimeographed), U. S. Dept. Agr. Bur. Agr. Econ., in coop. N. Y. City Port Authority, August, 1924.

79. Kantor, Harry, Factors affecting the price of peaches in the New York City market, *U. S. Dept. Agr. Tech. Bul.* 115, 1929.

80. Rauchestine, E., Economic aspects of the canteloupe industry, *Calif. College of Agr. Bul.* 419, 1928.

81. Hoos, Sidney, and S. W. Shear, Relation between auction prices and supplies of California fresh Bartlett pears, *Hilgardia*, Vol. 14, No. 5, pp. 233–319, January, 1942.

82. Foytik, Jerry, Characteristics of demand for California peaches, *Hilgardia*, Vol. 20, No. 20, April, 1951.

83. Mehren, G. L., and H. E. Erdman, An approach to the determination of intraseasonal shifting of demand, *Jour. Farm Econ.*, Vol. 28, No. 2, pp. 587–596, May, 1946.

84. Fox, Karl A., Factors affecting farm income, farm prices, and food consumption, U. S. Dept. of Agr., *Agr. Econ. Res.*, Vol. 3, No. 3, pp. 65–82, July, 1951.

85. ———, Changes in the structure of demand for farm products, *Jour. Farm Econ.*, Vol. 37, No. 3, pp. 411–428, August, 1955.

86. Waite, Warren C., and Henry C. Trelogan, *Agricultural Market Prices*, 2d edition 440 pp., John Wiley and Sons, and Chapman and Hall, 1951.

87. FAO Commod. Series No. 22, pp. 60–74, September, 1952.

88. Stone, J. R. N., *The Measurement of Consumers' Expenditure and Behaviour in the U. K.*, 1926–38, Vol. I, Demand for consumers goods behaviour, etc., and problems of the statistical analysis, 496 pp., Cambridge, 1954.
89. Wold, Herman, and Lars Juréen, *Demand Analysis, A Study in Econometrics*, John Wiley and Sons, New York, 1953.
90. FAO, *The State of Food and Agriculture*, 1957, Chapter III, espec. pp. 77–95, 1957.
91. Woollam, G. E., The influence of price on the relative consumption of beef and pork, Canad. Dept. Agr. Mktg. Serv., Econ. Div. Ottawa, June, 1953.
92. Ross. H. A., The demand side of the New York milk market, *Cornell Univ. Agr. Expt. Sta. Bul.* 459, 1927.
93. Bean, Louis H., Demand and supply curves on potatoes and cotton, unpublished ms., filed in Library, U. S. Agr. Mktg. Serv., 1929.
94. Viton, A., Economic aspects of cocoa consumption, *FAO Mo. Bul. Agr. Econ. and Stat.*, Vol. IV, No. 5, pp. 1–10, May, 1955.
95. Szarf, A., and F. Pignalosa, Factors affecting U. S. coffee consumption, *FAO Mo. Bul. Agr. Econ. and Stat.*, Vol. III, No. 10, pp. 6–10, October, 1954.
96. Goreaux, L., Long-range projections of food consumption, *FAO Mo. Bul. Agr. Econ. and Stat.*, Vol. 6, No. 6, pp. 1–18, June, 1957.
97. Metzdorf, Fr. Hans-Jürgen, Bestimmungsgründe des Trinkmilchverbrauchs, *Hefte für Landwirtschaftliche Marktforschung*, Heft 5, 85 pp., Verlag Paul Parey, Berlin-Hamburg, 1951.
98. Gollnick, H., Die Elasticität der Nachfrage nach Zucker, *Agrarwirtschaft* 4, No. 11, pp. 337–341, November, 1955.
99. Smith, Bradford B., Forecasting the acreage of cotton, *Jour. Amer. Stat. Assoc.*, Vol. 20, No. 149, pp. 31–47, 1925.
100. Bean, Louis H., The farmer's response to price, *Jour. Farm Econ.*, Vol. XI, No.3, pp. 368–385, July, 1929.
101. Elliott, Foster F., Adjusting hog production to market demand, *Univ. Ill. Agr. Expt. Sta. Bul.* 293, 1927.
102. Gans, A. R., Elasticity of supply of milk from Vermont plants, *Vt. Agr. Expt. Sta. Bul.* 269, 1927.
103. Reference 64, pp. 34–50.
104. Murray, K. A. H., Wheat prices and acreage, *The Farm Econ.*, Vol. 1, pp. 77–78, October, 1933.
105. Meinken, Kenneth W., The demand and price structure for wheat, *U. S. Dept. Agr. Tech. Bul.* 1136, 93 pp., November, 1955.
106. Foote, Richard J., Statistical Analyses relating to the Feed-Livestock Economy, *U. S. Dept. Agr. Tech. Bul.* 1070, 41 pp., June, 1953.
107. Foote, Richard J., John W. Klein, and Malcolm Clough, The demand and price structure for corn and total feed concentrates, *U. S. Dept. Agr. Tech. Bul.* 1061, 79 pp., October, 1952.
108. Armore, Sidney J., The demand and price structure for food fats and oils, *U. S. Dept. of Agr. Tech. Bul.* 1068, June, 1953.
109. Gollnick, Heinz, Die Nachfrage nach Nahrungsmitteln and ihre Abhängigkeit von Preis- und Einkommensänderungen, *Hefte für Landwirtschaftliche Marktforschung*, Heft 6, 110 pp. Verlag Paul Parey, Hamburg-Berlin, 1954.
110. Ezekiel, Mordecai, Statistical analyses and the "laws" of price, *Quart. Jour. Econ.*, Vol. XLII, pp. 199–225, February, 1928.
111. Working, Holbrook, Factors influencing price differentials between potato markets, *Jour. Farm Econ.*, Vol. VII, pp. 377–398, October, 1925.
112. Waugh, Frederick V., *Quality as a Determinant of Vegetable Prices*, pp. 39–45, Columbia Univ. Press, 1929.

113. Diedjens, V. A., W. D. Whitcomb, and R. M. Koon, Asparagus and its culture, *Mass. Agr. College Extens. Leaflet* 49, April, 1929.

114. Howe, Charles B., Some local market price characteristics which affect New Jersey egg producers; factors affecting the retail prices of eggs, *N. J. Agr. Expt. Sta. Bul.* 150, 1927.

115. Benner, Claude L., and Harry G. Gabriel, Marketing of Delaware eggs, *Del. Agr. Expt. Sta. Bul.* 150, 1927.

116. Kuhrt, W. J., A study of farmer elevator operation in the spring wheat area. Series of 1925–26, Part II. Analysis of the variation in the quality factors of the 1925 crop of spring wheat, and the relation to such variation to price received and premiums paid in 1925–26, U. S. Dept. Agr., Bur. Agr. Econ. preliminary report, October, 1927.

117. Schultz, Henry, *The Theory and Measurement of Demand*, Univ. Chicago Press, 1938.

118. Shepherd, Geoffrey S., *Agricultural Price Analysis*, 4th ed., Iowa State College Press, 1957.

119. Thomsen, Frederick Lundy, and Richard Jay Foote, *Agricultural Prices*, 509 pp., McGraw-Hill, New York, 1952.

120. Waite, Warren C., and Harry C. Trelogan, *Agricultural Market Prices*, 2d edition, 440 pp., John Wiley and Sons, and Chapman and Hall, New York and London, 1951.

121. Hearings before the Temporary National Economic Comm., Part 26, Iron and Steel Industry, Exhibit 1416, An analysis of steel prices, volumes and costs—controlling limitations on price reductions, pp. 14,032–82, Washington, 1940.

122. Wylie, Kathryn H., and Mordecai Ezekiel, The cost curve for steel production, *Jour. Pol. Econ.*, Vol. XLVIII, pp. 777–821, December, 1940.

123. Dean, Joel, Statistical cost curves in various industries, *Econometrica*, Vol. VIII, No. 2, p. 188–189, April, 1940.

124. Hearings before the Temporary National Economic Comm., Part 26, Iron and Steel Industry, A statistical analysis of the demand for steel, 1919–1938, pp. 13,913–13,942, Washington, 1940.

125. Roos, C. F., and Victor von Szeliski, Factors governing changes in domestic automobile demand, *The Dynamics of Automobile Demand*, General Motors Corp., New York, 1939.

126. Derksen, J. B. D., Long cycles in residential building: an explanation, *Econometrica*, Vol. VIII, No. 2, pp. 97–116, April, 1940.

127. Koopmans, T., *Tanker Freight Rates and Tankship Building*, Netherlands Economic Institute, London, 1939.

128. Chamberlin, Edward, *The Theory of Monopolistic Competition*, Harvard Univ. Press, Cambridge, 1936.

129. Ezekiel, Mordecai, Distribution of gains from rising technical efficiency in progressing economies, *Amer. Econ. Rev.*, Vol. XLVII, No. 2, pp. 361–375, 1957.

130. Douglas, Paul H., and Grace Gunn, The Production Function for American Manufacturing in 1919, *Amer. Econ. Review*, Vol. 31, pp. 67–80, March, 1941. See also Paul H. Douglas, *Theory of Wages*, Macmillan Co., 1934.

131. Opmerkingen naar aanleiding van het rapport, Het Amsterdamse taxivraagstuk, Nederlandsch Economisch Instituut, Rotterdam, August, 1953.

132. Keasberry, J. E., L. H. Klaassen, and J. Koopman, De invloed van de belastingdruk op het aantal personenauto's, *Wegen. de Vereniging Het Nederlandsche Wegencongres*, December, 1954.

133. Girschick, Meyer, and Ruth O'Brien, Children's body measurements for sizing garments and patterns, *U. S. Dept. Agr. Misc. Pub.* 365, 1940.

134. Cowan, Donald R. G., The commercial application of forecasting methods, *Jour. Farm Econ.*, Vol. XII., pp. 139–163, January, 1930.
135. Smith, Bradford B., The use of interest rates in forecasting business activity, *Proc. Management Week*, 1926, Ohio State Univ., Bur. of Business Res., 1926.
136. Crum, W. L., The statistical allocation of joint costs, *Jour. Amer. Stat. Assoc.*, Vol. XXI, pp. 9–24, March, 1926.
137. Hansen, Alvin, *Fiscal Policy and Business Cycles*, with Appendix by Paul A. Samuelson on "A statistical analysis of the consumption function", W. W. Norton and Co., N.Y., pp. 250–260, 1941.
138. Ezekiel, Mordecai, Statistical investigations of saving, consumption, and investment, Part I. Saving, consumption and national income; Part II. Investment, national income, and the saving-investment equilibrium, *Am. Econ. Rev.*, Vol. XXXII, pp. 22–49, 272–307, March and June, 1942.
139. Duesenberry, James, Some new income-consumption relationships and their implications, *Econometrica*, Vol. XV, pp. 162–63, April, 1947.
140. Radice, E. A., A dynamic scheme for the British trade cycle, 1929–37, *Econometrica*, Vol. VII, No. 1, pp. 47–56, January, 1939.
141. Tinbergen, J., *An Econometric Approach to Business-Cycle Problems*, Hermann, Paris, 1937.
142. Bean, Louis H., *Ballot Behavior*, Amer. Council on Public Affairs, Washington, 1940.
143. ———, *How to Predict Elections*, Alfred A. Knopf, New York, 1948.
144. Gosnell, Harold, *Machine Politics, Chicago Model*, Univ. Chicago Press, 1937.
145. Mensenkamp, L. E., Ability classification in ninth-grade algebra, *The Math. Teacher*, January, 1929.
146. Garrison, K., Correlation between intelligence test scores and success in certain rational organization problems, *Jour. Applied Psychol*, December, 1928.
147. Weeks, Angelina L., A vocabulary information test, *Arch. of Psychol.*, May, 1928.
148. Hull, Clark L., Prediction formulae for teams of aptitude tests, *Jour. Applied Psychol.*, Vol. VII, pp. 277–284, 1923.
149. Higbie, Edgar Creighton, *An Objective Method for Determining Certain Fundamental Principles in Secondary Agricultural Education*, published by the author, Madison, Wis., 1924.
150. Sutherland, H. E. G., The relationship between I.Q. and size of family, *Jour. Educ. Psychol.*, February, 1929.
151. Freeman, Frank S., Power and speed, their influence upon intelligence test scores, *Jour. Applied Psychol.*, December, 1928.
152. Chauncey, Marlin R., The relation of the home factor to achievement and intelligence test scores, *Jour. Educ. Res.*, Vol. XX, No. 2, pp. 88, September, 1929.
153. Winch, W. H., Accuracy in school children. Does improvement in numerical accuracy "transfer"? *Jour. Educ. Psychol.*, Vol. I, pp. 557–589, 1910.
154. Goodenough, F. L. The Kuhlman-Benet test for children of pre-school age, Univ. Minn. Institute of Child Welfare, *Mon. Series* 2, 1928.
155. Witmer, Helen Leland, *Attitudes of Mothers Toward Sex Education*, Univ. Minn. Press, 1928.
156. Allport, Gordon W., The composition of political attitudes, *Amer. Jour. Sociol.*, Vol. XXXV, No. 2, pp. 220–238, September, 1929.
157. Spearman, C., A footrule for measuring correlation, *Brit. Jour. Psychol.*, Vol. II, p. 89, 1906.
158. Kendall, Maurice G., *Rank Correlation Methods*, 160 pp., Charles Griffin, London, 1948.

159. Siegel, Sidney, *Non-parametric Statistics for the Behavioral Sciences*, McGraw-Hill Series in Psychology, pp. 312, 1956.
160. Stephenson, William, *The Study of Behavior, Q-Technique and its Methodology*, Univ. of Chicago Press, pp. 376, 1953.
161. Spearman, C., The factor theory and its troubles, I. Pitfalls in the use of probable errors, *Jour. Educ. Psychol.*, 1932; II. Garbling the evidence, *Jour. Educ. Psychol.*, October, 1933; III. Misrepresentation of the theory, *Jour. Educ. Psychol.*, November, 1933; IV. Uniqueness of G. *Jour. Educ. Psychol.*, February, 1934; V. Adequacy of Proof., *Jour. Educ. Psychol.*, April, 1934.
162. Thurstone, L. L., *The Vectors of Mind, Multiple-factor Analysis for the Isolation of Primary Traits*, Univ. Chicago Press, 1935.
163. Gelfand, Leonard, Bruce Quarrington, Harley Widemand, and Jean Brown, Inter-judge agreement on traits rated from the Rorschach, *Jour. Consult. Psychol.*, Vol. 18, No. 6, 1954.
164. Mighell, R. L., and R. H. Allen, Supply schedules—"long-time" and "short-time," *Jour. Farm Econ.*, Vol. XXII, No. 3, 1940.
165. Allen, R. H. Erling Hole, and R. L. Mighell, Supply responses in milk production in Cabot-Marshfield, Vermont, *U. S. Dept. Agr. Tech. Bul.* 709, 1940.
166. Jensen, Einar, Determining input-output relationships in milk production, *U. S. Dept. Agr. Farm Mgt. Reports* 5, January, 1940.
167. Ezekiel, Mordecai, A check on a multiple correlation result, *Jour. Farm Econ.*, Vol. XXII, No. 2, 1940.
168. ———, Agricultural situation and outlook work, national and international, *FAO Mo. Bul. Agr. Econ. and Stat.*, Vol. III, No. 6, pp. 18–28, June, 1954.
169. Collins, David N., The Engineering Applications of Statistics, *Industrial Math.*, Vol. 7, pp. 1–15, 1956.
170. *Het Verband tussen de Temperatuur en het Verbruik van Gas*, 10 pp., Nederlandsch Economisch Instituut, Rotterdam, July, 1953.
171. Ventura, M. Mateus, Velocidade ultra-sônica, parácoro, refração molar e índice de refração, *Escola de Agronomia do Ceará Pub. Tech.* 6, A. Forteleza, Brasil, May, 1951.
172. Maunder, A. H., Size and efficiency in farming, Univ. Oxford, *Institute for Res. in Agr. Econ. Occasional Papers*, IV. 23 pp., 1952.
173. Baum, E. L., Earl O. Heady, and John Blackmore, *Economic Analysis of Fertilizer Use Data*, Iowa State College Press, 218 pp., 1956.
174. Jensen, Einar, and others, Input-output relationships in milk production, *U. S. Dept. Agr. Tech. Bul.* 815, 88 pp., 1942.
175. Ferger, Wirth F., Measurement of tax shifting, economics, and law. *Nat. Tax Jour.*, May, 1940.
    ———, Role of economics in federal tax administration, *Nat. Tax Jour.*, June, 1948.
176. Bandeen, Robert A., Automobile consumption, 1940–50, *Econometrica*, Vol. 25, pp. 239–248, April, 1957.
177. Breimeyer, Harold F., On price determination and aggregate price theory, *Jour. Farm Econ.*, Vol. XXXIX, pp. 676–694, August, 1957.
178. Heady, Earl O., Robert McAlexander, and W. D. Shrader, Combinations of rotations and fertilization to maximize crop profits on farms in North-Central Iowa (an application of linear programming), *Iowa Agr. Expt. Sta. Res. Bul.* 439, 20 pp., 1956.
179. Nerlove, Marc, Estimates of the elasticities of supply of selected argicultural commodities, *Jour. Farm Econ.*, Vol. 38, No. 2, pp. 496–512, May, 1956.

180. George, James P., Job evaluation through the medium of multiple and partial correlation and regression analysis, *Jour. Amer. Stat. Assoc.*, Vol. 52, pp. 370–71, September, 1957.

181. Schumacher, Francis X., and Francisco dos Santos Hall, Logarithmic expression of timber-tree volume, *Jour. Agr. Res.*, Vol. 47, No. 9, pp. 719–734, Nov. 1, 1933.

182. McCorkle, Chester O., Jr., Linear programming as a tool in farm management analysis, *Jour. Farm Econ.*, Vol. XXXVII, pp. 1222–1235, December, 1955.

183. Hopp, Henry, and Richard J. Foote, A statistical analysis of factors that affect prices of coffee, *Jour Farm Econ.*, Vol. XXXVII, pp. 429–438, August, 1955.

184. Helmberger, Peter, and Willard W. Cochrane, A short-run supply relationship for eggs in Minnesota, *Jour. Farm Econ.*, Vol. XXXIX, pp. 532–39, May, 1957.

185. Ottoson, Howard W., Andrew R. Aandahl, and L. Burbank Kristjanson, A method of farm real estate valuation for tax assessment, *Jour. Farm Econ.*, Vol. XXXVII, pp. 471–483, August, 1955.

186. Hoos, Sidney, Weekly prices and retail margins—small, medium, and large stores—oranges, lemons, and grapefruit, Denver, August, 1948–July, 1949, Calif. Agr. Expt. Sta., mimeographed report 170, 150 pp., September, 1954.

187. ———, Pacific coast canned fruits, Calif. Agr. Expt. Sta., Mimeographed report 197, 30 pp., June, 1957.

188. ———, and R. E. Seltzer, Lemons and lemon products: Changing economic relationships, 1951–52, *Calif. Agr. Expt. Sta. Bul.* 729, 78 pp., 1952.

189. ———, and J. N. Boles, Oranges and orange products, *Calif. Agr. Expt. Sta. Bul.* 731, 68 pp., 1953.

190. Harberger, Arnold, Some evidence on the international price mechanism, *Jour. Pol. Econ.*, Vol. LXV, pp. 506–521, December, 1957.

191. Rojko, Anthony S., The demand and price structure for dairy products, *U. S. Dept. Agr., Agr. Mktg. Serv. Tech. Bul.* 1168, 252 pp., May, 1957.

192. Basu, D., and G. T. Jones, International pattern of demand for foodstuffs, *The Farm Econ.*, Vol. VIII, No. 9, pp. 1–15; No. 10, pp. 23–54, Oxford, 1957.

193. Hildreth, Clifford, and F. G. Jarrett, *A Statistical Study of Livestock Production Marketing*, John Wiley and Sons, New York, 1955.

194. Yeh, Martin H., Fertilizer demand functions for the United States and selected regions, Iowa State College, Prod. Econ. 18 (mimeographed).

195. Griliches, Zvi, Hybrid corn; an exploration in the economics of technological change, *Econometrica*, Vol. 25, pp. 501–520, October, 1957.

196. ———, Research costs and social returns: hybrid corn, with comparisons, *Univ. of Chicago Office of Agr. Econ. Res.*, Paper 5803, February, 1958.

197. Nerlove, Marc, *The Dynamics of Supply*, Johns Hopkins Press, Baltimore, 1958.

198. Clark, Colin, *The Conditions of Economic Progress*, 3rd edition, Macmillan & Co., 720 pp., 1957.

199. Köttl, Hans, Betriebswirtschaftliche und marktanlytische Untersuchengen den österreichischen Zuckerrüben- und Zuckerwirtachaft Teil I. 1, Production; II, Verarbeitung; III, Verbrauch; 78 *Arbeit des Österreichischen Kuratoriums für Landtechnik*, Wien, 224 pp., August, 1958.

200. Friend, Irwin, and Gordon Fulcher, Comment, *Amer. Econ. Rev.*, Vol. XXXII, No. 4, pp. 829–840, December, 1942.

201. Staehle, Hans, The Measurement of Statistical cost Functions: an appraisal of some recent contributions, *Amer. Econ. Rev.*, Vol. XXXII, No. 2, part 1, pp. 321–333, June, 1942.

202. Stephenson, Charles A., Methods of correlation analysis: Reply, *Amer. Econ. Rev.*, Vol. XXXIII, No. 4, pp. 902–903, December, 1943.

CHAPTER 26

Steps in research work, and the place of statistical analysis

Relation of Statistical Analysis to Research. Statistical analysis is only a tool to be used by the investigator. The analyst must be a worker in some field, or in several; he cannot use his statistical training except in analyzing problems any more than a carpenter can use his skill without lumber and something to be made. Now that the routine of statistical analysis has been discussed, and the types of problems to which it may be applied have been surveyed, it is pertinent to ask just what are the steps in research work and just where and how does statistical analysis fit into the picture.

The research worker must have an adequate knowledge of the facts, technical and otherwise, of the field in which he is to work. This knowledge is usually insured by the situation that in most cases the worker is a biologist, an economist, a psychologist, or an agronomist, first, and then a statistician only secondarily or in addition. When his training has been primarily in mathematics or statistics, however, the statistician must acquaint himself thoroughly with the facts and theories of the field involved before he can expect to do significant and substantial work on applied problems.

Stating the Objective. If adequate acquaintance with the field is given, the first step in a particular research problem is setting up the objective of the project. The objective can best be stated in the form of a direct question, such as "Why did sales of European automobiles in the United States increase from 1946 to 1958?" The more exact and specific the question can be made, the more clearly is the field of the investigation defined. Stating the objective as a question has the important effect of clarifying the issue, and so insuring that the worker knows what he is really trying to find out. It has the further effect of instantly challenging the attention and of instinctively calling forth mental answers which aid in the next step of the research.

Any research project which cannot be stated as a definite question has not been clearly defined. Starting out merely "to collect figures on automobile sales" would not constitute research. Clear formulation of the question to be answered is an essential prerequisite of good research work.

Developing an Hypothesis. The second step is a deductive analysis of the question raised to suggest possible answers. This analysis draws on all the theoretical and practical training and experience the worker has. In addition, he may study previous work along the same lines, ask questions of those concerned in the industry, or make brief reconnaissance studies to decide on the factors which may be involved and to judge of the probable relationships. This phase of the research should lead to the setting up of a definite hypothesis as to the elements which will be involved and of the ways in which they will be related. Thus in the automobile problem, the hypothesis might be that the relative prices of domestic and imported cars was one important factor determining the demand; that the relative economy in gas consumption and repairs were also important; that foreign cars were used to supplement American cars for shopping and other short trips, and foreign sports cars were purchased by wealthy young people; and that the readiness to buy foreign cars was influenced by the growth of facilities to service them, and by acquaintance with their use by other people.

The process of developing the hypothesis may be aided by breaking up the main question to be answered into a number of subquestions, each one of which may be further broken up. Thus the initial question might be broken up into such subquestions as "Do purchases vary because of increasing cheapness of foreign cars? Are purchasers concerned with price *per car* or price *per horsepower* or *per pound*? Are they influenced by relative economy in use of gas, tires, and upkeep? Does readiness to purchase European cars differ between families living in different kinds of areas, or with different incomes, or with different compositions of the family? Does the relative use of different cars vary between different geographical sections of the country?" And so on until the possible approaches to the problem have been thought out for every phase.

In setting up the hypothesis the investigator should also attempt to think through the probable nature of the relationships. Thus, should it be assumed that the influence of relative prices on purchases will be constant, and independent of other factors, or is the relation likely to change from time to time through the year, or with high or low levels of employment and prosperity?

In setting up his hypotheses, the investigator not only should rely on his own knowledge but also should draw upon all the skill and knowledge of others who have experience in the same field. This will involve not only a careful study of earlier investigations of the same problem but also discussions with practical men who are operating in the field to be studied. Thus the student of automobile sales should talk with automobile dealers selling at retail and at wholesale, with officials of automobile companies, and with individuals owning both American and foreign cars, to get their opinions of the factors involved. This will enable the student to check his hypothesis against the ideas of businessmen concerned with the same problem, and to check it with those concerned as consumers, and these people often may call to his attention elements in the situation which otherwise he might completely overlook.

Measuring the Factors.   Once the hypothesis has been set up, and the various factors involved have been considered with much care, the next step is to secure measurements of the various factors. This will involve deciding whether the data are to be taken from published records or other secondary sources, or whether they are to be secured first hand. In the auto problem, facts on imports, sales, and registrations of automobiles can be secured from official or trade association publications, and facts about individual car purchasers or owners, about the kinds of cars they drive, and about their families, might have to be obtained by direct enquiry. Will they be collected by direct observation, by enumerators, by schedules, by mail questionnaires? Advantages and disadvantages of each method, and the problems involved in laying out a record form, defining the units, securing the records, and checking or editing the reports are available in standard statistical textbooks and are not considered here.

In obtaining the basic data it is necessary to decide on the particular items to be measured to represent the hypothetical factors. Are prices to be listed retail prices, or net prices actually paid on purchase? Are taxes and other sales charges to be included? Will the cost of "extras" be included? Will the allowance for "trade-ins" be taken at face value, or will the net price be adjusted when a used car taken in trade is valued far above its current market value? What characteristics of the various types of cars will be considered—weight, length, speed, or fuel economy, etc.? What variables will be used to represent the status, location, and other characteristics of individual families? How will availability of services and repairs be measured? In collecting the data, thought must be given to whether a sample will be used or whether (as often happens in time-series analysis) the entire universe will be covered. If a sample is employed, will it be of the regression model, or of the correlation model, as discussed in Chapter 17? The type of sample must be considered with reference to the use which will be made of the results, and the extent to which it is intended to generalize from them as to the relations existing in the universe.

Studying the Apparent Relations.   After quantitative or qualitative

data are available for all the variables, the next step is to make a preliminary study of the apparent relationships before proceeding to more
elaborate analyses. The relation of the independent factors both to each
other and to the dependent factor must be studied. As has been pointed
out before, the relation of the dependent factor to an independent factor
that is not related to the others can be determined by simple correlation,
whereas otherwise multiple correlation might be necessary. (This does not
hold, however, if joint functions are present.) At this point the investigator
begins to test out the various elements in the hypothetical picture and to
compare the hypothesis with the observed facts. Some elements which
were thought to be of importance may prove unrelated, and other variables
which were thought of doubtful significance may show important relations.
This preliminary examination may even prove the entire hypothesis to be
wrong and necessitate a re-examination of the basic ideas and a reformu-
lation of the proposed explanation more in line with the facts as observed.

Setting up the Model for a Correlation or Regression Analysis

The preliminary examination will assist in setting up the final multiple
regression analysis, if such an analysis is found to be needed. This is not
the whole of the research project, but is merely that portion of it in which
the adequacy of the theoretical hypothesis is tested and in which the exact
relations, as expected in the hypothesis, are measured and determined.

The statement of the hypothesis in equation form, in one or more
equations, constitutes the "model" on which the investigation is to be
based. This involves consideration of whether single-equation or multi-
equation systems will be necessary to express the relations, as discussed in
Chapter 24, and the use of flow diagrams to represent the elements involved
and their logical relation to one another in time or in space, as was
illustrated in Figure 20.1, on page 345.

In the automobile illustration, for example, the problem might be
approached from three different angles, which could be expressed in three
different models.

(1) Changes in Time—Such as:

$$
\begin{pmatrix} \% \text{ of all car purchases} \\ \text{of foreign autos} \end{pmatrix} = f\begin{pmatrix} \text{average prices of} \\ \text{selected foreign} \\ \text{cars} \end{pmatrix} \Big/ \begin{pmatrix} \text{average prices of} \\ \text{selected domestic} \\ \text{cars} \end{pmatrix}
$$

$$
+ f\begin{pmatrix} \text{availability of repair} \\ \text{services for selected} \\ \text{foreign cars} \end{pmatrix} + f\begin{pmatrix} \text{miles per gallon of} \\ \text{gasoline,} \\ \text{foreign} \div \text{domestic} \end{pmatrix} + f(\text{time})
$$

(2) GEOGRAPHIC DIFFERENCES IN LOCATION OF SALES OR REGISTRATIONS:

$$\left.\begin{array}{l}\text{\% of all registered cars}\\ \text{foreign (in countries}\\ \text{or other local units)}\end{array}\right\} = f\left(\begin{array}{c}\text{proportion of}\\ \text{population rural}\end{array}\right) + f\left(\begin{array}{c}\text{proportion of}\\ \text{population sub-}\\ \text{urban}\end{array}\right)$$

$$+ f\left(\begin{array}{c}\text{income per}\\ \text{capita}\end{array}\right) + f\left(\begin{array}{c}\text{average cars}\\ \text{per family}\end{array}\right)$$

(3) BETWEEN INDIVIDUAL FAMILIES:

$$\left.\begin{array}{l}\text{Number of foreign}\\ \text{cars per family}\end{array}\right\} = f\left(\begin{array}{c}\text{total number}\\ \text{of cars}\\ \text{owned}\end{array}\right) + f\left[\left(\begin{array}{c}\text{number}\\ \text{of}\\ \text{adults}\end{array}\right)\left(\begin{array}{c}\text{number of}\\ \text{children}\end{array}\right)\right]$$
$$\text{(jointly)}$$

$$+ f\left(\begin{array}{c}\text{Location of}\\ \text{home}\end{array}\right) + f\left(\begin{array}{c}\text{average income}\\ \text{per capita}\end{array}\right)$$
$$\text{(distance}$$
$$\text{from nearest}$$
$$\text{urban center)}$$

Some aspects of these relations might be explored by correlation analysis, and other aspects might require other types of statistical treatment. Also, potentially available data on some aspects, such as (1), may cover so few observations as to preclude any elaborate statistical analysis. The relations under (3), might need to be studied separately for different types of cars, such as sports cars, station wagons, and other passenger cars.

Units in Which Variables Are Stated. Once the variables to be employed are decided on, the next problem is to decide in what units to state them. In studying land values, for example, the value of a given farm may be stated as total value, as value per acre of all land, or as value per acre of improved land. Which one to select depends on what other variables are included and how they are to be stated. The total value of the farm might be correlated with the value of the dwelling, the value of other buildings, the acres in cultivated land, the acres in pasture, etc. This would tend to show the contribution *per acre* of each of the acreage elements and should give a high correlation, since under normal conditions the value of the farm might be expected to approximate the value of the buildings *plus* that of the several tracts of land. In this case the simple or additive regression equation would be quite appropriate, for it would give

Farm value = value of dwelling + value of other buildings
+ (value per acre of cultivated land)(number acres of cultivated land)
+ (value per acre of pasture land)(number acres pasture land)
+ (value per acre of woodland)(number acres woodland)
+ etc.

But if it were desired to measure the influence of type of road, fertility of land, and distance from town on land value, they could not be so readily included in the same additive equation. For example, a 40-acre farm yielding 60 bushels of corn to the acre might be worth on the average $2,000 more than a farm of the same size yielding 45 bushels of corn per acre. Under the same conditions, it would not be reasonable to expect that a 160-acre farm yielding 60 bushels of corn to the acre would be worth only $2,000 more than a 160-acre farm yielding 45 bushels per acre. In the first case, the higher yield would add $50 per acre to the farm value, in the latter, only $12.50. Yet if yield of corn were added as a factor to the above equation, that would assume that a given increase in fertility would add the same amount to the value of the farm, no matter how large or how small the farm was.

If the value were stated as value per acre, that would partly solve the difficulty, for a given change in fertility, distance from town, or type of road would then be assumed to have the same influence upon value per acre no matter how large or how small the farm was. But that would introduce difficulty with other variables. The dwelling, for example, would not become larger in direct proportion to the size of the farm. Very large farms with good dwellings would have a very low "value-of-dwellings-per-acre," and small farms with poor dwellings would also have a low "value-of-dwellings-per-acre." Only some method of determining the effect of value of dwellings on land values separately for farms of different sizes would take care of this difficulty.[1]

Type of Equation to Be Fitted. The case mentioned also illustrates the need of something other than a simple additive regression equation to express certain cases.

Finally, if the effect of fertility upon land value be found to vary with location, say, and the effect of building value with size of farm, not even a logarithmic equation would be applicable. Instead, an equation of the joint-function type (as discussed in Chapter 21) might be used, such as

$$\text{Log (value per acre)} = f\text{(distance, fertility, roads)}$$
$$+ f\text{(value of dwelling, size of farm, value barns)}$$
$$+ \text{etc.}$$

One further consideration is the danger of false results or spurious correlation if the variables are improperly stated. Thus if an attempt were made to correlate the value of farms with three factors, (*A*) the percentage of land in corn, (*B*) the percentage of land in wheat, and (*C*) the percentage of land in all other uses, it would be impossible to solve

[1] Reference (59), pp. 39–54, of Chapter 25, gives an example of statistical treatment of this problem.

the problem, or else it would give a spurious result. The factors would add up to exactly 100 per cent, and after variation in (A) and (B) had been held constant there would not be any variation left in (C).[2] Only by dropping out one of the factors, say (C), would significant results be secured. The regressions on (A) and (B) would then also show the effect of (C); for example, the increase in value for each unit increase in (A) would mean the increase due to substituting one unit of (A) for one unit of (C); changing the sign would give the effect of substituting one unit of (C) for one of (A). The same principle would then apply as between (B) and (C); whereas the increase in the dependent variable for substituting one unit of (B) for one of (A) would be the difference between the two net regression coefficients.

After the variables to be examined and the nature of the regression function to be used have been decided upon, at least tentatively, it is necessary to decide whether curves are to be fitted. If mathematical regressions are to be used, this involves deciding what form of equation is to be used. (Note pages 70 to 80 of Chapter 6, and 205 to 210 of Chapter 14.) If curves are to be fitted by one of the graphic methods, limiting conditions to be applied in fitting the curves must be worked out, in the light of the hypotheses stated and of the technological and other knowledge of the relations. (See Chapter 6, and Chapter 14, pages 211 to 213.)

Steps in Carrying Through the Computations. After the variables and the form of the equation for the statistical analysis have been decided upon, the next step is actually carrying through the computation. This involves "coding" the numerical values of the variables; calculating the extensions; setting up and solving the normal equations; and calculating the standard error of estimate, the coefficient of multiple correlation, and the standard errors for the regression coefficients. Then if curvilinear regressions are desired or found necessary, they will be determined by mathematical or graphic methods. (A reconnaissance study by the short-cut graphic method is often useful as a preliminary test before such further work.) After the final curves are determined, the standard error of estimate for the curvilinear regression and the index of multiple correlation are computed. If joint functions are suspected, the residuals are grouped with respect to two or more variables, or studied with respect to compound variables, and a joint function is fitted graphically or algebraically, as found appropriate. As a final step, the standard error of each of the regression coefficients should be computed and indicated on the regression

---

[2] For an extended mathematical treatment of this problem, see Ragnar Frisch. Statistical confluence analysis by means of complete regression systems, *Oslo University Økonomiske Institutt. Publikazion* 5, 1934.

charts, to indicate the significance to be attached to the results. All through the process, the statistical relations found should be checked back against the hypothetical expectations. If the statistical results conflict with the hypothesis, both should be re-examined to see where the conflict lies, as discussed in more detail subsequently.

Meaning of Correlation and Regression Results

A statistical determination of the nature of any relation, no matter how complicated the methods used, tells nothing of the *reason* for the relation observed.

Thus the variation in potato yields with differences in early and late rainfall, as determined in Chapter 21, might be due to a large variety of causes. The plant requires certain conditions of soil moisture, nutrients, sunshine, maximum and minimum temperature, and relative humidity to achieve the best growth, and the factors used reflect certain of them. One set of conditions may be required while the plant is developing its leaves and top, and another set later on while it is developing the tubers; and the rainfall factors used might thus relate to the growth periods of the plant.

There are other possibilities, however. The peculiar relation of potato yields to early and late rainfall considered jointly, as shown in Figure 21.9, might reflect the relation of late rainfall to potato diseases. Heavy early rainfall may stimulate good growth of the top; then if heavy late rainfall should follow it might result in conditions favorable to the development of potato blight, and so reduce an otherwise promising yield.

A considerable range of specific technical information is necessary to interpret correctly the results of a regression analysis, and to develop the reasons for the particular relations which have been found to exist as well as to plan the investigations. The analysis itself can never provide the interpretation of cause and effect. It can only establish the *facts* of the relations—for the meaning of those facts, the investigator must look elsewhere.

The way in which correlation analysis establishes the facts of relationship and nothing else may be illustrated by a specific example. If the number of automobiles moving down Sixteenth Street in Washington, D.C., for each 15-minute period through a given 12 hours is related to the height of the water in the Potomac River during each of the same periods, a definite correlation will be obtained. On some days this correlation would be so high that its probable error would indicate that it would be very unlikely that it could have occurred by chance. However, if on the basis of this correlation a person were to attempt to forecast the flow of traffic from the height of the water, he would find his forecast sadly in error if he made it for another day when the water was high because of a flood, or when the moon was in a different phase. There is no direct causal relation between the two phenomena, yet there is real correlation between them because they both are influenced, though very remotely, by the same sequence of cosmic events. The rising and the setting of the sun have a very definite influence on the movements of persons and therefore on the flow of traffic, whereas the rising and the setting of the moon likewise have a definite influence on the height of the water. Washington has so low an elevation, that the Potomac River has a definite ebb and flood of tide. There is a certain specific though complex relation between the rising and setting of the sun and of the moon, changing constantly from day to day. This illustrates a case in which real and significant correlation between two variables reflects relation to a common factor or factors, yet gives no inference as to direct causal connections. Many similar cases are met with in practical work in which the correlation between two variables is due to both being influenced by common causes although neither may in any conceivable way influence the other. This illustrates again the need for clear, logical thinking and for a technological basis for the interpretation of the statistical results, which measure the relationships, but of themselves tell nothing of cause or effect.

Statement of Results of Correlation and Regression Analysis. Having completed the statistical analysis, the next step is to translate the statistical results to an intelligible non-technical statement. This may go only so far as simple regression charts or estimating tables of the type shown at the end of Chapter 14, or of carefully worked-out pictorial statements such as shown in Figure 25.3. After the results are reduced to intelligible form—intelligible, that is, at least to the investigator—they should be carefully compared with the original hypothesis. If hypothesis and the statistical results do not agree, the hypothesis must be carefully examined to see if it may logically be restated so as to be consistent with the facts as found; and the analysis must be studied to see if there are any loopholes in the way the facts are stated, or in the way the problem has been worked through, which may be responsible for the results. (The preliminary results, next-to-last paragraph, page 442, reflect such mis-statement of the variables.) If the hypothesis and results are found to be consistent, or if, without doing violence to either, they can be brought into reasonable agreement, the research may be regarded as completed. If such agreement is not obtained, the results may be announced as actual observations inconsistent with what was expected and subject to further study or independent checks before being accepted as scientific conclusions.

Finally, if forecasts of future events or estimates for new observations are to be made from the results of the analysis, the methods outlined in Chapter 19 should be used to help judge how much confidence can be placed in such estimates or forecasts.

When the hypothesis and the analysis are found to be in satisfactory agreement, all that remains is to interpret the results to those who will be interested in them. Many investigators fail to take into account the audience for which they are writing. In writing a technical paper for a scientific journal, a full discussion of the methods and techniques used will be quite in place, so that their fellows may pass on the adequacy of the work. In a general or a popular report for a lay audience, interested only in what they have discovered and what it means, details of statistical techniques may be as out of place as computations of stresses and strains would be in a magazine devoted to "The Home Beautiful." The plethora of technical terminology in some popular reports may lead readers to suspect that the investigator himself did not understand what his results really meant. Unless the conclusions can be translated back into "the King's English," and stated so simply that practical men dealing with the problem investigated can understand what they mean, the usefulness of the research may be largely wasted.

Summary

The place of statistical analysis in scientific research is no different from the place of any other technical aid the investigator may employ. It furnishes a means of measuring the elements that are involved and of examining the way in which they are related; but it does not of itself furnish an explanation of phenomena. The effort to state the hypothesis as a mathematical model, and to reduce the variables to specific numerical statement, definitely related, should force the investigator to think more clearly and definitely about his problem. The methods of analyzing complicated relations set forth in this book furnish the student with keen tools for their investigation; but, like all keen tools, they may yield unsatisfactory or misleading results if employed carelessly or heedlessly. Statistical analysis is not a substitute for careful thinking, technical knowledge, and skilled workmanship in research work; instead, it is an aid which may make that thought and skill even more productive of worthwhile results.

Glossary and important equations

Glossary

The Greek and Roman letters used as symbols in this text, and the most important of the other symbols, are as follows:

M_x	(Roman)	= arithmetic mean of X.
σ	(small *sigma*)	= standard deviation in the universe.
Σ	(capital *sigma*)	= sum of the items specified.
n	(Roman)	= number of observations in a sample.
b	(Roman)	= coefficient of regression.
$f()$	(Roman)	= function of the variable in the parentheses.
s_x	(Roman)	= standard deviation in a sample.
r	(Roman)	= coefficient of correlation in a sample.
i	(Roman)	= index of correlation (curvilinear relations).
S	(Roman)	= standard error of estimate.
m	(Roman)	= number of constants in the regression equation.
z	(Roman)	= residual, or difference between observed and estimated values of a dependent variable.
R	(Roman)	= coefficient of multiple correlation.
β	(small *beta*)	= "beta" coefficient of regression, in terms of standard deviation units; also universe coefficient of regression.
I	(Roman)	= index of multiple (curvilinear) correlation.
η	(small *eta*)	= correlation ratio.
θ	(small *theta*)	= function of (used here for the Bruce adjustment function).
Δ	(capital *delta*)	= arbitrary symbol.
π	(small *pi*)	= arbitrary symbol.
Φ	(capital *phi*)	= function of.
X, Y	(Roman)	= variables, as observed.

x, y	(Roman)	= variables, in terms of departures from their means.
d	(Roman)	= coefficient of determination.
k	(Roman)	= coefficient of alienation, also (eq. 17.9) number of variables in curvilinear multiple regression.
ρ_{yx}	(Greek *rho*)	= coefficient of correlation in the universe.
m	(Roman)	= number of variables in a linear multiple regression study.

List of Important Equations

For convenience in referring to the most important of the equations which are introduced from time to time in the text, all numbered equations are repeated here in numerical order.

$$M_x = \frac{\Sigma X}{n} \tag{1.1}$$

$$X - M_x = x \tag{1.2}$$

$$\text{Mean deviation} = \frac{\Sigma |x|}{n} \tag{1.3}$$

$$s_x = \sqrt{\frac{\Sigma x^2}{n}} \tag{1.4}$$

$$s_x = \sqrt{\frac{\Sigma X^2}{n} - M_x^2} \tag{1.5}$$

$$s_u = \sqrt{\frac{\Sigma (d^2 F)}{n} - \left[\frac{\Sigma (dF)}{n}\right]^2 - \frac{c^2}{12}} \tag{1.6}$$

$$\bar{s}_x = s_x \sqrt{\frac{n}{n-1}} \tag{2.1}$$

$$\bar{s}_x = \sqrt{\frac{\Sigma x^2}{n-1}} \tag{2.2}$$

$$\bar{s}_x = \sqrt{\frac{\Sigma X^2 - n M_x^2}{n-1}} \tag{2.3}$$

$$s_{M_x} = \frac{\bar{s}_x}{\sqrt{n}} \tag{2.4}$$

$$\frac{\sigma_s}{\sigma_x} = \frac{1}{\sqrt{2(n-1)}} \tag{2.5}$$

$$Y = a + bX \tag{5.1}$$

$$b = \frac{\Sigma(XY) - nM_xM_y}{\Sigma(X^2) - n(M_x)^2} \tag{5.2}$$

$$a = M_y - bM_x \tag{5.3}$$

$$\left.\begin{aligned} \Sigma(XY) - nM_xM_y &= \Sigma(xy) \\ \Sigma(X^2) - n(M_x)^2 &= \Sigma(x^2) \end{aligned}\right\} \tag{5.4}$$

$$\left.\begin{aligned} b &= \frac{\Sigma(xy)}{\Sigma(x^2)} \\ a &= M_y - bM_x \end{aligned}\right\} \tag{5.5}$$

$$\bar{S}_{y \cdot x}^2 = \frac{\Sigma z^2}{n-2} \tag{5.6}$$

$$Y = a + bX + cX^2 \tag{6.1}$$

(With X used for X, x for $X - M_x$, U for X^2, u for $U - M_u$, equation [6.1] becomes $Y = a + bX + cU$. These symbols are used in equations [6.2] to [6.4], inclusive.)

$$\left.\begin{aligned} (\Sigma x^2)b + (\Sigma xu)c &= \Sigma xy \\ (\Sigma xu)b + (\Sigma u^2)c &= \Sigma uy \end{aligned}\right\} \tag{6.2}$$

$$a = M_y - b(M_x) - c(M_u) \tag{6.3}$$

$$\left.\begin{aligned} M_x &= \frac{\Sigma X}{n} \quad M_u = \frac{\Sigma U}{n} \quad M_y = \frac{\Sigma Y}{n} \\ \Sigma x^2 &= \Sigma X^2 - nM_x^2 \\ \Sigma xu &= \Sigma XU - nM_xM_u \\ \Sigma u^2 &= \Sigma U^2 - nM_u^2 \\ \Sigma xy &= \Sigma XY - nM_xM_y \\ \Sigma uy &= \Sigma UY - nM_uM_y \end{aligned}\right\} \tag{6.4}$$

$$Y = a + bX + cX^2 + dX^3 \tag{6.5}$$

(With U for X^2, V for X^3, equation [6.5] becomes $Y = a + bX + cU + dV$. These symbols are used in equations [6.6] to [6.10] inclusive.)

$$(\Sigma x^2)b + (\Sigma xu)c + (\Sigma xv)d = \Sigma xy \left.\right\}$$
$$(\Sigma xu)b + (\Sigma u^2)c + (\Sigma uv)d = \Sigma uy \left.\right\} \quad (6.6)$$
$$(\Sigma xv)b + (\Sigma uv)c + (\Sigma v^2)d = \Sigma vy \left.\right]$$

$$a = M_y - b(M_x) - c(M_u) - d(M_v) \quad (6.7)$$

$$M_v = \frac{\Sigma V}{n} \left.\right\}$$
$$\Sigma uv = \Sigma UV - nM_u M_v \left.\right\}$$
$$\Sigma xv = \Sigma XV - nM_x M_v \left.\right\} \quad (6.8)$$
$$\Sigma v^2 = \Sigma V^2 - nM_v^2 \left.\right\}$$
$$\Sigma vy = \Sigma VY - nM_v M_y \left.\right]$$

$$Y = bX + cX^2 \quad (6.9)$$

$$\Sigma X^2 b + \Sigma(XU)c' = \Sigma(XY) \left.\right\}$$
$$\Sigma(XU)b + \Sigma(U^2)c' = \Sigma(UY) \left.\right\} \quad (6.10)$$

$$S_{y \cdot x}^2 = s_z^2 = \frac{\Sigma z^2}{n} \quad (7.1)$$

$$S_{y \cdot f(x)}^2 = s_{z''}^2 = \frac{\Sigma(z'')^2}{n} \quad (7.2)$$

$$\bar{S}_{y \cdot x}^2 = \frac{n s_z^2}{n - 2} = \frac{n S_{y \cdot x}^2}{n - 2} \quad (7.3)$$

$$\bar{S}_{y \cdot x}^2 = \frac{\Sigma(z^2)}{n - 2} = s_z^2 \left(\frac{n}{n - 2} \right) \quad (7.4)$$

$$\bar{S}_{y \cdot f(x)}^2 = \frac{n s_{z''}^2}{n - m} = \frac{n S_{y \cdot f(x)}^2}{n - m} \quad (7.5)$$

$$\bar{S}_{y \cdot f(x)}^2 = \frac{\Sigma(z''^2)}{n - m} = s_{z''}^2 \left(\frac{n}{n - m} \right) \quad (7.6)$$

$$r_{yx} = \frac{s_{y'}}{s_y} \quad (7.7)$$

$$i_{yx} = \frac{s_{y''}}{s_y} \quad (7.8)$$

$$d_{xy} = r_{xy}^2 \quad (7.9)$$

$$d_{y \cdot f(x)} = i_{yx}^2 \tag{7.10}$$

$$r_{yx} = \sqrt{b_{yx} b_{xy}} \tag{7.11}$$

$$r_{xy} = \frac{\Sigma(XY) - nM_x M_y}{\sqrt{[\Sigma(X^2) - nM_x^2][\Sigma(Y^2) - nM_y^2]}} \tag{8.1}$$

$$b_{yx} = \frac{\Sigma(XY) - nM_x M_y}{ns_x^2} = \frac{\Sigma(xy)}{ns_x^2} \tag{8.2}$$

$$r_{xy} = \frac{\Sigma(XY) - nM_x M_y}{ns_x s_y} = \frac{\Sigma(xy)}{ns_x s_y} \tag{8.3}$$

$$\bar{S}_{y \cdot x} = \sqrt{\frac{\Sigma(Y^2) - n(M_y)^2}{n - 2} (1 - r_{xy}^2)} \tag{8.4}$$

$$i_{yx}^2 = 1 - \frac{s_{z''}^2}{s_y^2} \tag{8.5}$$

$$\beta_{yx} = b_{yx} \left(\frac{s_x}{s_y} \right) \tag{9.1}$$

$$X_1 = a + b_2 X_2 + b_3 X_3 + \ldots b_m X_m \tag{10.1}$$

$$X_1 = a + b_2 X_2 + b_3 X_3 \tag{11.1}$$

$$\left. \begin{array}{l} \Sigma(x_2^2)b_2 \ + \Sigma(x_2 x_3)b_3 = \Sigma(x_1 x_2) \\ \Sigma(x_2 x_3)b_2 + \Sigma(x_3^2)b_3 \ = \Sigma(x_1 x_3) \end{array} \right\} \tag{11.2}$$

$$a = M_1 - b_2 M_2 - b_3 M_3 \tag{11.3}$$

$$X_1' = a + b_2 X_2 + b_3 X_3 \tag{11.4}$$

$$z = X_1 - X_1' \tag{11.5}$$

$$X_1 = a_{1.23} + b_{12.3} X_2 + b_{13.2} X_3 \tag{11.6}$$

$$X_1 = a_{1.234} + b_{12.34} X_2 + b_{13.24} X_3 + b_{14.23} X_4 \tag{11.7}$$

$$X_1 = a_{1.2345} + b_{12.345} X_2 + b_{13.245} X_3 + b_{14.235} X_4 + b_{15.234} X_5 \tag{11.8}$$

$$\left. \begin{array}{l} \Sigma(x_2^2)b_{12.34} \ + \Sigma(x_2 x_3)b_{13.24} + \Sigma(x_2 x_4)b_{14.23} = \Sigma(x_1 x_2) \\ \Sigma(x_2 x_3)b_{12.34} + \Sigma(x_3^2)b_{13.24} \ + \Sigma(x_3 x_4)b_{14.23} = \Sigma(x_1 x_3) \\ \Sigma(x_2 x_4)b_{12.34} + \Sigma(x_3 x_4)b_{13.24} + \Sigma(x_4^2)b_{14.23} \ = \Sigma(x_1 x_4) \end{array} \right\} \tag{11.9}$$

$$a_{1.234} = M_1 - b_{12.34} M_2 - b_{13.24} M_3 - b_{14.23} M_4 \tag{11.10}$$

$$\left.\begin{aligned}
\Sigma(x_2^2)b_{12.345} \quad &+ \Sigma(x_2x_3)b_{13.245} + \Sigma(x_2x_4)b_{14.235} \\
&\qquad\qquad\qquad + \Sigma(x_2x_5)b_{15.234} = \Sigma(x_1x_2) \\
\Sigma(x_2x_3)b_{12.345} &+ \Sigma(x_3^2)b_{13.245} \quad + \Sigma(x_3x_4)b_{14.235} \\
&\qquad\qquad\qquad +\Sigma(x_3x_5)b_{15.234} = \Sigma(x_1x_3)
\end{aligned}\right\} \quad (11.11)$$

Etc.

$$a_{1.2345} = M_1 - b_{12.345}M_2 - b_{13.245}M_3 - b_{14.235}M_4 - b_{15.234}M_5 \quad (11.12)$$

$$\bar{S}_{1.234}^2 = \frac{ns_{z1.234}^2}{n-m} \tag{12.1}$$

$$\bar{S}_{1.234\ldots m}^2 = \frac{\left\{\begin{aligned}\Sigma(x_1^2) - [b_{12.34\ldots m}(\Sigma x_1x_2) + b_{13.24\ldots m}(\Sigma x_1x_3) \\ + \ldots + b_{1m.23\ldots(m-1)}(\Sigma x_1x_m)]\end{aligned}\right\}}{n-m} \tag{12.2}$$

$$X_1' = a_{1.234} + b_{12.34}X_2 + b_{13.24}X_3 + b_{14.23}X_4 \tag{12.3}$$

$$R_{1.234} = \frac{s_{x_1'}}{s_1} \tag{12.4}$$

$$R_{1.234\ldots m}^2 = \frac{\left\{\begin{aligned}b_{12.34\ldots m}(\Sigma x_1x_2) + b_{13.24\ldots m}(\Sigma x_1x_3) + \ldots \\ + b_{1m.23\ldots(m-1)}(\Sigma x_1x_m)\end{aligned}\right\}}{\Sigma(x_1^2)} \tag{12.5}$$

$$R_{1.234\ldots m}^2 = 1 - \left(\frac{S_{1.234\ldots m}^2}{s_1^2}\right) \tag{12.6}$$

$$S_{1.234\ldots m}^2 = s_1^2(1 - R_{1.234\ldots m}^2) \tag{12.7}$$

$$r_{14.23}^2 = 1 - \frac{1 - R_{1.234}^2}{1 - R_{1.23}^2} \tag{12.8}$$

$$\beta_{12.34} = b_{12.34}\frac{s_2}{s_1} \tag{12.9}$$

$$\Sigma X_1 + \Sigma X_2 + \Sigma X_3 + \Sigma X_4 = \Sigma(\Sigma_0) \tag{13.1}$$

$$M_1 + M_2 + M_3 + M_4 = M_0 \tag{13.2}$$

$$\Sigma(X_1^2) + \Sigma(X_1X_2) + \Sigma(X_1X_3) + \Sigma(X_1X_4) = \Sigma(X_1\Sigma_0) \tag{13.3}$$

$$X_1 = a' + f_2(X_2) + f_3(X_3) + f_4(X_4) + \ldots \tag{14.1}$$

$$\left.\begin{aligned}
X_1 = a &+ b_2X_2 + b_{2'}(X_2^2) + b_3X_3 + b_{3'}(X_3^2) \\
&+ b_4X_4 + b_{4'}(X_4^2)
\end{aligned}\right\} \tag{14.2}$$

$$X_1 = a + b_2(X_2) + b_{2'}(X_2^2) + b_{2''}(X_2^3) + b_3(X_3) + b_{3'}(X_3^2)$$
$$+ b_{3''}(X_3^3) + b_4(X_4) + b_{4'}(X_4^2) + b_{4''}(X_4^3) \quad (14.3)$$

$$X_1'' = a_{1.234}' + f_2'(X_2) + f_3'(X_3) + f_4'(X_4) \quad (14.4)$$

$$a_{1.234}' = M_1 - \frac{\Sigma[f_2'(X_2) + f_3'(X_3) + f_4'(X_4)]}{n} \quad (14.5)$$

$$z'' = X_1 - X_1'' \quad (14.6)$$

$$X_1' = F_2(X_2) = f_2(X_2) - M_{f(2)} + M_1 \quad (14.7)$$

$$x_1' = F_2(x_2) = f_2(X_2) - M_{f(2)} \quad (14.8)$$

$$X_1' = F_2(x_2) + F_3(X_3) + F_4(x_4) + \ldots + F_k(x_k) \quad (14.9)$$

$$X_{1.f(2,3,\ldots,k)} = a + f_2(X_2) + f_3(X_3) + \ldots + f_k(X_k) \quad (15.1)$$

$$z_{1.f(2,3,\ldots,k)} = X_1 - X_1' \quad (15.2)$$

$$S_{1.f(2,3,\ldots,k)} = s_z \quad (15.3)$$

$$X_1 = a + b_2 X_2 + b_2' X_2^2 + b_3 X_3 + b_3' X_3^2 \quad (15.4)$$

$$S_{1.f(2,3,\ldots,k)}^2 = s_1^2(1 - R_{1.2,2',3,3',\ldots,k}^2) \quad (15.5)$$

$$\bar{S}_{1.f(2,3,\ldots,k)}^2 = S_{1.f(2,3\ldots,k)}^2 \left(\frac{n}{n-m}\right) \quad (15.6)$$

$$I_{1.23\ldots k}^2 = 1 - \frac{s_z^2}{s_1^2} = 1 - \frac{S_{1.f(2,3\ldots k)}^2}{s_1^2} \quad (15.7)$$

$$d_{1.f(2,3,\ldots,k)} = I_{1.23\ldots,k}^2 \quad (15.8)$$

$$i_{12.3\ldots k}^2 = r_{1(2,2^2).(3,3^2,\ldots,k,k^2)} \quad (15.9)$$

$$X_1 = a' + b_{12'.3'4'}[f_2(X_2)] + b_{13'.2'4'}[f_3(X_3)] + b_{14'.2'3'}[f_4(X_4)] \quad (15.10)$$

$$i_{12.34} = r_{12'.3'4'} \quad (15.11)$$

$$s_{b_{yx}} = \frac{\bar{S}_{y.x}}{s_x\sqrt{n}} \quad (17.1)$$

$$s_{b_{12.34\ldots m}} = \sqrt{\frac{\bar{S}_{1.234\ldots m}^2}{ns_2^2(1 - R_{2.34\ldots m}^2)}} \quad (17.2)$$

$$s_{M_{y'}} = \frac{\bar{S}_{y.x}}{\sqrt{n}} \quad (17.3)$$

$$s_{y'} = \sqrt{s_{M_{y'}}^2 + (s_{b_{yx}}x)^2} \quad (17.4)$$

$$s_{1.2,2^2}^2 = \frac{\bar{S}_{1.2,2^2}^2}{n} + (s_b x_2)^2 + 2s_b s_{b'} u x_2 + (s_{b'} u)^2 \quad (17.5)$$

$$s_{r_{yx}} = \frac{1 - r_{yx}^2}{\sqrt{n - 1}} \tag{17.6}$$

$$\rho_{yx}^2 = \frac{\sigma_{y'}^2}{\sigma_y^2} = \frac{\sigma_y^2 - \sigma_z^2}{\sigma_y^2} \tag{17.7}$$

$$\bar{r}_{yx}^2 = \frac{s_y^2\left(\dfrac{n}{n - 1}\right) - S_{y \cdot x}^2\left(\dfrac{n}{n - 2}\right)}{s_y^2\left(\dfrac{n}{n - 1}\right)} = 1 - \left[\frac{S_{y \cdot x}^2}{s_y^2}\right]\left(\frac{n - 1}{n - 2}\right) \tag{17.8}$$

$$\left.\begin{array}{c} \bar{i}_{12}^2 \\[12pt] \bar{R}_{1.23\ldots k}^2 \\[12pt] \bar{I}_{1.23\ldots k}^2 \end{array}\right\} = 1 - \left[\frac{S_{1.23\ldots k}^2}{s_1^2}\right]\left(\frac{n - 1}{n - m}\right) = \left\{\begin{array}{c} 1 - (1 - i_{12}^2)\left(\dfrac{n - 1}{n - m}\right) \\[12pt] 1 - (1 - R_{1.23\ldots k}^2)\left(\dfrac{n - 1}{n - m}\right) \\[12pt] 1 - (1 - I_{1.23\ldots k}^2)\left(\dfrac{n - 1}{n - m}\right) \end{array}\right\}$$

$$\tag{17.9}$$

$$s_{Y'-Y}^2 = s_{M_{y'}}^2 + (s_{b_{yx}}x)^2 + \bar{S}_{y \cdot x}^2 \tag{19.1}$$

$$s_{x'_{12.34}-x_1}^2 = \bar{S}_{1.234}^2\left[1 + \frac{1}{n} + c_{22}x_2^2 + c_{33}x_3^2 + c_{44}x_4^2\right.$$

$$\left. + 2c_{23}x_2x_3 + 2c_{24}x_2x_4 + 2c_{34}x_3x_4\right] \tag{19.2}$$

$$\left.\begin{array}{l} (\Sigma x_2^2)c_{22} + (\Sigma x_2x_3)c_{23} + (\Sigma x_2x_4)c_{24} = 1 \\ (\Sigma x_2x_3)c_{22} + (\Sigma x_3^2)c_{23} + (\Sigma x_3x_4)c_{24} = 0 \\ (\Sigma x_2x_4)c_{22} + (\Sigma x_3x_4)c_{23} + (\Sigma x_4^2)c_{24} = 0 \end{array}\right\} \tag{19.3}$$

$$\left.\begin{array}{l} (\Sigma x_2^2)c_{32} + (\Sigma x_2x_3)c_{33} + (\Sigma x_2x_4)c_{34} = 0 \\ (\Sigma x_2x_3)c_{32} + (\Sigma x_3^2)c_{33} + (\Sigma x_3x_4)c_{34} = 1 \\ (\Sigma x_2x_4)c_{32} + (\Sigma x_3x_4)c_{33} + (\Sigma x_4^2)c_{34} = 0 \end{array}\right\} \tag{19.4}$$

$$\left.\begin{array}{l} (\Sigma x_2^2)c_{42} + (\Sigma x_2x_3)c_{43} + (\Sigma x_2x_4)c_{44} = 0 \\ (\Sigma x_2x_3)c_{42} + (\Sigma x_3^2)c_{43} + (\Sigma x_3x_4)c_{44} = 0 \\ (\Sigma x_2x_4)c_{42} + (\Sigma x_3x_4)c_{43} + (\Sigma x_4^2)c_{44} = 1 \end{array}\right\} \tag{19.5}$$

$$s^2_{x'_{1,23\ldots m}-x_1} = \bar{S}^2_{1.23\ldots m}\left[1 + \frac{1}{n} + (c_2 x_2 + c_3 x_3 + \ldots + c_m x_m)^2\right] \Bigg\}$$ (19.6)

on condition that $(c_2 c_2) = c_{22}$, $c_2 c_k = c_{2k}$ etc.

$$r_a = \frac{\Sigma(z_t z_{t+1})}{\Sigma z_t^2}$$ (20.1)

$$\frac{\delta^2}{s_z^2} = \frac{\dfrac{\Sigma(z_{t+1} - z_t)^2}{(n-1)}}{\dfrac{\Sigma z_t^2}{n}}$$ (20.2)

$$X_1 = f(X_2, X_3)$$ (21.1)

$$X_1 = a + eX_3 + g(X_2 X_3)$$ (21.2)

$$X_1 = a + eX_3 + g(X_2 X_3) + h(X_2)$$ (21.3)

$$X_1 = f_{2,3}(X_2, X_3) + f_4(X_4)$$ (21.4)

$$X_1 = f_{2,3,4,k}(X_2, X_3, X_4, \ldots X_k)$$ (21.5)

$$X_1 = f_{2,3}(X_2, X_3) + f_{4,5}(X_4, X_5) + f_6(X_6)$$ (21.6)

$$X_1 = a + b_2 X_2 + b_3 X_3 + b_4 X_4 + b_5\sqrt{X_2} + b_6\sqrt{X_3}$$
$$+ b_7\sqrt{X_4} + b_8\sqrt{X_2 X_3} + b_9\sqrt{X_2 X_4} + b_{10}\sqrt{X_3 X_4}$$ (21.7)

$$\eta_{yx} = \sqrt{\frac{\Sigma[n_0(M_0^2)] - n(M_y)^2}{ns_y^2}}$$ (22.1)

$$s_{z^m}^2 = \frac{ns_{z^m}^2 - [\Sigma(n_0 M_0^2) - n(M_{z^m})^2]}{n}$$ (22.2)

$$t_b = \frac{\Sigma xy}{\bar{S}_{y\cdot x}\sqrt{\Sigma x^2}}$$ (23.1)

$$F = \frac{(\Sigma xy)^2}{\bar{S}_{y\cdot x}^2(\Sigma x^2)} = t^2$$ (23.2)

$$Q = a_1 + b_1 P + u$$ (24.1)

$$Q = a_2 + b_2 P + v$$ (24.2)

$$B = \frac{b_1\sigma_v^2 - (b_1 + b_2)\sigma_{uv} + b_2\sigma_u^2}{\sigma_v^2 - 2\sigma_{uv} + \sigma_u^2}$$ (24.3)

$$q = b_1 p + u$$ (24.4)

$$q = b_2 p + v \tag{24.5}$$

$$p = \frac{v - u}{b_1 - b_2} \tag{24.6}$$

$$q = \frac{b_1 v - b_2 u}{b_1 - b_2} \tag{24.7}$$

$$B = \frac{\Sigma qp}{\Sigma p^2} = \frac{\Sigma[(b_1 v - b_2 u)(v - u)]}{\Sigma(v - u)^2} \tag{24.8}$$

$$\left.\begin{array}{l}
\Sigma(x_2^2)c_{22} \ + \Sigma(x_2 x_3)c_{23} + \Sigma(x_2 x_4)c_{24} = 1 \\[4pt]
\Sigma(x_2 x_3)c_{22} + \Sigma(x_3^2)c_{23} \ + \Sigma(x_3 x_4)c_{24} = 0 \\[4pt]
\Sigma(x_2 x_4)c_{22} + \Sigma(x_3 x_4)c_{23} + \Sigma(x_4^2)c_{24} \ = 0
\end{array}\right\} \tag{A2.1}$$

$$\left.\begin{array}{l}
s_{b_{12.34}} = \bar{S}_{1.234}\sqrt{c_{22}} \\[4pt]
s_{b_{13.24}} = \bar{S}_{1.234}\sqrt{c_{33}} \\[4pt]
s_{b_{14.23}} = \bar{S}_{1.234}\sqrt{c_{44}}
\end{array}\right\} \tag{A2.2}$$

$$\left.\begin{array}{l}
s_{F_{1.234}}^2 = \bar{S}_{1.234}^2 \left\{ \dfrac{1}{N} + \dfrac{N}{d_{11}} \left[c_{22}''(X_2 - M_2)^2 + c_{33}''(X_3 - M_3)^2 \right.\right. \\[8pt]
\quad + c_{44}''(X_4 - M_4)^2 + 2c_{23}''(X_2 - M_2) \\[8pt]
\quad \times (X_3 - M_3) + 2c_{24}''(X_2 - M_2)(X_4 - M_4) \\[8pt]
\left.\left. \quad + 2c_{34}''(X_3 - M_3)(X_4 - M_4) \right] \right\}
\end{array}\right\} \tag{A2.3}$$

APPENDIX 2

Methods of computation

Coefficients of Correlation and Regression from a Double-Frequency Table. A short-cut method of calculating coefficients of simple correlation from double-frequency tables is given in previous editions (pages 455 to 457 in the second edition). It is omitted from this edition.

Coefficients of Multiple Correlation and Net Regression, Doolittle Method

Use of the Check Sum. Where a number of different variables are involved, every operation in making the extensions, computing the averages and corrections, and solving the normal equations through to the "back solution," can be verified by an automatic check known as the "check sum." The way in which the check sum is used will be illustrated by a small problem, carried through every step in turn, but it is equally applicable to any other other method of tabulation and is especially valuable with machine tabulation, where it serves as an overall control on the accuracy of the machine processes.

THE CHECK SUM AS A CHECK IN EXTENDING. The values in the following table (Table A2.1) may be used to illustrate the use of the check sum.

The values in the columns X_2, X_3, X_4, and X_1 are the three independent factors and the dependent factor, which are to be correlated. The values in the column headed "ΣX" are the arithmetic totals of the values for the four other variables, and are designated "the check sum."

As the first step, each of these five columns is added. Since, for each line,

$$X_2 + X_3 + X_4 + X_1 = \Sigma X$$

it also holds true that

$$\Sigma X_2 + \Sigma X_3 + \Sigma X_4 + \Sigma X_1 = \Sigma(\Sigma X)$$

489

Table A2.1

CALCULATION OF EXTENSIONS, USING THE CHECK SUM

Variables					Extensions with X_2				
X_2	X_3	X_4	X_1	ΣX	X_2^2	$X_2 X_3$	$X_2 X_4$	$X_2 X_1$	$X_2 \Sigma X$
0	136	106	103	345	0	0	0	0	0
1	140	103	108	352	1	140	103	108	352
2	86	108	102	298	4	172	216	204	596
3	115	102	111	331	9	345	306	333	993
4	115	111	95	325	16	460	444	380	1,300
12	161	91	109	373	144	1,932	1,092	1,308	4,476
13	235	109	118	475	169	3,055	1,417	1,534	6,175
14	304	118	123	559	196	4,256	1,652	1,722	7,826
15	224	123	108	470	225	3,360	1,845	1,620	7,050
16	185	108	100	409	256	2,960	1,728	1,600	6,544
17	108	100	88	313	289	1,836	1,700	1,496	5,321
18	193	88	109	408	324	3,474	1,584	1,962	7,344
19	175	109	103	406	361	3,325	2,071	1,957	7,714
134	2,177	1,376	1,377	5,064	1,994	25,315	14,158	14,224	55,691

Extensions with X_3				Extensions with X_4			Extensions with X_1	
X_3^2	$X_3 X_4$	$X_3 X_1$	$X_3 \Sigma X$	X_4^2	$X_4 X_1$	$X_4 \Sigma X$	X_1^2	$X_1 \Sigma X$
18,496	14,416	14,008	46,920	11,236	10,918	36,570	10,609	35,535
19,600	14,420	15,120	49,280	10,609	11,124	36,256	11,664	38,016
7,396	9,288	8,772	25,628	11,664	11,016	32,184	10,404	30,396
13,225	11,730	12,765	38,065	10,404	11,322	33,762	12,321	36,741
13,225	12,765	10,925	37,375	12,321	10,545	36,075	9,025	30,875
25,921	14,651	17,549	60,053	8,281	9,919	33,943	11,881	40,657
55,225	25,615	27,730	111,625	11,881	12,862	51,775	13,924	56,050
92,416	35,872	37,392	169,936	13,924	14,514	65,962	15,129	68,757
50,176	27,552	24,192	105,280	15,129	13,284	57,810	11,664	50,760
34,225	19,980	18,500	75,665	11,664	10,800	44,172	10,000	40,900
11,664	10,800	9,504	33,804	10,000	8,800	31,300	7,744	27,544
37,249	16,984	21,037	78,744	7,744	9,592	35,904	11,881	44,472
30,625	19,075	18,025	71,050	11,881	11,227	44,254	10,609	41,818
409,443	233,148	235,519	903,425	146,738	145,923	539,967	146,855	542,521

Adding the sums of the first four colums together gives the same value as the sum of the check sum column, which verifies all the totals.

The first set of extensions is made by multiplying the items in each line by the X_2 item in the first column of that line, giving the values shown under "Extensions with X_2." Since for each line

$$X_2 + X_3 + X_4 + X_1 = \Sigma X,$$

it also follows that

$$X_2^2 + X_2 X_3 + X_2 X_4 + X_2 X_1 = X_2 \Sigma X.$$

Then, adding each column, we find that the sums of the four other columns should total to the same value as the sum of the $X_2 \Sigma X$ column. Checking up, we see that $1{,}994 + 25{,}315 + 14{,}158 + 14{,}224 = 55{,}691$, verifying all the calculations.

The other extensions are made in similar fashion, and the sums of each column verified with the sum of the check-sum column, according to the relation

$$\Sigma X_2 X_3 + \Sigma X_3^2 + \Sigma X_3 X_4 + \Sigma X_3 X_1 = \Sigma(X_3 \Sigma X)$$

and the corresponding relation for the other extensions. It should be noted that in checking the "extensions with X_3," the value $\Sigma X_2 X_3$ is taken from the previous set of extensions; in checking the "extensions with X_4," the value for $\Sigma X_2 X_4$ is taken from the "extensions with X_2," and the value for $\Sigma X_3 X_4$ from the "extensions with X_4"; and so on for the remaining checks.

While the check sum would not disclose exactly compensating errors made in different columns the possibility of such errors is so remote that, after the arithmetic has been checked by the comparisons indicated, it may be assumed that no errors have been made either in making the multiplications or adding the columns.

THE CHECK SUM AS A CHECK IN CORRECTING TO THE MEANS. After the values for ΣX_2^2, $\Sigma X_2 X_3$, etc., have all been computed as indicated in Table A2.1, the process of making the corrections to get the values Σx_3^2, $\Sigma x_2 x_3$, etc., may be organized in regular fashion and checked by the check sum, as shown in Table A2.2.

The first line in Table A2.2 gives the sums of each of the variables, including the check sum. Dividing by the number of observations (13 in this case), gives the mean for each variable, as entered in the second line. Again, the entries for the first four columns total to equal the entry in the Σ column, checking the division.

The sums from the "extensions with X_2," of Table A2.1, are next entered in line 3. The sums for each variable, in line 1, are next multiplied by the mean of X_2 (10.30769), and the products entered in the corresponding column in line 4. Subtracting the entries in line 4 from those in line 3

Table A2.2

CALCULATION OF PRODUCT SUMS CORRECTED TO DEPARTURES FROM MEANS,
WITH CHECK SUM

	X_2	X_3	X_4	X_1	ΣX	Line
Sums	134.	2,177.	1,376.	1,377.	5,064.	1
Means	10.30769	167.46154	105.84615	105.92308	389.53846	2
Extensions with X_2	1,994.00	25,315.00	14,158.00	14,224.00	55,691.00	3
Corrections	1,381.23	22,439.84	14,183.38	14,193.69	52,198.14	4
Extensions with x_2	612.77	2,875.16	−25.38	30.31	3,492.86	5
Extensions with X_3	$\cdots$	409,443.00	233,148.00	235,519.00	903,425.00	6
Corrections	$\cdots$	364,563.77	230,427.08	230,594.54	848,025.23	7
Extensions with x_3	$\cdots$	44,879.23	2,720.92	4,924.46	55,399.77	8
Extensions with X_4	$\cdots$	$\cdots$	146,738.00	145,923.00	539,967.00	9
Corrections	$\cdots$	$\cdots$	145,644.30	145,750.15	536,004.90	10
Extensions with x_4	$\cdots$	$\cdots$	1,093.70	172.85	3,962.10	11
Extensions with X_1	$\cdots$	$\cdots$	$\cdots$	146,855.00	542,521.00	12
Corrections	$\cdots$	$\cdots$	$\cdots$	145,856.08	536,394.48	13
Extensions with x_1	$\cdots$	$\cdots$	$\cdots$	998.92	6,126.52	14

gives the values which are entered in line 5. These values are the extensions, expressed as departures from the means.

In column X_3, for example, the entry in line 3 is $\Sigma X_2 X_3$; and the entry in line 4 is $\Sigma X_3 M_2$.

The entry in line 5, then, is $\Sigma X_2 X_3 - \Sigma X_3 M_2$

$$= \Sigma X_2 X_3 - n M_3 M_2$$

$$= \Sigma x_2 x_3$$

Again, the values in the first four columns add to the same as the value in the check-sum column, verifying the work.

The rest of the table is entered in similar fashion. Lines 6, 9, and 12 are the extensions with X_3, X_4, and X_1, from Table A2.1. Lines 7, 10, and 13 are the values in the corresponding columns of line 1, multiplied by M_3, M_4, and M_1, respectively (from line 2). Lines 8, 11, and 14, obtained by subtracting the items in lines, 7, 10, and 13 from those in 6, 9, and 12, show the values corrected for departures from the means.

In verifying the sum of the other entries in line 8 by the check sum, the item $\Sigma x_2 x_3$ must be included, from column X_3, line 5, before comparing with the check sum; in checking line 11, $\Sigma x_2 x_4$ and $\Sigma x_3 x_4$, from column X_4, lines 5 and 8, must be included; and in checking line 14, the values $\Sigma x_1 x_2$, $\Sigma x_1 x_3$, and $\Sigma x_1 x_4$, from column X_1, lines 5, 8, and 11, must all be included. For line 11, the other items add to 3,962.11, as against the check sum of 3,962.10; and for line 14, they add to 6,126.55, as against the check sum of 6,126.52. In both cases the discrepancies are so small as to be readily due to raising and lowering in the last digit, and, therefore, may be disregarded.

Lines 5, 8, and 11 now give the values required to determine the regression coefficients by simultaneous solution, according to equations (11.9).

THE CHECK SUM AS A CHECK IN SOLVING THE NORMAL EQUATIONS. The solution of the simultaneous equations by the Doolittle method has already been illustrated in Chapter 11. The check sum may be used to verify each step in the computation, as shown in Table A2.3.

Table A2.3

SOLUTION OF NORMAL EQUATIONS BY THE DOOLITTLE METHOD, WITH CHECK SUM

Line	X_2	X_3	X_4	X_1	ΣX
I	612.77	2,875.16	−25.38	30.31	3,492.86
I′	−1.00000	−4.69207	0.04142	−0.04946	−5.70011
II	(2,875.16)	44,879.23	2,720.92	4,924.46	55,399.77
(−4.69206)(I)	(−2,875.16)	−13,490.45	119.08	−142.22	−16,388.74
Σ_2	...	31,388.78	2,840.00	4,782.24	39,011.03
II′	...	−1.00000	−0.09048	−0.15236	−1.24283
III	(−25.38)	(2,720.92)	1,093.70	172.85	3,962.10
(0.04142)(I)	(25.38)	(119.08)	−1.05	1.26	144.67
(−0.09048)(Σ_2)	...	(−2,840.00)	−256.96	−432.70	−3,529.72
Σ_3	...	...	835.69	−258.59	577.05
III′	...	...	−1.00000	0.30944	−0.69056

BACK SOLUTION

$b_{12.34}$	$b_{13.24}$	$b_{14.23}$
0.04946	0.15236	−0.30944
−0.01282	0.02800	
−0.84626		−0.30944
	0.18036	
−0.80962		

The values from line 5 of Table A2.2, including the check sum, are entered as line I of Table A2.3. Each item is divided by the first item of the line, with its sign changed (−612.77). The quotients are entered as line I′. The sum of the first four items checks to the value in the last column, the check sum.

The values from line 8 of Table A2.2 are entered as line II of Table A2.3, beginning with column X_3 (the values enclosed in parentheses will be explained later). Line I is next multiplied by the value in column X_3 of line I′ (−4.69207), and the products entered in the corresponding columns below line II. These two lines are then summed, giving line Σ_2. These operations are now verified by adding the items of line Σ_2 in columns X_3.

X_4, and X_1, and comparing the sum with the check sum in column ΣX, The three values add to 39,011.02, agreeing to 0.01 with the check sum, 39,011.03.

The values in line Σ_2 are next divided by the value in column X_3, with its sign changed ($-31,388.78$). The quotients are entered as line II'. Again the check sum verifies the computation.

The values from line 11, Table A2.2, are then entered as line III, beginning with column X_4. (Again disregard the figures in parentheses.) Line I is multiplied by the value in column X_4 of line I' (0.04142), and the products entered in the corresponding columns below line III; and line Σ_2 is multiplied by the value in column X_4 of line II', and the products entered in the corresponding columns in the next line. Line III and the two following lines are then summed, giving line Σ_3. The values in line Σ_3 are divided by the value in column X_4 of that line, *with its sign changed*. The quotients are entered as line III'. Again the check sum verifies the work. The values in line Σ_3 (before the check sum) add to 577.10, which agrees to 0.05 with the check sum of 577.05.

The values in lines I', II', and III' of column X_1, *with the signs changed*, are then entered at the foot of columns X_2, X_3, and X_4 (designated here $b_{12.34}$, $b_{13.24}$, and $b_{14.23}$). The value at the foot of the X_4 column, -0.30944, is the value for $b_{14.23}$. The item in column X_4, line I' (0.04142), is then multiplied by the last of these values (-0.30944), and the product (-0.01282) entered in the X_2 column; and the item in column X_4, line II' (-0.09048), is also multiplied by -0.30944, and the product entered in the X_3 column. The two entries at the foot of the X_3 column are then added, giving 0.18036 as the value for $b_{13.24}$. The item in column X_3, line I' (-4.69207), is then multiplied by 0.18036, and the product (-0.84626) entered below the other two entries at the foot of the X_2 column. The sum of these three entries, -0.80962, is then the value for $b_{12.34}$.

The way the check sum works in checking the operations may be seen by filling in the missing spaces in Table A2.3, as indicated by the entries enclosed in parentheses. Thus in line II, the first item, 2,875.16, is the same item, $\Sigma x_2 x_3$, as appears in line I, column X_3. If when line I had been multiplied by -4.69207, the operation had included the X_2 column also, the product would have been $-2,875.16$, or exactly the same as the item, in line I, column X_3, with the sign changed. This value, entered below line II in column X_2, exactly cancels the previous value when the two lines are added, leaving line Σ_2 still the same.

Similarly, the values -25.38 and 2,720.92, from lines I and II of column X_4, may be entered in parentheses, in columns X_2 and X_3 of line III. If the previous operations had been carried out in full, below them would appear 25.38 in column X_2 (column X_4, line I, times -1), and 119.08 and $-2,840.00$ ([column X_4 line I][-4.69206] and column X_4, line Σ_2, times -1). When the three lines are totaled to give line Σ_3, the items exactly cancel out, as before.

It should be noted that when all the items are entered in each line, including those in parentheses, the sum of the items in columns X_2 to X_1 exactly equals, line by line, the item in column ΣX. For that reason, if any error is found when one of the Σ lines is reached, the line in which the error occurred can be determined by adding the items line by line, and verifying the totals against the individual check sums. To do this it is not necessary to enter the missing items, as has been done in Table A2.5 (in parentheses); instead, the items left out can be picked out by going up the columns for the particular variable concerned. Thus all the missing terms for line III (extensions for X_4) and the next two lines appear above in the X_4 column. Once the location of the missing items in the previous work has been learned, they can be used to verify the computations line by line, and any error readily located.

The "back solution" is simply the solution, in regular form, of lines III′, II′, and I′ for b_4, b_3, and b_2. Thus line III′, if written out, is

$$-b_4 = 0.30944$$

Hence $b_4 = -0.30944$, the value at the foot of column X_4. Similarly, line II′, written out, becomes

$$-b_3 - 0.09048b_4 = -0.15236$$

Substituting the above value for b_4, and rearranging,

$$b_3 = 0.15236 - (0.09048)(-0.30944)$$

$$= 0.15236 + 0.02800$$

These last two values are the same as shown at the foot of column X_3, hence $b_3 = 0.18036$.

Similarly line I′, when written out in full,

$$-b_2 - 4.69207b_3 + 0.04142b_4 = -0.04946$$

Substituting values for b_3 and b_4, and rearranging,

$$b_2 = 0.04946 + (0.04142)(-0.30944) + (-4.69207)(0.18036)$$

$$= 0.04946 - 0.01282 - 0.84626 = -0.80962$$

exactly as shown at the foot of column X_2.

Having computed the values of the three regression coefficients, the final steps are (*a*) to check those values by substituting them in the *last* equation (line III, in full); (b) to compute the coefficient of multiple correlation; and (c) to compute the constant $a_{1.234}$ for the regression equation. These steps are all shown in Table A2.4.

Table A2.4

FINAL STEPS IN SOLUTION OF MULTIPLE CORRELATION PROBLEM

Variable	Regression Coefficient (1)	Equation III (2)	Check (3)	Equation X_1 (4)	Computation of R^2 (5)	Means (6)	Computation of a (7)
X_2	−0.80962	−25.38	20.55	30.31	−24.54	10.308	−8.35
X_3	0.18036	2,720.92	490.75	4,924.46	888.18	167.462	30.20
X_4	−0.30944	1,093.70	−338.43	172.85	−53.49	105.846	−32.75
Sums		172.85	172.87	998.92	810.15	105.92	−10.90

The first operation in Table A2.4 is the final checking of the entire solution, including the back solution. This is done by substituting the values found for the b's in the *last* equation of the normal equations. For this problem, that equation is:

$$\Sigma(x_2 x_4)b_2 + \Sigma(x_3 x_4)b_3 + \Sigma(x_4^2)b_4 = \Sigma x_1 x_4$$

The values of the 3 b's are entered in column 1 of the table, and the values of the corresponding coefficients of the unknowns, such as $\Sigma(x_2 x_4)$ etc., are entered in column 2. The product of each b with its coefficient is then computed, and entered in column 3. These add to 172.87, checking satisfactorily with the value of $\Sigma(x_1 x_4)$, 172.85, as shown at the foot of column 2.

The computation of the coefficient of multiple correlation, according to equation (46):

$$R_{1.234}^2 = \frac{b_2(\Sigma x_1 x_2) + b_3(\Sigma x_1 x_3) + b_4(\Sigma x_1 x_4)}{\Sigma(x_1^2)}$$

is shown in tabular form in columns 4 and 5.

The values $(\Sigma x_1 x_2)$, etc., as shown in Table A2.2, lines 5, 8, and 11 of column X_1, and $\Sigma(x_1^2)$, shown in line 14, are entered in column 4 of Table A2.4. Each product sum is multiplied by the corresponding b, shown in column 1, and the products entered in column 5. The sum of these products is then the numerator of the fraction in equation (46). The computation is then readily completed:

$$R_{1.234}^2 = \frac{810.15}{998.92} = 0.8110$$

$R = 0.9006$. With $n = 13$, and $m = 4$,

$$\bar{R}^2 = 1 - (1 - 0.811)\tfrac{12}{9} = 0.784, \text{ and } \bar{R} = 0.86$$

The standard error of estimate may also be readily computed

$$\Sigma(x_1^2) = ns_1^2 = 998.92$$

$$\Sigma[(x_1')^2] = ns_{x'_1}^2 = \underline{810.15}$$

then since $ns_1^2 - ns_{x'_1}^2 = ns_z^2$

$$\Sigma z^2 = ns_z^2 = 188.77$$

Since there are 13 cases and 3 independent variables,

$$\bar{S}_{1.234}^2 = \frac{ns_z^2}{n-m} = \frac{188.77}{9} = 20.97$$

and

$$\bar{S}_{1.234} = 4.58$$

The a for the regression equation is next computed. Using equation (11.10),

$$a_{1.234} = M_1 - b_2 M_2 - b_3 M_3 - b_4 M_4$$

we may arrange the work in tabular order as shown in columns 6 and 7 of Table A2.4. The means, from line 2 of Table A2.2, are entered in column 6, then multiplied by their respective b's, and the products entered in column 7. To complete the computation, following equation (11.10), the sum of this column is then subtracted from the mean of X_1.

$$a_{1.234} = 105.92 - (-10.90) = 116.82$$

This completes the computation of all the linear multiple correlation constants.[1] The results may be summarized:

$$X_1' = 116.82 - 0.810 X_2 + 0.180 X_3 - 0.309 X_4$$

$$\bar{R}_{1.234} = 0.86$$

$$\bar{S}_{1.234} = 4.58$$

Tables A2.1 to A2.4 and the computation following, have shown every arithmetic step in obtaining these results, arranged in the most convenient form for ready computation and checking.

In solving the equations in actual practice, only the items that are not enclosed in parentheses in Table A2.3 would be entered.

All the steps necessary for solving a 6-variable problem by this method are shown in Table A2.5, with the entries omitted which previously have been shown in parentheses.

For shorter methods of solution with modern computing machines, see pages 506 to 517.

---

[1] For the effects of rounding, see Note 7, Appendix 3.

Table A2.5

Doolittle Solution of Normal Equations for Six Variables

Line Designation	Column Designation							
	X_2	X_3	X_4	X_5	X_6	X_1	ΣX	

	Equations to be Solved							
Eq. I	100.00	23.32	19.86	25.69	10.64	40.17	219.68	
Eq. II	(23.32)	100.00	17.47	45.20	21.39	60.03	267.41	
Eq. III	(19.86)	(17.47)	100.00	26.28	0.33	23.79	187.73	
Eq. IV	(25.69)	(45.20)	(26.28)	100.00	29.89	68.07	295.13	
Eq. V	(10.64)	(21.39)	(0.33)	(29.89)	100.00	35.53	197.78	

	Front Solution							
I	100.0000	23.3200	19.8600	25.6900	10.6400	40.1700	219.6800	
I′	−1.000000	−0.233200	−0.198600	−0.256900	−0.106400	−0.401700	−2.196800	
II		100.0000	17.4700	45.2000	21.3900	60.0300	267.4100	
II−1		−5.4382	−4.6314	−5.9909	−2.4812	−9.3676	−51.2294	
Σ_2		94.5618	12.8386	39.2091	18.9088	50.6624	216.1806	
II′		−1.000000	−0.135769	−0.414640	−0.199962	−0.535760	−2.286130	
III			100.0000	26.2800	0.3300	23.7900	187.7300	
III−1			−3.9442	−5.1020	−2.1131	−7.9778	−43.6284	
III−2			−1.7431	−5.3234	−2.5672	−6.8784	−29.3506	
Σ_3			94.3127	15.8546	−4.3503	8.9338	114.7510	
111′			−1.000000	−0.168107	+0.046126	−0.094725	−1.216706	
IV				100.0000	29.8900	68.0700	295.1300	
IV−1				−6.5998	−2.7334	−10.3197	−56.4358	
IV−2				−16.2577	−7.8403	−21.0067	−89.6371	
IV−3				−2.6653	+0.7313	−1.5018	−19.2904	
Σ_4				74.4772	20.0476	35.2418	129.7667	
IV′				−1.000000	−0.269178	−0.473189	−1.742367	
V					100.0000	35.5300	197.7800	
V−1					−1.1321	−4.2741	−23.3740	
V−2					−3.7810	−10.1306	−43.2279	
V−3					−0.2007	+0.4121	+5.2930	
V−4					−5.3964	−9.4863	−34.9303	
Σ_5					89.4898	12.0511	101.5408	
V′					−1.00000	−0.134664	−1.134664	

Back Solution						Eq. V	Check	Eq. X_1	Computation of R^2
	$b_{12.3456}$	$b_{13.2456}$	$b_{14.2356}$	$b_{15.2346}$	$b_{16.2345}$				
	+0.401700	+0.535759	+0.094725	+0.473189	+0.134664				
	−0.014328	−0.026928	+0.006212	−0.036249	+0.134664	100.00	13.4664	35.53	4.7846
	−0.112250	−0.181173	−0.073453	+0.436940		29.89	13.0601	68.07	29.7425
	−0.005458	−0.003731	+0.027484			0.33	0.0091	23.79	0.6538
	−0.075540	+0.323927				21.39	6.9288	60.03	19.4453
	+0.194124					10.64	2.0655	40.17	7.7980
						Eq. V = 35.53	35.5299		62.4242

$$R^2 = \frac{62.4242}{s_1^2} = \frac{62.4242}{100.00} \qquad\qquad R = \sqrt{0.624242} = 0.790$$

Standard Errors of Partial Regression Coefficients and Standard Error of an Individual Estimate. The computation of standard errors of net or partial regression coefficients by equation (17.2), and of standard errors of an individual estimate, by equations (19.2) to (19.6), may be simplified by the following procedure:

$$\left.\begin{array}{l}
\Sigma(x_2^2)c_{22} + \Sigma(x_2x_3)c_{23} + \Sigma(x_2x_4)c_{24} = 1 \\
\Sigma(x_2x_3)c_{22} + \Sigma(x_3^2)c_{23} + \Sigma(x_3x_4)c_{24} = 0 \\
\Sigma(x_2x_4)c_{22} + \Sigma(x_3x_4)c_{23} + \Sigma(x_4^2)c_{24} = 0
\end{array}\right\} \quad \text{(A2.1)}$$

Solve simultaneously to obtain the values for c_{22}, c_{23}, and c_{24}. Then set up exactly the same set of equations, with c_{32}, c_{33}, and c_{34} as the unknowns, and with 0, 1, and 0 to the right of the equal signs, in the first, second, and third equations, respectively, and solve. Then set up again, with c_{42}, c_{43}, and c_{44} as the unknowns, and with 0, 0, and 1 to the right of the equal signs, and solve again. The standard errors of the regression coefficients may then be found by the following equations.

$$\left.\begin{array}{l}
s_{b_{12.34}} = \bar{S}_{1.234}\sqrt{c_{22}} \\
s_{b_{13.44}} = \bar{S}_{1.234}\sqrt{c_{33}} \\
s_{b_{14.23}} = \bar{S}_{1.234}\sqrt{c_{44}}
\end{array}\right\} \quad \text{(A2.2)}$$

Except for the values to the right of the equal sign, the coefficients of the equations are exactly the same as those required to obtain the values of $b_{12.34}$, $b_{13.24}$, and $b_{14.23}$. For that reason the values for c_{22}, c_{33}, and c_{44} may be most readily calculated by introducing as many new columns in the form of the Doolittle solution (Table A2.3) as there are independent factors, between the columns for X_1 and Σ. These columns will be

Line	Error b_2	Error b_3	Error b_4	Error b_5
(Eq. I)	1	0	0	0
(Eq. II)	0	1	0	0
(Eq. III)	0	0	1	0
(Eq. IV)	0	0	0	1
etc.				

These values can be included in the check sum, and the operations carried through for them just as for the other columns until the "back solution" to find the b's is reached. Then a separate "back solution" can be run for each set of "c" values, starting with the values in each "Error" column just as the back solution to find the b's started with the values in the X_1 column.[2]

Table A2.6 shows all the computations necessary to compute all the b's and c's from the product sums calculated in Table A2.2, except for the back solution on X_1, as shown in the lower section of Table A2.3. Table A2.6 thus replaces all of Table A2.3, except this last section. In practice, this back solution would be included in Table A2.6 ahead of the three back solutions on c_2, c_3, and c_4.

In computing Table A2.6, the work in the c columns is carried out to two more decimal places than in the other columns. This is necessary because of the small size of the values involved. It should also be noticed that in the back solution on c_3, only c_{34} and c_{33} are calculated directly. Since c_{32} is identical with c_{23}, the value previously calculated for the latter is inserted instead. Similarly, the back solution on c_{44} involves no additional calculating at all, since c_{44} is copied down (with the sign changed) from line III', c_{24} is written down for c_{42}, and c_{34} for c_{43}. Only the computation by substitution in the check equations is involved. Even that computation can be omitted for the c_4 values, since each of them has been checked earlier—c_{42} and c_{43} by substitution and c_{44} by the check sum in lines Σ_3 and III'.

As a result of these computations, the following values are secured:

$$c_{22} = 0.00259;\ c_{33} = 0.000042;\ c_{44} = 0.00120$$

Since $\bar{S}_{1.234} = 4.58$, the standard error of the b's may be readily calculated by equation (A2.2)

$$S_{b_{12.34}} = 4.58\sqrt{0.00259} = 0.233$$

$$S_{b_{13.24}} = 4.58\sqrt{0.000042} = 0.030$$

$$S_{b_{14.23}} = 4.58\sqrt{0.00120} = 0.159$$

The net regression coefficients may then be stated

$$b_{12.34} = -0.810$$
$$(0.233)$$

$$b_{13.24} = 0.180$$
$$(0.030)$$

$$b_{14.23} = -0.309$$
$$(0.159)$$

[2] For an explanation of why this process and equation (A2.2) gives the standard error of the b's see Note 13 in the second edition of this book. For other uses of the "c" constants see R. A. Fisher, *Statistical Methods for Research Workers*, 12th ed., pp. 129–166, Oliver and Boyd, Edinburgh and London, 1954.

Table A2.6

SOLUTION OF NORMAL EQUATIONS BY THE DOOLITTLE METHOD, TO CALCULATE
REGRESSION COEFFICIENTS AND THEIR STANDARD ERRORS

Line Designation	COLUMN DESIGNATION							
	X_2	X_3	X_4	X_1	c_2	c_3	c_4	$\Sigma(X+c)$
	EQUATIONS TO BE SOLVED							
Eq. I	612.77	2,875.16	−25.38	30.31	1	0	0	3,493.86
II	2,875.16	44,879.23	2,720.92	4,924.46	0	1	0	55,400.77
III	−25.38	2,720.92	1,093.70	172.85	0	0	1	3,963.09
	FRONT SOLUTION							
I	612.77	2,875.16	−25.38	30.31	1.00000	0	0	3,493.86
I′	−1.0000000	−4.6920704	0.0414185	−0.0494639	−0.0016319	0	0	−5.7017477
II		44,879.23	2,720.92	4,924.46	0	1.00000	0	55,400.77
I(−4.6920704)		−13,490.45	119.08	−142.22	−4.69207	0	0	−16,393.44
Σ_2		31,388.78	2,840.00	4,782.24	−4.69207	1.00000	0	39,007.33
II′		−1.0000000	−0.0904782	−0.1523551	0.0001495	−0.0000319	0	−1.2427157
III			1,093.70	172.85	0	0	1.0000	3,963.09
I(0.0414185)			−1.05	1.26	0.0414185	0	0	144.71
$\Sigma_2(−0.0904782)$			−256.96	−432.69	0.4245300	−0.0904782	0	−3,529.31
Σ_3			835.69	−258.58	0.4659485	−0.0904782	1.0000000	578.49
III′			−1.0000000	0.3094210	−0.0005576	0.0001083	−0.0011966	−0.6922304

BACK SOLUTION ON c_2				Eq. II—c_2	Check
	c_{22}	c_{23}	c_{24}		
	0.0016319	−0.0001495	0.0005576		
	0.0000231	−0.0000505	0.0005576	2,720.92	1.52
	0.0009384	−0.0002000		44,879.23	−8.98
	0.0025934			2,875.16	7.64
				0.00	0.00

BACK SOLUTION ON c_3				Eq. II—c_3	Check
	c_{32}	c_{33}	c_{34}		
		0.0000319	−0.0001083		
		0.0000098	−0.0001083	2,720.92	−0.29
		00.0000417		44,879.23	1.87
	−0.0002000			2,875.16	−0.58
				1.00	1.00

BACK SOLUTION ON c_4				Eq. II—c_4	Check
	c_{24}	c_{34}	c_{44}		
			0.0011966		
			0.0011966	2,720.92	3.26
		−0.0001083		44,879.23	−4.86
	0.0005576			2,875.16	1.60
				0.00	0.00

Just as in the illustrations discussed in Chapter 17, some of the net regression coefficients are much more reliable than are others. If we assume that the conditions of random sampling are fulfilled, there is some possibility that the regression for $b_{14.23}$ in the universe from which the sample was drawn is really positive instead of negative; but there is only a very slight chance that $b_{12.34}$ is really positive, and it is almost a certainty that $b_{13.24}$ is really positive, and above 0.1.

The computation of the standard errors of the net regression coefficients, by the method just presented is not a difficult one. It should be made an integral part of every multiple correlation solution, so that not only will the regression coefficients be obtained, but also the amount of confidence that can be placed in each value will be determined. Only if that is done can the regressions be interpreted with confidence.

The computations shown in Table A2.6 also give all the values needed to estimate the standard error of an individual estimate. Substituting these values in equation (19.2), and using the value for $\bar{S}_{1.234}$ previously calculated on page 497 (in practice the calculations on that page would all be made after Table A2.6 was calculated, including the back solution on X_1), we have:

$$(X) \quad s^2_{x'_1 - x_1} = 20.97\left[1 + \frac{1}{13} + .00259x_2^2 + .000042x_3^2 + .00120x_4^2 \right.$$

$$\left. + 2(-.00020)x_2x_3 + 2(.00056)x_2x_4 + 2(-.00011)x_3x_4\right]$$

The use of this equation may be shown as follows: Suppose we draw a new observation from the same universe as that from which the original sample (shown in Table A2.1) was drawn, and the new observation has values of 18 for X_2, 300 for X_3, and 90 for X_4. After we estimate the probable X_1 value from the regression equation, how much confidence can we place in that estimate?

The estimated value works out as follows:

The regression equation (from Table A2.4) is

$$X'_1 = -10.90 - 0.80962X_2 + 0.18036X_3 - 0.30944X_4$$
$$= -10.90 - 0.80962(18) + 0.18036(300) - 0.30944(90)$$
$$= 78.29$$

Before the values of X_2, X_3, and X_4 for this new observation can be substituted in equation (X), they must be put in the form of x_2, x_3, or x_4. Using the means shown in Table A2.2, we calculate

$$x_2 = X_2 - M_2 = 18 - 10.31 = 7.69$$
$$x_3 = X_3 - M_3 = 300 - 167.46 = 132.64$$
$$x_4 = X_4 - M_4 = 90 - 105.85 = -15.85$$

Substituting these values in equation (X), we have

$$s^2_{x'_1-x_1} = 20.97\left[1 + \frac{1}{13} + .00259(7.69)^2 + .000042(132.64)^2\right.$$
$$+ .00120(-15.84)^2 + 2(-.00020)(7.69)(132.64)$$
$$+ 2(.00056)(7.69)(-15.84)$$
$$\left. + 2(-.00011)(132.64)(-15.84)\right] = 45.8939$$

$$s_{x'_1-x_1} = 6.77$$

We can now say that our estimate of X_1, for the new observation with $X_2 = 18$, $X_3 = 300$, and $X_4 = 90$, is $X'_1 = 78.29 \pm 6.77$.

It is evident that the standard error of this particular estimate, 6.77, is larger than the standard error of estimate for the sample, 4.58. That is because the values of the independent variables for this new observation lay near the extremes of their several ranges in the sample. It is also evident that the value of $s_{x'_1-x_1}$ will vary with each new observation, depending on the combination of values of the independent variables in each observation.

Coefficients of Partial Correlation. Computation of the coefficients of partial correlation by equation (12.8) involves the calculation of the multiple correlation of the dependent variable with successive sets of the independent variables, with a different independent variable left out in each set. Thus, for the four-variable problem whose solution is shown in Table A2.3 (and A2.4) the three coefficients of partial correlation involve not only the value $R_{1.234}$, but also $R_{1.23}$, $R_{1.24}$, and $R_{1.34}$. These may be calculated readily by the same process shown in Table A2.3. It is not necessary to repeat the process three times, however, as the several columns may be rearranged with little additional calculation to omit each independent variable in turn. The first two stages in this process are illustrated in Table A2.7. The values for lines I and I' and Σ_2 and II' are copied from Table A2.3, as shown in the first four lines of Table A2.7. The columns X_4 and ΣX are dropped, however, as they are not needed at this step.

Lines I' and II' give all the information needed for the "back solution" with X_4 omitted. This is accordingly given in the second section of Table A2.7, using the same form as in the back solution of Table A2.3. The verification of the b's by substitution in equation II, and the calculation of $R_{1.23}$ by use of equation I are also shown, organized the same as in Table A2.3.

The next step is to enter the values necessary for the "front solution" with X_3 omitted. This is shown in the third block of Table A2.7. Lines I and I' are entered again, with the column for X_3 omitted. Lines III and

Table A2.7

DOOLITTLE SOLUTION OF NORMAL EQUATIONS, TO FIND COEFFICIENTS OF
PARTIAL CORRELATION, FOR THREE INDEPENDENT VARIABLES

LINE DESIG- NATION	COLUMN DESIGNATION						
	X_2	X_3	X_1				
I	612.77	2,875.16	30.31				
I′	−1.00000	−4.69207	−0.04946				
Σ_2		31,388.78	4,782.24				
II′		−1.00000	−0.15236				

BACK SOLUTION, X_4 OMITTED			Eq. II	Check	Eq. I	Compu- tation of $R^2_{1.23}$
	$b_{12.3}$	$b_{13.2}$				
	+0.04964	+0.15236				
	−0.71488	+0.15236	44,879.12	6,837.78	4,924.46	750.29
			2,875.16	−1,912.67	30.31	−20.16
	−0.66524		4,924.46	4,925.11	998.92	730.13

$$R^2_{1.23} = \frac{730.13}{998.92} = 0.730919$$

FRONT SOLUTION, X_3 OMITTED

	X_2	X_4	X_1
I	612.77	−25.38	30.31
I′	−1.00000	0.04142	−0.04946
III		1,093.70	172.85
(0.04142)(I)		−1.05	1.26
Σ_3		1,092.65	174.11
III′		−1.00000	−.15935

BACK SOLUTION, X_3 OMITTED			Eq. III	Check	Eq. I	Compu- tation of $R^2_{1.24}$
	$b_{12.4}$	$b_{14.2}$				
	0.04946	0.15935				
	0.00660					
		0.15935	1,093.70	174.28	172.85	27.54
	0.05606		−25.38	−1.42	30.31	1.70
			172.85	172.86	998.92	29.24

$$R^2_{1.24} = \frac{29.24}{998.92} = 0.029272$$

(0.04142)(I) are copied from Table A2.3 for columns X_4 and X_1. All that is necessary to complete the front solution is to add the new totals, Σ_3, and to divide col. X_1 by col. X_4, to get the new line $\Sigma III'$, and then to proceed with the back solution, as before. The check on equation III and the calculation of $R_{1.24}$ by substitution in equation I are also shown under the back solution on X_3.

The final step in computing the needed coefficients of multiple correlation involves calculating $R_{1.34}$. Since this involves rearranging Table A2.3 to omit X_2, which appears in the first column, it is necessary to carry through an entire new front solution, with X_2 omitted. This process is shown in Table A2.8. (The column ΣX is a new Σ, obtained by adding the values in columns X_3, X_4, and X_1 for lines II and III, and then using it as a check thereafter.)

In problems where there are four independent variables, this new back

Table A2.8

Doolittle Solution of Normal Equations, to Find Coefficients of Partial Correlation, for Three Independent Variables (*continued*)

Line Designation	Column Designation			
	X_3	X_4	X_1	ΣX
II	44,879.23	2,720.92	4,924.46	52,524.61
II	−1.00000	−0.06063	−0.10973	−1.17046
III		1,093.70	172.85	3,987.47
(−0.06063)(II)		−164.97	−298.57	−3,184.45
Σ_2		928.73	−125.72	803.01
		−1.00000	0.13537	−0.86463

Back Solution, X_2 Omitted

$b_{13.4}$	$b_{14.3}$	Eq. III	Check	Eq. I	Computation of $R^2_{1.34}$
0.10973	−0.13537				
0.00821	−0.13537	2,720.92	−368.33	172.85	−23.39
0.11794		44,879.23	5,293.06	4,924.46	580.79
		4,924.46	4,924.73	998.92	557.40

$$R^2_{1.34} = \frac{557.40}{998.92} = 0.558002$$

solution should be arranged in this column order X_4, X_5, X_3, X_2, X_1. After the entries were calculated through the front solution, two back solutions could then be run, one leaving out the X_2 column, and one the X_3 column. Where there are six independent variables, a third step could be used by repeating the last two steps of the front solution for the third independent variable to be dropped out; or a complete new solution could be run with X_3 and X_4 occupying the last columns before X_1. In many-variable problems, various other time-saving combinations can be worked out by the ingenious computer.

Tables A2.4, A2.7, and A2.8 provide all the values necessary for the calculation of the partial correlation coefficients, using equation (12.8). These calculations may be tabled as follows:

Variables	(1) R^2	(2) $1 - R^2$
1.234	0.8110	0.1890
1.23	0.7309	0.2691
1.24	0.0293	0.9707
1.34	0.5580	0.4420

$$r_{12.34}^2 = 1 - \frac{1 - R_{1.234}^2}{1 - R_{1.34}^2} = 1 - \frac{0.1890}{0.4420} = 0.5724 \qquad r_{12.34} = -0.76$$

$$r_{13.24}^2 = 1 - \frac{1 - R_{1.234}^2}{1 - R_{1.24}^2} = 1 - \frac{0.1890}{0.9707} = 0.8053 \qquad r_{13.24} = 0.89$$

$$r_{14.23}^2 = 1 - \frac{1 - R_{1.234}^2}{1 - R_{1.23}^2} = 1 - \frac{0.1890}{0.2691} = 0.2977 \qquad r_{14.23} = -0.55$$

The signs of the partial correlation coefficients are taken from the signs of the corresponding net regression coefficients, as shown in Table A2.3 or A2.4

Alternative Methods of Solving Normal Equations. ·The methods for solving normal equations and obtaining the various constants necessary in correlation analysis, which have been presented in Tables A2.3 to A2.8 inclusive, employ the so-called Doolittle method of solving equations, first developed by Dr. M. H. Doolittle, a computer in the Geodetic Survey.[3] His method involved slight modifications of the methods originally suggested by Gauss, the discoverer of the least-squares technique. (The solutions shown in Tables A2.7 and A2.8 involve short cuts added by the senior author.) The use of the 0–1–0, etc., method of calculating

[3] M. H. Doolittle, Adjustment of the primary triangulation between Kent Island and Atlanta base lines (Paper No. 3, Method employed in the solution of normal equations and the adjustment of a triangulation), *Report of the Superintendent*, Coast and Geodetic Survey, pp. 115–120, 1878.

error formulas (the reciprocal matrix), was also first developed by Gauss, and was revived by R. A. Fisher. Its further application to calculating the standard errors of an individual estimate was developed by the late M. A. Girshick of the U. S. Department of Agriculture, at the senior author's request.

A simple short cut in the solution of the normal equations has been suggested by P. S. Dwyer.[4] He points out that much of the "front solution" involves subtracting a series of products from, or adding them to, a given figure. In Table A2.5, for example, the item that appears in line Σ_4 of column X_5 is simply the value:

$$100.0000 + (25.6900)(-0.256900) + (39.2091)(-0.414640)$$
$$+ (15.8546)(-0.168107) = 74.4772$$

With modern computing machines this value can be computed directly without clearing the total dial, using the reverse lever whenever the product is to be subtracted instead of added. This method saves reading off and entering in the table the values that appear in line IV-1, IV-2, and IV-3. Whether this additional operation and possibility of error offset the other savings each computer can determine for himself.

Using this Dwyer short cut all the way through, the front solution of Table A2.3 would show only the lines I and I′, Σ_2 and II′, and Σ_3 and III′. Similarly Table A2.5 would show in the front solution only I and II′, Σ_2 and II′, Σ_3 and III′, Σ_4 and IV′, and Σ_5 and V′. Various other possible modifications of the Doolittle solution, all based on the same basic principle, are shown in Dwyer's article.

Obtaining all Multiple Correlation and Regression Coefficients by Matrix Solution.

Modern developments in calculation and matrix solution have been combined in one composite solution to give practically all constants desired, in a recent publication by Friedman and Foote.[5] Using the problem presented in Table 13.1, the successive steps in this method are summarized below.

Obtaining the Augmented Sums of Squares and Cross-Products. The first step is to compute the "augmented" sums of squares and cross

[4] P. S. Dwyer, The solution of simultaneous equations, *Psychometrika*, Vol. 6, No. 2, April, 1941.

[5] Joan Friedman and Richard J. Foote, Computational Methods for Handling Systems of Simultaneous Equations, U. S. Dept. of Agriculture, Agricultural Marketing Service, Agricultural Handbook No. 94, pp. 2–220, Nov., 1955. Acknowledgement is due to the authors for permission to reproduce here the text of this section of the handbook, adapted to the illustration shown in Chapter 13.

Table A2.9

COMPUTATION OF AUGMENTED SUMS OF SQUARES AND CROSS-PRODUCTS*

Item	X_1 (1)	X_2 (2)	X_3 (3)	X_4 (4)	Σ (5)
Sum	1,093.60	549.60	10,100.	1,315.900	13,059.100 √
Mean†	54.68	27.48	505.	65.795	652.955 √
1. Extensions with X_1:					
$\Sigma X_1 X_1$	60,180.04	29,952.36	554,069.	71,911.25	716,112.65 √
$N\Sigma X_1 X_1$	1,203,600.80	599,047.20	11,081,380.	1,438,225.00	14,322,253.00 √
$\Sigma X_1 \Sigma X_1$	1,195,960.96	601,042.56	11,045,360.	1,439,068.24	14,281,431.76 √
Difference	7,639.84	−1,995.36	36,020.	−843.24	40,821.24 √
(Diff.)(K_iK_j)‡	(.0001) 0.763984	(.0001) −0.199536	(.00001) 0.360200	(.0001) −0.084324	
2. Extensions with X_2:					
$\Sigma X_i X_2$		15,283.42	279,613.9	36,011.31	360,860.99 √
$N\Sigma X_i X_2$		305,668.40	5,592,278.0	720,226.20	7,217,219.80 √
$\Sigma X_i \Sigma X_2$		302,060.16	5,550,960.0	723,218.64	7,177,281.36 √
Difference		3,608.24	41,318.0	−2,992.44	39,938.44 √
(Diff.)(K_{ij})		(.0001) 0.360824	(.00001) 0.413180	(.0001) −0.299244	
3. Extensions with X_3:					
$\Sigma X_i X_3$			5,170,916.	665,096.3	6,669,695.2 √
$N\Sigma X_i X_3$			103,418,320.	13,301,926.0	133,393,904.0 √
$\Sigma X_i \Sigma X_3$			102,010,000.	13,290,590.0	131,896,910.0 √
Difference			1,408,320.	11,336.0	1,496,994.0 √
(Diff.)(K_iK_j)			(.000001) 1.408320	(.00001) 0.113360	
4. Extensions with X_4:					
$\Sigma X_i X_4$				87,461.61	860,480.47 √
$N\Sigma X_i X_4$				1,749,232.20	17,209,609.40 √
$\Sigma X_i \Sigma X_4$				1,731,592.81	17,184,469.69 √
Difference				17,639.39	25,139.71 √
(Diff.)(K_iK_j)				(.0001) 1.763939	

* The computations were performed with 9 decimal places, of which only 6 appear in the table; therefore, some of the computations may appear to be slightly in error. Where fewer decimals are shown than 6, they were all zeros.

† For this example, $N = 20$.

‡ The $k_i k_j$ value is shown in parentheses, as taken from Table A2.10.

products or moments. Use of augmented moments is suggested to avoid rounding errors involved in obtaining arithmetic means. As used in this connection, an augmented moment equals the actual moment such as Σx, x_2, multiplied by the number of observations in the sample, here designated as N. In working with augmented moments, the sums of squares and cross-products in terms of original values are cumulated directly on the calculating machine as for any problem of this type. The total for the observations included in the analysis is then multiplied by N, or the number of such observations. The correction factor for an augmented sum of squares equals the square of the sum of the series. The correction factor for an augmented cross-product equals the product of the sum for each series. Subtraction of the correction factor from the augmented sum gives the augmented sum in terms of deviations from the respective means. These computations for the 5-variable regression problem are illustrated in Table A2.9. It should be noted that X_1, the dependent variable, is written first.

A check column should always be carried. This is obtained by computing a "new" variable, Σ, for each year or observation which equals the sum of all of the variables for that observation. To check the computations involved in obtaining Σ, the sum for each of the variables, including the variable Σ, over all of the years included in the analysis is obtained. These sums constitute the first row of Table A2.9. The sum of these sums for the variables other than Σ should exactly equal the sum of Σ. If they do, the computations involved in obtaining Σ are correct. The second row (i.e., line) in Table A2.9—the means—is obtained by dividing the sums in the first row by N, the sample size, which, for this example, equals 25. Cross-products for Σ with the other variables in the analysis are obtained in the usual way and are shown in the last column. The check for each row is carried out by computing the sum of all the items in the row, except for the·item in the last, or Σ, column. For example, in the second row, the check is obtained by adding the means for X_1, X_2, X_3, and X_4. This should equal, except for rounding errors, the item in the Σ column, and if true, this is indicated by a $\sqrt{}$ placed next to that item.[6]

The terms in the lower left-hand part of the table are omitted. But in order to check the computations in all sections after the first, these omitted terms must be included. For example, the computation of the check for the first row of the third section of Table A2.9 is given by:

$$554{,}069 + 279{,}613.9 + 5{,}170{,}916 + 665{,}096.3 = 6{,}669{,}695.2$$

[6] Rounding errors are usually taken to mean a discrepancy in the final decimal place. In some computations, the number of significant figures in the items operated upon is a further consideration.

The terms omitted from this row in the table, 554,069 and 279,613.9, are obtained, respectively, from the first row of the first section and the first row of the second section of column (3), the column in which the first written term of the row, 5,170,916, appears. In general, for the ith row of any section, the omitted terms to the left of any given term, call it m, are obtained from the ith row of each section of the column in which m appears.

If a discrepancy due to a rounding error should occur, the sum across the row is considered as the correct figure and the figure originally shown in the Σ column is corrected accordingly. This corrected value is used in further computations. (For details of this adjustment, see the Friedman-Foote handbook, page 5.)

Since only a limited number of decimals are shown here, a $\sqrt{}$ is placed after *all* items in the Σ column that serve as checks. However, rounding errors do occur in some of these items. These result in part because the omitted figures were dropped without rounding. In such cases the wrong figure in the Σ column is crossed out, and the sum of the columns is substituted.

If an error is made in computing the sums of squares and cross-products, the following method is more efficient than a direct recomputation as a means of locating the error. Suppose that the checking operation indicates that an error has been made in obtaining the extensions with X_1. Continue to calculate the extensions with X_2 and, if the check for this indicates that no error has been made, we know that the augmented moment between X_1 and X_2 in the first section is correct. Similarly, if the extensions with X_3 check, we know that the augmented moment between X_1 and X_3 in the first section is correct. If all other extensions check, the mistake is in the computation of the sum of the squares for X_1. If one of the extensions does not check, recomputation of the corresponding element in the first section is indicated. A similar procedure is used if the initial error occurs in an extension other than with X_1.

Adjustments to Make the Sums of Squares Nearly Equal I. It is a great convenience in computations to have all the elements on the main diagonal close to 1. In making this adjustment we are concerned only with the last row in each section of Table A2.9. A set of values that are powers of 10, the k_i, where i is the variable to which it applies, is chosen such that when the sum of squares for the variable is multiplied by the square of the k_i the answer lies between 0.1 and 10. The value $(k_i)^2$ is referred to as the adjustment factor. The k_i for this example are shown in the second column of Table A2.10. They are determined in the following manner: In Table A2.9, note that the sums of squares for X_1, X_2, and X_4, respectively, lie between 1,000 and 20,000; therefore the adjustment factor equals $(0.01)^2$ and k equals 0.01 for X_1, X_2, and X_4. For X_3 however, the sum of the squares equals 1,408,320 and it must be multiplied by an adjustment factor of $(0.001)^2$; therefore k_3 equals 0.001. The adjustment factors for the cross-products, the $k_i k_j$, are obtained by multiplying the k's for the variables involved. For example, $k_3 k_4$, the adjustment factor for $\Sigma x_3 x_4$, equals $(0.01)(0.001) = 0.00001$. The $k_i k_j$ are shown in the right-hand section of Table A2.10.

Table A2.10
ADJUSTMENT FACTORS

Variable	Values of k_i	$k_i k_j$			
		X_1 (0.01)	X_2 (0.01)	X_3 (0.001)	X_4 (0.01)
X_1	0.01	0.0001	0.0001	0.00001	0.0001
X_2	0.01		0.0001	0.00001	0.0001
X_3	0.001			0.000001	0.00001
X_4	0.01				0.0001

Adjusted augmented moments are obtained by multiplying the augmented moment, the terms in the next to last row of each section of Table A2.9, by its appropriate adjustment factor from Table A2.10.

The adjustment process is important, naturally, only when the k_i differ considerably from 1. Particularly when working with logarithms or first differences of logarithms, all of the k_i normally are close to 1. Some computers may prefer to adopt a general rule that adjustments are made only when at least one of the k_i lies outside the range 0.1 to 10.

The steps involved in obtaining the adjusted augmented moments are exactly the same for single- and multiple-equation analyses.

Obtaining Multiple and Partial Regression and Correlation Measures. The method of determining multiple regression constants discussed in the following pages differs in these two important ways from that given in some of the standard statistical textbooks: (1) The use of D, the inverse of the complete moment matrix,[7] and (2) the computation of the inverse

[7] This approach, suggested to Friedman and Foote by Frederick V. Waugh, Director, Agricultural Economics Division, Agricultural Marketing Service, substantially reduces the number of calculations necessary for the estimation of the various multiple regression coefficients, particularly the partial correlation coefficients. The method explained in Chapter 19 for calculating the standard error of an individual estimate, is based upon the computation of the inverse of a matrix using only the moments for the independent variables. There the inverse of this matrix is referred to as the C matrix, the elements of which are the c_{ij}.

Table A2.11

CALCULATING PARTIAL AND MULTIPLE REGRESSION MEASURES FOR

Row	x_1 (1)	x_2 (2)	x_3 (3)	x_4 (4)	Σx (5)
(1) x_1	.763984	−.199536	.360200	−.084324	.840324
(2) x_2		.360824	.413180	−.299244	.275224
(3) x_3			1.408320	.113360	2.295060
(4) x_4				1.763939	1.493731
(1)	.763984	−.199536	.360200	−.084324	.840324
(1″) (1.308927935)	1.	−.261178	.471476	−.110374	1.099924 √
(2)		.360824	.413180	−.299244	.275224
(1) (.261178)		−.052114	.094076	−.022024	.219474
(2′)		.308710	.507256	−.321268	.494698 √
(2″) (3.239286061)		1.	1.643147	−1.040679	1.602468 √
(3)			1.408320	.113360	2.295060
(1) (−.471476)			−.169826	.039757	−.396193
(2′) (−1.643147)			−.833496	.527891	−.812862
(3′)			.404998	.681008	1.086005 6
(3″) (2.469147995)			1.	1.681510	2.681510 √
(4)				1.763939	1.493731
(1) (.110374)				−.009307	.092750
(2′) (1.040679)				−.334337	.514822
(3′) (−1.681510)				−1.145122	−1.826130
(4′)				.275173	.275173 √
(4″) (3.634077471)				1.	1.

CALCULATION OF PARTIAL COEFFICIENTS

(5) $b_{ij} = d_{ij}/d_{11}$	−1.	−1.849845	.831894	−.415085	−2.433036 √
(6) $d_{11}d_{jj}$	275.717092	1037.507961	211.617583	60.342871	1585.185506 7
(7) d_{ij}^2	275.717092	943.483898	190.809412	47.504750	1457.515152
(8) (6) − (7) (c_{ij}'')	0	94.024063	20.808171	12.838121	127.670355 √
(9) (8)(.000227)	0	.021343	.004723	.002914	.028981 0
(10) $sb_{ij} = \sqrt{(9)}$		.146092	.068723	.053981	
(11) $r_{ij}^2 = (7)/(6)$		.909375	.901671	.787247	
(11a) r_{ij}		.954	.950	.887	. . .

CALCULATION OF MULTIPLE COEFFICIENTS

$$(12)\ R_{1.234} = \sqrt{\frac{(16.604731)(.763984) - 1}{(16.604731)(.763984)}} = \frac{11.685749}{12.685749} = \sqrt{.921171} = .959777$$

CALCULATION OF DEADJUSTED COEFFICIENTS

	Variable	b	s_b	k_i	$k_i' = k_i/k_1$
(13)	X_1			.01	
(14)	X_2	−1.849845	.146092	.01	1.0
(15)	X_3	.831894	.068725	.001	.1
(16)	X_4	−.415085	.053981	.01	1.0

(17) $\bar{S}_{1.234} = (.013711/.01) = 1.3711$ $a = 54.630000 - (-.36.133814) = 90.813814$

(18) $X_1 = 90.813814 - 1.849845 X_2 + .083189 X_3 - .415085 X_4$
    (.146092)   (.006872)   (.053981)

A FOUR-VARIABLE PROBLEM

Row	I_1 (6)	I_2 (7)	I_3 (8)	I_4 (9)	ΣI (10)
(1)	1.	0	0	0	1.
(2)	0	1.	0	0	1.
(3)	0	0	1.	0	1.
(4)	0	0	0	1.	1.
(1)	1.	0	0	0	1.
(1″)	1.308928	0	0	0	1.308928 √
(2)	0	1.	0	0	1.
	.261178	0	0	0	.261178
(2′)	.261178	1.	0	0	1.261178 √
(2″)	.846030	3.239286	0	0	4.085316 √
(3)	0	0	1.	0	1.
	−.471476	0	0	0	−.471476
	−.429154	−1.643147	0	0	−2.072301
(3′)	−.900630	−1.643147	1.	0	−1.543777 √
(3″)	−2.223789	−4.057173	2.469148	0	−3.811814 √
(4)	0	0	0	1.	1.
	.110374	0	0	0	.110374
	.271802	1.040679	0	0	1.312481
	1.514418	2.762968	−1.681510	0	2.595876
(4′)	1.896594	3.803647	−1.681510	1.	5.018731 √
(4″)	6.892370	13.822748	−6.110738	3.634077	18.238457 √

CALCULATION OF D MATRIX

	(1)	(2)	(3)	(4)	Σ
(1)	(d_{11}) 16.604731	(d_{12}) 30.716183	(d_{13}) −13.813378	(d_{14}) 6.892369	$(d_{1\Sigma})$ 40.399905
(2)		(d_{22}) 62.482672	(d_{23}) −27.300263	(d_{24}) 13.822746	$(d_{2\Sigma})$ 79.721339 8
(3)			(d_{33}) 12.744415	(d_{34}) −6.110737	$(d_{3\Sigma})$ −34.479962 3
(4)				(d_{44}) 3.634077	$(d_{4\Sigma})$ 18.238457 5
...					$Sd_{jj} = 95.465895$

CALCULATION OF MULTIPLE COEFFICIENTS

$$R^2_{1.234} = \frac{d_{11}m_{11} - 1}{d_{11}m_{11}} \qquad \bar{S}^2_{1.234} = \frac{1}{NN'd_{11}} = \frac{1}{(20)(16)(16.604731)} = \frac{1}{5313.513920} = .000188$$

$$\bar{S}_{1.234} = .013711$$

CALCULATION OF DEADJUSTED COEFFICIENTS

Row	Deadjusted b	Deadjusted s_b	M_x	Deadjusted bM_x	
(13)			54.680		
(14)	−1.849845	.146092	27.480	−50.833741	
(15)	.083189	.006872	505.000	42.010445	
(16)	−.415085	.053981	65.795	−27.310518	
(17)				−36.133814	

using a variation of the Doolittle method that omits the conventional back solution.

Steps involved in the forward solution of the Doolittle method are given here in full detail. Experience demonstrates this as the easiest way to learn how to carry out these operations. Once the general approach is learned, many of the computations shown individually in Table A2.11 can be cumulated directly in the calculating machine. Use is made of all possible shortcuts of this kind in the so-called abbreviated Doolittle method. This is the method described by Klein.[8]

Computations involved in the forward solution of the Doolittle method are shown in Table A2.11 as follows:

In rows (1)–(4), columns (1)–(4), enter the adjusted augmented moments computed above. The reader will note that the X's are listed in numerical order; in the method used in the usual solution elsewhere in this book, X_1 is placed after the last independent variable. Computations involved in obtaining Table A2.10 and the adjusted augmented moments should be carefully checked as no automatic checks are available for these steps.

Additional columns, I_i, one for each variable in the analysis, are added in columns (6)–(9). The makeup of these is obvious from the table.

As an alternative, data shown in the upper section of Table A2.11 can be recorded directly as the first row of each subsequent section.

In this forward solution we carry two check columns: Σ_x, column (5), for that part of the solution concerning the x's; and Σ_I, column (10), for that part of the solution concerning the I's. For the upper section of Table A2.11, that is, rows (1)–(4), these columns are obtained in the following way: The element in the ith row[9] of the Σ_x column is obtained by adding together the elements in the ith row of columns (1)–(4), including the omitted elements. The element omitted in the ith row and jth column can be found in the jth row of the ith column: For example, the omitted element in row (4), column (3), is the element in row (3), column (4), namely, .113360, etc. The element in the ith row of the Σ_I column is obtained by adding the elements in the ith row of columns (6)–(10). Because of the makeup of the columns, however, each element in these rows of the Σ_I column equals 1. In the computations outlined below, Σ_x and Σ_I are treated as additional variables, with all the operations performed upon them.

Only the second row in the first section and the last two rows in each succeeding section of the solution are checked. This is done in two parts,

[8] Lawrence R. A. Klein, *Textbook of Econometrics*, Row, Peterson, and Co., Evanston Ill. and White Plains, N.J., pp. 151–155, 1953.

[9] Foote and Friedman, from whom this explanation is taken, use "row" where previously in this book "line" has been used.

one for the x's and one for the I's. In order to check the computations in either of these rows in the x part of the forward solution, sum all the elements *in that row* for the x columns and compare that sum with the element in the Σ_x column for that row. There is no question of omitted elements here. These figures should be identical, except for rounding errors. If they are identical, this is indicated by a $\sqrt{}$. Where a discrepancy occurs due to a rounding error, the sum across the row replaces the element in the Σ column and is used in further computations. The check on the computations in the I section is obtained in like manner; that is, sum the elements in the ith row, columns (7)–(10), and compare that sum with the element in the ith row of the Σ_I column.

FORWARD SOLUTION. We now consider computations involved in each row of the lower sections of the forward solution in Table A2.11. These computations are performed on both halves of the table at the same time, i.e., on all columns from (1) to (10).

Row (1): Copy row (1) from the upper section of Table A2.11.

Row (1″): Divide row (1) by its first term, that is by 0.763984, and perform the check. For computational purposes, it is more efficient to compute $1/0.763984 = 1.308928$, lock it in the calculating machine, and multiply each item of row (1) by it.

Row (2): Copy row (2) of the upper section of Table A2.11.

Row (1)(0.261178): Multiply row (1) by 0.261178. This factor is the element of row (1″), column (2), *with its sign changed*. Note that no figures are inserted in this section of the table in columns to the left of column (2).

Row (2′): Add row (2) and the following line and perform the check.

Row (2″): Divide row (2′) by its first term, that is, by 0.308710, and perform the check. Or, multiply row (2′) by $1/.308710 = 3.239286$.

Row (3): Copy row (3).

Row (1)(−.471476): Multiply row (1) by −0.471476. This factor is the element of row (1″), column (3), *with its sign changed*.

Row (2′)(−1.643147): Multiply row (2′) by −1.643147. This factor is the element of row (2″), column (3), *with its sign changed*.

Row (3′): Add row (3) and the two rows following it and perform the check.

Row (3″): Divide row (3′) by its first term, that is, by 0.404998, and perform the check. Or multiply row (3′) by $1/.404998 = 2.469148$.

Row (4): Copy row (4).

Row (1)(.110374): Multiply row (1) by 0.110374. This factor is the element of row (1″), column (4), *with its sign changed*.

Row (2′)(1.040679): Multiply row (2′) by 1.040679. This factor is the element of row (2″), column (4), *with its sign changed*.

Row $(3')(-1.681510)$: Multiply row $(3')$ by -1.1681510. This factor is the element of row $(3'')$, column (4), *with its sign changed*.

Row $(4')$: Add row (4) and the three rows following it and perform the check.

Row $(4'')$: Divide row $(4')$ by its first term, that is, by 0.275173 and perform the check. Or, multiply row $(4')$ by $1/0.275173 = 3.634077$.

This completes the forward solution.

Unfortunately, the checks do not guarantee that the correct multiplicand has been used; they only prove that the multiplications were carried out correctly. As a final check, it is suggested that the multiplicands shown in the stub be examined to make sure that the correct value was used and that these then be used to recheck the computations in the Σ_I column [column (10) in Table A2.11] Experience in our central computing unit has indicated that occasionally a statistical clerk is interrupted between the computations involved in the x and the I part of the table and that the wrong multiplicand is used in the latter set of computations. It seems unlikely, however, that a wrong multiplicand would be used in the x part of the table and the correct one in the I part. When the abbreviated Doolittle solution is used, this final check is not needed, as the computations are carried out on a column-by-column basis rather than a row-by-row basis.

D MATRIX. The D matrix is shown in Table A2.11 immediately following the I part of the forward solution, at the foot of columns (6)–(10). Its computation involves the terms in the last two rows of each section in the I part of the forward solution. The element in the ith row and jth column of the D matrix, d_{ij}, is obtained by the following formula:

$$d_{ij} = (1, I_i)(1'', I_j) + (2', I_i)(2'', I_j) + (3', I_i)(3'', I_j) + (4', I_i)(4'', I_j)$$

where the first term within the parentheses refers to the row and the second, to the column designation of the elements in the forward solution. Therefore

For column $\quad I_1, d_{11} = (1)(1.308928) + (.261178)(.846030)$
$$+ (-.900630)(-2.223789) + (1.896594)$$
$$\times (6.892370) = 16.604731$$
and for column $I_2, d_{12} = (1)(0) + (.261178)(3.239286) + (-.900630)$
$$\times (-4.057173) + (1.896594)(13.822748)$$
$$= 30.716183$$
$$d_{22} = (1)(3.239286) + (-1.643147)(-4.057173)$$
$$+ (3.803647)(13.822748) = 62.482672$$

and so on for each of the items.

These sums should be cumulated directly in the calculating machine. A check column, Σ, is also carried in this computation. The elements in the Σ column, $d_i\Sigma$, are computed in the same way as any other element in the D matrix. In the general formula given above, I_j becomes Σ_I. That the sum across the ith row of D is identical (except for possible rounding errors) with the element in the ith row of the Σ column is indicated by a check mark. This checks the computation of the ith row. It will be noted that the elements in the lower part of the D matrix have been omitted. These need not be computed, since $d_{ij} = d_{ji}$. In computing the check on the computations in rows after the first, however, these omitted elements must be included. For example, the check on the computation of the fourth row of the D matrix is given by:

$$6.892369 + 13.822746 - 6.110737 + 3.634077 = 18.238457\ 5$$

(here the last digit of the total comes out 5, so the Σ is changed to that.)

The next-to-last (fourth) column of the D matrix need not be computed, since it corresponds to the last row (4″) in the I part of the forward solution.

All the usual measures of partial regression and correlation can be obtained easily from the D matrix. These calculations are shown in rows (5)–(11) of columns (1)–(4); column (5) is a check column.

PARTIAL REGRESSION COEFFICIENTS. The calculation of the highest order partial regression coefficients, the "b's", is shown in row (6). This is done as follows:

Row (5): Divide each element of the first row of the D matrix, including the element in the Σ column, by the first element in the first row of D, and *change the sign* of the resulting quotient. Symbolically, $b_{1j} = -d_{1j}/d_{11}$, where j refers to the subscript of the x's and the column of the D matrix. $b_{12.34}$, the coefficient of x_2, therefore equals $-d_{12}/d_{11}$ or $-(30.716183)/(16.604731) = -1.849845$. $b_{13.24}$, and $b_{14.23}$, the coefficients on x_3 and x_4, respectively, are obtained in like manner. That the sum across row (5) is identical (except for possible rounding error due to carrying only 6 decimals) with the element in the Σ column is indicated by a check mark. This checks the computation of the b's. Since x_1 is the dependent variable, no coefficient is attached to it. The -1 in row (5), column (1), and the figures in the following rows are written in order to check the computations.

STANDARD ERRORS OF THE REGRESSION COEFFICIENTS. The calculation of the standard errors of the highest order partial regression coefficients is shown in rows (6)–(10). This is done as follows:

Row (6): Compute $d_{11}d_{jj}$, that is, the product of the element in the first row and first column of D with the successive diagonal elements of D. For example, the element in the first or x_1 column is obtained by squaring

d_{11}; the element in the second or x_2 column equals $d_{11}d_{22} = (16.604731)$ $(62.482672) = 1037.507961$; and the element in the Σ column is obtained by multiplying d_{11} by Sd_{jj}. Sd_{jj} is the sum of the diagonal elements of D and is shown in the last column of row (10). That the sum across row (6) is identical (except for rounding error) with the element in the Σ column is indicated by a check mark, or by correcting the last digit, as shown.

Row (7): Compute d_{1j}^2, that is, the square of each of the elements in the first row of the D matrix, excluding the element in the Σ column. For example, the item in the second column of row (7) is $d_{12}^2 = (30.716183)^2$. The element in the Σ column of row (7) is the sum across the row. The check on this row is one of recomputation.

Row (8): Subtract each element of row (7) from the element in the corresponding column of row (6), including those in the Σ column. That the sum across row (8) is identical (except for possible rounding error) with the element in the Σ column is indicated by a check mark. This checks the computation of row (8).

Row (9): Compute $1/(N - m)(d_{11}^2)$, where $(N - m)$ equals the sample size minus the total number of variables, and d_{11}^2 is the square of the element in the first row and first column of D. The value of d_{11}^2 is given in the first column of rows (6) and (7). In this example, $(N - m)$ equals $20 - 4$ and d_{11}^2 equals 275.717092; therefore $1/(N - m)(d_{11}^2) = 1/4,411.4735 = 0.000227$. Multiply each element in row (8) by 0.000227, including that in the Σ column. That the sum across row (9) is identical (except for possible rounding error) with the item in the Σ column is indicated by a check mark, or by correcting the last digit.

Row (10): Compute the square root of the element in the corresponding column of row (9), except the element in the Σ column. The elements in row (10) are the standard errors of the coefficients in the corresponding column of row (5). The check is one of recomputation.

Where the number of variables is larger or smaller than in the examples shown, the number of columns from (1)–(5) and (6)–(10), and the number of rows from (1)–(4) and in the D matrix, will change accordingly; but not, of course, the rows from (5)–(11).

COEFFICIENTS OF PARTIAL DETERMINATION. The calculation of the highest order coefficients of partial determination (the square of the partial correlation coefficient) is shown in row (11). This is done as follows:

Row (11): Divide each element in row (7) by the element in the corresponding column of row (6), except the Σ column. The elements in row (11) are the coefficients of partial determination. The element in row (11), column (2), for example, equals $r_{12.34}^2$. The check on this row is one of recomputation.

If the coefficients of partial correlation are desired, they can be obtained by taking the square root of the elements in row (11), as shown in row 11a.

COEFFICIENT OF MULTIPLE DETERMINATION. Row 12: $R^2_{1.234}$, the coefficient of multiple determination, is obtained by the following formula:

$$R^2_{1.234} = \frac{d_{11}m_{11} - 1}{d_{11}m_{11}}$$

where d_{11} is the element in the first row and first column of the D matrix, and m_{11} is the adjusted augmented moment of x_1 on x_1, which is found in the first row and first column of the upper part of Table A2.11. In this example d_{11} equals 16.604731 and m_{11} equals 0.763984. Therefore,

$$R^2_{1.234} = \frac{(16.604731)(.763984) - 1}{(16.604731)(.763984)} = 0.921171. \quad \text{The coefficient of}$$

multiple correlation, $R_{1.234}$, if desired, can be obtained by taking the square root of the coefficient of multiple determination.

STANDARD ERROR OF ESTIMATE. $\bar{S}_{1.234}$, the adjusted standard error of estimate, is obtained by the following formula:

$$\bar{S}_{1.234} = \sqrt{\frac{1}{(N)(N - m)d_{11}}}$$

where d_{11} is the element in the first row and first column of the D matrix, N is the sample size, and $N - m$ is N minus the total number of variables. In this example, d_{11} equals 16.604731, N equals 20, and $N - m$ equals 16. Therefore, $\bar{S}^2_{1.234} = 1/(20)(16)(16.60473) = 0.000188$. $\bar{S}_{1.234}$, the standard error of estimate, equals the square root of this value or 0.013711. The N in this formula is required because of the use of augmented moments.

REGRESSION EQUATION BASED ON DEADJUSTED DATA. Since the regression coefficients and their standard errors are computed on the basis of adjusted data, they must be deadjusted in order to apply to the original data. This deadjustment, carried out in rows (13)–(16), is as follows:

Column (1): Enter the variables in numerical order.

Column (2): Enter the regression coefficients, the b's, obtained in row (5). Note that no figure is entered for X_1.

Column (3): Enter the standard errors of the regression coefficients, obtained in row (10).

Column (4): Enter the appropriate values of k_i from Table A2.10.

Column (5): Compute $k'_i = k_i/k_1$. Thus, for X_2, this is $0.01/0.01 = 1.0$.

Column (6): The deadjusted b's are obtained by multiplying the b's, column (2), by their respective k'_i.

Column (7): The deadjusted standard errors of the b's are obtained by multiplying the s_b, column (3), by their respective k_i'.

Column (8): Enter the means of the variables from Table A2.9.

Column (9): Computations in this column are used in obtaining the constant for the equation. Multiply the deadjusted b's, column (6), by the mean in the corresponding row of column (8) and add the figures in column (9), or cumulate the products directly in the machine. The constant in the equation, a, is obtained by subtracting the cumulated product from the mean of X_1, the element in row (13), column (8). Hence, $a = 54.6800 - (-36.1338) = 90.8138$. This result can be recorded directly as the constant in the regression equation shown in row (18).

The final regression equation, in the following form, is shown in row (18):

$$X_1 = a + b_{12.34}X_2 + b_{13.24}X_3 + b_{14.23}X_4$$

The figures in the table within the parentheses are the standard errors of the respective regression coefficients.

Row (17): The standard error of estimate, $\bar{S}_{1.234}$, also must be deadjusted. This is done by dividing $\bar{S}_{1.234}$ by k_1. The latter is given in row (13), column (14). In our example, $k_1 = 0.01$; therefore, for this example, the standard error of estimate is $0.013711/0.01 = 1.371$. The indicated computation is shown in row (17).

The coefficient of multiple determination need not be deadjusted.

The check in this section is one of recomputation.

If all of the k_i' equal one, columns (6) and (7), for lines (14)–(16) can be omitted. In this case, column (2) is used in place of column (6) in obtaining the constant in the equation in column (9).

This completes the solution and computation of all .the multiple coefficients.

Eliminating or Adding Variables. If one or more variables are to be eliminated or added, the measures of correlation and regression can be obtained without rerunning the analysis.

ELIMINATING VARIABLES. Application of the formula given below, which applies if one variable is to be eliminated, yields elements of a similar D matrix, $D_{)k(}$, for all variables except the omitted one, x_k.[10] The elements of this matrix, the $d_{ij)k(}$, can be obtained by the formula:

$$d_{ij)k(} = \frac{d_{ij}d_{kk} - d_{ik}d_{jk}}{d_{kk}}$$

where the d's are the elements of D. These $d_{ij)k(}$ values are used in place of the corresponding d_{ij} values in the computations beginning with row (5) of Table A2.11.

---

[10] This formula was suggested by Frederick V. Waugh.

For example, if x_4 were to be dropped from the previous analysis, we would compute the first row of $D_{)4(}$, that is, $d_{11)4(}$, $d_{12)4(}$, and $d_{13)4(}$, by the formula:

$$d_{1j)4(} = \frac{d_{1j}d_{44} - d_{14}d_{j4}}{d_{44}}$$

If we consider the adjusted augmented moments of x_1 with the other variables given in the first row of Table A2.11 as m_{ij}, a check on the computation of the first row of $D_{)4(}$ is given by computing $m_{11}d_{11)4(} + m_{12}d_{12)4(} + m_{13}d_{13)4(}$. This sum should equal 1.

It is not necessary to compute the entire $D_{)4(}$ matrix. In addition to the first row, we need compute only the diagonal elements, that is, $d_{jj)4(}$, given by the formula:

$$d_{jj)4(} = \frac{d_{jj}d_{44} - d_{j4}^2}{d_{44}}$$

The partial regression coefficients can be obtained by:

$$b_{1j.)4(} = -\frac{d_{1j)4(}}{d_{11)4(}}$$

Their standard errors are given by:

$$s_{b_{1j.)4(}} = \sqrt{\frac{d_{11)4(}d_{jj)4(} - d_{1j)4(}^2}{N'd_{11)4(}^2}}$$

The coefficients of partial determination equal:

$$r_{1j.)4(}^2 = \frac{d_{1j)4(}^2}{d_{11)4(}d_{jj)4(}}$$

The coefficient of multiple determination equals:

$$R_{1.23}^2 = \frac{d_{11)4(}m_{11} - 1}{d_{11)4(}m_{11}}$$

The standard error of estimate is given by:

$$s_{1.23} = \sqrt{\frac{1}{NN'd_{11)4(}}}$$

Similarly, if both x_r and x_k were to be eliminated, we would compute the elements of $D_{)kr(}$, the $d_{ij)kr(}$, as follows:

$$d_{ij)kr(} = \frac{d_{ij)k(}d_{rr)k(} - d_{ir)k(}d_{jr)k(}}{d_{rr)k(}}$$

Thus, if more than one variable is to be eliminated, the computations must be done in steps by eliminating them one at a time.

Use of the above formula is easy if only one variable is to be eliminated; it becomes more difficult as additional variables are dropped. Sometimes the analyst knows fairly well in advance which variables may need to be eliminated. If so, he should use them as the highest numbered independent variables. Thus, in a five-variable problem, if X_4 and X_5 were to be eliminated, this could be done by dropping columns (4), (5), (10), and (11) and rows (4) and (5) and their corresponding sections in the forward solution. The D matrix then could be easily recomputed and the remaining computations carried out as in any three-variable analysis. New check sums for use in the computations beginning with row (6) probably would be advisable.

ADDING VARIABLES. In general, it is easier to drop variables than to add them. Hence, as many variables as are likely to be used should be incorporated in the initial analysis; some of these can then be dropped if this appears advisable. At times, however, a variable will need to be added. Assume that the added variable is X_5. Use can be made of all of the computations already made in the forward solution. Columns are added between the former columns (4) and (5) and between columns (9) and (10), and a row (5) and a corresponding section are added in the forward solution. Figures in these columns can be filled in by performing the same sort of computations that were done previously. An additional product from the new section (5) will need to be added to *each* of the elements in the original D matrix, and a new column (5) and row (5) should be added. These steps can be checked by recomputation or by use of a new check sum. All of the coefficients should be recalculated, making use of the new D matrix and of new check sums.

Standard Errors of the Function and of Forecasts. The standard error of a point on the regression equation, or function, relates to a point on the regression surface corresponding to specified values of the independent variables (note Chapter 19).

For a four-variable multiple regression problem, the square of the standard error of a point on the regression equation, or function, is given by:

$$s^2_{F_{1.234}} = \bar{S}^2_{1.234} \left[\frac{1}{N} + Nc_{22}(X_2 - M_2)^2 + Nc_{33}(X_3 - M_3)^2 \right.$$
$$+ Nc_{44}(X_4 - M_4)^2 + 2Nc_{23}(X_2 - M_2)(X_3 - M_3)$$
$$\left. + 2Nc_{24}(X_2 - M_2)(X_4 - M_4) + 2Nc_{34}(X_3 - M_3)(X_4 - M_4) \right]$$

where $\bar{S}^2_{1.234}$ is the deadjusted value of the square of the adjusted standard error of estimate obtained by squaring the deadjusted standard error of estimate, N is the number of observations on which the analysis is based,[11] and the c_{ij} are obtained from the elements of the D matrix shown in Table A2.11 by the formula:

$$c_{ij} = \frac{d_{11}d_{ij} - d_{1i}d_{1j}}{d_{11}}$$

If these values of c_{ij} are substituted in the formula for the square of the standard error of the function, N and d_{11} appear in each of the products within the brackets. [Compare with equation (19.2).] We can, therefore, rewrite the formula as:

$$
\begin{aligned}
s_{F_{1.234}}^2 = \bar{S}_{1.234}^2 \Bigg\{ \frac{1}{N} + \frac{N}{d_{11}} \Bigg[& c_{22}''(X_2 - M_2)^2 + c_{33}''(X_3 - M_3)^2 \\
& + c_{44}''(X_4 - M_4)^2 + 2c_{23}''(X_2 - M_2) \\
& \times (X_3 - M_3) + 2c_{24}''(X_2 - M_2)(X_4 - M_4) \\
& + 2c_{34}''(X_3 - M_3)(X_4 - M_4) \Bigg] \Bigg\}
\end{aligned}
\tag{A2.3}
$$

where $c_{ij}'' = d_{11}d_{ij} - d_{1i}d_{1j}$. Thus $c_{23} = d_{11}d_{23} - d_{12}d_{13}$; and $c_{22} = d_{11}d_{12} - (d_{12})^2$. The computed c_{ij}'' are shown in Table A2.12.

Table A2.12

c_{ij}'' FOR THE FOUR-VARIABLE MULTIPLE REGRESSION PROBLEM

Outline			Values		
c_{22}''	c_{23}''	c_{24}''	94.02414	−29.0186	−17.8135
	c_{33}''	c_{34}''		20.8082	−6.2598
		c_{44}''			12.8381

c_{22}'', c_{33}'', and c_{44}'' were computed in row (8) of Table A2.11, columns (2), (3), and (4), respectively. The other c_{ij}'' must be computed directly. For example, $c_{23}'' = d_{11}d_{23} - d_{12}d_{13}$. Substituting the values from the D matrix, we obtain: $c_{ij}'' = (16.6047)(-27.3003) - (30.7162)(-13.8134) = -29.0186$. The computation of the c_{ij}'' can be checked by computing the following sums of products:

(a) $c_{22}''m_{22} + c_{23}''m_{23} + c_{24}''m_{24}$

(b) $c_{23}''m_{23} + c_{33}''m_{33} + c_{34}''m_{34}$

(c) $c_{24}''m_{24} + c_{34}''m_{34} + c_{44}''m_{44}$

---

[11] N is required in all terms after the first one within the brackets because of the use of augmented moments in the computations.

where m_{ij} is the adjusted augmented moment of x_1 on x_j shown in rows (1)–(4), columns (1)–(4), of Table A2.11. Each of these sums of products should equal d_{11}, except for possible rounding errors. For example, to check the first row of the c''_{ij}, we compute (a): $(94.0241)(.360824) + (-29.0186)(.413180) + (17.8135)(-.299249) = 16.6047$. The second and third rows are checked by computing (b) and (c), respectively. These c''_{ij} are adjusted terms.

For use in the formula for the standard error of a function, the c''_{ij} must be deadjusted. This is done by multiplying c''_{ij} by $k_i k_j$, the appropriate adjustment factor from Table A2.10. For example, to deadjust c''_{24}, 17.8135, multiply by $k_2 k_4$, or 0.00001. Therefore, the deadjusted value of $c''_{24} = 0.00178$. The nature of the formula is such, however, that d_{11} is *never* deadjusted.

The means that are used in the formula are obtained from Table A2.11, rows (13)–(16) of column (8). These are given on a deadjusted basis.

Inserting the deadjusted standard error of estimate, c''_{ij}, means, and the adjusted d_{11} in equation (A2.3) gives the formula for the square of the standard error of the function.

The standard error of the function is obtained by inserting the specified values of X_2, X_3, and X_4 for any given observation and taking the square root of the result.

The standard error of a specified forecast is obtained from the following formula:

$$s_{x'_1} = \sqrt{s^2_{F_{1 \cdot 234}} + \bar{S}^2_{1 \cdot 234}}$$

$$= \sqrt{s^2_{F_{1 \cdot 234}} + 1.88}$$

where $\bar{S}^2_{1 \cdot 234}$ is on a deadjusted basis.

Use of an Alternative Variable as the Dependent One. All measures of regression and correlation given in preceding sections are based on the use of X_1 as the dependent variable. If, after the analysis is run, it seems desirable to have one of the other variables, X_i, as the dependent one, the various statistical measures can be obtained from the original D matrix by use of the following:

The partial regression coefficients equal:

$$b_{ij.} = -\frac{d_{ij}}{d_{ii}}$$

If, for example, X_2 is to be used as the dependent variable in the five-variable problem given above, we would compute: $b_{21.34}$, $b_{23.14}$, and $b_{24.13}$, where $b_{21.34} = -d_{21}/d_{22}$, etc.

The standard errors of the regression coefficients are given by:

$$s_{b_{ij}} = \sqrt{\frac{d_{ii}d_{jj} - d_{ij}^2}{(N - m)d_{ii}^2}}$$

For example,

$$s_{b_{21.34}} = \sqrt{\frac{d_{22}d_{11} - d_{21}^2}{(N - m)d_{22}^2}}$$

The coefficients of partial determination equal:

$$r_{ij.}^2 = \frac{d_{ij}^2}{d_{ii}d_{jj}}$$

For example,

$$r_{21.}^2 = \frac{d_{21}^2}{d_{22}d_{11}}$$

The coefficient of multiple determination is given by:

$$R_{i.}^2 = \frac{d_{ii}m_{ii} - 1}{d_{ii}m_{ii}}$$

For example,

$$R_{2.134}^2 = \frac{d_{22}m_{22} - 1}{d_{22}m_{22}}$$

The standard error of estimate equals:

$$\bar{S}_1 = \sqrt{\frac{1}{N(N - m)d_{ii}}}$$

For example,

$$\bar{S}_{2.134} = \sqrt{\frac{1}{N(N - m)d_{22}}}$$

It should be noted that when variables are eliminated, added, or interchanged, the regression coefficients, their standard errors, and the standard error of estimate must be readjusted before they can be applied to the original data. All of the formulas shown apply to adjusted values.

This completes the explanation of the Friedman-Foote method of solution.

Computing Residuals for Graphic Multiple Curvilinear Regressions

Where there are a large number of individual observations, the average residual around the net regression line may be computed from group averages, instead of calculated for each individual observation as described in Chapter 14. This may save much time in calculating the average residuals to obtain the first approximation regression curves.

After the net linear regression coefficients are computed, the observations are thrown into groups with respect to the first independent factor, say X_2, and averages of each factor are computed for the records falling in each group. If there are four groups, for example, there will be four sets of averages.

Value of X_2	Average X_2	Average X_3	Average X_1
0–9	M_{2-1}	M_{3-1}	M_{1-1}
10–19	M_{2-2}	M_{3-2}	M_{1-2}
20–29	M_{2-3}	M_{3-3}	M_{1-3}
30 and over	M_{2-4}	M_{3-4}	M_{1-4}

The average estimated value, $M_{x'}$, may then be calculated for each group by substituting the means for that group in the regression equation. Thus for the first group,

$$M_{x'} = a + b_2(M_{2-1}) + b_3(M_{3-1})$$

and

$$M_z = M_{1-1} - M_{x'}.$$

In a similar manner the average residual may be calculated from the group averages for each of the other groups, and then plotted as a departure from the net regression line, as illustrated in Figure 14.5. After the computation is completed for X_2, the records may be reclassified with respect to X_3, new means calculated for each variable for each group, and the process continued just as for X_2. The same steps are carried out for each other independent variable in turn. This method may be used to determine the net residuals around curvilinear regressions fitted by mathematical curves just as well as for linear regressions.

Once the first set of freehand approximation curves has been drawn, the remainder of the work has to be carried forward just as described in Chapter 14, as the average of values along a curve do not precisely represent that curve in the same way that the average of values along a straight line will represent that line.

Auxiliary Graphic Processes with the Short-Cut Graphic Method

The short-cut method of determining a net curvilinear regression, described in Chapter 16, may be materially aided by using graphic methods in transferring departures from one figure to another, and in calculating the averages of the values as plotted.

After the original observations are plotted and the first approximation to the regression line or curve is drawn (as in Figure 16.2), the departures from that line must be plotted against the next variable. A procedure for making those transfers graphically is shown in Figures A2.1 to A2.5.

The first step is to place an arrow in the middle of a strip of blank paper. Using the arrow to indicate the position of the regression line or curve, we mark off on the paper the vertical departures of each observation from that line, with each observation indicated by its number. Figure A2.1 shows this process just as the first observation (29) is marked on the strip.

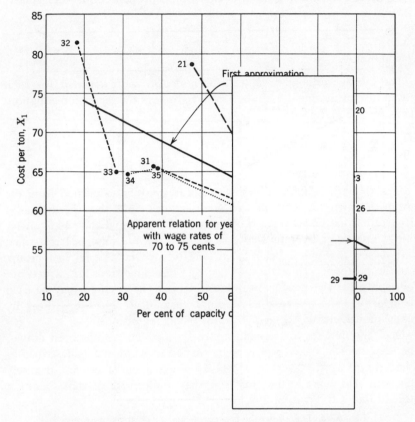

Fig. A2.1. This shows the start of the process of scaling off graphically the departures from the first approximation to the net regression line or curve.

Figure A2.2 shows it after several such values have been marked on the strip. The process is continued (with one or more strips of paper) until the vertical departures have been marked off for each observation.

The next step in the process is to transfer these departures to the next figure, Figure 16.3. After the chart form has been prepared, the arrow on the slip is centered on the zero line, and the departures marked on the figure, with the slip moved to the corresponding X_3 value as ordinate.

Figure A2.3 shows this step just after the value for the 1920 observation was entered on the chart. After the value is marked on the new figure, it is crossed off on the slip, to prevent confusion. Figure A2.4 shows the process completed, just as the last value on the slip—that for the 1933 observation—is entered. (It will be noticed that the values are transferred in sequence from top to bottom of the slip, to prevent confusion.)

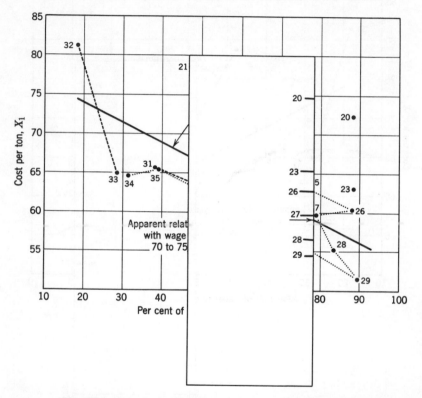

Fig. A2.2. This shows the process of scaling off the departures partially completed.

After the new curve is inserted on the chart, the next step is to transfer residuals from the new curve to the next figure. The departures can be scaled off from a curve as readily as from a line. Figure A2.5 shows the start of the next stage of the process, after the departures for the observations for 1920 and 1937 have been scaled off from the first approximation curve on Figure 16.3, and just as the value for 1936 is entered. The process is completed and carried on to the next chart (Figure 16.4) just as illustrated above. The same process is used in transferring the

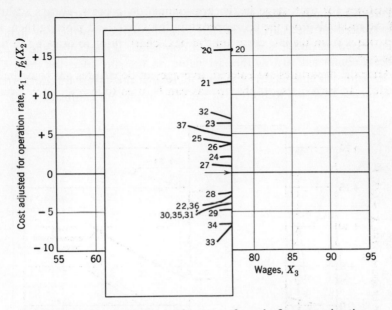

Fig. A2.3. Here the slip with the departures from the first approximation is moved to the next chart, ready to start transferring the departures to get the first approximation for the next variable. The first observation, for 1920, has been entered and checked off.

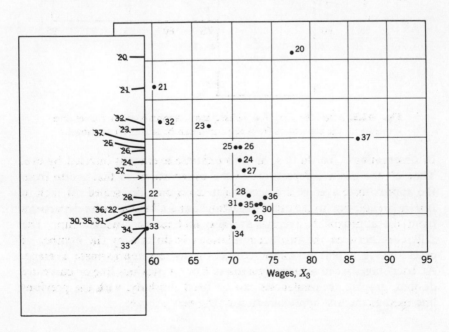

Fig. A2.4. The process shown at the start in Fig. A2.3 is here shown completed.

departures for each stage in the approximation process, always scaling off the residuals *from* the last approximation curve, and plotting them as departures from the last curve on the next chart, prior to drawing in the new curve.

After the departures are entered, averages of departures are sometimes needed. In such cases, graphic means can be used to average each group

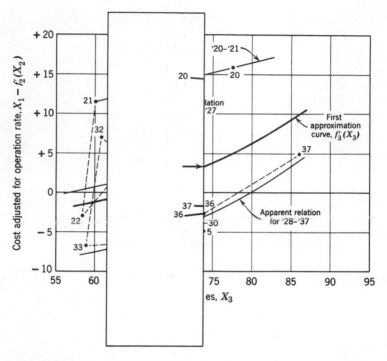

Fig. A2.5. After the first (or subsequent) approximation curves are drawn in, the departures from the curve can be scaled off as shown.

of observations. To do this, an approximate average is inserted by eye. Then all the positive departures of the observations in that group from the approximate average are accumulated on one slip, scaled off each in turn as an addition to the other departures, and all the negative departures from the approximate average are accumulated on another slip. The difference between the two accumulations is divided by the number of cases, giving a plus or minus correction to the approximate average. At later stages when average deviations from a previous line or curve are desired, graphic accumulations can be used similarly, with the previous line used as the first approximation to the new average.

APPENDIX 3

Technical notes

Note I. (Chapter 7). To prove for a very simple case that r_{yx}^2 measures the proportion of variance in Y explained by X. Let a, b, c, etc., be series of variables with $\sigma_a = \sigma_b = \sigma_c$, and with all intercorrelations such as r_{ab}, r_{ac}, etc., $= 0$.

Let $Y = a + b + c$, and $X = a + b$. Then $r_{yx}^2 = \dfrac{p_{yx}^2}{\sigma_x^2 \sigma_y^2}$

$\left(\text{Here the symbol } p_{yx} \text{ is used to represent } \dfrac{\Sigma yx}{n} \cdot \right)$

$$\text{each } (y)(x) = (a + b + c)(a + b)$$
$$= a^2 + 2ab + ac + b^2 + bc$$

Since
$$\Sigma(ab), \Sigma(ac), \Sigma(bc) = 0$$
$$\Sigma(y)(x) = \Sigma a^2 + \Sigma b^2$$

Similarly,
$$(y^2) = (a + b + c)^2$$
$$= a^2 + 2ab + 2ac + b^2 + 2bc + c^2$$
$$\Sigma(y^2) = \Sigma a^2 + \Sigma b^2 + \Sigma c^2$$

By similar proof,
$$\Sigma(x)^2 = \Sigma a^2 + \Sigma b^2$$

Hence
$$r_{yx}^2 = \frac{(\sigma_a^2 + \sigma_b^2)^2}{(\sigma_a^2 + \sigma_b^2 + \sigma_c^2)(\sigma_a^2 + \sigma_b^2)}$$
$$= \frac{\sigma_a^2 + \sigma_b^2}{\sigma_a^2 + \sigma_b^2 + \sigma_c^2}, \quad = \tfrac{2}{3} \ (\text{since } \sigma_a = \sigma_b = \sigma_c)$$

Similar results will be obtained for other combinations of elements.

Note 2. (Chapter 12). It can be proved that the coefficient of multiple determination (R^2) measures the percentage of variance ascribable to the several independent factors for certain simple cases. Thus assume four variables, A, B, C, D, with all intercorrelations equal to 0, and all σ equal. Let $Y = A + B + C$. Then correlate Y with A, B, and D. The regression equation will work out

$$Y = a + A + B[+0(D)]$$

531

Computing $R_{Y.ABD}$ by equation (12.5),

$$R^2_{Y.ABD} = \frac{b_{ya.bd}\Sigma ya + b_{yb.ad}\Sigma yb + b_{yd.ab}\Sigma yd}{\Sigma y^2}$$

Each

$$(y)(a) = (a + b + c)(a) = a^2 + ab + ac$$

$$\Sigma(ya) = \Sigma a^2 + \Sigma ab + \Sigma ac = \Sigma a^2 \text{ (Since } r_{ab} = r_{ac} = 0)$$

Similarly

$$\Sigma(yb) = \Sigma b^2; \ \Sigma(yd) = \Sigma d^2$$

And each

$$(y^2) = (a + b + c)^2 = a^2 + 2ab + b^2 + 2ac + 2bc + c^2$$

and

$$\Sigma(y^2) = \Sigma a^2 + \Sigma b^2 + \Sigma c^2$$

Hence

$$R^2_{Y.ABD} = \frac{(1)(\Sigma a^2) + 1(\Sigma b^2) + 0(\Sigma d^2)}{\Sigma a^2 + \Sigma b^2 + \Sigma c^2}$$

And since all σ's are identical,

$$R^2_{Y.ABD} = \tfrac{2}{3}$$

In this case, then, when Y, composed of three equally variable non-correlated elements, is correlated with two of those elements, and with one other equal element which is not represented in Y and which is not correlated with elements present in Y, the multiple determination of Y by the two elements (A and B) is found to be $\tfrac{2}{3}$.

Similar results will be secured for other experimental cases which may be set up.

Note 3. (Chapter 12). Coefficients of partial correlation are usually defined by the formula

$$r_{12.3} = \frac{r_{12} - r_{13}r_{23}}{\sqrt{1 - r^2_{13}}\ \sqrt{1 - r^2_{23}}}$$

For coefficients with more variables eliminated, such as $r_{12.345}$, for example, this becomes

$$r_{12.345} = \frac{r_{12.34} - r_{15.34}r_{25.34}}{\sqrt{(1 - r^2_{15.34})(1 - r^2_{25.34})}}$$

To determine the coefficients with several factors held constant by this method involves a lengthy process of elimination, variable by variable; and for that reason the method presented in the text is preferred as shorter, simpler, and more readily subject to checking.

Note 4 (Chapter 17). Reliability of observed correlations. Figures 17.2 to 17.5 provide a ready means of judging the probable minimum value for the correlation in the universe, with any observed value and any given size of sample. The chart is entered with the observed correlation as abscissa; the ordinate for the intersection of that abscissa with the curve for the given size of sample gives the probable minimum correlation. Thus if the coefficient of simple correlation, $r_{xy} = 0.65$, is obtained from a sample of 22 cases, the researcher will know from Figure 17.2 that, if he makes the statement that the true correlation in the universe is at least 0.38, he will be wrong in only 5 per cent of such statements, on the average. Figure 17.3 applies to $R_{1.234}$, Figure 17.4 to $R_{1.23456}$, and Figure 17.5 to $R_{1.2345678}$. Values for 2, 4, and 6 independent variables may be estimated by interpolation. The figures are based upon the researches of R. A. Fisher, summarized in his publication, The general sampling distribution of the multiple correlation coefficient, *Proceedings of the Royal Society*, A, Vol. 121, pp. 655–673, 1928.

Note 5 (Chapter 19). The standard error of an individual forecast is composed of the error of *points along the calculated regression* line plus that of *individual estimates around that line*. The standard error of the former is given by

$$s_{y'} = \sqrt{s_{M_{y'}}^2 + (s_{b_{yx}}x)^2} \qquad (17.4)$$

while that of the latter is the standard error of estimate, $\bar{S}_{y\cdot x}$ (equation 8.4).

Assuming that individual errors of calculated points along the line are uncorrelated with departures of individual forecasts from the line, the square of the standard error of an individual forecast is the sum of the squares of the standard errors of the two components, as follows:

$$s_{Y'}^2 = s_{M_{y'}}^2 + (s_{b_{yx}}x)^2$$

$$s_{\text{individual estimates}}^2 = \bar{S}_{y\cdot x}^2$$

hence

$$s_{y'-y} = s_{y'}^2 + \bar{S}_{y\cdot x}^2$$

or

$$s_{Y'-Y}^2 = s_{M_{y'}}^2 + (s_{b_{yx}}x)^2 + \bar{S}_{y\cdot x}^2 \qquad (19.1)$$

Note 6 (Chapter 24). The complete algebraic derivation of the reduced-form equations in the "just-identified" structural model for pork (page 419) is as follows:
The structural model is
Demand function: $\qquad\qquad Q = a_1 + b_1 P + c_1 Y + u$

Supply function: $\qquad\qquad Q = a_2 + b_2 P + c_2 Z + v$

where u and v are random disturbances, Y and Z are predetermined variables, and P and Q are jointly dependent endogenous variables. We must solve the structural equations for P and Q in such a way that each is expressed as a function only of Y, Z, u, and v.

Subtracting the first equation from the second, we have $0 = (a_2 - a_1) + (b_2 - b_1)P - c_1 Y + c_2 Z + (v - u)$. Transposing the term involving P to the left side of the equation and dividing through by the coefficient of P we obtain

$$P = \left(\frac{a_2 - a_1}{b_1 - b_2}\right) - \left(\frac{c_1}{b_1 - b_2}\right)Y + \left(\frac{c_2}{b_1 - b_2}\right)Z + \left(\frac{v - u}{b_1 - b_2}\right)$$

To solve for Q we multiply each term in the first equation by b_2 and each term in the second by b_1. Subtracting the first equation from the second, we have

$$(b_1 - b_2)Q = (a_2 b_1 - a_1 b_2) - c_1 b_2 Y + c_2 b_1 Z + b_1 v - b_2 u$$

Dividing through by the coefficient of Q, we obtain

$$Q = \left(\frac{a_2 b_1 - a_1 b_2}{b_1 - b_2}\right) - \left(\frac{c_1 b_2}{b_1 - b_2}\right)Y + \left(\frac{c_2 b_1}{b_1 - b_2}\right)Z + \left(\frac{b_1 v - b_2 u}{b_1 - b_2}\right)$$

Each of these equations now contains a single endogenous (logically dependent) variable; the values of Y and Z are given numbers, like independent variables in the "regression model" of least squares; and the terms in u and v have the same statistical properties as the random residuals of least-squares regression, being uncorrelated (by assumption) with Y and Z. Therefore each of the two equations, one in P and the other in Q, can be fitted separately by the method of least squares. These equations constitute the "reduced form" of the model and are called "reduced-form equations" to distinguish them from the structural equations.

The coefficients of the structural equations can be derived from those of the reduced forms by, in essence, reversing the process of derivation shown above. Thus, the coefficients of Y in the two equations differ only by the factor "b_2" in one of the numerators; the denominators are identical. We can therefore estimate b_2 as follows:

$$b_2 = \frac{-\left(\dfrac{c_1 b_2}{b_1 - b_2}\right)}{-\left(\dfrac{c_1}{b_1 - b_2}\right)}$$

Similarly, the coefficients of Z differ only by the factor "b_1", and b_1 can be estimated as

$$b_1 = \frac{\left(\dfrac{c_2 b_1}{b_1 - b_2}\right)}{\left(\dfrac{c_2}{b_1 - b_2}\right)}$$

Knowing these two values, b_2 and b_1, we can estimate c_1 as

$$c_1 = -\left(\frac{c_1}{b_1 - b_2}\right)[-(b_1 - b_2)]$$

and c_2 as

$$c_2 = \left(\frac{c_2}{b_1 - b_2}\right)(b_1 - b_2)$$

The values of a_1 and a_2 can be derived by multiplying the numerical values of $(a_2 - a_1)/(b_1 - b_2)$ and $(a_2 b_1 - a_1 b_2)/(b_1 - b_2)$ each by $(b_1 - b_2)$; then, calling the resulting values k_1 and k_2, we have two simultaneous equations in the two unknowns, a_1 and a_2:

$$a_2 - a_1 = k_1$$

$$b_1 a_2 - b_2 a_1 = k_2$$

Multiplying the first equation through by b_2 and subtracting it from the second equation, we have

$$a_2(b_1 - b_2) = k_2 - k_1 b_2$$

and

$$a_2 = \frac{k_2 - k_1 b_2}{b_1 - b_2}$$

where every element to the right of the equality sign is a known number. Substituting the numerical value of a_2 in the first equation, we obtain

$$a_1 = a_2 - k_1$$

Our estimates of the coefficients of the structural equations are now complete.

Note 7 (Appendix 2). In computing means from rounded-off data, errors due to rounding tend to compensate for each other, so that in large samples the means may be carried out to more significant digits (or decimal places) than are given in the individual items from which they are computed. In division and multiplication, however, this is not true, so that in general the product or dividend will not be accurate to as many decimal places as the numbers from which it was calculated. These principles can be readily demonstrated by arithmetic examples or mathematical models. For these reasons, slight differences in values of various statistics may result merely from differences in how many decimal places they are carried out to in the computations. Where the computations are carried out to four to five significant digits, the differences generally tend to be insignificant compared to the standard errors of the statistics to which they relate. While it is customary to carry elaborate or long series of calculations (such as those shown on pages 507 to 513) out to six to ten decimal places, to minimize errors due solely to rounding, the final results are not usually shown to more than two or three, to avoid giving a spurious impression of exact accuracy.

Similarly, in working with squares and square roots, the squares must be carried out to twice as many decimal places as the square roots, to maintain the same degree of accuracy in the computations.

Note 8 (Chapter 17). Beta coefficients, like coefficients of correlation, can be strongly influenced by purposeful selection of the sample values of one or more of the independent variables. This similarity is suggested by the fact that β_{12} and r_{12} are identical in the case of simple linear regression. If beta coefficients are calculated from a random sample drawn from a "bivariate" or "multivariate" normal population, they qualify as estimates of corresponding parameters of this population. But if values of one or more of the independent variables are fixed by the investigator, as in a controlled experiment, the beta coefficients have sampling significance only in relation to a very special "universe" in which, roughly speaking, the standard deviation of each independent variable is held perfectly constant for all possible samples.

Ordinary simple and net regression coefficients are not subject to these limitations. Thus beta coefficients are of doubtful value except in the so-called *correlation model* involving random sampling from a normally distributed "natural" universe.

Author Index

Aandahl, Andrew R., 467
Aitchison, J., 186
Albert, W. W., 460
Allan, D. H. W., 185
Allen, R. H., 466
Allport, Gordon W., 465
Anderson, R. L., 338, 347
Armore, Sidney J., 324, 463
Armstrong, David T., 460
Attridge, R. F., 185

Bandeen, Robert A., 466
Barnicoat, C. R., 460
Bartlett, M. S., 334, 347
Basu, D., 467
Baum, E. L., 466
Bean, L. H., inventor of short-cut graphic
  method, 254
  leader in correlation use, 435
  on cotton supply-demand relations,
  462
  on farmers' response to price, 463
  on freehand versus algebraic curves,
  108
  on graphic short-cut method, 277, 278
  on potato and cotton supply and demand, 463
  on predicting elections, 465
  on voting patterns, 465
Bean method, basis for, 169

Beckman, F. S., 186
Been, Richard O., on validity of multiple
  regression results, 324
Bell, G. D. H., 460
Benner, Claude L., 464
Black, John D., input-output study, 459
  inspirer of research, 435
  on graphic method, 108, 273, 277
  study of creamery organization, 461
Blackmore, John, 466
Boles, J. N., 467
Breimyer, Harold F., 466
Brounoff, P. I., 459
Brown, C. J., 461
Brown, J. A. C., 185, 186
Brown, Jean, 466
Brown, William G., 333, 377, 397, 408,
  409
Bruce, Donald, 377, 481
Burmeister, Gustave, 459
Burtis, Edgar L., 324
Butler, O. D., 460

Calzada, José, 459
Cartwright, T. C., 460
Casady, R. B., 460
Castle, Elizabeth J., 460
Castle, M. E., 460
Cavin, James, 449
Chamberlin, Edward, 464

537

Subject Index

Introductory Note. Subjects are classified when they relate to regression and correlation issues, or when they have been used as examples in exercises. The many subjects of individual studies mentioned in Chapter 25 are not indexed, however.